I0882171

PLANNING LAW AND PRACTICE IN NORTHERN IRELAND

Each of the jurisdictions within the United Kingdom is constantly refining the operational characteristics of its planning system and while there are some common practices, there are also substantive divergences. In each territory the planning template is shaped within a dynamic political and legal context and thus students and practitioners require an accessible, in-depth and up-to-date literature dealing with this matter. The multi-disciplinary contributors to this expanded Second Edition of *Planning Law and Practice in Northern Ireland* explore the progression of planning within the region and discuss prominent facets of contemporary development management, development plans, environmental law, property law and professional practice. Consideration is given to the consequences of Brexit for planning in Northern Ireland, devolved government institutional structures for planning, and the post-2015 emergent performance of local authorities in this arena. The book makes an important contribution to the wider literature in this field and, with its extensive citing of statutes and cases, provides an essential resource for students, planning practitioners and researchers.

Stephen McKay is Reader in Planning Law and Practice and was Head of Planning at Queen's University Belfast. He is a Chartered Town Planner. He was Co-Director of the 2021 award winning UK Knowledge Transfer Project *VU* City that has created an interactive 3D digital model to help inform development management decision making.

Michael Murray, prior to retirement, was Reader in Spatial Planning at Queen's University Belfast and where, as Honorary Associate, he continues his writing on international planning, rural development and planning law. His recent scholarship includes *The Camino de Santiago: Curating the Pilgrimage as Heritage and Tourism* published by Berghahn Books in 2021.

PLANNING LAW AND PRACTICE IN NORTHERN IRELAND

Second Edition

Edited by Stephen McKay and Michael Murray

Routledge
Taylor & Francis Group
LONDON AND NEW YORK

Cover image: Martin Siepmann | Getty Images.

2nd Edition published 2023
by Routledge
4 Park Square, Milton Park, Abingdon, Oxon, OX14 4RN

and by Routledge
605 Third Avenue, New York, NY 10158

Routledge is an imprint of the Taylor & Francis Group, an informa business

© 2023 selection and editorial matter, Stephen McKay and Michael Murray; individual chapters, the contributors

The right of Stephen McKay and Michael Murray to be identified as the authors of the editorial material, and of the authors for their individual chapters, has been asserted in accordance with sections 77 and 78 of the Copyright, Designs and Patents Act 1988.

All rights reserved. No part of this book may be reprinted or reproduced or utilised in any form or by any electronic, mechanical, or other means, now known or hereafter invented, including photocopying and recording, or in any information storage or retrieval system, without permission in writing from the publishers.

Trademark notice: Product or corporate names may be trademarks or registered trademarks, and are used only for identification and explanation without intent to infringe.

First edition published by Routledge, 2017

British Library Cataloguing-in-Publication Data
A catalogue record for this book is available from the British Library

Library of Congress Cataloging-in-Publication Data
Names: McKay, Stephen, editor. | Murray, Michael, 1953- editor.
Title: Planning law and practice in Northern Ireland / edited by Stephen McKay and Michael Murray.
Description: 2nd edition. | Abingdon, Oxon [UK] ; New York, NY : Routledge, 2022. | Includes bibliographical references and index. |
Identifiers: LCCN 2022017455 (print) | LCCN 2022017456 (ebook) | ISBN 9781032110745 (hardback) | ISBN 9781032110738 (paperback) | ISBN 9781003218319 (ebook)
Subjects: LCSH: Regional planning--Law and legislation--Northern Ireland.
Classification: LCC KDE135 .P57 2022 (print) | LCC KDE135 (ebook) |
DDC 346.41604/5--dc23/eng/20220831
LC record available at https://lccn.loc.gov/2022017455
LC ebook record available at https://lccn.loc.gov/2022017456

ISBN: 9781032110745 (hbk)
ISBN: 9781032110738 (pbk)
ISBN: 9781003218319 (ebk)

DOI: 10.4324/9781003218319

Typeset in Bembo
by Deanta Global Publishing Services, Chennai, India

To the memory of John H. Caldwell, Senior Lecturer in Town and Country Planning at Queen's University Belfast from 1971 to 1998

CONTENTS

ILLUSTRATIONS

Figures

Photographs

TABLES

CONTRIBUTORS

Joseph Onwona Ansong is a PhD graduate from Ulster University, Belfast, Northern Ireland and a Marine Planner with Howell Marine Consulting, Morpeth, England.

Peadar Davis is Senior Lecturer in Property Appraisal and Management at Ulster University, Belfast, Northern Ireland.

Wesley Flannery is Reader in Environmental Planning at Queen's University Belfast, Belfast, Northern Ireland.

Neil Galway is Lecturer in Planning at Queen's University Belfast, Belfast, Northern Itreland.

Brian Jack, prior to retirement, was Senior Lecturer in Environmental Law at Queen's University Belfast, Belfast, Northern Ireland.

Sean MacIntyre is Senior Lecturer in Environmental Health at Ulster University, Belfast, Northern Ireland.

John McCord is Lecturer in Law at Ulster University, Belfast, Northern Ireland.

Michael McCord is Reader in Property Market Research at Ulster University, Belfast, Northern Ireland.

Gary McGhee is Partner and Head of the Planning and Environmental Law team at Carson McDowell, Solicitors, Belfast, Northern Ireland.

Jane McKay is Solicitor with ByrneWallace LLP, Dublin, Ireland and New York, USA.

Stephen McKay is Reader in Planning Law and Practice at Queen's University Belfast, Belfast, Northern Ireland.

Thomas Murphy is Senior Lecturer in Land, Property and Housing Law at Ulster University, Belfast, Northern Ireland.

Michael Murray, prior to retirement, was Reader in Spatial Planning at Queen's University Belfast, Belfast, Northern Ireland.

Brendan Murtagh is Professor of Urban Planning at Queen's University Belfast, Belfast, Northern Ireland.

Diane O'Neill is Senior Commissioner with the Planning Appeals Commission for Northern Ireland, Belfast, Northern Ireland.

William Orbinson QC is Senior Counsel and Member of the Bar of Northern Ireland, Belfast, Northern Ireland.

Mura Quigley is a Consultant in Sustainable Placemaking and Design and is a Doctoral Researcher at Ulster University, Belfast, Northern Ireland.

Gavan Rafferty is Lecturer in Spatial Planning and Development at Ulster University, Belfast, Northern Ireland.

Heather Ritchie is Lecturer in Spatial Planning and Energy Policy at Ulster University, Belfast, Northern Itreland.

Roisin Willmott OBE, FRTPI, is Director of the Royal Town Planning Institute for Northern Ireland and Wales.

FOREWORD

There are many similarities in the planning systems across the United Kingdom but the differences, which reflect the context of the different jurisdictions, have always interested me. My role in the Royal Town Planning Institute puts me in a privileged position, enabling me to delve into the detail of the planning systems in both Northern Ireland and Wales (and frequently indulging me to look at other experiences and practices, too). Inevitably this leads me, at least at a cursory level, to compare and contrast the approaches taken and, interestingly, when the same process or mechanism is put in place, how it is implemented differently; it stimulates the sleeping (and unqualified) anthropologist in me.

Those professionals operating in the United Kingdom's devolved planning systems – in Northern Ireland, Scotland and Wales – appreciate that their planning systems are far from identical. Finding authoritative material focused on the Northern Ireland planning system is rare, which is why *Planning Law and Practice in Northern Ireland* is such a valuable resource for those entering the profession, as well as for those who are more established practitioners. It provides an unencumbered reference for understanding planning in the region.

All policy and practice undergoes change: adapting to societal transformation, new technology and wider circumstances. It is for this reason that I am delighted to see the Second Edition of this valuable reference published, thus keeping it in tune with amendments in legislation, policy and practice over the period from when the First Edition was published in 2017, shortly after the current planning governance framework came into operation. There has been a significant change in the interim, some of it being a subtle evolution, some more substantive. The plan-led system has been establishing itself, with an increased understanding of what this means through the preparation of Local Development Plans. Placemaking and the steps needed to achieve this have acquired momentum. The approach to planning in Northern Ireland continues to evolve, alongside developing principles and approaches to address climate change. There have been changes induced by technology, accelerated by the COVID-19 pandemic, regarding how planning services can be delivered and how the built environment itself responds to digitisation and changes in working, living, education and leisure patterns. The spatial challenges arising from Brexit add a further layer of implications for planning. Accordingly, this Second Edition, by including updates to all the chapters of the

First Edition, helps with mapping the progress made over recent years. The addition of new chapters on development plans, developer contributions, placemaking, cross-border spatial planning and renewable energy reflect this steady evolution of planning in Northern Ireland. The journey will continue and thus the inclusion of these topics will support the progression of the planning agenda and the embedding of these matters in professional practice.

This Second Edition of *Planning Law and Practice in Northern Ireland* provides a timely compilation of contemporary planning themes for students and practitioners; it should be regarded as the seminal reference for anyone with an interest in the discipline in Northern Ireland.

Roisin Willmott OBE, FRTPI
The Royal Town Planning Institute

PREFACE TO FIRST EDITION

The academic year 2016–2017 marks the fiftieth anniversary of the commencement of planning education at Queen's University Belfast. Over the intervening period additional planning programmes have been introduced at Ulster University and Belfast Metropolitan College and thus the planning profession is currently well served in terms of educational opportunities in Northern Ireland. In addition, the discipline has moved over time from an essentially architecture based curriculum to one more linked to the broad sweep of the social sciences, while the substantive focus has been sharpened to demonstrate the possibilities for mediation between environmental protection and business investment interests, increasingly within an international context. Notwithstanding the emergence of alternative theoretical discourses, new challenges for planning practice and innovative methodologies in the planning toolkit, the legal framework within which the profession operates has remained a constant theme of academic inquiry and student learning. The importance of planning law and practice in planning degree programmes at Queen's, at both undergraduate and postgraduate levels, was quite recently highlighted by a comprehensive review of teaching content that involved intensive discussion with students, graduates, practitioners and professional institutes. A key message coming through from those conversations was the primacy of understanding and competency in this area. This book seeks to make a contribution to the wider literature that all planning students across the three local institutions and further afield are directed to, as well as acting as a source of reference for planning practitioners and researchers. Its jurisdictional scope is essentially constructed around the Northern Ireland planning system, but inevitably there is an operational embeddedness within the broader United Kingdom and European Union dimensions of planning law and policy. The content of the book recognises this geographical interdependence, but within the contemporary governance preference for devolution, it seeks to reveal the distinctive personality of planning practice in the region.

On 23 June 2016 United Kingdom citizens voted by referendum to discontinue membership of the European Union. A withdrawal will require lengthy consideration by Parliament of the need to maintain, amend or repeal legislation over the years ahead and will bring into focus the European Communities Act 1972, European Union directives and regulations, as well as the decisions emanating from the Court of Justice of the European Union. There are implications here for planning law and practice as illustrated by some of the chapter content in this book. In that regard these matters constitute serious challenges for the effective reshaping of the United Kingdom planning systems during the forthcoming transitional period and beyond, and it will be for a later volume of essays to review how this has fared in Northern Ireland.

Our initial interest, as editors of and contributors to this book on planning law and practice, owes much to the stimulating lectures in this sphere provided by John Caldwell, Alan Dowling, Brian Garrett and David Trimble during our time as students on the then two years postgraduate programme in planning at Queen's. That trajectory has been further developed through a close professional collegiality with those working in the planning and legal professions, along with our teaching and research activities at Ulster University and Queen's University Belfast. It is appropriate that we acknowledge these formative influences and express our thanks for the institutional support that has sustained our engagement over many years in talking and writing about the legal and policy environments of planning.

We extend our gratitude to Valerie Rose at Ashgate for her early enthusiasm for this book and, following a very helpful peer review process of our proposal, her support in moving the project to contract stage. Subsequently, Grace Harrison and Sade Lee at Routledge / Taylor & Francis provided valuable editorial support in bringing the book through the production process. We are also grateful to Libby Mulqueeny for her work on some of the illustrations.

Finally, we convey our deep thanks to colleagues and friends who contributed to this book. We appreciate the generosity shown in sharing research-based and practice insights, and in responding graciously to our requests and queries in the run up to publication. We wish to also thank a number of readers for their valuable comments on draft material – they shall remain anonymous, but they know who they are.

Stephen McKay and Michael Murray
Queen's University Belfast
2016

PREFACE TO SECOND EDITION

The First Edition of *Planning Law and Practice in Northern Ireland* was published by Routledge in 2017 and was set to a baseline of the previous year. The book has been well received by planning students, by those teaching in the discipline, by planning practitioners and by planning decision-makers. The intervening period has seen the bedding-down of our new two-tier planning system in Northern Ireland, the emergence of fresh challenges linked to the outworking of Brexit, legislative and case law progression and ongoing policy refinement. It is against that multidimensional backcloth that a Second Edition of this book was commissioned in the Spring of 2021.

All thirteen chapters in the First Edition have been revisited and significant updates have been inserted to give greater currency. Moreover, five new chapters have been added to the Second Edition that take account of the bedrock instruments and ever-widening compass of our profession. The list of judicial decisions and the index have been extended to facilitate further research and easy access to the content of the book.

We are delighted that all contributors to the First Edition responded enthusiastically to our proposal for an update and we greatly appreciate the willingness of colleagues to add new chapter content. The academic and planning practice environments over the period from 2020 have been deeply affected by COVID-19 and we have been fully aware that routines of teaching, research and writing have been interrupted by constant uncertainty and new workplace protocols. We would like to place on record our deepest thanks to all our contributors who stepped up to the plate and gave of their time and expertise during these difficult circumstances. Our special thanks are extended to Roisin Willmott at the Royal Town Planning Institute for penning a generous Foreword to the book.

Our thanks also go to Kate Schell, Senior Editor, Planning, Landscape, and Urban Design at Routledge for her enthusiasm in commissioning this Second Edition and guiding our proposal through the peer review and contracting processes, and to Megha Patel, Editorial Assistant, Planning, Landscape, and Urban Design at Routledge who steered the book through its production schedule.

Finally, our thanks to several reviewers of our draft proposal whose thoughtful commentaries have helped shape the structure and content of this Second Edition of *Planning Law and Practice in Northern Ireland*. Again, we respect their anonymity but they know who they are.

Stephen McKay and Michael Murray
Queen's University Belfast
2022

1

THE EVOLVING NORTHERN IRELAND PLANNING SYSTEM

Stephen McKay and Michael Murray

Introduction

Over time the planning discipline has acquired a series of labels which seek to better represent its professional ambitions. Among the earliest and most longstanding is *town and country planning*, which, from a traditional British public policy perspective, harks back to the social idealism of Ebenezer Howard and the need to address unsanitary and overcrowded living conditions in the 19th-century city by the establishment of garden villages. The regulation of urban sprawl into the countryside, the protection of good farmland for food production, the building of new towns and the protection of rural landscapes subsequently became key responsibilities for planning in the lead up to and following the Second World War. The passage of the Town and Country Planning Act 1947 for England and Wales provided the significant statutory foundation for the instruments of development plans and development control to help achieve these goals. The momentum provided by environmentalism gained prominence during the 1980s in the context of an emerging global debate around sustainable development, climate change and renewable energy. When linked to the perceived collapse of the international modernist project of urban redevelopment in the 1960s and 1970s and the desire for new ways to build communities, the second big label of *environmental planning* chimed particularly well with the search by the planning profession at that time for a new set of credentials. In the immediate post-2000 era, the favoured phrase emerged as *spatial planning*, which accepts the logic of planning the interaction of places with the multiple spaces within which they are situated and that links with approaches which are, at once, strategic, integrated and locally sensitive. The influence of the European Spatial Development Perspective, agreed at the informal Council of Planning Ministers in Potsdam in 1999, was pivotal here. In the 2020s *placemaking* has become the moniker of a profession that proudly asserts its appositeness to urban design, strategic thinking and local engagement. This ongoing reinvention of planning has not abandoned its previous mission; rather, it has resulted in adding new layers of activity linked to emergent societal issues at varied scales of geography. All these themes are rehearsed within this book.

DOI: 10.4324/9781003218319-1

The Northern Ireland planning system has not been unaffected by these wider trends, changing circumstance and the need to adapt in order to better serve the public interest. This chapter explores how planning has fared from its earliest application to public policy in the region, its shifting operation within the complexities of regional government structures, including lengthy periods of direct rule, and the substantive challenges currently being addressed in a dynamic trajectory of institutional and practice reform. In this regard the Northern Ireland planning system is far removed from being an exact mirror image of the other United Kingdom planning systems, although it has drawn considerably over time on legal principles and practice in England and Wales. But there are differences in context and content and thus the purpose of this book is to offer a more nuanced insight into how the business of planning is conducted in Northern Ireland. This chapter, put simply, sets the scene by way of an unfolding chronology of planning succession and, in so doing, identifies some of the key legislative, administrative and policy provisions that have helped to shape planning governance and to determine the scope of its regulatory intervention.

The early years

The establishment of Northern Ireland on 3 May 1921, under the Government of Ireland Act 1920, allowed for its administration by a legislature separate from Westminster, while at the same time remaining within the United Kingdom. The region thus acquired under devolution many of the characteristics of an independent state with its own Parliament, Civil Service, High Court and Court of Appeal, and a system of executive control comprising a Prime Minister and Cabinet. The neoclassical splendour of the Parliament Buildings at Stormont, opened in 1932, represented at that time an architectural symbol of this new dispensation. Regional government, however, enjoyed the trappings more of pomp than circumstance since its effectiveness, as argued by Buckland (1981), was hamstrung by inadequate powers, especially in relation to finance, and by the key role assigned to local authorities. Political tensions operated on two distinct axes: outwards to the Treasury in London, with which there were frequent engagements by the Ministry of Finance and downwards to a mosaic of local councils in regard to service delivery. And, of course, in the background there was the border, the outworking of partition and its effects on everyday life. Nash et al. (2013: 36) observe that while cross-border cooperation did occur, 'it was always problematic'. Cross-border rail and road travel was impacted, postal services required reorganising and, for example:

> barristers and solicitors working along the border were suddenly faced with the problem that practicing law in Northern Ireland and the Free State required different systems of qualification. There were similar problems for other professionals such as pharmacists and teachers.
>
> *(ibid. 36)*

It was within these evolving circumstances that a fledgling system of town and country planning emerged but which trundled along with a great degree of operational impotence, prompting legislative, administrative and policy reform in the 1960s. This reform agenda led to a more robust planning system based on new legislation that came into effect in 1973 with the handing over of planning responsibilities to professional Civil Service planners within central government departmental and agency structures.

The legal context for town and country planning was initially provided by the Planning and Housing Act (Northern Ireland) 1931, whereby Article 2 invited local authorities to prepare and adopt planning schemes for lands within their districts. The second Schedule of the Act identified the matters that may be dealt with in a planning scheme as comprising built development, infrastructure provision, private and public open space and the preservation of natural beauty. However, given the discretionary nature of the legislation, no planning schemes were brought forward by local councils with the result that much of the development carried out in Northern Ireland was to a large extent uncontrolled between 1931 (and the decade prior) and subsequent legislation in the form of the Planning (Interim Development) Act (Northern Ireland) 1944. This statute made it obligatory for each local authority to become involved in planning activity. The Act placed all land under statutory planning control and provided that within three months of its passing all land which was not the subject of a planning resolution under the 1931 Act was deemed to be subject to such a resolution. However, as noted by Newman (1965), it proved impossible to prepare the statutory schemes with the very limited number of planning staff available at the time in Northern Ireland, and, as a result, only outline advisory plans were introduced as an interim measure to guide prospective developers and planning authorities. Some fifty-one of these plans were produced, but, in many instances, they were nothing more than a general extension of land use zoning together with illustrative road improvement proposals. They made little reference to future population targets and were not bound to specific programmes of public expenditure. Moreover, effective development control was hamstrung by compensation difficulties which could be traced back to an interpretation of the Government of Ireland Act 1920, whereby planning restraint could be deemed contrary to the property rights of individual landowners. Only with the 1959 House of Lords judgment in the landmark case *Belfast Corporation v OD Cars Ltd* (1960 1 All ER 65) was some respite afforded; it held that compensation was not an entitlement to a landowner denied the right to develop land through the operation of the planning system (Walsh 2015).

A multiplicity of local authorities across Northern Ireland at that time placed added constraints on the operation of the planning system. Local government was derived from the Local Government (Ireland) Act 1898. Following partition of the island of Ireland, Northern Ireland inherited a two-tiered mosaic of county boroughs and county councils below which were urban district councils and rural district councils. There were varying functions across health, education, welfare, housing, sanitary services and roads. The Planning Acts Amendment Act (Northern Ireland) 1944 enabled district councils to transfer their planning powers to county councils and in the event all the rural district councils did so together with six of the urban district councils, thus establishing thirty-seven planning authorities, including Belfast Harbour Commissioners and Belfast Corporation. Hayes (1967: 94) has commented on that outcome as follows:

> Town and country planning suffers from the number of small areas and the rivalry between county, rural and urban authorities. A small urban district in the middle of a county cannot properly plan its own development because that depends also on the development of the surrounding countryside and it may impede the development of the whole area by its individual existence.

Hayes, in particular, was strongly critical of the arrangements in place for the planning of Belfast and argued that the whole of the Belfast metropolitan area should be administered by a single authority taking in, as he suggested, the suburban areas of Holywood, Castlereagh, Lisburn, Newtownabbey and Carrickfergus. One consequence, though, of such a rational configuration would unquestionably have been the embedding of a deeply contested power relationship between Stormont and a very expansive metropolitan local authority, which the former would never have countenanced. At the commencement of the 1960s, therefore, the evidence available would suggest that any political commitment, thus far, to town and country planning at central and local government levels had been limited. The time was right for reform as Northern Ireland sought to expand its economy through major international inward investment and a modernisation agenda linked to improved infrastructure, better housing and extended third-level education campus provision.

The initial stimulus was the commissioning by the Northern Ireland government of Robert Matthew (Photograph 1.1), an Edinburgh-based planner, to prepare a planning report for the Belfast region. While the scope of his planning proposals was wide-ranging across Northern Ireland and included the creation of new and expanded towns linked to the containment of Belfast, his report emphasised that effective implementation depended upon suitable administrative and legislative arrangements, hitherto absent, being in place. His advocacy for the creation of a central planning authority operating under the auspices of a newly established Ministry of Planning and Development and led by a government Minister was key. He also suggested that the option of a smaller number of local planning authorities might be considered as an additional measure. And, finally, Matthew argued that an adequate and highly skilled planning profession would be essential for success.

At that time, the staffing of local planning authorities in Northern Ireland was a low priority conditioned by the often-modest budgets commanded by local government. Even as late as 1965 only forty-six practitioners were listed in the membership records of the

PHOTOGRAPH 1.1 Sir Robert Matthew, 1963. *Credit*: Henk Snoek / RIBA Collections Ref No: RIBA63274

regional branch of the Town Planning Institute; only six local authorities were serviced by full-time planning staff, with the remainder of the thirty-seven planning authorities drawing upon those services available at county level or upon part-time planning consultants. In October 1964 the Northern Ireland Government established a high-level Working Party on the Recruitment and Training of Town and Country Planning Staff, which reported back to the Minister of Development, William Craig, the following August. The Working Party convened twelve meetings during its period of operation and established two subcommittees to specifically examine the demand for planners in Northern Ireland and the possibility of setting up an additional course for the training of planning technicians. It estimated the immediate need for between 120 to 140 planners in public service in Northern Ireland. Thus a two-pronged strategy emerged: to recruit and retain planners drawn from other parts of the United Kingdom and to facilitate the training of Northern Ireland people to become planners by offering Great Britain-based planning school bursaries to local candidates as well as developing a new capability for planning education at degree level within the region based at Queen's University Belfast. Subsequently, in the autumn of 1966 a two-year postgraduate programme in Town and Country Planning commenced at Queen's with an initial enrolment of thirteen students and headed by an endowed Chair appointment (see Houston et al. 2017)

Meanwhile, following the publication in 1964 of a Command Paper (Cmd 456) on the administration of town and country planning, the government constituted a Ministry of Development in 1965 with responsibility for planning. Government also proposed that development control would become a function of a Central Development Office and operate in consultation with local authorities. The latter failed to generate a political consensus, as Aitkin (1967: 12–13) observed:

> public reaction was disappointingly thin, although favourable. The reactions of local authorities, on the other hand, were quite strong and, in the main, unfavourable … where the strongest opposition came was in the matter of development control by the central authority. It is apparent that this is the strongest focus of local interest for councils and councillors.

Faced with this impasse, the Government moved to a different strategy, one that sought to achieve fundamental reform of the structure of local government itself. Consideration of any future planning system was to be set within the context of reshaping local administration, the guiding parameters of which were held to be efficiency, economy and effective representation of local aspirations. In December 1967, a White Paper (Cmd 517) on local government reform was published. The concept of twelve to eighteen single-tier, all-purpose authorities was commended for consideration and while the framework for the preparation of a range of plans (regional, area, sub-area and action plans) was related to appropriate levels of central and local government, the thorny issue of development control responsibilities was purposely omitted. Following consultation, a further set of proposals (Cmd 530) for the reshaping of local government was published in July 1969. An administrative geography of seventeen area councils was proposed, which would have precluded the retention of the traditional counties but whose primary functions of health, education and welfare were in any event under review. In the sphere of town and country planning, a Central Development Office was to assume responsibility for regional policy, area plan preparation and approval of sub-area plans prepared by local authorities. Development control was to remain a council function, but with call-in powers given to the Ministry for certain classes of major development. Critically for the events which were to follow, housing was also to remain a local authority function.

The 1960s in Northern Ireland were set to close, therefore, as a decade of enquiry into the institutional context for town and country planning. In considering the future form of local administration at this time, policymakers in the region were moving in tandem with those elsewhere in the United Kingdom. In February 1966 the United Kingdom government announced the appointment of the Royal Commission on Local Government in England and Wales and its counterpart in Scotland. Reports from both were delivered in 1969, but, as noted by Alexander (1982), there was still no consensus about the principles upon which the new system should be based. In Northern Ireland, however, certainly in regard to town and country planning, many of the technical criticisms of the planning system and the accompanying political sensitivities had apparently been addressed. Local government was ensured a continuing executive role in planning within a one-tier council structure, and, with the promise of additional legislation to give effect to this final set of proposals, a political compromise between central and local interests had been effected. But the outbreak of civil violence in 1969 changed this utterly, not least because these governance structures were precariously balanced in the context of the divided society of Northern Ireland.

At the root of the problem was the intense political rivalry between Nationalists and Unionists affecting about two-thirds of the councils (Elliott and Wilford 1987). Nationalists claimed that local electoral areas were subject to a Unionist gerrymander and that the longstanding alteration of the franchise by incorporating property ownership as a voting qualification put them at a disadvantage. This had been compounded by the replacement of proportional representation with the simple majority method of election under the Local Government Act (Northern Ireland) 1922. Alleged discrimination in the allocation of public sector housing and employment gave increased force to the dissatisfaction expressed by the minority population. Indeed, the Cameron Commission (Cmd 532), established to report on the civil unrest in 1969, found well-documented cases of the manipulation of local government electoral boundaries and the distortion of council housing policy for political ends. When combined with the announcement through a joint communiqué issued by the Northern Ireland and United Kingdom Governments in October 1969 that a central housing body was to be established for Northern Ireland involving a phased transfer of housing functions from local councils, a more far-reaching reform of local government than was previously considered became inevitable. The outcome was the publication of the Macrory Report (Cmd 546) in June 1970.

While reform of policy on public sector housing provision constituted the essential departure point for the deliberations of this review committee, the need to guarantee provision of land supply and supporting infrastructure dictated that the future of town and country planning should form an essential element of the working agenda. Macrory, accordingly, recommended that central government should be responsible for a suite of regional services and that, in particular, forward planning, roads, traffic management, water and sewerage functions should be vested in the Ministry of Development. The Minister in charge was to be accountable to the Northern Ireland Parliament, thus ensuring satisfactory administration. At a local level the establishment of not more than twenty-six district councils was recommended with a range of ceremonial, consultative, representative and executive functions. These would include services which the councils would operate in their own right by virtue of powers given to them by the Northern Ireland Parliament (recreation, environmental health and cleansing), as well as a range of activities carried out on an agency basis for central government. Close reading of the Macrory Report points to development control being operated on these lines when it is stated: 'We certainly envisage the District Councils should handle local planning applications within the overall development plan for which the Ministry of Development would be responsible' (Macrory 1970: 43).

The recommendations of the Macrory Report were accepted by the Government as the basis for the reorganisation of local government and enabling legislation was passed under the Local Government Act (Northern Ireland) 1972. Council elections followed in 1973 (the voting system having been changed to proportional representation) and this process brought into play a new, single-tier structure of twenty-six district councils that was to endure through to 1 April 2015. As Macrory had envisaged, the attendant range of functions which local government had previously enjoyed was reduced, but, in contrast to what he had proposed, all planning responsibilities were assumed by the Ministry of Development within its Town and Country Planning Service. Local government was given only a consultative role in regard to planning. With the Northern Ireland Government and Parliament having been suspended by the United Kingdom Prime Minister, Edward Heath, in March 1972 in the face of domestic political instability and escalating street violence, it was left to Westminster to approve enabling planning legislation in the form of the seminal Planning (Northern Ireland) Order 1972. For the first time, Northern Ireland had received comprehensive town and country planning legislation, which, as already noted, came into effect in October 1973.

Planning under direct rule

The arrangements for town and country planning in Northern Ireland that placed executive responsibility within a central government department and reduced the role of local authorities to mere consultation are very different from what has prevailed elsewhere in the United Kingdom. In England, Scotland and Wales planning functions are divided between each tier, and local government takes a strong lead in plan-making and development management in line with the imperative of local political accountability. However, in Northern Ireland, the centralisation of planning, along with many other aspects of public services delivery including the creation of a substantial non-departmental/quango sector, also became intertwined with the additional and unforeseen circumstances of lengthy 'direct rule' in effect to 1999. Blackman (1984: 12), in writing about that early period of public administration, has observed:

> With the introduction of Direct Rule in 1972, the relegation of local government and concentration of power in unelected government agencies, a mini Orwellian state took shape. For Northern Ireland, it seems as if 1984 arrived back in 1972. Its people lost their democratic rights in the confusion of identity and the conflict brought about by the nature of the state in Northern Ireland.

More specifically in regard to town and country planning, Hendry (1984: 38) has argued:

> This final shift of powers now effectively precludes that element of local control which has remained an essential component of the planning system in Britain. Once physical planning is removed from the control of both local and regional government, it can be

> represented primarily as a restriction on the rights of the individual by some remote bureaucratic institution.

Direct rule was intended to be only a temporary expedient and much of the subsequent wrangling in establishing stable regional government reflected the seemingly intractable problems of a divided society grappling with the wicked problems of constitutional positioning and power-sharing. In the event, as noted by O'Leary et al. (1988), it was to become a more enduring feature of the political system as successive efforts to restore devolved government foundered. Initial Northern Ireland Assembly elections took place in 1973 and led to the appointment of a multiparty executive coalition that took office in January 1974. It collapsed in May of that year and the Assembly was formally dissolved in March 1975. The convening of a Constitutional Convention, headed by former Lord Chief Justice of Northern Ireland Sir Robert Lowry, followed in 1975 as an attempt to break through the political impasse. It was dissolved in March 1976. A round table conference of constitutional parties in 1980, under the auspices of Secretary of State Humphrey Atkins, lasted only three months. A new Northern Ireland Assembly with only deliberative and scrutiny functions was elected in 1982. The committees established as part of a hybrid regional governance infrastructure allowed locally elected politicians to interact with direct rule Ministers and make recommendations to government, but it was dissolved in 1986. Elections to a Northern Ireland Forum followed in 1996 and this body was charged with making progress on the structural issues of Northern Ireland regional governance within the context of political relationships on the island of Ireland and with the rest of the United Kingdom. This work culminated in the signing of the intergovernmental Belfast/Good Friday Agreement in April 1998 which, inter alia, paved the way for the establishment of a new, 108-member, Northern Ireland Assembly and Executive Committee of Ministers in December 1999 following a shadow or pre-devolution period of institutional preparatory work. The engagement of the post-1998 institutions with town and country planning in Northern Ireland will be explored in the next section of this chapter but, prior to that, some brief comment is appropriate on the operation of the planning system during the previous years of direct rule.

The provisions of the Northern Ireland Act 1974 introduced alternative arrangements to the devolved administration of the region. Legislation would henceforth be brought forward within Westminster by Orders in Council, thus maintaining Northern Ireland's separate statute book and avoiding the increased time that would be necessary in introducing Northern Ireland bills as primary legislation in the House of Commons (Birrell 2007a). Government departments in discharging their functions would operate under the direction and control of the Secretary of State for Northern Ireland, but, in practice, the outworking of direct rule saw increased Civil Service influence over policy matters. In this vein, as noted by Carmichael (2002), the term 'Minister' was only a courtesy title and many direct rule Ministers adopted a 'hands off' approach (Carmichael and Osborne 2003). The business of administering the region thus adopted a style of technocratic proceduralism devoid of party-political ideology and was denoted by a reliance on professional expertise and senior civil servant managerialism to develop and implement policy. The essence of the approach was, simply, to keep the system going (Birrell 2013). It was within this context of governance adjustment that the Northern Ireland planning system was fashioned.

In the early years of direct rule, planning responsibilities rested with the Department of Housing, Local Government and Planning; however, by 1976, this had merged with the

Department of the Environment to create a new entity that endured for some forty years. For the greater part of that time, planning operated under the organisational title of the Town and Country Planning Service located within the Department of the Environment. It became a United Kingdom 'Next Steps' agency in 1996 with the chief executive designated as an accounting officer additional to the Department's Permanent Secretary, and it was tasked with responding to parliamentary questions and general queries about the activities of the agency. Renamed the Planning Service, the agency within the Department of the Environment continued to function by way of a Headquarters Branch in Belfast, which was responsible in the main for legislation, policy guidance, regional planning, major development proposals and research. Six divisional planning offices, with two sub-offices, looked after development plan preparation and routine development control. With a central government department now taking all planning decisions, the appellate function which it previously enjoyed was deemed inconsistent with this new situation. Accordingly, an independent appellate body, the Planning Appeals Commission, was established under statute to decide appeals and report on a variety of matters under planning and other legislation. In essence, therefore, professional planners, employed as civil servants, were charged with shaping planning policy and making planning decisions.

The legacy of that era can be appreciated by the fact that Northern Ireland widened its coverage of adopted development plans in line with the content of a new *Regional Physical Development Strategy 1975–1995* and its commitment to a district towns approach of multiple mini-growth centres. In 1993 the Department of the Environment published *A Planning Strategy for Rural Northern Ireland*, although this was deemed something of a misnomer by being more a collection of statements on individual aspects of planning policy and capable of being used 'as a rule book in order to find out restrictions on construction or use of buildings' (Northern Ireland Affairs Committee 1996: xxxiii). These documents were accompanied by the publication of supplementary planning guidance by way of *Development Control Advice Notes* and an emergent suite of *Planning Policy Statements* to replace the much-criticised *Planning Strategy for Rural Northern Ireland* and to assist both plan-making and development control. The 1972 Planning Order was rolled forward, notably by the passage of the Planning (Northern Ireland) Order 1991 and served to set town and country planning on a more complete statutory footing. However, it must be remembered that some of these technical accomplishments could well have been secured under the centralised operation of the planning system alone and cannot be attributed to the impact of direct rule per se. The main limitation was a democratic deficit, which, Birrell (2009) has argued, meant a denial of self-government over transferred matters as the decisions were not taken by local politicians with local knowledge. The absence of regional government with accountability to Northern Ireland citizens, with powers of close scrutiny and charged with steering the formulation and implementation of planning policy, fits with that assessment.

The Northern Ireland Assembly and planning

Following the 1998 political settlement, the New Northern Ireland Assembly was elected on 25 June 1998 and, during a pre-devolution or shadow period, it appointed a First Minister (designate) and Deputy First Minister (designate) to lead an Executive Committee of Ministers. In addition to the Office of the First Minister and Deputy First Minister, agreement was reached on the establishment of a further ten departments, which Wilford (2000: 590) has labelled 'an exercise in parity of ministerial esteem'. A Department of Justice was subsequently added to the range of transferred powers on 12 April 2010 following the successful outcome of Northern Ireland political party negotiations with the United Kingdom and Irish Governments linked to the St Andrews Agreement in 2006 and the Hillsborough Agreement in February 2010. From the outset this arrangement was designed to accommodate the sharing of ministerial portfolios across the major Northern Ireland parties in line with the pursuit of a consociational model of governance in a divided society where majority rule is unworkable. Thus, the broad planning functions of the Department of the Environment were, in part, hived off to a newly established Department for Regional Development (regional planning) and Department for Social Development (urban regeneration). Finally, statutory departmental committees were approved to provide scrutiny and policy advice and were given the power to initiate primary legislation. The configuration of department Ministers and committee chairpersons and deputy chairpersons was agreed at the end of November 1999. With devolution restored on 2 December 1999, bringing to an end some twenty-seven years of direct rule, the Northern Ireland Assembly was finally in charge of transferred matters that included planning legislation and policy.

Notwithstanding this promising start, the operation of the Northern Ireland Assembly was short-lived. Due to a bitter political fallout in October 2002, it was suspended that month and eventually dissolved on 28 April 2003. A further lengthy period of direct rule followed to 2007 during which Assembly elections were held in November 2003 and a non-legislative Assembly, made up of members from that 2003 election, was created by the Secretary of State. Its remit was to again make preparations through committee deliberations for the restoration of devolved government to Northern Ireland. The post-2007 period did witness continuity of the Northern Ireland Assembly, although there were prolonged interludes of deep anxiety regarding the functioning of the Executive linked, for example, to division over budgetary matters and welfare reform. The issue of paramilitary activity surfaced in 2015, prompting ministerial withdrawal and the suspension of meetings by the coalition Executive Committee of Ministers. The Assembly was not suspended by Secretary of State Theresa Villiers, and it did see out its full term through to March 2016 with planned elections following in May 2016 for a new mandate. However, controversy surrounding the implementation of a Non-domestic Renewable Heat Incentive Scheme (see McBride 2019; Coghlin et al. 2020) led to the collapse of the Executive Office of Northern Ireland and the dissolving of the Northern Ireland Assembly in January 2017, to be followed by a fresh election in March of that year. There was no immediate return to ministerial decision making and, with the United Kingdom Government choosing not to reintroduce direct rule, the Northern Ireland Civil Service assumed responsibility for public administration. This lacuna in Executive Committee decision making did have implications for planning, as discussed below, and prompted litigation in the High Court and Court of Appeal in regard to planning approval for a major waste disposal incinerator within the urban fringe of Belfast. Nonetheless, following prolonged multiparty negotiations that involved the United Kingdom and Irish Governments and the subsequent signing of the *New Decade, New Approach Deal*, the appointment of a new five-party Northern Ireland Executive in January 2020 ended this period of political deadlock. The interfacing of planning policy and the high politics of devolution in Northern Ireland over the period from 1998 to 2021 can be illustrated by the cases of regional planning and rural planning.

The case of regional planning

The fresh start to the governance of Northern Ireland was no better symbolised than by the commitment written into the 1998 Belfast/Good Friday Agreement regarding the preparation of a new regional planning framework. In fairness, the need for a coherent regional plan or strategy had emerged previously as a key recommendation from the review of the Northern Ireland planning system by the House of Commons Northern Ireland Affairs Committee in 1996. Enabling legislation was provided by the Strategic Planning (Northern Ireland) Order 1999, made in March of that year and just prior to devolution in December 1999. The strategy in final form was endorsed by the Northern Ireland Assembly in 2001 and published with the very positive slogan *Shaping Our Future* (Department for Regional Development 2001). This was substantially revised in 2012, not least to accommodate the economic slowdown and with the long-term time horizon shifting out to 2035 (Department for Regional Development 2012). It sets out a long-range vision for the physical development of the region and provides a framework within which to locate subsequent statutory development plans in regard to key settlement designations, housing growth indicators (refreshed periodically), infrastructure and economic development. Compared with the 2001 version, the latter strategy places more emphasis on the potential of Belfast and Derry/Londonderry to act as engines for regional growth. Under the terms of the Strategic Planning (Northern Ireland) Order 1999, government departments must have regard to the regional development strategy in exercising any functions in relation to development. The preparation of Local Development Plans and development schemes by local authorities must take account of the regional development strategy and that strategy is material to decisions on individual planning applications and appeals.

Significantly, this strategy, at the time of its introduction, stood in the vanguard of a new European 'spatial planning' approach that shortly afterwards saw equivalent plans launched for Wales, Scotland and the Republic of Ireland (see Healey 2004; Davoudi and Strange 2008). A key element of the European influence on strategic spatial planning is cross-border cooperation and, while the 2001 *Regional Development Strategy for Northern Ireland* largely represented Northern Ireland as a place apart, subsequent years have witnessed the refining of a more meaningful island of Ireland planning perspective dealing with shared interests in infrastructure connectivity and harnessing the economic and social potential of the Irish borderlands for mutual benefit (see Chapter 14 in this book for a further discussion). Thus, in February 2011 Northern Ireland Minister for Regional Development Conor Murphy and Ireland Minister for the Environment, Heritage and Local Government Éamon Ó Cuív, jointly issued a draft framework for public consultation titled *Spatial Strategies on the Island of Ireland – Framework for Collaboration* (Department for Regional Development and Department for the Environment, Heritage and Local Government 2011). It is noteworthy that this politically very sensitive approach to planning had initially been brokered some years previous by interests outside government, but with the support of the two government departments mentioned above, in the preparation and publication of a report with a largely similar title (InterTradeIreland 2006). The non-statutory framework was eventually approved by the Northern Ireland Executive on 27 June 2013 (Department for Regional Development and Department of the Environment, Community and Local Government 2013). This cross-border engagement is delicately poised as evidenced by the title of the final document *Framework for Cooperation: Spatial Strategies of Northern Ireland and the Republic of Ireland*, which underlines the existence of two planning strategies and by an inserted east-west constitutional positioning into the text that regular reports on spatial planning work streams will be prepared for the British Irish Council meetings. Moreover, as noted by Peel and Lloyd (2015: 19) the publication is denoted by 'a nuanced shift in nomenclature from collaboration to cooperation, and the launch involved officers from the two jurisdictions rather than Ministers, potentially undermining the status of the document'.

The case of rural planning

Rural planning policy in Northern Ireland has long been one of the most contested areas of town and country planning, pitting community-led rural development interests against environmental protection advocates. The content and application of policy have ranged over time from very strict control against new housing in the countryside to more liberal approaches to development control. It is probably the case that no other sphere of planning practice has had as many policy reviews over the period from 1973 involving multiple iterations of public consultation. Thus, in June 2004 during a period of direct rule, the Department for Regional Development embarked on yet another quest for policy revision by publishing an Issues Paper designed to capture stakeholder perspectives at an early stage. A total of eighty-six responses were received, indicating a broad 50/50 split on the question regarding a presumption in favour of development in the open countryside. In September 2005 officials from the Department for Regional Development circulated an internal paper that set out four options on the way forward:

1 Ongoing extensions to green belt and countryside policy areas through the development plan process while maintaining current planning policies.
2 Ongoing extensions to green belt and countryside policy areas through the development plan process and review the content of current planning policies.
3 The use of personal information related to kinship beyond green belts and country side policy areas thus filtering out urban generated rural housing demand.
4 Immediate application of green belt/countryside policy area designations to all of rural Northern Ireland, thus removing the presumption in favour of development.

All options were oriented towards a considerable tightening of rural planning policy with Option 4 being the most severe. The Department's internal paper noted its perceived impact in the following manner:

> It would immediately impact on applications and significantly reduce the number of approvals. Approvals would be limited to exemptions such as agricultural need, retiring farmer, special personal or domestic circumstances and replacement dwellings. It would generate significant political and community opposition in that it would be perceived as a draconian measure undermining the long-term sustainability and viability of indigenous rural communities. There would be every likelihood that, if devolution were restored, this policy (if it were to be adopted) would be reversed. Even if the exemptions were reviewed and extended, opposition would remain.
>
> *(Committee on the Programme for Government 2007: 29)*

Accordingly, officials recommended Option 3, but prior to its consideration by the Minister, the Special Adviser to the Secretary of State was consulted. In his response to officials, the

Adviser stated: 'I actually feel that we should now go for Option 4 and bring us into line with the rest of United Kingdom. I am particularly opposed to the "kinship" addition to current practice for wider political reasons' (Committee on the Programme for Government 2007: 30). The published report of the Committee on the Programme for Government within the Northern Ireland Assembly indicates that no further information on what was meant by 'wider political reasons' was given, but that following a meeting between the Special Adviser and officials at which the Adviser indicated 'a particular direction might be appropriate in the present circumstances', he wrote by email on 9 November 2005 that he had spoken to the Secretary of State and the Minister and that both agreed to proceed with the most restrictive policy, in line with the wider sustainability agenda. A memorandum submitted to the Minister on 17 November 2005 highlighted a potential adverse reaction to the adoption of Option 4 from all the local political parties. Despite this advice, direct rule Ministers decided to proceed with Option 4, which introduced a presumption against development in the countryside. On 16 March 2006 the Department for Regional Development issued *draft Planning Policy Statement 14: Sustainable Development in the Countryside*, with immediate implementation. The consultation process generated 8,513 written responses of which 95 percent expressed opposition to the strict policy stance being adopted and, as further evidence of this outcry, the draft planning policy statement was challenged by way of judicial review. In its deliberations on the matter between November 2006 and January 2007 the Northern Ireland Assembly Committee on the Programme for Government:

> considered the decision-making process which led to the introduction of Draft PPS14 and reached the conclusion that it had been seriously flawed. Members expressed serious concerns that the detailed work over a two to three year period under taken by professional staff within the Department for Regional Development, the Department of Agriculture and Rural Development and the Department of Environment had effectively been dismissed in a matter of weeks by the Secretary of State's Special Adviser. Members were appalled that such an important decision was taken because 'of wider political reasons' and that 'a particular direction might be appropriate in the present circumstances' rather than on the basis of sound planning policy reasons.
>
> *(Committee on the Programme for Government 2007: 34)*

What makes this vignette especially interesting is that it reveals the tensions instilled in the interaction at that time between ongoing direct rule and restored devolution. Arguably, the subliminal message being advanced was that if direct rule decisions are unpopular, then it is for devolved government to take action. The case of rural planning needs to be set alongside other contentious decisions taken contemporaneously by the direct rule administration in pursuit of that political agenda, including the abolition of the '11 plus' school transfer test and the introduction of water charges (Scannell and Turner 2009). Additionally, the controversial nature of public administration reform at that time, discussed further in the next section of this chapter, can also be viewed as goading Northern Ireland political parties to respond positively to a power-sharing executive and the restoration of devolution (Knox 2012). In the event, devolution returned on 8 May 2007, a draft replacement for Planning Policy Statement 14 was issued by the Department of the Environment on 25 November 2008, and on 11 December 2009 the locally appointed Minister indicated to the Northern Ireland Assembly that a final policy statement had been submitted to the Executive Committee of Ministers for consideration. On 1 June 2010 the Minister, Edwin Poots, announced to the Assembly that *Planning Policy Statement 21: Sustainable Development in the Countryside* was operational.

But, as indicated previously, policy formulation in this sphere is iterative, and on 7 March 2016 the Department of the Environment launched yet another review of strategic planning policy for development in the Northern Ireland countryside. Subsequently, in August 2021 the Department for Infrastructure issued a Planning Advice Note on *Implementation of Strategic Planning Policy on Development in the Countryside*. Rural interests, including the Ulster Farmers Union, expressed concern that this would have an adverse impact on rural communities. On 15 October 2021 the Planning Advice Note was withdrawn by the Department and four days later in Private Members' Business in the Northern Ireland Assembly the question was resolved:

> That this Assembly acknowledges that the planning system must be sustainable while also working for our rural communities … and calls on the Minister for Infrastructure to withdraw immediately the Planning Advice Note and to engage meaningfully with local councils regarding the very real concerns of people in our rural communities.

The reconfiguration of planning governance in Northern Ireland

Four prominent themes command deeper consideration in this preliminary examination of the Northern Ireland planning system: first, reshaping central government departments; second, evolving planning initiatives; third, decision making by the Executive Committee of Ministers; and fourth, redefining the role of local government.

Central government departments

The reshaping of Northern Ireland government departments was quietly addressed by political parties and the Northern Ireland Assembly over the period 2012 to 2016. The Executive's Programme for Government 2011–2015 had included a commitment to agree to changes to the structures of government which would operate in the next Assembly mandate. Again, the background can be traced to the 1998 Belfast/Good Friday Agreement, which contained a provision that executive authority would be discharged on behalf of the Assembly by a First Minister and Deputy First Minister and up to ten Ministers with departmental responsibilities. The subsequent Northern Ireland Act 1998 introduced a degree of upward flexibility for a greater number to be ordered by the Secretary of State.

As already noted, ten departments were established at that time (replacing the six departments operating prior), along with the Office of the First Minister and Deputy First Minister, and a further department was added in 2010. Clearly, the political capital to be gained by power-sharing outweighed administration inefficiencies. This arrangement points to a difference in the devolution experience in Northern Ireland, compared with other parts of the United Kingdom, where its focus was (and continues to be) securing accommodation between unionism and nationalism (Knox and Carmichael 2005). The Stormont House Agreement of 23 December 2014 is illustrative of that imperative in that, while passing mention is made of a reduction in government departments from twelve to nine, the document mainly addresses agreed measures on the very contentious headline-grabbing issues of flags, parades and the troubled past of Northern Ireland. The subsequent publication by the

Northern Ireland Executive of *A Fresh Start: The Stormont Agreement and Implementation Plan* on 17 November 2015 presented further information on a nine department model and indicated that the planning functions of the Department of the Environment, the Department for Regional Development, the Department for Social Development and the Office of the First Minister and Deputy First Minister would be spread across new departments of (1) Agriculture, Environment and Rural Affairs, (2) Infrastructure, (3) Justice, and (4) Communities. Thereafter the legislative process moved along quite rapidly, with broad political support weighing in on the need for a change in structures. A Departments Bill was introduced in the Northern Ireland Assembly on 30 November 2015 and received Royal Assent on 29 February 2016 to be known as the Departments Act (Northern Ireland) 2016. It provides for the new nine departments structure by renaming seven departments and abolishing three others (Table 1.1). The Departments (Transfer of Functions) Order (Northern Ireland) 2016 received a necessary affirmative resolution by the Northern Ireland Assembly on 15 March 2016, the last day of Assembly business before dissolution to facilitate elections.

Political balance aside, this reorganisation of departments and functions endeavours to meet aspirations to secure financial savings, to group functions within organisations of manageable size, to provide a better service to the public and to avoid an overlap in authority. But possible implications for planning practice arise from fragmentation across departments. Indeed the Royal Town Planning Institute (with some 650 members in Northern Ireland in 2021 drawn from central government departments, district councils, the Planning Appeals Commission, consultancies, the development industry and the education sector), while supporting a reduction in the number of departments, had commented by way of stakeholder

TABLE 1.1 Government departments in Northern Ireland

Pre-devolution 1998	*1998–2016*	*Post-2016 Assembly elections*
	Office of the First and Deputy First Minister	The Executive Office
Department of Agriculture	Department of Agriculture and Rural Development	Department of Agriculture, Environment and Rural Affairs
Department of Economic Development	Department of Enterprise, Trade and Investment	Department for the Economy
Department of Education	Department of Education	Department of Education
Department of Finance and Personnel	Department of Finance and Personnel	Department of Finance
Department of Health and Social Services	Department of Health, Social Services and Public Safety	Department of Health
	Department of Justice	Department of Justice
	Department for Regional Development	Department for Infrastructure
	Department for Social Development	Department for Communities
Department of the Environment	Department of the Environment	(dissolved)
	Department of Culture, Arts and Leisure	(dissolved)
	Department for Employment and Learning	(dissolved)

input to the deliberations of the Assembly and Executive Review Committee: 'It is vital to the delivery of a fit for purpose planning system that these functions are not split in order to ensure a smooth and joined up approach that will avoid unnecessary delays and enhance accountability' (Assembly and Executive Review Committee 2012: 252). While the post-1998 superficial, albeit legally significant, split between regional planning and strategic planning policy preparation has been addressed by each moving, respectively, from the Department for Regional Development and the Department of the Environment to the Department for Infrastructure, there is still an incomplete representation of planning activities in one department. Thus, for example, powers to list buildings, to designate conservation areas and to exercise listed building enforcement under the Planning Act (Northern Ireland) 2011 have been transferred from the Department of the Environment to the Department for Communities, with most other central government planning functions under that legislation located in the Department for Infrastructure. Meanwhile, environmental management, including waste and the emergent policy area of marine spatial planning, is within the compass of the Department of Agriculture, Environment and Rural Affairs. Interdepartmental collaboration continues to be required as these post-2016 structures do business and which is well illustrated by the establishment of the Strategic Design Group jointly by the Departments for Infrastructure and Communities and whose aim is 'to promote successful, inclusive, well designed places which inspire civic stewardship and have an enduring positive impact on people's lives'.

Planning initiatives

The Northern Ireland planning system is denoted by evolving episodes of legislation, policy initiative, public announcement and consultation that continuously seek to stimulate reform, modernisation and responsiveness to pressing challenges, which not least in recent times are related to the consequences of Brexit vis-à-vis the implementation of EU environmental protocols and the impact of COVID-19 on planning procedures. The post-devolution lengthy reform agenda in Northern Ireland (Table 1.2) has been aptly described by Lloyd (2008: 5) as being primarily concerned with improving efficiencies in decision making, securing greater effectiveness in the delivery of societal and planning objectives and promoting more active engagement by stakeholders in its various processes. Such considerations can be set alongside the much-aired criticisms of planning coming, on occasion, from issue advocates and coalitions holding very different ideological perspectives. The planning system has been attacked for its tardy performance in making development plans and delivering decisions on planning applications, in being hostile to the introduction of third-party planning appeals, in frustrating the aspirations of the development industry, in being weak in regard to environmental protection and building design, in operating within a piecemeal and poorly defined regional planning context and in having produced a confusing panoply of planning policies which on occasion are misunderstood, disregarded or not applied with consistency. And there has been no shortage of documented scrutiny: in June 2020 academic research on perceptions of planning in Northern Ireland highlighted issues around public trust; in October 2020 the Northern Ireland Confederation of British Industry published its review and recommendations related to major planning applications; in 2022 the Northern Ireland Audit Office completed its examination of the Northern Ireland planning system with special reference to the period since 2015 when local government assumed greater responsibilities in this sphere. This was followed up by a report from the Northern Ireland Assembly Public Accounts Committee (2022) that, in its press

TABLE 1.2 Evolving planning initiatives

Date	*Key planning element*
13 February 2002	Modernising planning processes: a consultation paper (Sam Foster MLA, Environment Minister, and Planning Service)
3 February 2003	Modernising Planning Processes Implementation Plan (Angela Smith MP, Parliamentary Under-Secretary of State, and Planning Service)
19 August 2004	Reforming Planning: Proposals to Amend Primary Planning Legislation in Northern Ireland public consultation paper (Angela Smith MP, Parliamentary Under-Secretary of State, and Planning Service)
28 October 2005	Consultation commences on draft Planning Reform (Northern Ireland) Order 2006 and Explanatory Memorandum (Department of the Environment)
9 May 2006	Planning Reform (Northern Ireland) Order 2006 made at Westminster
29 November 2007	Announcement of planning reform process (Arlene Foster MLA, Environment Minister)
April 2008	Planning reform in Northern Ireland: independent report to the Minister for the Environment (Professor Greg Lloyd)
6 July 2009	Reform of the planning system in Northern Ireland: your chance to influence change (Edwin Poots MLA, Environment Minister, and Planning Service)
October 2009	Reform of the planning system in Northern Ireland: your chance to influence change. Report of the independent consultation events undertaken on behalf of the Planning Service (Ms Lisa Copeland)
25 November 2009	The performance of the Planning Service (Northern Ireland Audit Office, 2009)
25 February 2010	The NI Executive agrees to the final policy proposals for reform of the planning system and next steps to bring forward legislation and guidance
March 2010	Reform of the Planning System in Northern Ireland: your chance to influence change. Government Response to Public Consultation, July–October 2009 (Planning Service)
6 December 2010	Planning Bill introduced to Northern Ireland Assembly (Edwin Poots MLA, Environment Minister)
4 May 2011	Planning Act (Northern Ireland) 2011 receives Royal Assent
14 January 2013	Planning Bill introduced (Alex Attwood MLA, Environment Minister)
22 October 2013	Planning Bill withdrawn (Mark H Durkan MLA, Environment Minister)
4 February 2014	A Strategic Planning Policy Statement for Northern Ireland – Planning for sustainable development – public consultation draft (Mark H. Durkan MLA, Environment Minister)
28 May 2014	Phase 1 consultation on planning reform and transfer to local government: proposals for subordinate legislation (Mark H. Durkan MLA, Environment Minister)
18 September 2014	Living places - an urban stewardship and design guide for Northern Ireland published
31 October 2014	Phase 2 consultation on planning reform and transfer to local government: proposals for subordinate legislation (Mark H. Durkan MLA, Environment Minister)
1 April 2015	Transfer of planning functions to local government becomes operational
28 September 2015	A Strategic Planning Policy Statement for Northern Ireland: 'Planning for sustainable development' (Mark H. Durkan MLA, Environment Minister)
27 September 2016	Announcement of intention to review strategic planning policy for renewable energy and development in the countryside (Chris Hazzard MLA, Infrastructure Minister)

(*Continued*)

TABLE 1.2 (Continued)

Date	*Key planning element*
8 December 2016	Public consultation on removal of permitted development rights for oil and gas exploration within the Planning (General Permitted Development) Order (NI) 2015
15 December 2016	Public consultation on proposals to transpose the requirements of the revised EIA Directive (2014/52/EU) into the Planning (EIA) Regulations (NI) 2017
1 April 2017	Direction amending the notification arrangements when Councils intend to grant listed building consent
17 May 2018	Best practice place profiles published
1 November 2018	Northern Ireland (Executive Formation and Exercise of Functions) Act 2018 comes into force
28 November 2018	Planning (Environmental Assessments and Miscellaneous Amendments) (EU Exit) (Northern Ireland) Regulations 2018 (S.I. 2018 No. 1235) in preparation for when the United Kingdom leaves the EU and dealing with Environmental Impact Assessment, Management of Extractive Waste and Hazardous Substances Consent formally laid in Westminster, in the absence of a functioning NI Assembly
1 May 2020	The Planning (Development Management) (Temporary Modifications) (Coronavirus) Regulations (Northern Ireland) 2020 come into force and suspend the requirement to hold a public event as part of the pre-application community consultation process for major planning applications
20 October 2020	Announcement of the Planning Engagement Partnership to enhance the quality and depth of community engagement in the planning process at both the regional and local planning levels (Nichola Mallon MLA, Infrastructure Minister)
2 November 2020	The Planning Act 2011 (Review) Regulations (Northern Ireland) 2020 come into force
21 December 2020	Statutory Rule: The Planning (General Permitted Development) (Amendment) Order (Northern Ireland) 2020 comes into force, providing extended PDR for electric vehicle charging points, removing PDR for petroleum exploration, expanding PDR for mobile network operators and allowing for the extension of premises
1 April 2021	The Planning (Development Management) (Temporary Modifications) (Coronavirus) (Amendment) Regulations (Northern Ireland) 2021 extend the temporary removal for the requirement for public events as part of the pre-application process for major planning applications
2 August 2021	Publication by Department for Infrastructure of Planning Advice Note on Implementation of Strategic Planning Policy on Development in the Countryside
15 October 2021	The Department for Infrastructure withdraws Planning Advice Note on Implementation of Strategic Planning Policy on Development in the Countryside
27 January 2022	The Department for Infrastructure Minister, Nichola Mallon, publishes the first review report of the Planning Act (Northern Ireland) 2011

release, headlined a commentary that the Northern Ireland planning system is not fit for purpose. The establishment of a Commission to identify immediate improvements in performance is called for. Such considerations, of course, are not unique to Northern Ireland. Mutations in planning practice can be observed more widely across the United Kingdom. Nonetheless, the purpose of reform is to move forward the complexity that is planning, in line with a political steer that is lawful, proportionate and respectful of the public interest. In that regard the reshaping of planning can be briefly considered in regard to legislative and policy succession.

The Planning Reform (Northern Ireland) Order 2006 confirmed in law the importance of the development plan in determining planning applications and indicated that determination must be in accordance with the development plan unless material considerations indicate otherwise. Furthermore, the making, alteration or repeal of development plans must be done in accordance with the objective of contributing to sustainable development, and, in considering objections to development plan proposals, the legislation provided for the convening of an independent examination with a right for anyone who has made representations to appear and be heard. This more inquisitorial approach to the scrutiny of development plans has been carried forward to the new post-2015 two-stage procedures set out in the Planning Act (Northern Ireland) 2011, which provides for an initial independent examination of the Local Development Plan strategy and a follow-on independent examination of the local policies plan to test for soundness. The important point that emerges, however, is that while local government is charged with driving the development plan process, a central government department will have a statutory oversight and directing role to ensure, for example, general conformity with the regional development strategy, not least in relation to the quantum of land zoning vis-à-vis housing growth indicators.

The Planning Act (Northern Ireland) 2011 expressly states that the Department must formulate and coordinate policy for securing the orderly and consistent development of land and the planning of that development. However, unlike the 1991 Order, the Department is now required to further the objective of sustainable development and promoting or improving wellbeing. But with the publication of the *Strategic Planning Policy Statement for Northern Ireland* (SPPS) in 2015, it is noteworthy that it is the planning system that is given the purpose of furthering sustainable development and improving wellbeing. Furthermore, sustainability is defined as a trinity of social, economic and environmental considerations and, in that regard, the tone of the document speaks very much to measured sentiments around the management and encouragement of development when it states:

> A key dimension of sustainable development for Northern Ireland is economic growth. This requires the planning system to continue to provide protection to the things we cherish most about our built and natural environment, including our heritage assets while unlocking development potential, supporting job creation and aiding economic recovery for the benefit of all our people.
>
> *(Department of the Environment 2015a: 10)*

Quite clearly, the planning system in Northern Ireland is being directly challenged to lend its support to the investment and economic development ambitions of government. The provisions of the SPPS must be taken into account in the preparation of Local Development Plans and are also material to all decisions on individual planning applications and appeals.

The Executive Committee of Ministers

The Executive Committee of Ministers exercises executive authority on behalf of the Northern Ireland Assembly, and it takes decisions on significant issues and matters which cut across the responsibility of two or more Ministers. It also agrees to proposals put forward by Ministers for new legislation in the form of executive bills for consideration by the Assembly and is responsible for drawing up a Programme for Government and an agreed budget for approval by the Assembly. Ministers of the Executive are nominated by the political parties in the Northern Ireland Assembly. The number of Ministers which a party can put forward is determined by its share of seats in the Assembly. The First Minister and Deputy First Minister are nominated by the largest and second largest parties, respectively, and act as joint chair-persons of the Executive. Each Executive Minister has responsibility for a specific Northern Ireland government department (www.northernireland.gov.uk/index/your-executive.htm, accessed 15 April 2016). The Ministerial Code sets out the rules and procedures to which all Ministers must adhere. It includes the Pledge of Office, the Ministerial Code of Conduct, the principles of public life and provisions governing the functioning of the Executive Committee. In his analysis of the Northern Ireland Ministerial Code, McGrath (2005) has argued that a Ministerial Code can be viewed 'as a map of the institutional framework of any particular system of government'; it comprises 'an overview of the processes through which public policy is arrived at, and through which a government operates as a cohesive unit' (pp. 107–108). The Northern Ireland Executive Committee of Ministers is charged with:

> (i) the discussion of, and agreement on, issues which cut across the responsibilities of two or more Ministers; (ii) prioritising executive proposals; (iii) prioritising legislative proposals; (iv) recommending a common position where necessary; (v) agreement each year on (and review as necessary of) a programme incorporating an agreed budget linked to policies and programmes (Programme for Government); (vi) discussion of and agreement upon significant or controversial matters that are clearly outside the scope of the agreed programme referred to in paragraph 20 of Strand One of the (Belfast/Good Friday) Agreement; and (vii) discussion of and agreement upon significant or controversial matters that the First Minister and Deputy First Minister acting jointly have determined to be matters that should be considered by the Executive Committee.
>
> *(Northern Ireland Executive 2007: 5–6)*

These functions are derived from Strand One of the (Belfast/Good Friday) Agreement 1998, the Northern Ireland Act 1998 and the Northern Ireland (St Andrews Agreement) Act 2006 and they impose an obligation on Ministers to bring cross-cutting issues and, subject to particular conditions, significant or controversial matters to the Executive for agreement. As noted by Treacy J in *The Minister of Enterprise, Trade and Investment's Application* (2016 NIQB 26):

> the purpose behind these provisions was to avoid a Minister making a unilateral decision which affected another Minister, or to which another Minister objected, or which was properly a matter to be decided by the Executive as a whole, (10)

and furthermore:

> In my view it plainly follows from the above provisions that a Minister has no power to take a decision in violation of the Ministerial Code relating to the obligation to bring to the attention of the Executive Committee any matter that requires to be considered

> by it for discussion and agreement by reason of being cross-cutting, significant or controversial. (11)

This judicial review case concerned the purported adoption of the Belfast Metropolitan Area Plan on 3 September 2014 in accordance with the instructions of the Minister with responsibility for the Department of the Environment. The plan included policy content restricting the expansion of Sprucefield Regional Shopping Centre except for bulky goods trade, and other Ministers had expressed objections to the adoption of the plan in its then current form. In line with the provisions of the Ministerial Code of Conduct and its related legislative base, the High Court found that the respondent had acted *ultra vires* his powers and that adoption of the plan was in fact a matter for the Executive Committee to discuss and agree upon, rather than for the Minister to act unilaterally. Subsequent litigation in the Court of Appeal before Weatherup LJ, Weir LJ and McBride J regarding a Remedies Order issued at the conclusion of the High Court case prompted the following observation within the judgment (2017 NICA 28):

> it is no longer a matter of dispute by any principal, successor or notice party that the Minister's decision to direct the formal adoption of BMAP was, for all the reasons articulated by Treacy J, *ultra vires*. The purported adoption was accordingly of no force or effect; the draft BMAP remains in its entirety unadopted. (7)

A further complexity in regard to Minister and Executive Committee decision making arose *In the matter of an application by Colin Buick for judicial review and in the matter of a decision of the Department for Infrastructure made on 13 September 2017 under Section 26 of the Planning Act (Northern Ireland) 2011* (2018 NIQB 43). This concerned a decision to grant planning permission for a major waste disposal incinerator on the edge of Belfast in the Hightown / Mallusk area. At that stage Minister oversight and the Executive Committee had ceased functioning and thus it was contended that the impugned decision had firstly, been taken in the absence of the Minister with responsibility for the Department for Infrastructure and, second, given that the decision also cut across two departments (the Department for Infrastructure and the Department for Agriculture, Environment and Rural Affairs) it could not lawfully be taken in the absence of an Executive Committee. The applicant for judicial review argued that, in those circumstances, civil servants had no legal power to make the decision. While the Court indicated its understanding of 'the frustration within the Civil Service that important decisions need to be made' (19), Keegan J concluded:

> In my view the role of a department in decision making in relation to this type of planning decision is clear. The Department has decision making functions conferred upon it. However, the Department cannot be entirely detached from the Minister who heads it up. The level and extent of direction and control may vary depending on the circumstances, however the Minister remains in overall charge … the requirement that the functions of the Department shall at all times be exercised subject to the direction and control of the Minister is clearly part of the statutory regime … I fail to see how it is possible to have democratic accountability without a Minister in place and a functioning Executive and in particular a working Assembly … I do not consider that the exigencies of the current situation are an adequate justification for the course

> that has been taken. The commendable motivation and aims I refer to cannot override the proper construction of the statutory regime which this case requires … decisions have to be taken in a lawful manner otherwise the whole structure of government is undermined. (17–19)

The matter was then progressed to the Court of Appeal (2018 NICA 26) by the Department for Infrastructure on the basis that this High Court decision had ramifications for the conducting of government business through all the Northern Ireland Departments given the prolonged absence at that time of Ministers in charge. The Office of the First Minister and Deputy First Minister had been vacant since 9 January 2017 and no Ministers had been in place since 2 March 2017. In agreeing that the appeal must be dismissed, Morgan LCJ, Stephens LJ and Treacy LJ considered:

> that this was a significant and controversial matter which again required determination by the Executive Committee. It would be contrary to the letter and spirit of the Agreement and the 1998 Act for such decisions to be made by departments in the absence of a Minister. We are reinforced in these views by our recognition of the constitutional position of civil servants. That role is to advise Ministers and be accountable to them. The appellant's submissions would effectively turn civil servants into Ministers. Such a remarkable constitutional change would require the clearest wording and we do not consider that the Northern Ireland Budget Act 2018 provides any basis for the implication of such a major departure from established constitutional principles. That is sufficient to deal with the appeal. We have not in this appeal heard argument on the precise limits of any power of the departments to take decisions but it follows from our analysis of the constitutional position of civil servants that any decision which as a matter of convention or otherwise would normally go before the Minister for approval lies beyond the competence of a senior civil servant in the absence of a Minister. (18)

However, this was not the conclusion to the controversy surrounding the governance of departmental decision making in Northern Ireland. The limitations exposed by the Courts led to the passage of enabling legislation in Westminster that makes provision for the exercise of governmental functions in, or in relation to, Northern Ireland in the absence of Northern Ireland Ministers. The Northern Ireland (Executive Formation and Exercise of Functions) Act 2018 contains three significant subsections:

> 3(1) The absence of Northern Ireland Ministers does not prevent a senior officer of a Northern Ireland department from exercising a function of the department during the period for forming an Executive if the officer is satisfied that it is in the public interest to exercise the function during that period.
>
> 3(4) The absence of Northern Ireland Ministers is not to be treated as having prevented any senior officer of a Northern Ireland department from exercising functions of the department during the period beginning with 2 March 2017 and ending when this Act is passed.
>
> 3(5) The fact that a matter connected with the exercise of a function by a Northern Ireland department has not been discussed and agreed by the Executive Committee

> of the Northern Ireland Assembly is not to be treated as having prevented the exercise of that function as mentioned in subsection. (1) or (4)

Guidance published by the Secretary of State for Northern Ireland (2018) in regard to the exercise of departmental functions under this Act sets out three principles that senior officers should take into account when taking decisions:

> (a) the principle that where decisions to be taken materially cut across or impact upon the responsibilities of another NI Department, discussions with the relevant Department(s) should take place and due regard should be given to the views of the relevant Department(s). (b) the principle that where decisions are deferred, the public interest consequences of a deferral should be kept under review. This recognises that, with the passage of time, there may be the need to take decisions on matters previously deferred. (c) the principle that where decisions are not taken, NI Departments may, however, continue to advance preparatory work as far as possible until such time as decisions can be taken by Ministers.

For purposes of development management, the outcome of this protracted period of political difficulty, litigation and new enabling legislation was, finally, a lawful capability for senior civil servants, in the absence of a Minister, to exercise the functions of the Department for Infrastructure when they are satisfied that it is in the public interest to exercise those functions. The Department was thus able to progress planning applications to a point where a decision could be made and, having regard to the Act and its accompanying guidance, consider whether to issue a decision. In November 2019 the Department reported that, to 31 August 2019, it had issued two decisions (one refusal and one approval) and determined fourteen applications which would be decided on a case-by-case basis as to issuing a decision. In January 2020 devolved government was restored in Northern Ireland.

Notwithstanding the legislative response to the Court of Appeal decision in *Buick*, it would seem that the Department for Infrastructure remained concerned about the potential impact of that judgment, with its potential to hinder planning application decision making even when devolved government had resumed with a functioning Executive. In a case (2021 NIQB 93) that concerned Minister approval of planning applications for the North-South Electricity Interconnector, Scoffield J commented that the phrase 'cross-cutting' was perceived to have been given too wide a meaning by the Court of Appeal and:

> since many planning decisions will in some way, engage the interests of other departments, the Department for Infrastructure was keen to restrict cross-cutting decisions to those involving only the legal 'responsibilities' of other departments … the Department had understandable concerns about the effect of the Executive Committee becoming the *de facto* planning authority for major applications. That would remove the decision-making power from the Minister who was statutorily responsible for making these decisions … it would add significantly to the work of the Executive and both eat into the Executive's time to deal with other pressing matters and seriously slow down the planning decision-making process. In light of these concerns, the Department quite understandably took the view that it should seek to narrow the occasions on which planning decisions may require Executive approval and press for legislative intervention in this regard.

That legislation change came in the form of an amendment to the Northern Ireland Act 1998 by the Executive Committee (Functions) Act (Northern Ireland) 2020, which increased exceptions to Executive decision making in the following terms:

> (7) Decisions may be made by the Department for Infrastructure or the Minister in charge of that Department in the exercise of any function under –
>
> (a) the Planning Act (Northern Ireland) 2011 (except for a function under section 1 of that Act); or
>
> (b) regulations or orders made under that Act, without recourse to the Executive Committee.

In short, the amended legislation now allows the Department for Infrastructure Minister to make operational decisions without recourse to the Executive. However, as noted by Scoffield J in the Interconnector case, while the Minister now has discretion to determine such matters, the insertion of the word 'may' means that 'the Minister is not precluded from referring a matter to the Executive for discussion and agreement where it would otherwise fall to them to consider. The relevant amendment simply removes the requirement … to do so'. The approach of Scoffield J in *Safe Electricity* was subsequently followed by Humphreys J *In the Matter of an Application by Mooreland and Owenvarragh Residents' Association for Judicial Review* (2022 NIQB 40) concerning the redevelopment of Casement Park stadium in Belfast.

Local government

Planning governance in Northern Ireland has comprised a complex and changing alchemy of direct rule, devolved government, central government departments and agencies, quasi-autonomous public bodies and local government. From 1973 to 2015 councils did not have any executive function in regard to development plans and development control and were involved in the planning system only as statutory consultees. The Planning (Northern Ireland) Order 1972 introduced mandatory consultation on development plans and planning applications (under Article 4(4) and Article 14(2a), respectively). Accordingly, each council would be asked to contribute its ideas on how the council area should develop, be invited to comment on thematic papers prepared as part of the plan preparation process and, in later years, be given confidential viewing of draft proposals for settlement development limits and zoning before publication. Councils would be consulted on planning applications when an opinion to approve or refuse had been formulated by central government department planners, and councillors could request deferral and reconsideration before any decision would issue. However, these arrangements fall far short of executive authority and its corollary of public accountability, and, in that regard, the consideration of the return of planning powers to local government became part of a wider process of public sector reform in Northern Ireland, branded locally as the Review of Public Administration.

The Review of Public Administration was formally announced by the Northern Ireland Executive in June 2002 with the remit of examining the arrangements for the accountability, development, administration and delivery of public services in Northern Ireland. It was carried forward by officials within the Office of First Minister and Deputy First Minister during the period of direct rule commencing in October 2002 and then returned again to direct oversight by devolved government from May 2007. The need for reform had, in fact, been mooted earlier than 2002 as illustrated by the content of the first draft Programme for Government, launched in 2000 by First Minister David Trimble and Deputy First Minister Seamus Mallon. The document identified:

> the need to consider the rationalisation of public administration so that resources can be used best to serve the public rather than maintaining bureaucratic structures. We have inherited from the last 30 years a wide range of public bodies. Their organisation and structure reflected the needs of those times. They helped maintain services at a time of very limited public accountability. But now that devolution has been achieved, there is a need for change that will provide not only greater accountability, but should ensure that the organisations that deliver many key services throughout Northern Ireland are much more coherently organised. It is therefore important that we set about a major process of reform in central government.
>
> *(Northern Ireland Executive 2000: 61–62)*

At that stage very little of substance was aired about the future of local government, which was merely applauded for knowing the needs of its respective areas and having the capacity to lead and coordinate. In the following year the draft Programme for Government, while still very centralist in content, hinted more fully at local government being part of the public sector reform process:

> The Executive remains committed to improving the efficiency and effectiveness of and accountability for, the administration and delivery of public services in Northern Ireland. We recognise the need for different structures under devolution, taking account of new relationships between local and regional government, as well as the full range of other bodies that function within the wider public sector. We are committed to a comprehensive and strategic review of all aspects of the public sector.
>
> *(Northern Ireland Executive 2001: 63)*

By February 2002, prior to its launch in June, the position of local government in the review process had visibly changed. It had become key to the optics of the review and, in opening a consultation process with the Northern Ireland Assembly on its terms of reference, the First Minister and Deputy First Minister stated: 'Among the important issues, which will be addressed by the Review, include the structure, accountability and responsibilities of local government, Non-Departmental Public Bodies and Next Steps Agencies' (Northern Ireland Assembly Research and Library Services 2002: 1). Interestingly, there is no specific mention of central government departments in that agenda.

Much has been written about the wider elements of the analytical process, consultation on the size and number of councils, the wrangling between competing political interests and the underwhelming outcome of the review (see for example: Birrell 2007b, 2008, 2013; Knox 2008, 2009, 2012; Knox and Carmichael 2006, 2015). The decisions on the reforms relating to local government have had to be mediated through a political process that was always going to favour regional centralism with political parties keen to hold on to their share of ministerial appointments and a powerful Civil Service unwilling to see their departments substantially diluted (Knox 2009). But what is significant in the context of this chapter on the Northern Ireland planning system is that the number of councils was reduced from twenty-six to eleven. New functions, partially transferred from the Department of the Environment, requiring the preparation of Local Development Plans and the operation of development management, planning enforcement and community planning have been added to their portfolios of executive responsibility, taking effect from 1 April 2015. But then, perhaps one commentator, writing in 2001, was wisely prescient when he suggested that, with the exception of town and country planning, there is little enthusiasm in local government to reacquire responsibility for direct provision of additional services. Rather, he argued 'the principal desire within local government appears not to (re)gain overall control but to progress beyond just a consultative role in shaping the determination of policy by those in authority' (Carmichael 2001). Nonetheless, since 2015 local government has embraced its new responsibilities to deliver on the determination of applications for planning permission, planning enforcement and the preparation of Local Development Plans and Community Plans. In addition, councils are also responsible for determining applications (subject to particular Department for Infrastructure notification requirements) for listed building consent, conservation area consent, hazardous substances consent, display of advertisement consent and applications for works to trees protected by Tree Preservation Orders (Department for Infrastructure 2022: 18).

A frequently voiced criticism of the Northern Ireland planning system has been the length of time taken to decide planning applications. Data published annually by the Department for Infrastructure offer insight into the functioning of councils in dealing with major and local planning applications over the period from April 2015. The data in Table 1.3 demonstrate a clear difference in performance between major and local development applications. The average processing time is in excess of the target of thirty weeks for the former category and the number of councils achieving the target is modest. In contrast, compliance with the processing target of fifteen weeks for local development applications has improved significantly over the five years to March 2020 with the average processing time for all councils standing at 14.0 weeks for the year 2019–20 and with seven councils meeting the target of fifteen weeks.

An additional consideration relates to local government being a champion for localism and which received strong endorsement by the Northern Ireland Executive in its draft Programme for Government Framework 2016–2021, published for public consultation in May 2016. In a joint-ministerial *Foreword* Arlene Foster and Martin McGuinness pointed to a new direction being necessary for the business of government:

> We believe a different approach is needed and so this new approach focuses on the impact on our people rather than the actions we take within Government. We recognise that for this to work effectively, we need a cohesive Executive working to deliver for all. We also need a system of Government that works across boundaries, organisations, groups and communities for the common good. One of the significant changes in this Programme is that we will be engaging outside Government to develop plans which recognise the various parts that many different organisations and sectors have to play. We want to work with local government, the private sector, voluntary and community sectors and beyond to maximise what we can achieve collectively.
>
> *(Northern Ireland Executive 2016: 5)*

TABLE 1.3 Major development and local development planning applications and processing times performance 2015/2016–2019/2020

Year	*Applications for major development* received*	*Major development applications decided / withdrawn*	*Average processing time to 30 weeks target*	*Number of councils under the 30 weeks target*
April 2015–March 2016	145	272 /14	46.4	1
April 2016–March 2017	142	192/19	68.6	0
April 2017–March 2018	161	176/13	50.2	1
April 2018–March 2019	137	137/11	59.0	3
April 2019–March 2020	149	129/11	52.8	2
Year	*Applications for local development** received*	*Local development applications decided / withdrawn*	*Average processing time to 15 weeks target*	*Number of councils under the 15 weeks target*
April 2015–March 2016	12,069	10,753 /444	19.4	3 (an additional council at 15.2 weeks)
April 2016–March 2017	12,893	12,764/637	16.2	5
April 2017–March 2018	12,770	12,137/612	15.2	5
April 2018–March 2019	12,404	12,019/618	14.8	5
April 2019–March 2020	12,058	11,616/542	14.0	7

Source: Department for Infrastructure Northern Ireland Planning Statistics Annual Statistical Bulletins, 2016-2020.

Notes: * Major developments have important economic, social and environmental implications. The majority of applications for major developments are dealt with by councils and are subject to pre-application consultation with the community.

**Local developments comprise all other developments (other than permitted development) that do not fall within the classes described for major or for regionally significant developments. They comprise the vast majority of residential and minor commercial applications to be received and determined by a council.

The fresh ideas being hinted at comprise an outcomes-based approach linked to the overarching theme of 'societal wellbeing', which, as observed by Wallace (2019), might create a common language that could sit above political decisions and constitutional struggles in a post-conflict society (113). However, as noted above, the Executive collapsed in January 2017, leaving the Northern Ireland Civil Service with responsibility for carrying forward work on the delivery of a cross-departmental Outcomes Delivery Plan against the previously agreed objective of 'Improving wellbeing for all - by tackling disadvantage and driving economic growth'. The January 2020 *New Decade, New Approach Deal* reiterated that commitment to societal wellbeing as the kernel of the Government's approach to responsive public spending. However, from the perspective of local government, the important point here is that the availability of Community Plans and Local Development Plans has the potential to integrate

TABLE 1.4 District Council Community Plans, 2022

District council	*Community Plan title*	*Time horizon*
Antrim and Newtownabbey	*Love Living Here*	2030
Ards and North Down	*The Big Plan*	2032
Armagh, Banbridge, Craigavon	*Connected*	2030
Belfast	*Belfast Agenda*	2035
Causeway Coast and Glens	*A Better Future Together*	2030
Derry City and Strabane	*Inclusive Strategic Growth Plan*	2032
Fermanagh and Omagh	*Fermanagh and Omagh 2030 Community Plan*	2030
Lisburn and Castlereagh	*Nurture and Reap Rewards*	2032
Mid and East Antrim	*Putting People First*	2032
Mid Ulster	*Our Community Plan*	2027
Newry, Mourne and Down	*Living Well Together*	2030

with and contribute to the delivery of this high-level goal. In its practice guidance notes, the former Department of the Environment (2015b) indicated that the duty of community planning requires a council to lead the process of creating a long-term vision for the social, environmental and economic wellbeing of its area and citizens, in partnership with community and service providers. The resultant Community Plan has a statutory linkage with the fifteen-year Local Development Plan under Section 77 of the Local Government Act (Northern Ireland) 2014, in that the preparation of the latter must take account of the content of the former. Accordingly, in 2022 it can be reported that all eleven district councils have published their agreed Community Plans deploying a range of time horizons and comfortable titles that speak to ambitions of inclusiveness and progressive optimism (Table 1.4). Progress continues to be made on the preparation of Local Development Plans, albeit it has been suggested that it will be 2028 before these are in place in each district council area. Identified issues include complexity of the work required, insufficient specialist planning expertise and staffing pressures that prioritise the processing of planning applications (Northern Ireland Audit Office 2022: 23–24).

Conclusion

This introductory chapter demonstrates the inextricable linkage between governance and practice in the Northern Ireland planning system. The filtering of the planning system through the sieve of political accommodation, in a manner that is very different from the rest of the United Kingdom, sets the Northern Ireland situation apart and is a narrative worth recounting. Subsequent chapters in this book will explicate the operational personality of planning in this jurisdiction and will give further attention to development management and lawful decision making.

Acknowledgement

The content of this chapter dealing with the early years of planning in Northern Ireland draws, in part, from Murray, M. 1991. *The Politics and Pragmatism of Urban Containment.* Aldershot: Avebury.

Legislation

Departments Act (Northern Ireland) 2016
Executive Committee (Functions) Act (Northern Ireland) 2020
Government of Ireland Act 1920
Local Government (Ireland) Act 1898
Local Government Act (Northern Ireland) 1922
Local Government Act (Northern Ireland) 1972
Local Government Act (Northern Ireland) 2014
Northern Ireland Act 1974
Northern Ireland Act 1998
Northern Ireland Budget Act 2018
Northern Ireland (Executive Formation and Exercise of Functions) Act 2018
Planning Act (Northern Ireland) 2011
Northern Ireland (St Andrews Agreement) Act 2006
The Departments (Transfer of Functions) Order (Northern Ireland) 2016
The Planning Act 2011 (Review) Regulations (Northern Ireland) 2020
The Planning Acts Amendment Act (Northern Ireland) 1944
The Planning and Housing Act (Northern Ireland) 1931
The Planning (Development Management) (Temporary Modifications) (Coronavirus) Regulations (Northern Ireland) 2020
The Planning (Development Management) (Temporary Modifications) (Coronavirus) (Amendment) Regulations (Northern Ireland) 2021
The Planning (Environmental Assessments and Miscellaneous Amendments) (EU Exit) (Northern Ireland) Regulations 2018
The Planning (General Permitted Development) (Amendment) Order (Northern Ireland) 2020
The Planning (Interim Development) Act (Northern Ireland) 1944
The Planning (Northern Ireland) Order 1972
The Planning (Northern Ireland) Order 1991
The Planning Reform (Northern Ireland) Order 2006
The Strategic Planning (Northern Ireland) Order 1999
Town and Country Planning Act 1947

Cases cited

Belfast Corporation v OD Cars Ltd (1960 1 All ER 65)
The Minister of Enterprise, Trade and Investment's Application (2016 NIQB 26)
In the Matter of an Application by the Minister of Enterprise, Trade and Investment for Judicial Review and in the Matter of a Decision by the Minister for the Environment Dated 3 September 2014 Between: Belfast City Council Appellant / Notice Party and the Minister of Enterprise, Trade and Investment Respondent / Applicant (2017 NICA 28)
In the Matter of an Application by Colin Buick for Judicial Review and in the Matter of a Decision of the Department for Infrastructure made on 13 September 2017 under Section 26 of the Planning Act (Northern Ireland) 2011 (2018 NIQB 43)
In the Matter of an Application by Colin Buick as Chairperson of NoARC21 for Judicial Review. In the Matter of a Decision by the Department for Infrastructure Dated 13 September 2017 (2018 NICA 26)

In the Matter of an Application by Mooreland and Owenvarragh Residents' Association for Judicial Review (2022 NIQB 40)

In the Matter of an Application by SAFE Electricity A & T Limited and Patrick Woods for Judicial Review and In the Matter of Decisions of the Minister for the Department for Infrastructure and the Executive Office (2021 NIQB 93)

References

Aitkin, J. 1967. 'Regional Planning in Northern Ireland', in *Report of the Town and Country Planning Summer School*. London: Town Planning Institute, pp. 5–21.

Alexander, A. 1982. *Local Government in Britain Since Reorganisation*. London: George Allen and Unwin.

Assembly and Executive Review Committee. 2012. *Review of the Number of Members of the Northern Ireland Legislative Assembly and of the Reduction in the Number of Northern Ireland Departments: Part 2 – Reduction in the Number of Northern Ireland Departments*. Belfast: The Stationery Office.

Birrell, D. 2007a. 'Northern Ireland Business in Parliament: The Impact of the Suspension of Devolution in 2002', *Parliamentary Affairs* 60(2): 297–312.

Birrell, D. 2007b. 'The Role of Local Government in the Review of Public Administration in Northern Ireland: The Challenges of Enhancement, Devolution and Modernisation', *Local Government Studies* 33(5): 657–75.

Birrell, D. 2008. 'The Final Outcomes of the Review of Public Administration in Northern Ireland. Tensions and Compatibility with Devolution, Parity and Modernisation', *Public Administration* 86(3): 779–93.

Birrell, D. 2009. *Direct Rule and the Governance of Northern Ireland*. Manchester: Manchester University Press.

Birrell, D. 2013. *Quango Reform in Northern Ireland: Challenging Devolution, Sustaining Democratic Deficiencies and Enhancing Centralisation*. Paper presented at Political Studies Association Annual Conference. Cardiff.

Blackman, T. 1984. 'Planning in Northern Ireland – The Shape of Things to Come?', *The Planner* 70(1): 12–15.

Buckland, P. 1981. *A History of Northern Ireland*. Dublin. Gill and Macmillan.

Carmichael, P. 2001. 'Where Next for Local Government and Quangos in Northern Ireland?', *Belfast Telegraph*, 15 October.

Carmichael, P. 2002. 'The Northern Ireland Civil Service: Characteristics and Trends Since 1970', *Public Administration* 80(1): 23–49.

Carmichael, P. and R. Osborne. 2003. 'The Northern Ireland Civil Service Under Direct Rule and Devolution', *International Review of Administrative Sciences* 69: 205–17.

Coghlin, P., U. O'Brien and K. MacLean. 2020. *The Report of the Independent Public Inquiry into the Non-domestic Renewable Heat Incentive (RHI) Scheme*. Belfast: Department of Finance.

Committee on the Programme for Government. 2007. *Report on the Review of Public Administration and Rural Planning*. Belfast: TSO Ireland.

Davoudi, S. and I. Strange (eds). 2008. *Conceptions of Space and Place in Strategic Spatial Planning*. London: Routledge.

Department for Infrastructure. 2022. *Review of the Implementation of the Planning Act (Northern Ireland) 2011, Report*. Belfast: Department for Infrastructure.

Department for Regional Development. 2001. *Regional Development Strategy for Northern Ireland 2025 – Shaping Our Future*. Belfast: Department for Regional Development.

Department for Regional Development. 2012. *Regional Development Strategy 2035 – Building a Better Future*. Belfast: Department for Regional Development.

Department for Regional Development and Department for Environment, Community and Local Government. 2013. *Framework for Cooperation: Spatial Strategies of Northern Ireland and the Republic of Ireland*. Belfast: Department for Regional Development and Dublin: Department for Environment, Community and Local Government.

Department for Regional Development and Department for the Environment, Heritage and Local Government. 2011. *Spatial Strategies on the Island of Ireland – Framework for Collaboration* (Consultation document). Belfast: Department for Regional Development and Dublin: Department for the Environment, Heritage and Local Government.

Department of the Environment. 2015a. *A Strategic Planning Policy Statement for Northern Ireland: 'Planning for Sustainable Development'*. Belfast: Department of the Environment.

Department of the Environment. 2015b. *Introduction: Context for Local Development Plans Development Plan Practice Note 1*. Belfast: Department of the Environment.

Elliott, S. and R.A. Wilford. 1987. 'Administration', in R.H. Buchanan and B.M. Walker (eds) *Province, City and People: Belfast and its Region*. Antrim: Greystone Books, pp. 285–304.

Hayes, M. 1967. 'Some Aspects of Local Government in Northern Ireland', in E. Rhodes (ed.) *Public Administration in Northern Ireland*. Londonderry: Magee University College.

Healey, P. 2004. 'The Treatment of Space and Place in the New Strategic Spatial Planning in Europe', *International Journal of Urban and Regional Research* 28(1): 45–67.

Hendry, J. 1984. 'The Development of Planning in Northern Ireland', in Bannon, M. and J. Hendry (eds) *Planning in Ireland: An Overview*. Occasional Papers in Planning No 1. Belfast: Queen's University Belfast.

Houston, D., S. McKay and M. Murray. 2017. *Celebrating Our Past - Shaping the Future: 50 Years of Planning Education at Queen's University Belfast*. Belfast: Queen's University Belfast.

InterTradeIreland. 2006. *Spatial Strategies on the Island of Ireland: Development of a Framework for Collaborative Action*. Newry: InterTradeIreland.

Knox, C. 2008. 'Policy Making in Northern Ireland: Ignoring the Evidence', *Policy and Politics* 36(3): 343–59.

Knox, C. 2009. 'The Politics of Local Government Reform in Northern Ireland', *Local Government Studies* 35(4): 435–55.

Knox, C. 2012. 'The Reform of Public Administration in Northern Ireland: A Squandered Opportunity', *Administration* 60(1): 117–38.

Knox, C. and P. Carmichael. 2005. 'Devolution – The Northern Ireland Way: An Exercise in 'Creative Ambiguity', *Environment and Planning C: Government and Policy* 23: 63–83.

Knox, C. and P. Carmichael. 2006. 'Bureau Shuffling? The Review of Public Administration in Northern Ireland', *Public Administration* 84(4): 941–65.

Knox, C. and P. Carmichael. 2015. 'Local Government Reform: Community Planning and the Quality of Life in Northern Ireland', *Administration* 63(2): 31–57.

Lloyd, G. 2008. *Planning Reform in Northern Ireland: Independent Report to the Minister for the Environment*. Belfast: Northern Ireland Assembly.

Macrory, P.A. 1970. *Review Body on Local Government in Northern Ireland*. Cmd 546. Belfast: Her Majesty's Stationery Office.

McBride, S. 2019. *Burned: The Inside Story of the 'Cash for Ash' Scandal and Northern Ireland's Secretive New Elite*. Newbridge: Merrion Press.

McGrath, C. 2005. 'The Northern Ireland Ministerial Code', *Irish Political Studies* 20(2): 107–25.

Nash, C., B. Reid and B. Graham. 2013. *Partitioned Lives: The Irish Borderlands*. Farnham: Ashgate.

Newman, C.F.S. 1965. 'A Short History of Planning in Northern Ireland', *Journal of the Town Planning Institute* 51(2): 47–53.

Northern Ireland Affairs Committee. 1996. *The Planning System in Northern Ireland (Volume 1)*. London: HMSO.

Northern Ireland Assembly Public Accounts Committee. 2022. *Planning in Northern Ireland. Report :NIA 202/17-22*. Belfast: Northern Ireland Assembly.

Northern Ireland Assembly Research and Library Services. 2002. *Review of Public Administration (Research Paper 07/02)*. Belfast: Northern Ireland Assembly.

Northern Ireland Audit Office. 2009. *The Performance of the Planning Service*. Belfast: Northern Ireland Audit Office.

Northern Ireland Audit Office. 2022. *Planning in Northern Ireland*. Belfast: Northern Ireland Audit Office.

Northern Ireland Executive. 2000. *Draft Programme for Government*. Belfast: Northern Ireland Executive.

Northern Ireland Executive. 2001. *Draft Programme for Government*. Belfast: Northern Ireland Executive.

Northern Ireland Executive. 2007. *Ministerial Code*. Belfast: The Stationery Office Limited.

Northern Ireland Executive. 2016. *Draft Programme for Government Framework 2016–21*. Belfast: The Northern Ireland Executive.

O'Leary, C., S. Elliott and R.A. Wilford. 1988. *The Northern Ireland Assembly 1982–1986: A Constitutional Experiment*. London: C. Hurst & Company.

Peel, D. and M. G. Lloyd. 2015. 'Towards a Framework for Cooperation: Spatial Public Diplomacy on the Island of Ireland', *European Planning Studies* 23(11): 2210–26.

Scannell, Y. and S. Turner. 2009. 'A Legal Framework for Sustainable Development in Rural Areas of the Republic of Ireland and Northern Ireland', in McDonagh, J., T. Varley and S. Shortall (eds) *A Living Countryside: The Politics of Sustainable Development in Rural Ireland*. Farnham: Ashgate, pp. 25–52.

Secretary of State for Northern Ireland. 2018. *Guidance on Decision-Making for Northern Ireland Departments During the Period for Northern Ireland Executive Formation*. Cm 9725. Belfast: Northern Ireland Office.

Wallace, J. 2019. *Wellbeing and Devolution: Reframing the Role of Government in Scotland, Wales and Northern Ireland*. Cham: Springer Nature / Palgrave Macmillan.

Walsh, R. 2015. 'Setting the Parameters for Restricting Use', in Douglas, S., R. Hickey and E. Waring (eds) *Landmark Cases in Property Law*. Oxford and Portland, OR: Hart Publishing, pp. 227–252.

Wilford, R. 2000. 'Designing the Northern Ireland Assembly', *Parliamentary Affairs* 53, pp. 577–90.

2

DEVELOPMENT AND PLANNING APPLICATIONS

Stephen McKay and Michael Murray

Introduction

At first sight, a chapter devoted to development, planning applications and their determination might seem superfluous given the many contributions to the wider planning literature on this topic. These interconnected matters are addressed, for example, in Cullingworth et al. (2014) *Town and country planning in the UK*, Willmore and Blackhall (2016) *Planning law and practice*, Duxbury (2018) *Telling and Duxbury's planning law and procedure*, and Bowes (2019) *A practical approach to planning law*. While the principles set out in these influential texts have relevance to Northern Ireland, there is, as argued in the previous chapter, the need for a more nuanced appreciation that the business of planning in this jurisdiction has developed over time a set of particular practices that somewhat place it apart from the other UK planning systems. Additionally, a corpus of case law has emerged from the Northern Ireland courts that has relevance to planning practice within the region and which, in the main, is omitted from discussion in other scholarly work.

This chapter responds to that lacuna by rehearsing some of the key provisions in planning law and procedure that are steering the trajectory of planning practice in Northern Ireland. Key elements of this statutory framework for development management are set out in the Planning (General Development Procedure) Order (Northern Ireland) 2015 and the Planning (Development Management) Regulations (Northern Ireland) 2015, which specifically deal with the relationship between different types of development proposal and the processes around applying for planning permission, improvements to public involvement and the determination of planning applications. It is within this context and rationale that the chapter considers the following matters: the meaning of development, pre-application procedures, the planning application, planning application notices, neighbour notification, material planning considerations, prematurity, the contribution of a public local inquiry and a hearing, Planning Committee and delegation procedures, planning application decisions, planning appeals and, finally, Completion Notices.

DOI: 10.4324/9781003218319-2

The meaning of development

In any discussion of the development management process, the significant departure point is consistently regarded as being the meaning of development. Section 23 of the Planning Act (Northern Ireland) 2011 defines development as the 'carrying out of building, engineering, mining or other operations in, on, over or under land, or the making of any material change in the use of any buildings or other land'. Building operations include the demolition of buildings, rebuilding, the structural alteration of buildings and other operations normally undertaken by a person carrying on business as a builder. The definition seeks comprehensiveness in its coverage, but to secure efficiencies of regulatory oversight, a number of operations and uses of land are identified as not comprising development. These include interior works to a building that do not materially affect its external appearance, infrastructure inspection, repairs and renewal that involve the breaking open of a street, the use of any buildings or other land within the curtilage of a dwelling-house that is incidental to the enjoyment of that dwelling-house, changes of use within the same use class and the use of land for the purposes of agriculture and forestry.

The two limbs of development which are explicit in the statute definition above comprise operational development (for example, construction work related to new buildings, alterations and additions to buildings) and material changes in use. The Planning (Use Classes) Order (Northern Ireland) 2015 provides amplification for the latter. This specifies classes of use of buildings or other land and provides that a change of use is not to be regarded as involving development where the former use and the new use are both within the same class. The range of classes is set out by way of a schedule as summarised in Table 2.1 below.

No class specified in the schedule includes use:

- as an amusement arcade or centre, or a funfair;
- as a betting office;
- for the purposes of a funeral undertaker;
- as a hostel where a significant element of care is provided;
- as a hotel;
- as a house in multiple occupations;
- for the sale of fuel for motor vehicles;

TABLE 2.1 Classes of use set out in the Planning (Use Classes) Order (Northern Ireland) 2015

Part A: Shopping and financial and professional services	Class A1: Shops
	Class A2: Financial, professional and other services
Part B: Industrial and business uses	Class B1: Business (other than use in Class A2)
	Class B2: Light industrial
	Class B3: General industrial
	Class B4: Storage or distribution
Part C: Residential uses	Class C1: Dwelling-houses
	Class C2: Guest houses
	Class C3: Residential institutions
	Class C4: Secure residential institutions
Part D: Community, recreation and culture	Class D1: Community and cultural uses
	Class D2: Assembly and leisure

- for the sale or display for sale of motor vehicles;
- for the sale of food or drink for consumption on the premises or of hot food for consumption off the premises;
- as a scrapyard, or a yard for the storage or distribution of minerals or the breaking of motor vehicles;
- as a swimming bath, skating rink, gymnasium or area for other indoor or outdoor sports or recreations, including those involving motorised vehicles or firearms;
- for a taxi business or business for the hire of motor vehicles;
- for or in connection with public worship or religious instruction;
- for the carrying out of any prescribed process which requires an authorisation under Article 6 of the Industrial Pollution Control (Northern Ireland) Order 1997 or for the operation of any installation or mobile plant which requires a permit under regulation 9 of the Pollution Prevention and Control (Industrial Emissions) Regulations (Northern Ireland) 2013;
- as a waste management facility for the collection, transport, treatment, recovery, recycling, transfer and disposal of waste.

Such uses are *sui generis*, and, because of their potential impact implications for amenity, they require particular consideration by way of a planning application.

An additional statutory rule, the Planning (General Permitted Development) Order (Northern Ireland) 2015, also amplifies the meaning of development and the implications for planning permission. It deals with permitted development which grants a general planning permission for certain, often minor, non-contentious types of development, removing the need to apply for planning permission. A total of thirty-seven categories of development are set out in the schedule format to which these provisions apply. In this regard an interesting point of law has been tested by way of judicial review that examines whether a condition attaching to the grant of planning permission could, by implication, take away rights set out within the General Development Order (GDO). In the case *R v Planning Appeals Commission ex parte Bailie* (1995 Queen's Bench Division (Crown Side)), planning permission had been granted on appeal in 1986 for a harness track at Ballymacbrennan to be confined to the private exercise and training of horses; the Planning Appeals Commission imposed conditions limiting the finished surface of the track to a maximum of 3 metres in width, prohibiting the use of the track for any commercial purposes, limiting its use solely for the training and exercising of horses and expressly prohibiting its use for private or public races or meetings associated with equine sports or any other sporting activity. Subsequent to 1986 a 10-metre track was constructed. In July 1993 harness racing took place on four occasions and the Department of the Environment then issued an Enforcement Notice requiring cessation of use. A planning appeal was lodged wherein it was submitted that the 1986 planning appeal decision failed to make express reference to the Planning (General Development) Order (Northern Ireland) 1973 and that in the absence of any consideration of the appellant's GDO rights and reasoned decision withdrawing those rights, that they continued to subsist and thus the lands could be used for any purpose on not more than twenty-eight days in any calendar year. The Planning Appeals Commission dismissed the appeal in June 1994, concluding that the 1986 decision was unequivocal in the limitations on the use of the development and that, by implication, it, therefore, removed GDO rights in relation to the use of the track. In his judicial review determination that dismissed the application, Kerr J stated:

> In the present case, the relevant condition expressly confines the sole use of the track to the training and exercising of horses and explicitly forbids its use for private or public races or meetings associated with equine sports or any other sporting activities. In my view the condition could not be more categoric in prohibiting the activity which the applicant claims Mr Dean is entitled to carry on under the General Development Order. While express reference to these rights was not made and while the inspector and PAC may not have been alert to Mr Dean's possible entitlement under the Order, it is beyond question, in my view, that neither intended or contemplated that Mr Dean would in any circumstances be allowed to carry on races at the track. It is clear that the grant of planning permission should be construed by reference to other documents which it incorporates … When one considers the report of the inspector, detailing, as it does, the frequently expressed concern about the very activity which it is now claimed is Mr Dean's right, the assurances made on his behalf that such would not occur, the invitation to the inspector to impose conditions which would prohibit it and the latter's acknowledgement of the reality of and justification for the fears expressed, it is abundantly clear that, by the conditions which he imposed, and which were adopted by the PAC, it was intended that such activity should never take place on these lands. Such an intention cannot be consistent with the continued existence of GDO rights which would, in effect, allow Mr Dean, if he chose, to arrange a race meeting every fortnight throughout the year. I do not accept the suggestion implicit in the case made for the applicant that a planning authority must specifically identify possible GDO rights and articulate precisely the reasons for imposing the condition which has the effect of withdrawing those rights by express reference to them. Provided the rationale of the condition is sufficiently clear and its intention is unambiguous that development or use of the lands which might be permitted under the General Development Order should not take place, advertence to or discussion of those rights in the grant of planning permission is not required, in my opinion.

In May 2016 the Department for Infrastructure issued a consultation paper dealing with potential amendments to the Planning (General Permitted Development) Order (Northern Ireland) 2015 relating specifically to development by electronic communications code operators, along with non-domestic roof-mounted solar photovoltaic panels, shops and financial and professional services establishments, and electric vehicle charging points. In December 2016 a further consultation took place to assist consideration of options to remove petroleum exploration permitted development rights. All measures were deemed to be in the best interests of removing regulatory barriers and protecting environmental amenity. The revisions were subsequently carried forward by the enactment of The Planning (General Permitted Development) (Amendment) Order (Northern Ireland) 2020, which provides for extended permitted development rights for electric vehicle charging points, expanded permitted development rights for mobile network operators, amendments to allow for the extension of premises related to shop, financial and professional services establishments, along with extensions to loading bays. The Order also removed permitted development rights for the drilling of boreholes for oil and gas exploration. The Department for Infrastructure decided, however, not to take forward amendments to the permitted development rights for non-domestic roof-mounted solar photovoltaic panels.

Pre-application determination procedures

The Planning Act (Northern Ireland) 2011 distinguishes between 'major developments' and 'local developments', and the classes of development belonging to 'major development' are set out in the schedule appended to the Planning (Development Management) Regulations (Northern Ireland) 2015 using threshold-based criteria. The categories comprise development coming under The Planning (Environmental Impact Assessment) Regulations (Northern Ireland) 2015 (as amended), energy infrastructure, transport infrastructure, waste infrastructure, minerals, housing, retailing/community/recreation/culture, business/industry/storage/distribution and all other development. The important factor here is the scale of development in terms of capacity, length, area of the site, number of units and floor space. Operationally, planning applications for local and major development are determined by a council, whilst applications for regionally significant development and called-in applications are determined by the Department for Infrastructure. Regionally significant development is development which, if carried out, would (a) be of significance to the whole or a substantial part of Northern Ireland or have significant effects outside Northern Ireland, or (b) involve a substantial departure from the local development plan for the area to which it relates. Before submitting an application for planning permission that may fall into this category, a developer must enter into discussions with the Department for Infrastructure to determine if the proposed development is regionally significant. Should the conclusion be that it is not regionally significant development, the planning application should be submitted to the relevant district council. Called-in planning applications apply in situations when a council is of the opinion to grant planning permission and (a) a government department or statutory consultee has raised a significant objection to a major development application, (b) a major development application would significantly prejudice the implementation of the Local Development Plan's objectives and policies, (c) a major development application would not be in accordance with a marine plan adopted under the Marine Act (Northern Ireland) 2013, (d) planning applications are for major and local development in which the council has an interest and the proposal would be significantly contrary to the development plan for its district, or (e) where a council proposes to grant planning permission for petroleum development. Requests submitted to the Department for Infrastructure from interested parties may also result in a planning application being called-in. Once notified of these circumstances the Department for Infrastructure will consider whether or not to call-in the planning application for its own determination or allow the council to continue and determine the application.

Before submitting a planning application for major development, a prospective applicant, under Section 27 of the Planning Act (Northern Ireland) 2011, must give formal notice, known as a Proposal of Application Notice, to the appropriate council, or, as the case may be, the Department for Infrastructure, that an application for planning permission for the development is to be submitted. A period of at least twelve weeks must elapse between serving the Notice and submitting an application for planning permission. The developer may lawfully meet with the planning authority officials before and after the submission of the Proposal of Application Notice. If then the developer decides to proceed further, there is a statutory requirement (Section 28) that a pre-application community consultation must take place. The Planning (Development Management) Regulations (Northern Ireland) 2015 specify that this necessitates at least one public engagement event, to facilitate comments from members of the public, be convened within the locality of the proposed development and that prior to the event there is an advertisement placed in a locally circulating newspaper giving details of the proposal, arrangements regarding the public consultation venue, and how and when members of the public can submit their observations on the proposal to the prospective applicant for planning permission. Development Management Practice Note 10 (Department of the Environment 2015a) indicates that the carrying out of a pre-application community consultation is the responsibility of the applicant and that all community comments should be directed to the applicant. A council, or as the case may be the Department for Infrastructure, cannot accept representations or objections to a proposal at this stage since no planning application has been received. Moreover, it is suggested that consultees in this process may include local community groups, individual residents within the local area and particularly neighbours likely to be notified of the planning application, landowners/businesses within the vicinity of the site, the district council(s), government departments, public bodies, local environmental groups and voluntary groups. Following the pre-application community consultation, a pre-application community consultation report must be written up to accompany the planning application. The matters to be covered are again set out in Development Management Practice Note 10 as comprising:

- details of any community or residents' steering group(s) established;
- what consultation was undertaken, when and with whom;
- dates, venues and numbers attended for all consultation events and meetings;
- details of how public events were made as accessible as possible to all members of the community;
- evidence of the consultation, e.g., dates and copies of advertisements, reference to material made available at events, minutes of meetings, samples of letters/leaflets if applicable;
- the comments made by the general public and those consulted;
- whether and how these comments have been taken into account in the development proposals;
- summary of the main issues/comments raised through the consultation and how they have been responded to;
- details of any amendments to the proposal as a result of the consultation; and
- details of any existing liaison arrangements, agreements with the local community or monitoring that is proposed during the construction and operational phases.

If, after considering the report, the council or as the case may be, the Department for Infrastructure, concludes that the applicant has failed to carry out the required steps, they may decline to determine the planning application. But before deciding whether an application must be declined, the council or, as may be the case, the Department can request additional information in order to decide whether to decline the application. Section 50 of the Planning Act (Northern Ireland) 2011 provides that a council or the Department *must* decline to determine a planning application where in its opinion the pre-application community consultation provisions of the legislation have not been met. The operation of these provisions has generated litigation by way of judicial review.

A determination in 2018 by the Department for Infrastructure that it would not decline to determine a planning application for a gold mine that was deemed compliant with the pre-application community consultation requirements of the 2011 Planning Act was judicially considered *In the Matter of an Application by Greencastle Rouskey Gortin Concerned Community Limited for Judicial Review v Department for Infrastructure* (2019 NIQB 12/2019 NIQB 24). In 2016 the developer, Dalradian, had undertaken a series of public information events and lodged its planning application in November of the following year, accompanied by a pre-application community consultation report. The plaintiff, comprising a company of some forty local residents, initiated judicial review proceedings, which essentially argued that the project proposals upon which Dalradian had consulted were not sufficiently advanced and too embryonic in nature. Accordingly, the Proposal of Application Notice process had been undertaken too early and consequently the decision of the Department for Infrastructure was defective. The court, however, identified no material operative legal defect and in dismissing the judicial review challenge, McCloskey J ruled:

> The new statutory arrangements do not require attainment of the theoretically perfect. Rather, their effect is to notionally stop the clock at a particular moment in time and to require the putative planning applicant to conduct a public engagement exercise which gives the community a fair and reasonable opportunity to express its views relating to the 'general terms' of the project then in contemplation. Neither a completed project concept nor a highly advanced one is required by the statute or the associated common law principles. The new statutory regime accommodates the possibility of what some might, subjectively, later consider to be imperfections or inadequacies or unacceptable omissions in the development proposal published at the PAN (Proposal of Application Notice) consultation stage. The section 27 stage is, by definition, a preliminary, or intermediate, one. It does not require the developer to publish and consult upon a completed project concept. At all stages the familiar public law juridical phenomena of balancing, margins, latitude and evaluative judgment shine brightly. Furthermore, it is the very essence of the new statutory regime that alterations, ranging from the minimal to the more significant, may legitimately be made to a planning project in the wake of a section 27 PAN consultation exercise.

During the COVID-19 emergency commencing in 2020 the Department for Infrastructure issued guidance under The Planning (Development Management) (Temporary Modifications) (Coronavirus) Regulations (Northern Ireland) 2020 that temporarily suspended the requirement for a pre-application community consultation public event. Other forms of community participation in the planning process were identified in the accompanying guidance, including web-based engagement, use of social media, non-face-to-face community surveys, information sheets and newsletters in order that prospective applicants for major development planning permission may comply with the requirement to prepare a pre-application community consultation report.

The important point here is that these procedures place emphasis on front-ending the planning application process with consultation in order to improve both the quality of development and give greater rigour in decision making. In this regard the potential of non-statutory pre-application discussions (PAD) with council or Department for Infrastructure staff for all development proposals is now canvassed as a valuable element of contemporary planning practice in Northern Ireland. Development Management Practice Note 10 (Department of the Environment 2015a) suggests that an engagement with PAD can help to:

- identify potential policy constraints and other material issues which need to be addressed at an early stage in the process;
- facilitate discussions with key consultees (where appropriate) at an early stage, especially where an environmental statement is likely to be required;
- identify related technical issues and allow for discussion with a view to resolving such matters;
- ensure that engagement with appropriate stakeholders takes place at an early stage in the planning process;
- offer an opportunity for informed amendments and improvements to be made to schemes prior to formal planning applications being submitted, thereby potentially reducing the time taken for an application to move through the planning system;
- improve the content and quality of planning applications;
- enhance the quality of a development scheme;
- speed up the statutory decision-making process; and
- ensure active case management.

During the interim Councils have now published their individual Statements of Community Involvement that include mention of PAD protocols. A key and consistent expected outcome is the identification of material issues relevant to the preparation of a planning application. Case law explication has taken this further by placing emphasis on procedural fairness and which *In the Matter of an Application by Belfast City Council for Judicial Review v The Planning Appeals Commission* (2018 NIQB 17) was defined by McCloskey J:

> The ingredients of procedural fairness are intrinsically fact specific and context sensitive, varying from one case to another. In the particular context of planning decision making, it seems to me harmonious with well-established principle that the developer should be treated on a 'cards face up' basis throughout the process. This entails, fundamentally, knowing the case which he has to meet. Being taken by surprise in any material respect is antithetical to this principle. This is so not merely because it is procedurally unfair vis-à-vis the developer. It also offends against the overarching requirement of transparency to which all planning decision making has aspired for many years.

In this case Council planning staff had recommended to the Planning Committee refusal of a planning application for a student apartment accommodation scheme based *inter alia* on the unacceptability of the proposed height, scale and massing of the development. But during the judicial review the developer argued that matters of scale and massing had not been raised during pre-application discussions as a possible reason to refuse this application and which on the evidence prompted McCloskey J to rule:

> I have no hesitation in concluding that the failure of the Council's officials to alert the IP's agents to their obviously profound concerns about the scale and massing of the proposed development, considered in its full context, was procedurally unfair. Its effect was that the IP and its agents were unfairly taken by surprise. The 'ambush' element was that they had insufficient time and opportunity to respond and rectify. This should never have occurred.

In essence, therefore, the principle of procedural fairness must watermark the PAD process that brings together the agents of developers and planning officials within Councils and the Department for Infrastructure. As a final word on this matter McCloskey J provides salutary advice:

> No planning application can lay claim to a guaranteed successful outcome. Every planning application entails varying degrees of unpredictability and uncertainty. This applies to a context in which the putative developer must invariably make a predominantly economic decision having substantial financial consequences. The fee for a planning application is measured by reference to a set formula. In the present case it cost the IP £33,512.00 to lodge his planning application. Furthermore, with effect from 01 April 2017 the fee payable in return for the Council's willingness to engage in the 'PAD' exercise would have been £1,500 plus VAT Basic fairness surely dictates that in the process of assessment, examination, consideration and, ultimately, determination which followed he was entitled to be treated with basic procedural fairness.

The planning application

The Planning (General Development Procedure) Order (Northern Ireland) 2015 sets out the information to be submitted in an application for planning permission as including:

- a written description of the proposed development;
- the postal address of the land to which the development relates, or if none exists;
- a description of the location of the land; and
- the name and address of the applicant and, in the case of an agent acting on behalf
- of the applicant, the name and address of that agent.

Additionally, the application must be accompanied by plans to identify the land to which the development relates, appropriate plans to describe the development, a plan denoting any neighbouring land owned by the applicant, the appropriate property interest certificate, a pre-application community consultation report where the proposed development is in the category of a major development, an access statement or a design and access statement as appropriate and where required, and any fee payable under the Fees Regulations. An application may be made for outline planning permission with a follow-on application for reserved matters if the outline application is approved, or alternatively for full planning permission; the former approach may be used to test the principle of development possibility and can avoid abortive expenditure on professional fees, but it is not suitable for change of use proposals. Reserved matters can deal with siting, design, external appearance, means of access and the landscaping of the site. A reserved matters application, however, is not an application for planning permission since that has been dealt with at the outline planning application stage, which provides the lawful consent context for this subsequent application.

Both a council and the Department have discretionary powers to decline to determine subsequent repeat applications for planning permission and overlapping, twin tracking applications. The main characteristics requiring judgement in these circumstances are the degree of similarity between different planning applications and any material changes in circumstance.

There is no right of appeal to the Planning Appeals Commission when an application for planning permission is declined to be determined.

The matter of the planning application has generated litigation in Northern Ireland. There has, for example, been a legal challenge around the inaccuracy of the land ownership certificate submitted by a developer seeking planning permission for housing on land in Londonderry. *In the matter of an Application by Callan for Judicial Review* (1997 Queen's Bench (Crown Side)) it was confirmed by the Department of the Environment at a pre-decision public inquiry that it had been aware that the certificate was inaccurate but expressed the view that this did not render the planning application itself invalid. Moreover, at the inquiry the Commissioner expressed the view that there did not appear to be any prejudice and declared his intention to proceed. The basic ground, therefore, upon which relief was sought by way of judicial review, was that the certificate was so defective as to render it invalid and that, accordingly, the Department of the Environment had no jurisdiction to entertain the application, and the Commissioner was wrong in law in permitting the inquiry to proceed after the alleged defects in the certificate had been drawn to his attention. At the court hearing it was accepted by all the parties that the statute provisions were mandatory in form and that the defective certificate constituted a breach thereof. Coghlin J took the view that the real issue of significance was, therefore, whether, in the circumstances of that particular case, the Department of the Environment had been deprived of jurisdiction to entertain the planning application. In his examination of the issue, Coghlin J stated:

> Once it is established that a provision is mandatory in form it becomes necessary to consider whether it is also mandatory in substance in that it was the intention of Parliament that, in order to achieve the effective purposes of the legislation, non-observance of such a provision should render subsequent proceedings dependent thereon a nullity. I have already referred to the importance of the 'notice' provisions within the structure of the 1991 Order and, as a matter of construction, it seems to me that the intention of Parliament was to ensure that planning applications should not proceed unless all practicable steps were taken to ensure that those persons who might be entitled to possession of the land or any part thereof had effective notice of the application. Therefore, in formal parlance, the breach may be designated mandatory in substance as well as in form.

Nonetheless, it was acknowledged that the court retained the discretion to withhold relief and in that regard counsel for the applicant 'was unable to point to any different result or outcome which might have been achieved had the certification been perfected'. In quoting observations by Lord Wilberforce (*Malloch v Aberdeen Corporation* (1971) 1 WLR 1578), Coghlin J noted: 'A breach of procedure, whether called a failure of natural justice, or an essential administrative fault, cannot give him (the appellant) a remedy in the Courts, unless behind it there is something of substance which has been lost by the failure'. In this case, given that the defective certification related to the land interest of the Water Executive, a subdivision of the body to which the planning application had been made, and that it was fully aware of the planning application, Coghlin J concluded:

> Setting the benefit to the public interest of ensuring formal compliance with Article 22 against the considerable demands upon both human and financial resources which

> would inevitably be required for a resubmission of the application and a rerun of the public inquiry, in the absence of any established prejudice on the part of any individual or body, after careful consideration, I have come to the conclusion that this is not a case in which relief should be granted and, accordingly, I refuse the application.

The relationship between a reserved matters planning application and a previous outline planning permission has also been the subject of judicial review. In *Re Central Craigavon Limited* (2008 NIQB 46) the applicant sought to quash a decision of the Department of the Environment whereby it had determined that an application for reserved matters permission by the applicant was invalid and returned the application to the applicant. The application for reserved matters approval contained proposals for a development of 14,787 square metres gross of retail floor space, spread over two retail areas and with internal and external car parking spread over two levels. In contrast the approved outline planning application indicated a total gross internal floor space of the shopping centre extension as 6,295 square metres on one level with single-level parking below. The outline permission granted on appeal by the Planning Appeals Commission had included a condition (Number 2) that the development be carried out in accordance with the approved plans, though the quantum of retail floor space was not included as a condition. Gillen J, in his conclusions, commenced by finding in principle:

> that an application which is made for approval of a reserved matter, must be within the ambit of the outline planning permission and must be in accordance with any conditions annexed to the outlined permission. An applicant cannot at his own instance modify the permission. Only the planning authority can do that. If the applicant desires to depart in any significant respect from the outline permission or the conditions annexed to it, he must apply for a new planning permission.

In dismissing this application Gillen J stated:

> In this case I consider that it was within the reasonable exercise of the Department's discretion to conclude that the proposed enlargement of floor space from 6297 square metres to 14787 square metres was so substantial as to merit a conclusion that it was out with the ambit of the OPP [outline planning permission] and in particular condition 2. The decision cannot be set aside as unreasonable. If members of the public viewed this OPP and observed the plans attached to condition 2 which record plainly on their face that the retail floor space is 6297 square metres, I consider they would be entitled to say that the proposed reserved matters were not what they understood the OPP to have granted. To accede to the reserved matters with a vastly increased floor space would indicate to the public at large that they had not been able to rely on the permission and conditions granted. It may not be uncommon for applicants to submit plans, as in this case, with the OPP application for 'illustrative purposes' or for there to be an expected measure of flexibility to be allowed in proposed reserved matters. However where those proposals are so far in excess of what is contained in plans attached as a condition to the actual OPP, I consider that the Department is entitled to conclude that the proposals are outside the ambit of the permission granted and to determine this at an early stage without wasteful and unnecessary further enquiry …

> It seems to me that the decision of the Department that these reserved matters were not in 'general compliance' with the conditions imposed was one that could be fairly made on the facts and thus their decision to return the application was lawful and not *Wednesbury* unreasonable.

Planning application notices

Article 8 of the Planning (General Development Procedure) Order (Northern Ireland) 2015 requires the council or, as the case may be, the Department to publish notice of the application in at least one newspaper circulating in the locality in which the land to which the application relates is situated, and on the council or Department website. A council or the Department for Infrastructure may not lawfully determine a planning application before the expiration of 14 days from the date on which the Notice is first published. In the case of a planning application accompanied by an environmental statement under the provisions of The Planning (Environmental Impact Assessment) Regulations (Northern Ireland) 2017 the council, or the Department for Infrastructure, must publish a Notice to the effect that the application for planning permission or subsequent application is accompanied by an environmental statement;

- in the case of a subsequent application, that a copy of the planning permission and supporting documents for the development in respect of which the application has been made may be inspected by members of the public at all reasonable hours at the relevant office of the council or, as the case may be, the Department for Infrastructure;
- the Notice contains a postal address (within the locality in which the land proposed to be developed is situated) at which copies of the environmental statement may be obtained from the developer, so long as stocks last, and if a charge is to be made for a copy, state the amount of the charge.

Under these environmental impact assessment regulations, the council or the Department for Infrastructure must allow the public a period of four weeks from the date on which the Notice is first published to make representations.

On occasion, the need for a readvertisement of a planning application may arise and the advice in Development Management Practice Note 14 (Department of the Environment 2015b) is that this is a matter at the discretion of the council or the Department of Infrastructure having regard to the nature of the application, the scale of the amendment and the level of public representations It is suggested that readvertisement should occur when:

- there are material amendments made to the proposal, for example, a new description to the proposal, significant changes in layout, design, number of buildings (including dwellings), amended access points, changes in landscaping proposals or other site works such as retaining walls which have an effect in the external appearance of the building;
- significant additional and new information is provided in support of a proposal including traffic assessments, retail impact assessments, concept statements, noise reports, and contaminated land remediation measures;
- an amendment, or additional information, relates directly to the grounds of objection.

Other matters relating to Notices include publicity in regard to a planning application hearing before the Planning Appeals Commission, public notices indicating that in circumstances of a high volume of correspondence received in relation to a planning application it is not feasible to notify all written representations individually, and announcements related to the granting or refusal of planning permission, including notification of decisions by An Bord Pleanála to approve/refuse planning permission for development in the Republic of Ireland that may have an impact within Northern Ireland.

In the often-cited case of *Morelli v DOE (NI) 1976* (NI 159), Murray J ruled that the advertisement of a planning application is designed to bring home to the mind of the reasonably intelligent and careful reader the nature of the development proposed. At issue here was the description of the proposed development whereby an application lodged on 7 March 1975 for 'Structural alterations to existing café to amusement arcade' had in fact been initially advertised on 15 March 1975 as 'Structural alterations to existing dwelling'. Planning permission was granted. A second application for an extended area then followed on 4 July 1975, which was advertised on 19 July 1975 as 'Change of use from restaurant to amusement area'. Murray J noted that 'unlike the first Notice, the second Notice produced a crop of objections and adverse comments from members of the public'. Headquarters Branch of the Planning Service then intervened and issued instructions that the initial planning decision was unlawful and that the first planning application would need to be readvertised and considered afresh. This was subsequently advertised on 30 August 1975 as 'Change of use from dwelling and take-out bar to amusement arcade' with the additional note appended: 'This is a correction of the notice published on 15 March 1975'. What is important here is that the Court ruled that the first application Notice did not comply with the requirements of Article 15(a) of the Planning (Northern Ireland) Order 1972 and that such non-compliance invalidated the subsequent planning permission:

> A Notice which omits any reference to a change in use sought is, in my view, a Notice of part only of the relevant objection and is not a Notice of the relevant objection … In the present case it is highly significant that when the first Notice was published no objection was received but when the second notice appeared a crop of objections came in … The requirement of a proper Notice is in my view clearly a mandatory one since the giving of such notice is an essential part of the scheme of the 1972 Order … people living in the locality of the land to which a planning application relates will be informed in advance in a local newspaper of the substance of what is proposed on that land, and will have a reasonable opportunity – if they so desire – of following the matter up and making representations … in this case I hold in effect that no Notice was published of the relevant application.

In drawing on this judgement, McCloskey J in the later *Belfast City Council Application* (2018 NIQB 17) commented more widely:

> the burden on developers and their agents of ensuring maximum accuracy and particularisation in every development proposal application. In this sphere one is dealing with well qualified experts of whom the highest standards are to be expected. The only qualification, juridical in nature, is that the law recognises a degree of latitude in these matters. In so doing it does not purport to espouse a charter for the slipshod or inefficient.

Neighbour notification

From 1985 the Northern Ireland planning system, by way of Planning Policy Statement 1 (PPS1), operated a non-statutory neighbour notification scheme which brought planning applications to the specific attention of those individuals who might be most directly affected by them. The Planning (General Development Procedure) Order (Northern Ireland) 2015 now places this element of public consultation in the planning process on a statutory footing. Applicants are asked to list on the application form the address of any occupier on neighbouring land. Identified occupier means the occupier of premises within a ninety-metre radius of the boundary of the application site and neighbouring land means land which directly adjoins the application site or which would adjoin it but for an entry or a road less than 20 m in width. As with readvertisement during the processing of a planning application, it may be necessary to repeat the neighbour notification procedure in order to advise members of the public that amendments have been made to a proposal and to invite further representations.

The significance of robust neighbour notification is particularly well illustrated by *Warner Chilcott UK Ltd and Caridian BCT NI Ltd's Application* (2011 NIQB 137) wherein it was accepted by the Planning Service of the Department of the Environment that it had failed, in breach of the neighbour notification scheme, to inform the applicants for judicial review of a planning application for an abattoir; the applicants were carrying on business in adjacent premises and were manufacturing pharmaceutical products and a range of sterile fluids. Treacy J underlined the wider advantages of formal neighbour notification in the face of this non-compliance when he stated:

> Formal notification by a public authority, in this case the Planning Service, of the details of a planning application is capable of being one of the most effective ways of ensuring that those who may be adversely affected by a particular planning application are put on notice. The solemnity of formal notification by the public authority with sufficient accompanying detail is much more likely to ensure that its potential significance registers at the appropriate level within a company. The obvious importance of what is at stake for everyone in the present case merely serves to emphasise how the failure to comply with PPS1 can have potentially far reaching consequences. Had the policy been complied with the applicants would have had the opportunity to furnish details of the nature of their operations and the potential impact on their businesses if planning permission were granted. Certainly no one can say or has said that had this opportunity been afforded and availed of that the outcome of the planning application must have been the same. Casual or informal conversation is no substitute for formal neighbour notification. Formal communication especially by a public authority is much more likely to fully engage (at the requisite level) corporate and commercial antennae and lead to an appropriate formal response. The fog of memory and the time consuming attempt to reconstruct the truth would have been displaced by the officially communicated and recorded notification. The advantage of PPS1 compliance and the consequences of non compliance have I believe been made manifest by this case. The other side of this coin is that if the applicants had been notified and had not taken prompt steps thereafter to challenge the grant of planning permission they would likely face an uphill struggle in obtaining relief in judicial review proceedings in the absence of some very compelling reason.

Material planning considerations

The scope of material considerations

The identification of all relevant material considerations and the giving of appropriate weight to each consideration lie at the core of the determination of planning applications. Needless to say, it is a matter that has generated a considerable body of authorities with the departure point commonly found in the much-cited case of *Stringer v Minister of Housing and Local Government* (1971 1 All ER 65) when Cooke J said that any consideration that related to the use and development of land was capable of being a planning consideration, its materiality depending on the individual circumstances of the case. In *Westminster City Council v Great Portland Estates* (1985 AC 661) the House of Lords held that the test of what is a material consideration in the planning context was whether it served a planning purpose relating to the character of the use of land. In *Tesco Stores Ltd v Secretary of State for the Environment* (1995 2 All ER 636), the House of Lords took the view that for something to be a material consideration, it must serve a planning purpose and have a connection with the proposed development which is not *de minimis*. In short, a material consideration should be related to the purpose of planning and should reasonably relate to the particular planning application. Development Management Practice Note 16 (Department of the Environment 2015c) identifies material considerations as including the Local Development Plan, planning policy, planning history, need, public opinion, consultation responses, existing site uses and features, layout, design and amenity matters, precedent, alternative sites and planning gain. Wider ranging government policy has also been deemed a material consideration, as illustrated *In the Matter of an application by Meekatharra (Northern Ireland) Limited for Judicial Review* (1995 Queen's Bench Division (Crown Side)). This concerned a proposal for an electricity interconnection with Scotland whereby the applicant challenged the degree to which the deliberations linked to the reporting of a public inquiry should be informed by current government policy. In his judgement, Kerr J stated: 'Planning decisions belong to the real world, however, and they must not ignore relevant government policy'. But there are also matters that would not be considered to be material considerations, including the protection of private interests (for example, boundary disputes or impact on a private view), moral considerations, political considerations and issues covered by other legislation. In short, all relevant matters must be taken into account and irrelevant matters must not be deemed relevant.

Statutory consultees

The inputs of statutory consultees constitute a vital part of the planning system and represent an especially important material consideration in decision making. In this vein, a research report by the Department for Infrastructure (2019) acknowledges that consultation with bodies that have an expertise in certain topics is an essential feature of the planning process as planning authorities may not have the necessary in-house expertise to decide on the technical and specialist aspects of a planning application's merits (pp. 8–9). Councils and the Department for Infrastructure (DfI) must consult with statutory consultees in line with the nature of the planning application and which include DfI Roads, DfI Rivers, other government departments and their agencies (for example, the Northern Ireland Environment Agency within the Department of Agriculture, Environment and Rural Affairs), Northern Ireland Water, the Health and Safety Executive for Northern Ireland, airports and the Northern Ireland Housing

Executive. Planning applications being determined by the Department for Infrastructure also require council consultation. McCloskey J *In the Matter of an Application by James Hugh Allister and Robert Edwin Agnew for Judicial Review v Causeway Coast and Glens Borough Council* (2019 NIQB 79) has more fully explained the rationale for this wider engagement:

> The legal framework is straightforward. Planning authorities consult with specified agencies for three reasons, a mixture of the factual and the legal. The factual driver is that while planning authority officials who advise decision makers are presumptively well qualified in the principles, theory and practice of town and country planning, their expertise in certain land use areas – environment, ecology, habitats, air quality, noise impact, together with roads and traffic issues (inexhaustively) – is limited. Lacking the necessary in-house skills and expertise, it is necessary for planning authorities to fill this void. The second reason is a legal one. It is rooted in the planning authorities' inter-related public law duties to take all material matters into account, to disregard matters extraneous and to avoid the prohibited territory of *Wednesbury* irrationality. In reduced terms, there is a legal obligation to engage fully with the issues thrown up by every planning application. In cases where a planning application raises issues requiring expert input and assessment which the planning authority is unable to provide it must take appropriate steps to rectify this lacuna in its inbuilt skills and expertise. Finally, there is a free standing duty imposed by statute to consult in prescribed cases.

The substantive responses from consultees will state contentment with the application, no comment to make, a reference to standing advice or particular advice to enable the determination of the application.

The invaluable contribution of this expert input into planning decision making has been underlined by further judicial comment. *In the Matter of an Application by Christine Alexander for Judicial Review v Causeway Coast and Glens Borough Council* (2018 NIQB 55), the granting of planning permission for a caravan park at Portrush was contested on grounds, inter alia, of road safety and traffic volumes. Evidence was considered from a Traffic Consultant and from DfI Roads (the roads authority for Northern Ireland) and regarding which McCloskey J commented:

> One of the recognised principles of planning decision making is that decision makers and their advisers are entitled, within certain limits, to place reliance on the contributions and input of presumptively expert consultees. I consider that, as a matter of legal principle, such limits are to be viewed from the perspective of the *Wednesbury* principle. This was uncontroversial as between the parties. In a case of the present kind, in which objections on traffic and road safety grounds grew as the decision- making process advanced, the court will be astute to ascertain whether, at each turn, the decision making authority turned to the expert consultee concerned for further input. This, demonstrably, occurred in the present case and reflects well on the planning officials concerned. Irrationality can only be evaluated by reference to the information available to the decision makers and this, in turn, throws up questions relating to the decision making process. Assessed in this way, and having scrutinised the relevant evidence critically, while remaining alert that the court's role is one of supervisory superintendence, I consider that the Council's decision withstands challenge on this particular ground.

This reliance by a planning authority on the advice received from statutory consultees, alongside other considerations, is further illustrated *In the Matter of an Application by William Donnelly for Leave to Apply for Judicial Review* (2020 NIQB 35) whereby the applicant sought to challenge the failure of Fermanagh and Omagh District Council to take enforcement action in respect of alleged unlawful development comprising boreholes at a site for underground mining that itself had a previous planning permission. The boreholes were deemed by the Council as a breach of a condition attached to that permission but decided that it would not issue an enforcement notice. In the judgement on this case, Humphreys J acknowledges that in arriving at this decision, the Council took into account:

> (i) The passage of time; (ii) The absence of any evidence of physical damage as a result of the boreholes; (iii) The opinions of other relevant statutory agencies; and (iv) The absence of any satisfactory remedy. Prior to arriving at this decision, the Council consulted the Northern Ireland Environment Agency and the Shared Environmental Services. Both these agencies commented that they had no concerns about the boreholes from an environmental perspective. The Council 'reluctantly' concluded that enforcement action was not expedient in the circumstances of this case.

In concluding that the applicant had not established an arguable case and thus dismissing the application for judicial review, the Court held:

> As in any proportionality challenge, it must be recognised that the law affords a margin of appreciation or latitude to the decision maker. I am satisfied in this case that the Council took into account all material considerations and that the decision not to pursue enforcement action was a proportionate one in the all the circumstances … In the instant case, there was no harm which could be remedied or activity ceased by the service of an Enforcement Notice … Given that the unlawful drilling took place over 4 years ago, and has not occurred since, and the lack of any physical damage to be remedied, it is difficult to see what purpose an Enforcement Notice would serve … the Council was entitled to arrive at the decision which it did, namely that it was not expedient in all the circumstances to issue an enforcement notice. The claim that such a decision was irrational in the *Wednesbury* sense was plainly unarguable.

In short, it is necessary for the planning authority to take into account the responses from the statutory consultees. In citing from Treacy J in *Newry Chamber of Commerce and Trade's Application* (2015 NIQB 65), McBride J *In the Matter of an Application by William Donnelly* (2017 NIQB 84) draws attention to the comments of Beatson J in *Shadwell Estates v Breckland DC* (2013 EWHC 12 (Admin)): a decision-maker should give the views of statutory consultees … 'great' or 'considerable' weight. A departure from those views requires 'cogent and compelling reasons'.

The input of statutory consultees is a vital part of the planning application decision-making process and in speeding-up it up much depends on statutory consultees providing timely responses. Performance on that front has attracted criticism from the Northern Ireland Audit Office (2022:60-63) but with slow consultation replies being linked to increasing requests, greater case complexity, staffing shortages, incomplete planning application information and the need for multiple consultations with some statutory consultees.

Pre-determination hearings

Councils are required to hold a pre-determination hearing in instances where the Department for Infrastructure has decided not to call-in a planning application that was notified to it by the council. A council also has the discretion to hold a pre-determination hearing for any planning application that it determines (Department of the Environment 2015d; Department for Infrastructure 2022). This procedure is designed to deal with material considerations that have surfaced during the consideration of a planning application and can be used by elected representatives to hear the views of interested parties and seek clarification around issues arising. It is thus a further input into the decision-making process and is designed to deepen public involvement and enhance quality determinations. The mandatory requirement for a pre-determination hearing mentioned above has been deemed as potentially duplicating earlier work by a council and causing delay, uncertainty and increased costs in the planning process. The Department for Infrastructure (2022: 52) has indicated that it will bring forward proposals to make all pre-determination hearings discretionary for councils.

Planning policy

Within Northern Ireland, planning policy, in particular, as a material consideration has generated a number of applications for judicial review. In *O'Hare's application* (2016 NIQB 20) the Department of the Environment had conceded in correspondence that policy prohibiting unsatisfactory piecemeal development on sites zoned for housing required that an application for partial development should demonstrate how this could be achieved in a manner which enabled the development of the entire site in a way which was satisfactory to the Department. These matters, it was acknowledged, were not properly taken into account 'in the sense that proper consideration was not given to the meaning and requirements of the policy or to the question of whether the decision to grant planning permission would be in accordance with policy or a departure'.

In the *Department of the Environment's Application* (2014 NIQB 4) the Department of the Environment challenged a decision of the Planning Appeals Commission granting planning permission for the unauthorised use of land at Aldergrove as a commercial car park which had not become time-barred from enforcement. In this instance the contended statutory planning policy framework comprised Planning Policy Statement 3 (PPS3) *Access, movement and parking* (Department of the Environment 2005) and Planning Policy Statement 21 (PPS21) *Sustainable development in the countryside* (Department of the Environment 2010). The applicant took the view that the planning authority had failed to give planning policy its single objectively correct meaning and had erred in law. In effect, it was argued that the Planning Appeals Commission had held that the only policy test which needed to be satisfied lay with PPS21, when, in fact, PPS21 states that the policies of that statement must be read together and in conjunction with other planning policy statements. This would include the provisions of PPS3. In contradistinction to this analysis, the respondent argued that a proposal which meets the policy requirements of PPS21 does not also have to meet the policy requirements of PPS3 in order to be acceptable in the countryside. It was submitted that once the Planning Appeals Commission was satisfied that planning permission should be granted in principle for the farm diversification proposal (the commercial car park), and there were no site-specific objections, then permission was required to be granted. In his discussion of the evidence,

Treacy J commenced by drawing on *Tesco Stores Ltd v Dundee City Council* (2012 UKSC 13), where, in regard to the development plan, Lord Reed said:

> It has long been established that a planning authority must proceed upon a proper understanding of the development plan … The need for a proper understanding follows, in the first place, from the fact that the planning authority is required by statute to have regard to the provisions of the development plan: it cannot have regard to the provisions of the plan if it fails to understand them.

The judgement in this Tesco case, in turn, cites *City of Edinburgh Council v Secretary of State for Scotland* (1998 SC(HL) 33, 1997 1 WLR 1447) where Lord Clyde observed:

> it will obviously be necessary for the decision-maker to consider the development plan, identify any provisions in it which are relevant to the questions before him, and make a proper interpretation of them. His decision will be open to challenge if he fails to have regard to a policy in the development plan which is relevant to the application or fails properly to interpret it … The development plan is a carefully drafted and considered statement of policy, published in order to inform the public of the approach which will be followed by planning authorities in decision-making unless there is good reason to depart from it. It is intended to guide the behaviour of developers and planning authorities. As in other areas of administrative law, the policies which it sets out are designed to secure consistency and direction in the exercise of discretionary powers, while allowing a measure of flexibility to be retained. Those considerations point away from the view that the meaning of the plan is in principle a matter which each planning authority is entitled to determine from time to time as it pleases, within the limits of rationality. On the contrary, these considerations suggest that in principle, in this area of public administration as in others … policy statements should be interpreted objectively in accordance with the language used, read as always in its proper context. That is not to say that such statements should be construed as if they were statutory or contractual provisions. Although a development plan has a legal status and legal effects, it is not analogous in its nature or purpose to a statute or a contract. As has often been observed, development plans are full of broad statements of policy, many of which may be mutually irreconcilable, so that in a particular case one must give way to another. In addition, many of the provisions of development plans are framed in language whose application to a given set of facts requires the exercise of judgment. Such matters fall within the jurisdiction of planning authorities, and their exercise of their judgment can only be challenged on the ground that it is irrational or perverse … Nevertheless, planning authorities do not live in the world of Humpty Dumpty: they cannot make the development plan mean whatever they would like it to mean.

The reference to the writing of Lewis Carroll in *Through the looking-glass* in this judgement is very important in that Humpty Dumpty (Figure 2.1) stated: 'When I choose a word, it means just what I choose it to mean, neither more nor less'.

Thus, while the Tesco and City of Edinburgh cases concerned themselves with the proper interpretation of the development plan, Treacy J in the Aldergrove commercial car park case has taken the view that the same considerations apply to 'the carefully drafted and considered

FIGURE 2.1 Humpty Dumpty and Alice. *Source*: Illustration by John Tenniel in Lewis Carroll (1871).

statements of policy embodied in Planning Policy Statements'. The court in this instance ruled the relevant policy content of PPS21 (i.e., policy CTY1) must be read together with and in conjunction with the relevant contents of development plans and other planning policy publications. The court indicated that it was not satisfied that other material considerations relating to the content of PPS3 had been conscientiously taken into account when assessing the planning merits of the development proposal and accordingly in quashing the decision, concluded:

> The starting point must surely be that the PAC [Planning Appeals Commission] erroneously concluded that CTY1 is a self-contained policy which, since it was satisfied, must lead to the grant of planning permission. In the first instance it seems to me that ordinarily one must expect the decision maker to have correctly identified the relevant policy(s) and the relationship between them, correctly interpreted and applied them. That, for the reasons given, did not occur in the present case.

This case, however, invites further reflection on the long-accepted principle that a planning authority need not 'slavishly' follow a policy. It is lawful for a planning authority to choose to disapply a policy, but as indicated in *Gransden & Co Ltd v Secretary of State for the Environment* (1987 54 P & CR 86) regarding the status, interpretation and application of planning policy, it is both exceptional and conditional on clear reasons being given and it is essential that the policy is properly understood by the determining body. A key criterion is stated as: 'If the body making the decision fails to properly understand the policy, then the decision will be as defective as it would be if no regard had been paid to policy'. Thus, in *Lamont's (David John Stewart and Elaine) Application* (2014 NIQB 3) the applicant contended that, in granting planning permission for an additional farm dwelling outside Lisburn, the Department of the Environment did not interpret and apply the relevant policy in PPS21 in reaching its decision and had arbitrarily disregarded it. In this instance the court found as part of the process of determining the planning application:

> that while there was consideration of a policy the minds of those present were not clearly directed to the actual policy in question. They either did not consider the policy (as they did not in the main consider the actual written requirements of the policy) or they did not understand the policy.

In granting the relief applied for, Treacy J concluded that the policy was, therefore, not lawfully departed from.

The final matter that arises for discussion in this section of the chapter relates to the weight to be given to different material planning considerations in determining a planning application. Girvan J *In the Matter of an Application for Judicial Review by: (1) Bow Street Mall Limited (2) Lisburn Chamber of Commerce (3) Multi-development UK Limited (4) Central Craigavon Limited(5) Belfast City Council (6) Belfast Chamber of Trade and Commerce and in the matter of a planning decision made by the Department of the Environment* (2006 NIQB 28) has rehearsed the relevant legal principles:

> It is settled principle that matters of planning judgment are within the exclusive province of the local planning authority or the relevant minister … The planning decision-maker's powers include the determination of the weight to be given to any particular contention. He is entitled to attach what weight he pleases to the various arguments and contentions of the parties. The courts will not entertain a submission that he gave underweight to one argument or failed to give any weight at all to another.

These sentiments have been qualified by Treacy J in *Ulster Architectural Heritage Society's Application* (2014 NIQB 21), where, in reviewing the decision of the Department of the Environment to grant planning permission for the demolition of a building in a conservation area in Belfast, he addresses the matter of policy presumption and the weight to be accorded it:

> In exercising the supervisory function of the Judicial Review Court the court is only empowered to review whether the Department applied its policy lawfully and fairly, I am not empowered to interfere with the expert planning judgement of the Department. To this end, questions of the weight to be attached to the material considerations which weighed in the balance of the final decisions are within the sole province of the

> Department … However, where there is a presumption in the policy, the Department are obliged to have regard to that presumptive weight and cannot in their decision undermine that presumption. Presumptions in policy are there to protect the overriding objectives of the policy itself and to undermine them would be to nullify the policy.

In this case, the policy framework relating to new development in conservation areas in Northern Ireland expressly states that the Department of the Environment will operate a presumption in favour of retaining any building that makes a positive contribution to the character or appearance of a conservation area. What makes the Ulster Architectural Heritage Society's application noteworthy is that the court, in line with its jurisdiction, did not say what weight the presumption in favour of retention should command, but that the Department had failed to take into account the presumption-respecting option:

> When weighing up an application for development, and particularly for demolition consent, in a conservation area, the Department must have regard to the presumption in favour of retaining. Against this presumption, it must weigh all other material considerations. The Department enjoys discretion as to the weight to be given to these other considerations. In the group decision of 25 May 2011 the group explicitly give 'determining' weight to the fact that the proposals for retention and repair were not economically viable. That is, the fact that the repair options submitted by the owner are insufficient to create a profit for the developer is considered sufficient to outweigh the presumption in favour of retaining the building. The problem with this decision is that the Department is not obliged to change the status quo. The instant application is not the end of the line for the building. It is not the only option. If this application were refused, the building would continue in use as it is at the moment for the time being. There would be no necessary detriment to the area. Nothing bad will immediately happen if the building continued in its present use save the effect of time on the fabric of the building which it is estimated will eventually render the building unusable in 30 years in the absence of alternative proposals. To my mind where the Department is obliged to have regard to a presumption, and there are two options available to it, one which respects the presumption, and one which does not, the existence of the presumption-respecting option is a very material consideration and one which must be considered and given its full presumptive weight. Key documents leading to the ultimate decision overlooked the existence of this presumption, and none of the documents gave any or proper consideration to the presumption-respecting option that was open to the Department. For this reason I hold that the Department failed to properly take into account a very relevant consideration.

To conclude, case law indicates that all relevant material considerations must be fully considered and that appropriate weight be given to each by way of planning judgement.

Prematurity

The Planning Act (Northern Ireland) 2011 requires that the determination of planning applications be made in accordance with the development plan unless material considerations indicate otherwise. In this vein, an issue arises in development management regarding the relationship of a planning application, the extant development plan and a plan that is under preparation which may bring forward proposals that are different from the content of the extant plan. At stake here is the possibility that planning decisions made on the basis of the existing plan may have the potential to compromise the emerging plan. Within Northern Ireland this matter has contributed to a series of housing-related planning appeals for non-determination and applications for judicial review, notably over the period from 2001 when the Regional Development Strategy (RDS) for Northern Ireland was approved, although there had been earlier litigation on this theme (see, *Re Lisburn Development Consortium's Application for Judicial Review* (2000 Queen's Bench Division (Crown Side)). The 2001 regional planning framework introduced a new calculus for housing provision that comprised aggregate housing growth indicators at district level that would, in turn, be interpreted and applied by the follow-on plan-making process across individual settlements. It heralded a shift from the previous practice of substantial over-zoning in development plans to one which more closely seeks to align land supply and demand. This approach, labelled 'Plan-Monitor-Manage', had potential to lead to, and indeed resulted in, the dezoning of land deemed unnecessary in meeting projected housing requirements within the stated planning time horizon. The Strategic Planning Policy Statement for Northern Ireland (Department of the Environment 2015e) gives summary guidance on the approach to the determination of planning applications when such considerations more generally arise vis à-vis forthcoming Local Development Plans (LDPs):

> Where a new LDP is under preparation or review it may be justifiable, in some circumstances, to refuse planning permission on the grounds of prematurity. This may be appropriate in respect of development proposals which are individually so substantial, or whose cumulative effect would be so significant, that to grant planning permission would prejudice the outcome of the plan process by predetermining decisions about the scale, location or phasing of new development which ought to be taken in the LDP context. A proposal for development that has an impact on only a small area would rarely come into this category, but refusal might be justifiable where a proposal would have a significant impact on an import ant settlement, or a substantial area, with an identifiable character.

The contentious nature of the prematurity ground in determining planning applications can be illustrated, first, by reference to two planning appeals. In a non-determination planning appeal for ninety-eight dwellings at Scarva (*PAC reference 2003/A229*), the Planning Appeals Commission in its 2004 decision noted that the then draft Banbridge/Newry and Mourne Area Plan had not been published (it appeared in 2006), that the appeal site was within the development limit of the existing plan and that significant weight should be attached to the existing statutory plan. The Commission found that 'the provisions of the existing plan should prevail and that prematurity in the context of the emerging area plan which must be in general conformity with the RDS should not be determining'. Key here was the absence of any draft plan in the public domain. In a subsequent non-determination planning appeal for a housing development on lands at Draperstown (*PAC reference 2004/A080*), the Commission noted that the appeal site lay within the development limit of Draperstown in the then existing Magherafelt Area Plan, but it was excluded from the village limits delineated in the published draft Magherafelt Area Plan. Given this conflict between the existing and draft plans the issue of prematurity was deemed to arise. Important considerations here revolved around the nature of objections to the land zoning for Draperstown in the draft plan, including an objection on the appeal site and the desirability of addressing all these matters at a subsequent public inquiry convened as part of the plan-making process, as well as the creation of a precedent. In its decision the Commission stated:

> the Commission accepts that Draperstown is not a significant settlement and that the proposal is not individually substantial within the context of housing allocations within the plan area as a whole. However, having regard to the nature and extent of the objections referred to above the Commission concludes that to allow the appeal would predetermine decisions about the scale and location of development, including the village form of Draperstown, and also the allocation of housing throughout settlements in the plan area. The public inquiry into the draft MAP is the appropriate forum for addressing these matters when objectors and other interested parties will be able to put forward their views, thereby enabling issues to be considered on a comprehensive basis. The Commission concludes therefore, that it would be premature to grant permission to this development as it would prejudice the outcome of the development plan process and this consideration outweighs the fact that the appeal site presently lies within the development limits of Draperstown as set out in existing the Magherafelt Area Plan. Reference was made by the appellant to previous Commission decisions and the Commission acknowledges that these establish that the particular circumstances of some cases may limit precedent to the extent that determining weight may not be attached to the issue of prematurity. The Commission considers that the issue of precedent does not apply to other sites identified by the Department which have been excluded from the Draperstown development limit in the dMAP as there is no evidence of objections relating to these sites, thereby distinguishing them from the appeal site. However, given the objections to the general housing allocation within the plan area and the insufficient provision for housing land in various settlements, planning applications are likely to be submitted for other sites which, like the appeal site, have been the subject of objections to their exclusion from the development limit of settlements. The Commission agrees with the appointed Commissioner, however, that the particular circumstances identified by the appellant in respect of the appeal site are unlikely to distinguish it sufficiently from other sites, thereby giving rise to a level of precedent that would support concerns over the cumulative effect of the proposal. Accordingly, the Commission concludes that the Department has sustained the second reason for refusal based on prematurity.

The approach taken by the Planning Appeals Commission on prematurity has found strong support in subsequent unsuccessful applications for judicial review. *In the Matter of an Application by The Logan Rodgers Partnership for Judicial Review and In the matter of a Decision by the Planning Appeals Commission Dated 24 March 2004* (2005 NIQB 4) concerning a proposed development of 200 houses on lands at Ballyhalbert which were within the development limit of the extant and statutory North Down Area Plan 1984–1995, but outside the limit of the draft Ards and North Down Plan 2015, Girvan J in his consideration stated:

> The rationale lying behind a refusal on grounds of prematurity is that if planning per mission were granted to a proposal it would prejudice the outcome of a planning process

> by pre-determining issues related to scale, location or phasing of new development which ought properly to be taken in the development plan context. If a proposal has merely a small incidental impact then the risk of prejudicing the outcome of the development plan inquiry does not arise. If however a planning permission has a wider impact (for example by substantially affecting an important settlement or a substantial area with an identifiable character) then the risk of prejudice will be much greater and a refusal on that ground may be justified. The temporal issue of the timing of the new development plans is clearly a relevant factor because if the plan is at a very early stage the plan may take a very long time to come to finality and the proposals may change or develop or be dropped with the passage of time. Refusal in the meantime could sterilise development land which would not be in the interests of land management and proper planning. The earlier the stage of the development the less certain the final form of the proposals will be. The decision maker is bound to have regard to the timescale of events in relation to the development proposal which would be a material consideration. In this case the PAC clearly proceeded on the basis that the emerging development plans were at a sufficiently advanced stage to trigger the prematurity issue … The PAC took account of the potential impact of the granting of permission for this type and size of development and on the wider area in the context whether it could prejudice the outcome of the inquiry which is the proper place to focus on the detail of such issues relating to the various sites and locations in the area. The approach adopted by the PAC was legally permissible in the circumstances – bearing in mind that in a planning application where an issue of prematurity arises the PAC must perforce approach the question in a more broad-brush way than would be the case in a long running and detailed inquiry. A prematurity decision must focus on the questions of whether and to what extent it was appropriate and fair to hold the fort in the meantime. I am not persuaded that the PAC's approach was flawed.

The issue of prematurity and the need to avoid undermining the integrity of a public inquiry dealing with objections to a draft area plan also surfaced *In the matter of an Application by FP McCann Developments Limited and Maghera Presbyterian Church for Judicial Review* (2005 NIQB 78). This involved two sites for residential development in Tobermore and one site in Maghera with a combined potential for 138 housing units. The very relevant draft Magherafelt Area Plan had been published in April 2004 and was a material consideration in the proceedings. In concluding that the Planning Appeals Commission did not misdirect itself as to the concept of prematurity, Weatherup J opined:

> What is under consideration is the balance between the making of the decisions on individual proposals and decisions made in the process of review of a development plan. Prematurity of decisions on individual proposals may arise from significant cumulative effect creating prejudice by pre-determining, for example, the location of development. The harm that arises from such prematurity is harm to the process of evaluating the proposals in the inquiry into the draft plan. This harm is the prejudice to decisions to be made in the inquiry, for example, as to whether there should be development within an area, when permission has recently been given on a proposal for development at a location within that area. Cumulative effect may arise from existing grants of permission, present proposals and future proposals. Evidence of future proposals will

> involve reasonable inferences being drawn in all the circumstances. Where such reasonable inferences may be drawn the decision maker will not be acting out of mere fear or generalised concern. The character of the matter in issue will be taken into account, for example, cumulative effect may differ in its impact in relation to the scale of development, which may involve a matter of degree, as opposed to whether there should be any, or any further, development within a particular area. There may or may not be existing grants of permission and/or other proposals for the location but in any event reasonable inferences may be drawn in all the circumstances as to the effect of an approval at the location on the process that has yet to determine the future location of development. This is a planning judgment.

The contribution of a public local inquiry and a hearing

The facility by the former Department of the Environment to cause a public local inquiry to be held prior to the determination of a planning application has long been a feature of the Northern Ireland planning system. In the Planning (Northern Ireland) Order 1991, for example, Article 31 provided for a special procedure for major planning applications to be the subject of a public local inquiry whereby the permission, if granted, would involve a substantial departure from the development plan for the area to which it relates, be of significance to the whole or a substantial part of Northern Ireland, affect the whole of a neighbourhood, or have particular road-based implications. The Planning Appeals Commission would then, if instructed by the Department to do so, convene a public local inquiry and report back to the Department, which would then take into account these 'hear and report' findings and recommendations in its decision making. The Planning Act (Northern Ireland) 2011 carries forward this element of planning practice and for developments of regional significance the Department, under Section 26, may cause a public local inquiry at which the proposed development, if carried out, would be of significance to the whole or a substantial part of Northern Ireland or have significant effects outside Northern Ireland, or involve a substantial departure from the Local Development Plan for the area to which it relates. Significantly, the criterion of affecting the whole of a neighbourhood has been deleted, perhaps in line with concern voiced as far as back as 2002 that too many applications were being dealt with by this procedure and that a high proportion fell into that category (Planning Service 2002). Planning applications that are the subject of the call-in procedure to the Department may also be the subject of a public local inquiry under Section 29 of the Act. In both instances, the current legislation provides for a public local inquiry to be held by either the Planning Appeals Commission or a person appointed by the Department for that purpose. The use of a person appointed by the Department would seem to be for simply expediency reasons as the Strategic Planning Policy Statement for Northern Ireland (Department of the Environment 2015e) indicates that the Planning Appeals Commission will be contacted in the first instance to check on the availability of a suitable date; if a suitable date is not available, the Department may avail of that alternative. Clearly, the findings of a public local inquiry comprise a material consideration in planning decision making and the statute requires that the report emanating from such proceedings must be taken into account in determining an application for planning permission. Thus, by way of illustration, Minister of the Department for Infrastructure Nichola Mallon MLA announced in June 2020 that the planning application for the Dalradian gold mine project in County

Tyrone, which had attracted in excess of 40,000 representations, would be the subject of this procedure:

> Before any decision is taken I consider it important to have a Public Inquiry and report which has independently considered the views of stakeholders, including the local community and other Government Departments, and which will robustly scrutinise the information provided by all interested parties.
>
> *(Department for Infrastructure 2020b)*

The Department is not bound, however, to accept the recommendation that emerges from a public local inquiry. This principle has been succinctly aired in another context by Weatherup J in *Tesco Stores Limited's Application* (2010 NIQB 15) regarding the relationship between a public examination into a draft plan, the resultant report of the Planning Appeals Commission to the Planning Service and subsequent decision making by the Planning Service, when he said: 'It is important to recognise that the adoption by Planning Service of recommendations in a PAC report need not amount to the adoption of every aspect of that report'.

In *Re an Application by Bangor Flagship Developments Ltd* (1991 Queen's Bench Division (Crown Side)), for example, it is noted by way of background that the Planning Appeals Commission had recommended that outline planning permission be refused for a shopping centre beyond the town centre of Bangor. The Department decided, however, to allow the application and granted outline planning permission with conditions that sought to control the scale of activity. Carswell J in *R v Secretary of State ex parte Ulster Estates Limited* (1990 Queen's Bench Division (Crown Side)) has elegantly expressed the underlying principle in this forum of decision making:

> Town and country planning involves decisions on matters of judgement and opinion. If a matter is sufficiently debatable to go to a public local inquiry, then it is in my opinion perfectly possible and proper for different persons making such judgement to come to different conclusions. The fact that the Inspector and the Planning Appeals Commission have come to one conclusion on the evidence does not point to any arguable case of unreasonableness on the part of the Department in coming to the opposite conclusion. It seems to me a perfectly sustainable position that the Department may have favoured the development in the first place, but having seen and considered the more voluminous evidence brought forth by the public local inquiry came to a different conclusion on reconsideration.

The different roles of the Planning Appeals Commission as an appellate body under Articles 24 and 32 of the Planning (Northern Ireland) Order 1991 or as a public inquiry convenor under Article 31 vis-à-vis the Department of the Environment have been explicated by Kerr J in *R v Secretary of State ex parte Blair* (1995 Queen's Bench Division (Crown Side)):

> The role of the PAC, therefore, is to hold an inquiry which provides a forum for the making of representations. The decision on the planning application is to be taken by the Department. That decision need not be based exclusively on the matters raised at the inquiry. Indeed, there may be highly material planning considerations which are not raised in the inquiry. It is important, therefore, not to confuse the role of the

> PAC under Article 31 with its adjudicatory functions under Articles 24 and 32. Under Article 31 the task of the PAC is to provide the opportunity, through the holding of an inquiry, for representations to be made and to report to the Department on those representations. The argument that the Department may only depart from the recommendations contained in such a report if they are perverse is, in my opinion, misconceived.

When in respect of a regionally significant or called-in planning application the Department for Infrastructure decides not to hold a public local inquiry, the Department must serve on the applicant and the council a Notice indicating the decision it proposes to make, whether to grant or refuse. This Notice is known as a Notice of Opinion. If within a period of not less than twenty-eight days from the Notice being served, the applicant or the council so requests, the Department for Infrastructure must afford to each of them an opportunity of a hearing before the Planning Appeals Commission or a person appointed for the purpose. If no hearing is requested, the decision can issue. As with a public local inquiry, a hearing is also a public forum and non-participatory observers are welcome to attend. The report of the hearing must be taken into account in determining the application for planning permission and the decision issuing from the Department for Infrastructure shall be final with no right of appeal to the Planning Appeals Commission. Interested parties may, however, pursue leave to apply for judicial review although this avenue of redress is one of supervisory superintendence and is not a merits appeal. In line with the comments of Carswell LCJ *In the Matter of an application by UK Waste Management Limited for Judicial Review* (1999 CARF2798) in the Court of Appeal, the Notice of Opinion cannot be regarded as the final determination of the planning application. Moreover, in this case, the court was unable to support a conclusion that the Department could only frame that final decision in accordance with the terms of the notice, so leaving it no room to change its mind, whether because of supervening events or other sufficient reason. The court determined:

> We consider, however, that it must still be open to the Department to change its mind for sufficient reasons and give a different final decision on the application. There must always be room, within certain limits, for a public authority to change tack in its administration of matters with which it has to deal … If the Department were debarred from changing its view and giving a different final decision, there would be no room for amendment of its proposal if supervening events … were to make it desirable in the public interest that the original proposal should not be adopted.

Planning Committee and delegation procedures

The reshaping of local government in Northern Ireland, as discussed in Chapter 1, has put councils centre stage in the determination of planning applications. This new responsibility is articulated in each council's Operating Protocol, which essentially provides a rubric of governance in regard to the working of its Planning Committee. The key aims of an Operating Protocol are to ensure that the Planning Committee makes decisions in a sound, lawful and transparent way and in a timely and efficient manner and which has more fully been rehearsed by McCloskey J *In the Matter of an Application by James Hugh Allister and Robert Edwin Agnew for Judicial Review v Causeway Coast and Glens Borough Council* (2019 NIQB 79):

> the Protocol prescribes a series of solemn formalities. Due observance of these has the twin merits of reminding Planning Committee members of the important duty which they are discharging and prescribing steps and procedures which will indisputably enhance the quality of the decision making process and its outcome. Simultaneously the values of transparency and accountability will be promoted. A further, free standing aspect of the Protocol is that those who elect councillors to office in local government have a right to expect that instruments of this kind will be respected and observed. It is not for this court, absent detailed argument on the issue, to determine that what the law recognises as a substantive legitimate expectation to this effect is engendered by the Protocol. However, the analysis that it operates as a species of representation, promise or assurance to the audience concerned – developers, objectors and other ratepayers alike – seems to me uncontentious. The solemnly and formally adopted protocols of every council belong to the realm of their Standing Orders and are to be treated with respect. These are mechanisms which curb the decision making powers of elected representatives and promote the well-recognized public law values of procedural fairness, proper motive, taking everything relevant into account and disregarding the immaterial ... The Protocol is an instrument to be observed by planning officers, Council officials and Planning Committee members alike. It is not an 'opt-in/opt-out' mechanism. It is no free-wheeling palm tree.

McCloskey J *In the Matter of an Application by Belfast City Council for Judicial Review v the Planning Appeals Commission* (2018 NIQB 17) provides a helpful summary of the follow-on Planning Committee decision-making process:

> The planning committees of councils do not compose essays documenting matters such as their understanding, insight, assessment of material considerations, evaluation of relevant planning policies and the reasons for their decisions. Rather, in brief compass, their decision making involves the receipt of a planning case officer's report, the consideration of the case papers (usually, one assumes), the possibility of oral presentations at their public meetings, debate and discussion in the same forum – albeit constrained by the requirements and limitations of Standing Orders (SOs) - the receipt of legal and other advice if considered appropriate and a site visit if so advised. Ultimately their decisions are taken by vote, the manifestation being a show of hands. This is followed by a relatively formulaic letter informing the developer of the outcome.

Four matters emerge from this commentary that require brief exploration: the relationship between the council as the local planning authority and its Planning Committee, the planning officer's report, the purpose of a site visit and internal delegation of decision making to a council's planning officers.

The guidance issued to councils relating to the operation of their Planning Committees (Department of the Environment 2015f) strongly recommends that the decisions of the Planning Committee should be taken under full delegated authority, meaning that the decisions of the Committee will not go to the full council for ratification. However, it is also acknowledged in this advice that if councils decide to retain the option of referring certain major applications to the full council for determination, it is recommended that these should be restricted to applications which are significantly contrary to the Local Development Plan or are contentious cross-boundary applications (p. 4). Again *In the Matter of an Application by Belfast City Council for Judicial Review v the Planning Appeals Commission* (2018 NIQB 17) McCloskey J has expanded on the practical and legal implications of full delegated authority to a council's Planning Committee:

> Stated succinctly, the Planning Committee operates as the (corporate) Council's *alter ego* as regards all planning – and related – decisions made by it pursuant to the Scheme of Delegation. It follows from the above analysis that the possibility of (and procedure for) amending the minutes and proceedings of any of the Council's Standing Committees, which is provided for and regulated by the Standing Orders, has no application to the Planning Committee. Thus the (corporate) Council has no power to amend, vary or revoke the decisions of its Planning Committee.

The planning officer's report is crucial to informing the Planning Committee's decision making and typically will comprise details of the proposed development with a list of supporting drawings and documentation, statutory body consultations and issues raised, details of a pre-application community consultation process if applicable, the objections and representations of support received following public advertisement of a planning application, the on-site and off-site planning history, an assessment of the application against statutory regulations, planning policies and planning studies, and finally a recommendation with attached reasons. Judicial comment (2019 NIQB 79) identifies that the specific responsibility of planning officers in this context is:

> to bring to the attention of the decision makers, the Planning Committee members, the relevant policy requirements in sufficiently detailed and accurate terms to ensure that they were properly understood by the reader, thereby facilitating the twofold exercise of relating them to the material aspects of the development proposal and forming a judgement accordingly.

Accordingly, the planning officer's report carries a heavy onus of completeness by not leaving out of account material considerations and of avoiding materially misleading elected representatives on any matter bearing upon their decision. Expressed more pithily, it is incumbent on planning officers to present to Planning Committees 'a fully rounded and balanced picture' (2018 NIQB 49). Their conduct in discharging their duties is to do so 'conscientiously and in good faith and without any conscious or subconscious alien motive or predisposition in favour of the developer' (2019 NIQB 79).

Prior to the determination of a planning application, the Planning Committee may decide to conduct a site visit and again there is formal advice and judicial comment surrounding this aid to decision making. Site visits are to be undertaken when the benefit is expected to be of considerable value but are not to be used as a lobbying opportunity and, ideally, should be attended by the full Planning Committee. McCloskey J (2019 NIQB 79) commends this practice in the following terms:

> Self evidently, this is to inform and educate Planning Committee members in advance of performing their solemn duty of determining the planning application in question. The rationale underpinning the Protocol must include the consideration that the broad

> array of visual aids, illustrations and depictions available in today's high tech world and which are frequently provided to Planning Committee members before and/or in the course of the meetings at which key decisions are made are considered not necessarily an adequate substitute for visiting the site. A further aspect of the rationale must be the consideration that at its monthly meetings every Council's Planning Committee is required to consider multiple planning applications. This is, self-evidently, a heavy and challenging responsibility. Advance preparation is a matter of obvious importance. Maximum familiarity with the substance, details and nuances of every planning application on the agenda of the forthcoming meeting is obviously essential. Furthermore Planning Committee members, as elected representatives, have a broad range of duties extending beyond those involved in their membership of this particular committee. The proposition, therefore, that in certain cases an advance site visit will enhance the prospects of a properly informed and balanced decision, encompassing all appropriate matters of planning judgement, is in my view incontestable. Site visits by Planning Committee members also promote transparency and accountability, two of the values underlying the recent major reforms in Northern Ireland transferring planning decision making responsibilities to democratically elected councillors.

Finally, in this section of the chapter, mention is required about the internal delegation of decision making from the Planning Committee to a council's planning officers. Under Section 31 (1) of the Planning Act (Northern Ireland) 2011 councils are required to produce a Scheme of Delegation whereby certain types of planning application are determined by a planning officer. This procedure is designed to facilitate speedier decision making and improved efficiency (Department of the Environment 2015g). A Scheme of Delegation cannot include major, regionally significant development or an application which has council involvement. There is a latitude of discretion in this procedure which allows elected representatives to determine a planning application which otherwise would have been done so by a planning officer when, for instance, issued are raised by a councillor or an objector which warrant consideration by the Planning Committee. Guidance suggests that councils should aim to have between 90 and 95 percent of planning applications dealt with under a Scheme of Delegation. Accordingly, a Planning Committee should consider only the most significant or complex applications. However, evidence points towards the type of applications being considered by Planning Committees as not always being appropriate with some appearing to be excessively involved in decisions for new single dwellings in the countryside and highlighting 'a disproportionate use of committee time and focus on these applications' (Northern Ireland Audit Office: 2022: 38–9)

Planning application decisions

The Planning (General Development Procedure) Order (Northern Ireland) 2015 specifies that councils shall give an applicant notice of its decision, determination or referral of the application to the Department within a period of sixteen weeks for major development and eight weeks for other development commencing when the application in writing has been lodged with the council, any certificates or documents required by the 2011 Planning Act have also been lodged with the council and any fee required to be paid in respect of the application has been paid to the council. These statutory provisions facilitate agreement in writing between an applicant and councils for an extended time period to determine, decide or refer the planning application and also allow an applicant for planning permission to take the matter to the Planning Appeals Commission as a deemed planning refusal. It is important that these legislative time periods are not confused with the provisions contained in the Local Government (Performance Indicators and Standards) Order (Northern Ireland) 2015, which specify performance targets related to the processing of local development decisions within an average of fifteen weeks, major development decisions within an average of thirty weeks and processing 70 percent of enforcement cases within thirty-nine weeks.

The Planning Act (Northern Ireland) 2011 allows the council or the Department, as the case may be, to grant planning permission unconditionally or with conditions, or to refuse planning permission. Where the Planning Appeals Commission deals with an appeal arising from a decision by a council or any other body, the Commission may confirm, reverse or vary the decision. Conditions attached to planning permission can deal with regulating the development or use of any lands under the control of the applicant or requiring the removal of any buildings or works authorised by the permission or the discontinuance of any so authorised use of land by a specified period and carrying out of works for the reinstatement of that land at the end of that period. The granting of planning permission for wind turbines, for example, might commonly have a condition that gives meaning to these provisions:

> The permission shall be for a limited period expiring 25 years from the date on which electricity from the turbine is connected to the grid. Within 6 months of the cessation of electricity generation at the site, or upon expiration of this permission, whichever is sooner, all structures and access tracks shall be removed and all land affected by the development restored in accordance with a decommissioning scheme submitted to and approved by the planning authority prior to the commencement of any works or in accordance with any variation to the scheme to which the local planning authority subsequently agrees in writing.
>
> *(PAC reference 2014/A0259)*

With exceptions, say in regard to outline planning permission or permission granted for a limited period, planning permission shall be granted subject to the condition that the development to which it relates must be begun within five years of the date on which the permission is granted or such other period, whether longer or shorter, deemed appropriate having regard to the provisions of the local development plan and to any other material considerations. To reiterate, a time limit on the commencement of development is not applied to an outline planning permission since, by itself, it does not permit development operations to proceed; what is important here is the submission of a reserved matters application within three years of the date of the grant of the outline planning permission, and that, if approved, the development to which the permission relates (i.e., the outline permission) must be begun by whichever is the later of the following dates: the expiration of five years from the date of the grant of outline planning permission or the expiration of two years from the final approval of reserved matters; again, these periods may be varied as appropriate. Once the time limit for applications for approval of reserved matters has expired, however, no applications for such an approval can be made (Development Management Practice Note 20, Department of the Environment 2015h). Planning permission goes with the land to which it relates and to those who have an estate in the land; the land may change ownership but the planning permission endures with the land which is the subject of the planning permission.

As noted above, planning permission can be granted with conditions and in borrowing from well-versed case law, they must be necessary, relevant to planning, relevant to the development to be permitted, enforceable, precise and reasonable in all other respects (see for example, *Newbury DC v Secretary of State for the Environment* (1981 AC 578)). As far back as 1958, the courts have indicated that a planning authority is not at liberty to use its powers for an ulterior objective, no matter how much that objective may seem to them to be in the public interest (*Pyx Granite Co Ltd v Ministry of Housing and Local Government* (1958 1 QB 554)). Additionally, it is necessary that conditions attached to a planning permission be accompanied by reasons in the same way that a refusal of planning permission contains explicit reasons. In *South Buckinghamshire County Council v Porter* (2004 1 WLR 1953), Lord Brown commented:

> The reasons for a decision must be intelligible and they must be adequate. They must enable the reader to understand why the matter was decided as it was and what conclusions were reached on the 'principal important controversial issues', disclosing how any issue of law or fact was resolved. Reasons can be briefly stated, the degree of particularity required depending entirely on the nature of the issues falling for decision. The reasoning must not give rise to a substantial doubt as to whether the decision-maker erred in law, for example by misunderstanding some relevant policy or some other important matter or by failing to reach a rational decision on relevant grounds. But such adverse inference will not readily be drawn. The reasons need refer only to the main issues in the dispute, not to every material consideration. They should enable disappointed developers to assess their prospects of obtaining some alternative development permission, or, as the case may be, their unsuccessful opponents to understand how the policy or approach underlying the grant of permission may impact upon future such applications. Decision letters must be read in a straightforward manner, recognising that they are addressed to parties well aware of the issues involved and the arguments advanced. A reasons challenge will only succeed if the party aggrieved can satisfy the court that he has genuinely been substantially prejudiced by the failure to provide an adequately reasoned decision.

In applying this dictum to *Tesco Stores Limited's Application* (2010 NIQB 15), Weatherup J commented as follows on the planning decision arrangements in place in Northern Ireland:

> Planning decisions issued by the Planning Appeals Commission reflect one such process and planning decisions issued by Planning Service reflect a different process. The latter do not appear in the form of a judgment. Nevertheless they must allow interested parties with knowledge of the facts and the issues and with access to the necessary papers and reports to understand the basis of the decision and to mount a challenge to that decision if so advised.

Accordingly, transparency and fairness remain the hallmarks of sound decision making and can put any subsequent planning appeal against reasons for refusal or unacceptable conditions on an approval, or judicial review action, on a fully informed footing.

With local authorities now involved in deciding planning applications, a new issue has emerged in regard to planning practice in Northern Ireland and which surfaces when a Planning Committee decides not to follow the recommendation contained within a planning officer's report. The matter comes into sharp focus when applications recommended for refusal by a planning officer are approved by elected representatives. Data published by the Department for Infrastructure (2020c) indicate that over the period April 2018 to March 2019 a total of 114 planning applications were within that category. The Northern Ireland Audit Office (2022: 40) has reported that over the period 2018–19 and 2019–20 almost 40 percent of the decisions made against officer advice related to single dwellings in the countryside whereby in all cases the officer recommendation to refuse planning permission was overturned and approved by a Planning Committee.

Decision making in this arena has generated litigation when *In the Matter of an Application by Stuart Knox for Judicial Review v Causeway Coast and Glens Borough Council* (2019 NIQB 34) full planning permission was granted for the conversion and alteration of a historic vernacular building to provide a new detached dwelling. The planning officer's report had recommended refusal and the challenge was pursued on grounds that there was a failure to provide adequate reasons for the impugned decision. The Protocol for the Operation of the Causeway Coast and Glens Borough Council Planning Committee states: 'The reasons for any decisions which are made contrary to the Planning Officer's recommendation must be formally recorded in the minutes and a copy placed on file'. In this instance, the reasons cited in the Planning Committee resolution to approve were: 'Consider that the conversion of the barn is acceptable in principle; no policy requirement for private amenity space; proposed development would be an improvement to what currently on site'. The applicant did not contend that these twenty-nine words did not amount to reasons but the question for the court was whether these were legally adequate supporting reasons. In his judgement, McCloskey J ruled in favour of the council but also sounded an important note of advice to all councils in Northern Ireland:

> However, having regard to the full evidential context, I conclude, by an admittedly narrow margin, that the recorded reasons pass muster in law. I thus conclude because considered in their full evidential and juridical contexts they are imbued with sufficient clarity, coherence and intelligibility. I am bound to add, however, that the Council should not have found itself in the position of defending the legal sustainability of the reasons for its impugned decision before this court. These proceedings were pre-eminently avoidable. The recorded reasons have been found to satisfy the legal minima. However, the court trusts that every Council in Northern Ireland will not satisfy itself with the bare minimum. Judicial review is designed inter alia to encourage and promote the highest standards of decision making in the realm of public law. The net effect of the authorities by which this court is bound is to erect a relatively high threshold for judicial intervention in a judicial review reasons challenge. The Applicant has failed, narrowly, to overcome this threshold.

In the Matter of an Application by Colum Sands for Judicial Review v Newry, Mourne and Down District Council (2018 NIQB 80) the court added the succinct observation that 'a challenge of this genre requires the demonstration of a defect of substance, to be contrasted with the trivial or technical'. In this case, the Planning Committee rejected the relevant refusal reason of the planning officer and, in his judgement, McCloskey J ruled that the Planning Committee members 'were not merely entitled, but obliged, to critically question the Case Officer's recommendation if they considered this appropriate and to form their independent view'.

Finally, there remains the question of the legally effective date of a planning decision and which prompted McCloskey J in *Belfast City Council's Application* (2018 NIQB 17) to opine:

> The comprehensive new legislative regime which came into operation in 2015 is surprisingly unclear with regard to this elementary and self-evidently important issue … there are two statutory provisions, namely section 59 of the 2011 Act and Article 22 of the Planning (General Development Procedure) Order (NI) 2015 (the 'General Development Order') which, in the final analysis, point most clearly to the correct answer … I turn to the first of the two statutory provisions to which I have already drawn attention. Section 58(3) of the 2011 Act specifies a time limit for appeals by the planning applicant to the PAC against refusal decisions and conditional approval decisions. Notice of appeal must be served 'within four months from the date of notification of the decision to which it relates'. This provision points strongly to the conclusion that the date of notification of the Council's decision is the date upon which it takes effect in law. The second statutory provision which illuminates this discrete issue is Article 22 of the General Development Order. This provides: 'The council or, as the case may be the Department, shall give notice of a decision or determination in writing and on an application for planning permission or for approval of reserved matters, where a permission or approval is granted subject to conditions or the application is refused, the Notice shall state the reasons for the refusal or for any condition imposed'. In my judgment, Article 22 combines with section 58(3) in impelling to the conclusion that the legally effective date of any decision or determination of the Council within the ambit of the planning legislative framework is that upon which it is notified to the planning applicant.

In the Belfast City Council litigation, the Council's Planning Committee considered the planning application at the heart of this case and voted to refuse planning permission on 19 January 2016. A Decision Notice dated 20 January 2016 was issued to the developer on 22 February 2016. Amended drawings were submitted to the Council on 22 January 2016 but were returned by the Council to the developer on 25 January 2016 advising that these amendments could not be considered as a decision had been made on 19 January 2016. The court ruled in this matter that having regard to statutory provisions and a wider principle espousing the importance of notice of a decision to give legal effect to a determination:

> The effect of this analysis is that the operative date in law of the Council's planning decision in the present case was 22 February 2016.

Planning appeals

The Planning Act (Northern Ireland) 2011 provides that where an application for planning permission, made to a council, is granted subject to conditions, or is refused, the applicant may appeal to the Planning Appeals Commission, an agency under the auspices of the Department of Justice. The legislation also provides an opportunity for an appeal against a failure to take a planning application decision. There are set time periods within which an appeal may be brought forward comprising four months from the date of the notification of the planning authority decision (for planning, listed building and advertisement appeals) and for non-determination appeals it is also four months from the date on which a decision should have been given, unless an extended date was agreed in writing between the applicant and the council. The Planning Appeals Commission may deal with the application as if it had been made to it in the first instance. However, before determining any appeal the Planning Appeals Commission must, if either the applicant or the council so desires, afford each of them an opportunity of appearing before and being heard by the Commission. A party to such proceedings is not permitted to raise any matter which was not previously aired at the time the decision appealed against was made unless that party can demonstrate to the satisfaction of the Planning Appeals Commission that the matter could not have been raised before that time or that its not being raised before that time was a consequence of exceptional circumstances. Finally, while Section 204 of the 2011 Act deals with procedural matters governing the determination of an appeal and with mention being made of the use of hearings and written representations, there is little by way of definitive explication of the different methods of processing appeals. Section 204(5) empowers the Planning Appeals Commission to draw up its procedures in this sphere and, at present, there are four modes by which this business can be conducted: a formal hearing, an informal hearing, written representations with an accompanied site visit and written representations with a Commissioner site visit. Table 2.2 gives insight into the mode and number of planning appeals in Northern Ireland for five annual reporting years 2015–16 to 2020–21, with the volume of hearings and accompanied site visits for the latter year noticeably constrained by COVID-19 protocols. While the available information in Table 2.2 masks the distinction between formal and informal hearings, earlier time-series data indicate that while four formal hearings were convened in 2010–11, no subsequent formal hearings were held through to the year 2014–15 (Planning Appeals Commission and Water Appeals Commission 2015). The essential point here is that the use of the adversarial formal hearing procedure has been replaced, in effect, by the informal hearing, which is more inquisitorial and arguably less intimidating in style for participants; the majority of appeals are decided by written representations. As each appeal is decided on its own particular facts, the Commission considers that there is no connection between the appeal outcome and the processing method selected.

TABLE 2.2 Mode and number of planning appeals in Northern Ireland

	2015/16	*2016/17*	*2017/18*	*2018/19*	*2019/20*	*2020/21*
Hearing	120	128	116	128	99	43
Written representations with an accompanied site visit	98	92	101	92	109	17
Written representations with a Commissioner site visit	76	76	59	58	69	124
Total	294	296	276	208	277	184

Source: The Planning and Water Appeals Commission (2020, 2021)

Nonetheless, the use of the informal hearing procedure has generated litigation, as illustrated *In the Matter of an Application by James Stewart for Judicial Review* (2002 NIQB 38) and which, following its dismissal in the High Court, was subsequently taken before the Court of Appeal in Northern Ireland (2003 NICA 4), which again dismissed the application. In terms of planning practice, the case is noteworthy because it raised issues related to Section 6 of the Human Rights Act 1998 and Article 6 of the European Convention on Human Rights that give expression to individual civil rights for a fair and public hearing before an independent tribunal. It was argued by the applicant that the informal appeals procedure was unlawful and contravened these obligations. The background to the case was the granting of full planning permission by the Planning Appeals Commission in June 2001 for the erection of a building for a motor factors business on land beside the applicant's dwelling in Carrickfergus. The Department had previously refused planning permission in June 2000 on the grounds that included the detrimental impact on neighbouring residents by reason of noise, nuisance and general disturbance given the close proximity of the proposed service yard and the affected residential property. The developer appealed to the Planning Appeals Commission, which decided to adopt the informal hearing procedure to deal with the appeal. In his judgement before the High Court, Gillen J, having rehearsed in some detail the rubric of the informal hearing procedure and cited content from the affidavit of the designated Commissioner, concluded:

> The applicant in this case relied on the absence of an adversarial procedure as constituting a breach of Article 6(1). I do not consider that such a proceeding is necessary in a planning application. In any event the DOE [Department of the Environment] in this instance opposed the application before the designated member. Essentially the DOE was on the same side as the applicant. All parties, including the present applicant, were given an opportunity to participate in discussion and could have asked informal questions through the designated member. There was no formal questioning or cross-examination. The designated member, at paragraph 7 of her affidavit, indicates that no requests were made for cross-examination at the hearing. One must remember that presentation of arguments in this case is facilitated by the requirement to lodge statements of case which are submitted and exchanged in advance of the hearing. All parties are, therefore, well aware of the issues. The views of the parties are, therefore, invited at all stages. Doubtless there is a heavy onus on the Commissioner to ensure that a proper inquiry is conducted, but this is what happens in a Commissioner-led discussion. I see nothing to suggest that a fair and thorough investigation was not carried out by the Commissioner in this case. She had distributed an agenda for discussion at the hearing at the commencement of the hearing followed by an explanation by her of the procedures. Thereafter the Commissioner directed the parties to the issues requiring further exploration and, at the end of the hearing all parties indicated their satisfaction that all issues had been fully explored. I find nothing in the affidavit on behalf of the applicant to suggest the contrary. The applicant submits that the failure to permit rebuttal evidence constitutes a breach of his rights under Article 6. Such formalities do not appear to me to be necessary, given the nature of the proceedings. Formal rebuttal evidence (oral or written responses) may not be introduced by any party at the informal hearing. The designated member, at paragraph 6 of her affidavit, gets to the heart of the matter when she records:

> but relevant comments (including observations which contradict, mitigate or contextualise) on statements of the case submitted by others may be made in the course of the Commissioner-led discussion. Parties are given the opportunity to discuss the agenda which is distributed at the hearing and to add matters to it prior to the Commissioner-led discussion … All parties to informal hearings are given the opportunity to participate in discussion and may ask informal questions through the Commissioner.

Accordingly, during the course of the Commissioner-led discussion, any matter can be raised with the Commission and, therefore, if rebuttal is considered necessary, it can always be raised through the Commissioner. Similarly, I accept entirely the submission of Mr Larkin QC that if the applicant wished to call a witness that application will be considered by the Commissioner on its merits. I have concluded that the informal procedure in general, and this informal procedure in particular, did not deny this applicant the right to any of the guarantees implicit in Article 6 of the Convention.

What emerges from this judgement is the watermarking of proceedings by the lead role of the presiding Commissioner, openness of engagement, and the exploration of all issues with rigour. The Court of Appeal, in its follow-on determination, usefully enlarged on this principle of rigour, while also sounding a note of caution, when Carswell LCJ stated:

> In our opinion the issues in planning decisions lie at the judgment and discretion end of the spectrum. We do not consider that the use of the informal hearing procedure is in itself unfair or a breach of the Article 6 rights of an objector. There may be cases where there is such a need to establish the correct facts in a conflict of evidence or test the validity of certain types of evidence that an informal hearing would not suffice to satisfy Article 6, but this was not in our opinion such a case. In the absence of such factors, we do not regard the resort to informal hearings as being *per se* in breach of the parties' Article 6 rights. It is necessary, however, to sound a note of caution. Where a person or body has the function of conducting an inquisitorial type of proceeding of the type of the informal hearing of planning appeals, it is of particular importance that care is taken to ensure that all reasonable expressions of opinion are received and that sufficient opportunity is given to the participants in the proceeding to present their case in an effective fashion. We would draw attention to the remarks of Pill LJ in Dyason v Secretary of State for the Environment [1998] JPL 778, in the course of which he said:
>
>> 'The hearing must not become so relaxed that the rigorous examination essential to the determination of difficult questions may be diluted. The absence of an accusatorial procedure places an inquisitorial burden on an inspector'.

The procedures adopted by the Planning Appeals Commission were also reviewed by the court *In the Matter of an Application by Belfast City Council for Judicial review v The Planning Appeals Commission* (2018 NIQB 17). While it was accepted that the Planning Appeals Commission did not require that parties inform it or each other if they intend to be legally represented at a hearing, it was contended in this case that the failure of the Planning Appeals Commission to require parties to give prior notification that they will be legally represented at an appeal hearing was in breach of the common law duty of fairness and thus placed the Council at a substantial disadvantage in regard to preparations and representations around legal argument. This ground of challenge was rejected by McCloskey J in a ruling that also points to the potential for improved practice on this matter:

> The nub of this ground of challenge is that the PAC's procedures should be such as to alert all participating parties in advance to the fact of legal representation on the part of any of the parties concerned. The PAC's existing procedures do not do this. I find nothing unlawful in this at the general level. However, to cater for the possibility, identified above, that inequality of arms giving rise to material procedural unfairness could result from the extant practice, I invite the PAC to consider the economic and practical feasibility of revising its pre-hearing procedures so as to ensure that (a) every party who intends to be legally represented is required to signify this fact and (b) any alteration occasioned by subsequent developments is also notified to the PAC in the first place and, thereafter, by the PAC to the other participating parties.

Under the terms of the Planning Act (Northern Ireland) 2011 and since 2015 the Planning Appeals Commission is permitted to make orders as to the costs of the parties in a range of appeals that include planning appeals, listed building consent appeals, conservation area consent appeals and hazardous substance appeals. The costs awards may be made in favour of and against any party to an appeal, but it is important to note that costs will only be considered by the Commission when claimed by any party. Costs awards are intended to provide redress when one party to an appeal has put another party to unnecessary expense due to unreasonable behaviour. Guidance (Planning Appeals Commission 2016) indicates that four criteria are to be met:

- a party has made a timely claim; and
- the claim relates to an eligible appeal type; and
- the party against whom the award is sought has acted unreasonably; and
- the unreasonable behaviour has caused the party claiming costs to incur unnecessary or wasted expense.

Unreasonable behaviour is defined as causing an unnecessary appeal, prolonging proceedings, abandoning or not pursuing part of a case, and withdrawing or conceding an appeal. Unnecessary or wasted expenses can include those connected to submitting an appeal, employing staff to prepare evidence, appearances at hearings and costs linked to the services of professional experts. The Commission will not determine the amount of costs to be awarded in a successful claim for costs. It places the onus on the parties to reach an agreement on the amount pursuant to the issuing of a costs order for a full or partial award. If an agreement cannot be reached, the matter can be referred to the High Court for detailed assessment. Additionally, a costs decision may be challenged by applying to the High Court for judicial review. It is customary to make separate decisions, issued simultaneously, on an appeal and related costs claim, except in a withdrawal case when no appeal decision would be forthcoming. Costs award claims in the year 2019–20 totalled forty-eight, of which six were decided as a partial award and eight were decided as a full award (Planning and Water Appeals Commissions 2021).

Completion Notices

Section 64 of the Planning (Northern Ireland) Act allows a council to issue a Completion Notice to require a development which has a time-bound planning permission, and which has been begun, to be completed. A council must give at least one year for the completion. Notices can be withdrawn by a council if appropriate. Under Section 65 Completion Notices issued by a council must be confirmed by the Department before they take effect. The person on whom it is served can request a hearing before the Planning Appeals Commission, as can a council. Once it takes effect the planning permission expires at the end of the period allowed for the development's completion. Section 66 of the Act also allows the Department to issue Completion Notices which have the same effect as those issued by a council, but it must consult a council before doing so.

Within Northern Ireland this longstanding provision has seldom been deployed, but it does have potential relevance when matters of visual amenity are at stake and when developments have lain dormant in regard to partial construction for some considerable period of time. A notable illustration of the persuasive value of a Completion Notice is found on Strand Road, Portstewart, where a developer had partially constructed seashore apartments on the basis of a planning permission granted in the late 1990s (Photograph 2.1). In October 2012 the Department of the Environment issued a Completion Notice on the owner who then appealed to the Planning Appeals Commission. However, prior to the date of the hearing in April 2013 a third party demolished the building and removed the rubble (Photograph 2.2). The Completion Notice was subsequently withdrawn by the Department, and the planning appeal did not proceed. In oral answers to questions in the Northern Ireland Assembly on 14 May 2013 in the wake of that outcome (www.theyworkforyou.com/ni/?id=2013-05-14.4.3, accessed 28 March 2016), Minister for the Environment Alex Attwood indicated that Completion Notice action would be taken on other unfinished or partially developed sites.

PHOTOGRAPH 2.1 Completion Notice site in Portstewart, 2012. *Source*: Photograph by authors.

PHOTOGRAPH 2.2 Completion Notice site in Portstewart, 2014. *Source*: Photograph by authors.

However, in the Portstewart case, given that the Completion Notice was withdrawn, it would seem that the planning permission has been kept alive and the site value retained given the completion of underpinning works apparent in Photograph 2.2. But significantly, the overarching public interest objective of enhancing visual amenity has in the main been achieved by the removal of the unfinished apartments.

Conclusion

This chapter has ranged widely across the development management situation within Northern Ireland and has combined elements of legislation, planning policy and case law to offer a succinct account of how planning applications are framed and how they are determined. The chapter clearly identifies the prominent role played by the courts in giving interpretation to the statutes and guidance to planning practice and in subsequent chapters of this book we will return to that arena. In the following chapter we move on to consider special forms of planning control which once again brings together an interaction between statutes, planning policy, planning authorities and judicial supervision.

Legislation

European Convention on Human Rights
Human Rights Act 1998
Marine Act (Northern Ireland) 2013
Planning Act (Northern Ireland) 2011
The Industrial Pollution Control (Northern Ireland) Order 1997
The Local Government (Performance Indicators and Standards) Order (Northern Ireland) 2015

The Planning (Development Management) Regulations (Northern Ireland) 2015
The Planning (Development Management) (Temporary Modifications) (Coronavirus) Regulations (Northern Ireland) 2020
The Planning (Environmental Impact Assessment) Regulations (Northern Ireland) 2015
The Planning (Environmental Impact Assessment) Regulations (Northern Ireland) 2017
The Planning (General Development) Order (Northern Ireland) 1973
The Planning (General Development Procedure) Order (Northern Ireland) 2015
The Planning (General Permitted Development) Order (Northern Ireland) 2015
The Planning (General Permitted Development) (Amendment) Order (Northern Ireland) 2020
The Planning (Northern Ireland) Order 1972
The Planning (Northern Ireland) Order 1991
The Planning (Use Classes) Order (Northern Ireland) 2015
The Pollution Prevention and Control (Industrial Emissions) Regulations (Northern Ireland) 2013

Cases cited

Appeal by Mr. Edward Burns against the Non-determination of a Planning Application for a Housing Development on Lands North of 26–44 Magherafelt Road, Draperstown, County Tyrone (PAC reference 2004/A080)
Appeal by W F Properties against the Non-determination of a Planning Application in respect of the Erection of 98 Dwellings at Main Street, Scarva (PAC reference 2003/A229)
Appeal by Redslave Ltd against the Refusal of Full Planning Permission for the Erection of 225kW Wind Turbine with 31m Momo-pole and 29m Rotar Diameter at Prospect Hill, Brookeborough, Co. Fermanagh (PAC reference 2014/A0259)
City of Edinburgh Council v Secretary of State for Scotland (1998, SC(HL) 33, 1997, 1 WLR 1447)
Department of the Environment's Application (2014 NIQB 4)
Dyason v Secretary of State for the Environment (1998 JPL 778)
Gransden & Co Ltd v Secretary of State for the Environment (1987 54 P & CR 86)
In the Matter of an Application by Belfast City Council for Judicial Review v The Planning Appeals Commission (2018 NIQB 17)
In the matter of an Application by Callan for Judicial Review (1997 Queen's Bench (Crown Side))
In the Matter of an Application by Christine Alexander for Judicial Review v Causeway Coast and Glens Borough Council (2018 NIQB 55)
In the Matter of an Application by Colum Sands for Judicial Review v Newry, Mourne and Down District Council (2018 NIQB 80)
In the matter of an Application by FP McCann Developments Limited and Maghera Presbyterian Church for Judicial Review (2005 NIQB 78)
In the Matter of an Application by Greencastle Rouskey Gortin Concerned Community Limited for Judicial Review v Department for Infrastructure (2019 NIQB 24).
In the Matter of an Application by James Hugh Allister and Robert Edwin Agnew for Judicial Review v Causeway Coast and Glens Borough Council (2019 NIQB 79)
In the Matter of an Application by James Stewart for Judicial Review (2002 NIQB 38)

In the Matter of an Application by James Stewart for Judicial Review (2003 NICA 4)
In the Matter of an Application by Stuart Knox for Judicial Review v Causeway Coast and Glens Borough Council (2019 NIQB 34)
In the Matter of an Application by The Logan Rodgers Partnership for Judicial Review and In the matter of a Decision by the Planning Appeals Commission Dated 24 March 2004 (2005 NIQB 4)
In the Matter of an Application by William Donnelly (2017 NIQB 84)
In the Matter of an Application by William Donnelly for Leave to Apply for Judicial Review (2020 NIQB 35)
In the Matter of an Application for Judicial Review by: (1) Bow Street Mall Limited (2) Lisburn Chamber of Commerce (3) Multi-development UK Limited (4) Central Craigavon Limited(5) Belfast City Council (6) Belfast Chamber of Trade and Commerce and in the matter of a planning decision made by the Department of the Environment (2006 NIQB 28)
In the Matter of an Application by Meekatharra (Northern Ireland) Limited for Judicial Review (1995 Queen's Bench Division (Crown Side))
Lamont's (David John Stewart and Elaine) Application (2014 NIQB 3)
Malloch v Aberdeen Corporation (1971 1 WLR 1578)
Morelli v Department of the Environment (1976 N.I. 159)
Newbury DC v Secretary of State for the Environment (1981 AC 578)
Newry Chamber of Commerce and Trade's Application (2015 NIQB 65)
O'Hare's Application (2016 NIQB 20)
Pyx Granite Co Ltd v Ministry of Housing and Local Government (1958 1 QB 554)
R v Planning Appeals Commission ex parte Bailie (1995 Queen's Bench Division (Crown Side))
R v Secretary of State ex parte Blair (1995 Queen's Bench Division (Crown Side))
R v Secretary of State ex parte Ulster Estates Limited (1990 Queen's Bench Division (Crown Side))
Re an Application by Bangor Flagship Developments Ltd (1991 Queen's Bench Division (Crown Side))
Re Lisburn Development Consortium's Application for Judicial Review (2000 Queen's Bench Division (Crown Side))
Re Central Craigavon Limited (2008 NIQB 46)
Shadwell Estates v Breckland DC (2013 EWHC 12 (Admin)):
South Buckinghamshire County Council v Porter (2004 1 WLR 1953)
Stringer v Minister of Housing and Local Government (1971 1 All ER 65)
Tesco Stores Limited's Application (2010 NIQB 15)
Tesco Stores Ltd v Secretary of State for the Environment (1995 2 All ER 636)
Tesco Stores Ltd v Dundee City Council (2012 UKSC 13)
Ulster Architectural Heritage Society's Application (2014 NIQB 21)
Warner Chilcott UK Ltd and Caridian BCT NI Ltd's Application (2011 NIQB 137)
Westminster City Council v Great Portland Estates (1985 AC 661)

References

Bowes, A. 2019. *A Practical Approach to Planning Law*. Oxford: Oxford University Press.
Carroll, L. 1871. *Through the Looking-Glass*. London: Macmillan.
Cullingworth, B., V. Nadin, T. Hart, S. Davoudi, J. Pendlebury, G. Vigar, D. Webb and T. Townsend. 2014. *Town and Country Planning in the UK*. London: Routledge.

Department for Infrastructure. 2019. *A Discussion Paper Examining the Role of Statutory Consultees in the Planning Process in Northern Ireland*. Belfast: Department for Infrastructure.

Department for Infrastructure. 2020a. *Review of Permitted Development Rights: Departmental Response to Public Consultation*. Belfast: Department for Infrastructure.

Department for Infrastructure. 2020b. 'Mallon: Dalradian Gold Mine Project Will Be Subject of Public Inquiry', *DfI Latest News*. Belfast: Department for Infrastructure.

Department for Infrastructure. 2020c. *Planning Applications Recommended for Refusal by Planning Officer and Approved by Planning Committee, by Council, April 2018 to March 2019*. Belfast: Department for Infrastructure.

Department of the Environment. 2005. *Access, Movement and Parking*. Planning Policy Statement 3 (PPS3). Belfast: Department of the Environment.

Department of the Environment. 2010. *Sustainable Development in the Countryside*. Planning Policy Statement 21 (PPS21). Belfast: Department of the Environment.

Department of the Environment. 2015a. *Pre-Application Community Consultation and Pre-Application Discussions*. Development Management Practice Note 10. Belfast: Department of the Environment.

Department of the Environment. 2015b. *Publicity Arrangements and Neighbour Notification*. Development Management Practice Note 14. Belfast: Department of the Environment.

Department of the Environment. 2015c. *The Determination of Planning Applications*. Development Management Practice Note 16. Belfast: Department of the Environment.

Department of the Environment. 2015d. *Pre-Determination Hearings*. Development Management Practice Note 17. Belfast: Department of the Environment.

Department of the Environment. 2015e. *Strategic Planning Policy Statement for Northern Ireland*. Belfast: Department of the Environment.

Department of the Environment. 2015f. *Best Practice Protocol for the Operation of Planning Committees*. Belfast: Department of the Environment.

Department of the Environment. 2015g. *Councils' Schemes of Delegation*. Belfast: Department of the Environment.

Department of the Environment. 2015h. *Use of Planning Conditions*. Development Management Practice Note 20. Belfast: Department of the Environment.

Duxbury, R. 2018. *Telling and Duxbury's Planning Law and Procedure*. Oxford: Oxford University Press.

Northern Ireland Audit Office. 2022. *Planning in Northern Ireland*. Belfast: Northern Ireland Audit Office.

Planning Appeals Commission. 2016. *Cost Awards Guidance*. Belfast: Planning Appeals Commission.

Planning Appeals Commission and Water Appeals Commission. 2015. *Annual Report of the Chief Commissioner, Planning and Water Appeals Commissions*. Belfast: Planning Appeals Commission and Water Appeals Commission.

Planning Service. 2002. *Modernising Planning Processes: A Consultation Paper*. Belfast: Planning Service.

The Planning and Water Appeals Commission. 2020. *Annual Report of the Chief Commissioner 2019–20*. Belfast: Planning and Water Appeals Commission.

The Planning and Water Appeals Commission. 2021. *Annual Review 2020/1*. Belfast: The Planning and Water Appeals Commissions.

Willmore, C. and C. Blackhall. 2016. *Planning Law and Practice*. London: Cavendish.

3

AMENITY AND SPECIAL FORMS OF DEVELOPMENT MANAGEMENT

Stephen McKay and Michael Murray

Introduction

In seeking to balance conflicting interests regarding the use and development of land, the planning systems of the United Kingdom have embedded the concept of amenity in policy guidance and practice. The meaning of amenity in the context of English case law is reviewed by Mynors (2009: 54) as referring to 'pleasant circumstances or features, advantages' and as the 'visual appearance (of land) and the pleasure of its enjoyment'. The Planning Portal for England and Wales (2021) suggests that amenity can be 'a positive element or elements that contribute to the overall character or enjoyment of an area. For example, open land, trees, historic buildings and the inter-relationship between them, or less tangible factors such as tranquillity'. In Northern Ireland, *Planning Policy Statement 1: General Principles* (Planning Service 1998: 7), cancelled in 2015, set out an approach to planning which is comparable to other jurisdictions by stating that 'it must secure economy and efficiency as well as amenity in the use of land'. The subsequent *Strategic Planning Policy Statement (SPPS) for Northern Ireland*, subtitled *Planning for Sustainable Development* (Department of the Environment 2015a: 17), more particularly specifies the remit of amenity protection as safeguarding residential and work environs in regard to health and wellbeing. Specific reference is made to visual intrusion, general nuisance, loss of light and overshadowing. Amenity and heritage conservation are inextricably linked and require responsive decision making (see Historic Environment Division 2021). The concept of amenity is consequently a key lens through which regulatory constraints on changes within the local built and natural environment are viewed as necessary. Nonetheless, it is a concept that is not without its difficulties in interpretation. As noted by Cullingworth et al. (2014: 159) 'the phrase "injurious to the interests of amenity" has become part of the stock-in-trade jargon of the planning world'. They acknowledge that there is considerable scope for disagreement on the preservation of amenities, and, in that regard, planning authorities can exercise their powers to reject proposals perceived as harmful to amenity. This chapter specifically relates the concept of amenity to special forms of development management in line with the provisions of the Planning Act (Northern Ireland) 2011, other legislation and the 2015 *Strategic Planning Policy Statement for Northern*

DOI: 10.4324/9781003218319-3

Ireland. It comprises a discussion of advertisements, Conservation Areas, Areas of Townscape and Village Character, Listed Buildings, Scheduled Monuments, landscape amenity and Tree Preservation Orders.

Advertisements

The advertising industry embraces multi-media formats and crosses sectors that include TV and radio, out-of-home, newsprint and magazine, cinema, digital and direct mail formats. The total UK advertising expenditure in 2020 was £23.5bn, and, with COVID-19 lockdowns pushing greater purchasing of goods and services from home, there was a marked shift by advertisers into online video, social media and search markets. The Advertising Association (2021) contends that for every pound invested in advertising, six pounds are generated in GDP. From the perspective of regulatory planning, outdoor advertising is becoming one of the most dynamic contributors to the changing scene in our cities and towns; it is truly a global business with local impact. Key criteria for advertisers' expenditure in this sphere are added impact and standout, which are continuously pushing the boundaries of conventional campaigns by embracing innovative technologies, new styles of display and wider site choices. *The Economist* (Economist 2011) has summed up this situation in the following way:

> Roadside billboards, posters on buses and subway escalators, ads in airport terminals – a type of publicity known as out-of-home advertising – used to be the dull end of the industry. No more. The falling price and improving quality of flatscreen displays mean that static posters printed on paper are being replaced by snazzy digital commercials with moving pictures, sound and sometimes inter – active features. As some advertising media, especially newspapers, see their audiences fade, streets, airports and other public spaces are becoming crowded with more potential viewers than ever, as people continue moving to cities and travel more.

The excesses of outdoor billboard advertising have, on occasion, drawn international academic scrutiny (see, for example, Gomez 2013). Nonetheless, the advantages of outdoor advertising can be summarised as providing twenty-four-hour exposure with high frequency and relatively low cost; it can serve as a point of purchase reminder to customers in the buying process, it extends brand identity for established products, it offers a localised approach for national advertising campaigns and it can give direction to local retail outlets (Lane et al. 2011). The brand package is a key part of the advertising process and, as argued by Lewis (1996: 138), its most fundamental aspect relates to the establishing of 'character or personality, building the elements which make the brand recognisable, visible and individual packaging which says, "Trust me, buy me, I am for you"'.

These comments have considerable resonance with the case of *BP Amoco PLC and John Kelly Ltd and Glenshane Tourist Services Limited* (2001 NICA 3) heard in the Northern Ireland Court of Appeal. In this matter, BP claimed that the respondents had infringed BP's UK registered trademarks by using a distinctive green colour on the whole or a substantial part of the exterior surfaces of their TOP service stations and that they had thereby committed the tort of passing off their TOP branded service stations and premises and oils, fuels and lubricants for those of BP. Finding in favour of BP on the first point, Carswell LCJ stated:

> We are prepared to hold that BP had by the time the present proceedings were commenced established sufficient consumer association [and that] in consequence of the use which had been made of it, it had after registration acquired a distinctive character … we are of opinion that it has been established that there exists a likelihood of confusion on the part of the public. We therefore conclude that BP has made out a case of infringement on the part of the respondents of its registered marks.

However, in dealing with the second aspect of the case related to passing off, the Court of Appeal declared:

> We consider that it is a necessary ingredient of the tort that the customer is deceived into making the purchase by reason of the confusion engendered by the defendant's use of a get-up similar to that of the plaintiff. As Lord Jauncey said in the *Jif Lemon* case at page 417, 'Mere confusion which does not lead to a sale is not sufficient'. If the customer can see sufficiently clearly when he gets close to the station that the product sold is not that of BP, he does not buy the petrol under the mistaken impression that he is getting BP petrol into the tank of his vehicle. In these circumstances the tort of passing off has not in our view been committed.

It is within this broad context of market share protection and growth that regulatory planning exercises outdoor advertisement control, principally in relation to the linked matters of large and small formats, siting and transport. The discussion in this section of the chapter now considers the statutory definition of advertisements and the range of development management provisions in place.

The definition of advertisements

Section 250 (1) of the Planning Act (Northern Ireland) 2011 defines an advertisement as:

> any word, letter, model, sign, placard, board, notice, awning, blind, device or representation, whether illuminated or not, in the nature of, and employed wholly or partly for the purpose of, advertisement, announcement or direction and includes any hoarding or similar structure used or designed, or adapted for use and anything else principally used, or designed or adapted principally for use for the display of advertisements.

The definition follows very closely the content of the English regulations, which, as noted by Blackhall (2005: 323), specifically include the phrase 'principally used, or designed or adapted for use' in order to deal with awnings and blinds on which an advertisement is displayed and structures such as rotating poster panels. These additional matters were initially addressed locally by the Planning (Amendment) (Northern Ireland) Order 2003 in order to ensure control over modern forms of advertising. The definition of an advertisement is, accordingly, very wide-ranging and includes the mundane (traffic signs, street name signs), the functional (shop fascia signs, estate agent boards) and the quirky (flyposting and graffiti). As noted above, advertisements are designed to provide information and seek to be noticed; they can also be highly dramatic, as evidenced in Photograph 3.1, which relates to an off-licence premises in rural Australia.

PHOTOGRAPH 3.1 Off-licence advertisement in rural Australia. *Source*: Photograph by authors, 2009.

Regulating outdoor advertisements

Notwithstanding the breadth of meaning in the definition of advertisements, planning control is exercised through a variety of regulatory modes with varying degrees of intervention. The Planning (Control of Advertisements) Regulations (Northern Ireland) 2015 specifies that powers are to be operated only in the interests of amenity and public safety. In so doing, Article 3 of these regulations enjoins councils to take into account the Local Development Plan in so far as material and any other relevant factors. The factors relevant to amenity include the general characteristics of the locality, including the presence of any feature of historic, archaeological, architectural or cultural interest. Any advertisements being displayed nearby can be disregarded if the planning authority thinks it fit to do so. *Planning Policy Statement 17: Control of Outdoor Advertisements* (Planning Service 2006: 11–12), retained by the Department for Infrastructure under the transitional arrangements of the SPPS, expands on this legislative framework with the following commentary:

> In relation to advertisements the term amenity is usually understood to mean its effect upon the appearance of the building or structure or the immediate neighbourhood where it is displayed, or its impact over long distance views. A well designed and sensitively sited advertisement, where thought has been given to size, colours, siting and levels of illumination, can contribute positively to the visual qualities of an area. All too often, however, advertisements are added to a building or placed in a location as an afterthought. The result is that a good building, neighbourhood or sensitive location can be easily spoiled by poorly designed advertising, which appears over-dominant, unduly prominent or simply out of place. A large number of advertisements on a building or along a road can create clutter and be disruptive to the appearance and character of an area. When

> preparing designs for new signage or advertisements, the opportunity should be taken to rationalise the number of signs on a building or in an area and remove those which are redundant or excessive ... The amenity of the countryside is particularly important and there is a need to protect its unique qualities from the negative effects of advertising. The only advertisements likely to be acceptable in the countryside are those proposed on-site and which relate to existing or approved commercial enterprises. These should be small in scale and not detract from the quality and character of the local landscape.

While each case is fact sensitive, the attention given to visual amenity in the development management of advertisements is illustrated, nonetheless, by a planning appeal (2015/A0017) against the refusal of consent for the retention of two banner-type advertisements and fixings at the Odyssey Arena in Belfast, a major entertainment venue latterly renamed as the SSE Arena. In this decision that granted consent to display, the presiding Commissioner stated:

> The host building is large in scale and it reads with the adjacent Odyssey Pavilion which increases its overall scale. It is a visually prominent feature in the landscape from the critical views (CVs) identified. Given the size and scale of the host building relative to the size of the advertisements, I do not consider the visual impact of both to be unduly prominent from the CVs. The visual impact is even less when the advertisements are viewed in the context of the overall Odyssey complex. The advertisements are also partially screened by the linear tree belt along sections of both approaches on the Sydenham Road and intervening vegetation aids with screening when travelling north on the M3. Furthermore, drivers would be concentrating their views towards the north and not in the direction of the advertisements. While the advertisements can be appreciated when travelling east on the M3, the overall distance within which they can be seen is diminished as drivers are generally travelling at speed. This combined with their scale in relation to the host building and the wider complex reduces their overall visual impact. The guidance indicates, inter alia, that the signs should not detract from any architectural feature. While I do not consider there to be any architectural feature of particular merit on the host building, the advertisements are made of a light canvas material and the orange tiles of the Arena can be partially seen through them. The subject matter of the advertisements relates to the primary use of the host building as they advertise upcoming events. In so doing, they relate to the premises upon which they are displayed and read as part of the building. The advertisements are not illuminated and they do not project above the eaves or parapet of the host building in accordance with the guidance. Furthermore, for the reasons set out above and given their simplistic design and positioning upon the frontage, the signs do not cause visual clutter. The surrounding land use is commercial and mixed use in nature and advertisements are to be expected in this context in accordance with policy. The subject advertisements do not therefore look out of place or detract from the character of the area.

Relevant factors in regard to public safety are cited in the regulations as including personal safety for users of any road, railway, waterway (including coastal waters), docks, harbour or airfield, as well as whether any display of advertisements is likely to obscure or hinder the interpretation of any road traffic sign, railway sign or aid to navigation by water or air (Article 3.-(2)(a) and (b)). It is noteworthy that the complementary Planning Policy Statement identifies drivers, cyclists and pedestrians as coming under the rubric of public safety. It also sets out the main types of advertisement which are likely to pose a threat to public safety:

> (a) those which obstruct or impair sight lines at corners, bends or at a junction or at any point of access to a road; (b) those which, by virtue of their size or siting, would obstruct or confuse a road user's view or reduce the clarity or effectiveness of a traffic sign or traffic signal, or those which would be likely to distract road users because of their unusual design; (c) signs which leave insufficient clearance on or above any part of the road or footpath, or insufficient lateral clearance for vehicles on the carriageway; (d) those which are located so as to impair the safety of any person looking at them because there is no protection from moving vehicles or where the footpath is narrow at the point where the public stop to look at them; (e) illuminated signs: where the means of illumination is directly visible from any part of the road; which, because of their colour, could be mistaken for, or confused with, traffic lights or any other authorised signals; and which, because of their size or brightness, could result in glare or dazzle, or otherwise distract road users especially in wet or misty weather; (f) signs which incorporate moving or apparently moving elements in their display, especially where the whole message is not displayed at one time therefore increasing the time taken to read the whole message; (g) those which resemble traffic signs because of their colour or content or those which embody directional or other traffic elements and which could therefore cause confusion with traffic signs; (h) signs sited or designed primarily to be visible from a motorway or other special road; and (i) those which cause possible interference with a navigational light or an aerial beacon.
>
> *(Planning Service 2006: 12–13)*

Again, the importance of public safety as a material consideration in the determination of planning consent for advertisements can be illustrated by a decision of the Planning Appeals Commission (2019/ A0067) involving the replacement of 2 × 48 sheet standard display boards with 2 × 48 sheet standard display boxes on vacant ground at Bridge End in Belfast. In dismissing the appeal, the presiding Commissioner ruled:

> Whilst the in-situ signs are externally illuminated, that lighting is directed down onto the advertising panel itself, which limits its effect on the legibility of the signals, as well as in terms of being a potential distraction in itself. The proposed internally illuminated signs would be backlit, with the light directed from behind the panel, giving a much a larger area of light source adjacent to the traffic signals and pelican crossing. During night time hours, or inclement weather conditions, when ambient light would be reduced, the proposed signs would become a more noticeable feature in the wider streetscape than during the day. Again, I am not persuaded that the existing street lighting in the vicinity would mitigate this effect. In my judgement the internally illuminated signs would prove more visually distracting to drivers approaching the pelican crossing in either of the two right hand lanes, even if the illumination levels comply with the guidance from the Institute of Lighting Engineers. Even if the colours of the adverts on the signs did not blend with the traffic signals, they would still reduce the overall legibility, clarity and consequently effectiveness of the traffic signals for road users in the two right hand lanes. I am not persuaded that the black and white trim around the edge of the traffic

signals would be sufficient to differentiate the signal head from the advertisements in the background to offset this reduction in legibility. The internally illuminated signs would have the potential to distract approaching motorists from signal changes at the pelican crossing ... The appeal signs would be more visually distracting than the in-situ signs and would attract road users' attention to a greater degree, reducing their due care and attention when driving towards this pedestrian crossing.

It is commonly accepted that planning authorities cannot exercise a censorship role in regard to advertisement content. However, the legislation and practice guidance set out one exception to this: the subject matter, content or design of what is to be displayed can be restricted in the interests of amenity or public safety. Content regulation, more generally, falls within the remit of the Advertising Standards Authority, which is empowered to take action against misleading, harmful or offensive advertisements.

Turning now to the categories of advertisement set out in the planning legislation, there are, first, nine classes of advertisement regarding which there is 'no requirement for consent' from the planning authority provided certain conditions are adhered to, as summarised in Table 3.1. So, for example, the site relating to an advertisement on a tethered balloon shall not be within an Area of Outstanding Natural Beauty or a Conservation Area, or an advertisement displayed on enclosed land shall not be readily visible from outside the land (see Department of the Environment 2015b).

TABLE 3.1 Classes of advertisements where, on condition, no consent is required

Class of advertisement	*Description*
Class A	The display on a site of an advertisement on or consisting of a balloon not more than 60 metres above ground level for not more than ten days in any calendar year with the exception of an Area of Outstanding Natural Beauty or Conservation Area.
Class B	An advertisement displayed on enclosed land and where the advertisement is not readily visible from outside the land.
Class C	An advertisement displayed in or on a vehicle where the vehicle is not normally employed except as a moving vehicle and is not used principally for the display of advertisements.
Class D	An advertisement incorporated into the fabric of a building; an advertisement fixed to or painted on a building is not regarded as incorporated in its fabric.
Class E	A non-illuminated advertisement not exceeding 0.1 square metre in area displayed on an article for sale or its container.
Class F	An advertisement required to be displayed by standing orders of either House of Parliament or by any enactment, subject to what is necessary in terms of size, height, number and time period.
Class G	Traffic signs provided for under the Road Traffic Regulation (Northern Ireland) Order 1997.
Class H	The national flag of any country displayed on a single flagstaff and with no additional subject matter.
Class I	An advertisement displayed inside a building where the advertisement is not illuminated and the building is not used principally for the display of advertisements.

Source: Adapted from Schedule 2, the Planning (Control of Advertisements) Regulations (Northern Ireland) 2015.

Within this grouping of advertisements, one category has long been problematic across the United Kingdom – the unlawful display of advertisements on vehicles. The regulations provide for the display of advertisements on moving vehicles not principally used for the display of advertisements. Buses, taxis and commercial vehicles, for example, come within these parameters, and, in the case of public transport, the advertising revenue generated from side, rear and full wrap options contributes to service budgets (see, for example, PML Group NI 2015). However, as stated in *Planning Policy Statement 17* (26):

> The display of an advertisement stationed on a trailer or other mobile equipment, which is principally used, or designed or adapted principally for use, for the display of advertisements, without the express consent of the Department is a breach of advertisement control … Such unauthorised mobile advertising is often sited close to the roadside and can have serious implications for road safety. It can cause distraction to drivers and impede visibility. In addition it is generally visually intrusive and can significantly detract from amenity.

Effective enforcement action can be difficult as the trailers parked on the roadside, in lay-bys or in fields can be moved to alternative locations, and the persons responsible for the trailers can be difficult to identify. Public consultation on draft PPS17 in 2004 had identified the control and enforcement of mobile advertisements as the issue raised most by respondents, including local authorities, government agencies and voluntary bodies, and, in that vein, the planning response has been to pursue, where expedient, enforcement action against the interests being given publicity by the advertisement.

The broader complexities within this sphere of planning management are well illustrated by the case *Calderdale Metropolitan Borough Council v Windy Bank Dairy Farm Ltd and another* (2010). This was an appeal against a decision from the Magistrates Court in England that had dismissed summonses relating to advertisements displayed by a farmer on the sides of either lorries or lorry trailers on land adjacent to a motorway. The respondent had claimed that loaned vehicles were, in fact, used as lockable mobile barns in which to store equipment and silage and that their principal use was to assist the farmer in tending his fields, feeding his animals and maintaining the perimeters of his land. In its determination, the appeal court held that the farmer and the vehicle satisfied the test of exemption from the requirement to obtain consent for the display of these advertisements. However, the decision is finely balanced vis-à-vis other potential cases as pointed out by Munby J in his judgment:

> The task for the Justices was one of judicial evaluation, to determine, in the light of the primary facts as they had found them, whether the vehicles were, within the meaning of the Schedule, 'normally employed as a moving vehicle' and, if so, whether or not they were being 'used principally for the display of advertisements'. Were the Justices entitled to conclude as they did? In my judgment they were. This was a case which on any view was pretty near the line. And in similar cases, where the evidence comes out in a slightly different way, Justices might well be justified in coming to the opposite conclusion. But the Justices here were, as I have said, entitled to find the primary facts as they did, and, having done so, and in circumstances where I can detect no error of law or principle in their approach, they were entitled to conclude as they did. Other Justices might have taken a different view. So, for all I know, might this court. But that is not the point. The

> Justices came to a conclusion which in my judgment was open to them on the evidence they had accepted, and that is the end of the appeal.

Within Northern Ireland the flying of multiple flags, which do not comprise the national flag of any country displayed on a single flagstaff, is a commonplace occurrence that can also give rise to a loss of visual amenity (Photograph 3.2). Other types of flags of a political and proscribed organisation nature are not included in the exemption provided by Class H of Schedule 2 of the Planning (Control of Advertisements) Regulations (Northern Ireland) 2015. Invariably, flag displays seek to celebrate cultural identity and mark out festive events and thus are not 'normally an issue' (Northern Ireland Human Rights Commission 2011). However, as in the case of some wall murals, flags can be interpreted as externally aggressive and internally cohesive by seeking to denote boundaries, mark territory and indicate the presence of proscribed organisations; in this regard the use of flags 'for other more sinister purposes' has been denoted as 'unacceptable in a peaceful and tolerant society' (Police Service of Northern Ireland et al. 2005). In its 2008 response to a consultation paper on proposed amendments to the advertisements legislation, the Alliance Party had, for example, called for appropriate planning enforcement in this matter. But again the enforcement issue here is identifying who may be responsible for erecting unlawful flags. Moreover, the matter is deeply sensitive; it goes well beyond regulatory planning, and at best it can be dealt with only through delicate community

PHOTOGRAPH 3.2 Flags mounted on telephone cable posts in Northern Ireland. *Source*: Photograph by authors, 2014.

mediation across local stakeholders. The 2005 *Joint Protocol in Relation to the Display of Flags in Public Areas* between the Police Service of Northern Ireland, the Department for Social Development, the Department for Regional Development (Roads Service), the Department of the Environment (Planning Service), the Northern Ireland Housing Executive and the Office of the First Minister and Deputy First Minister set out both ambition and principles on a thorny issue particular to Northern Ireland:

> the removal of all flags and emblems from arterial routes and town centres; the removal of all paramilitary flags and displays; the control of displays of flags and emblems in particular areas: e.g. mixed and interface areas and near buildings such as schools, hospitals, places of worship and community halls; flag flying should be limited to particular times and particular dates; and that where flags are displayed for a festive or other occasion, that the display is reasonably timebounded and that flags, including plastic ties, tape and poles, should be removed by the community after the agreed period; to encourage communities to accept that flags displayed which are tattered and torn or discoloured do not enhance the environment and should be removed.
>
> *(Police Service of Northern Ireland et al. 2005)*

Nonetheless, the flags problem in Northern Ireland remains complex and there is a belief by the Commission on Flags, Identity, Culture and Tradition (2021: 101) that 'the 2005 Protocol has not been effective' with an ongoing need for improved interagency working that includes district council participation and support for a possible Code of Practice constructed around appreciation for respect.

The second category of advertisement encompasses those with 'deemed consent' whereby permission is automatically granted simply because it would be if a formal application for express consent had to be made to the planning authority. The provisions set out in Table 3.2 are designed to ease the regulatory burden, but they do not operate without conditions. There are controls, for example, on the number of advertisements on a site, their duration, their size and height above ground, the size of characters and symbols, the nature of illumination and distance from site entrance in the case of residential development.

Arising from the 2015 regulations is the need for a site owner to prove that the site has been in continuous use for the display of advertisements without express consent for the past ten years from the date of any enforcement challenge by a planning authority. This rolling forward period in Class 12 has been carried across from the Planning (Control of Advertisements) (Amendment) Regulations (Northern Ireland) 2009, which, at the time of its mooted introduction in 2008, was perceived as a way to simplify the enforcement regime, bring Northern Ireland more into line with the legislative provisions in England and provide a measure of deregulation of the advertisement control regime. Prior to this provision becoming law, an advertiser facing enforcement proceedings had to establish that an advertisement had been in place before comprehensive advertisement control was introduced in Northern Ireland in 1973. With the passage of time, it had become difficult to establish if a site had been in continual use since that year. This provision allows for deemed consent for advertisements which have been displayed on the same site for the past ten years or more without any required express consent.

Finally, there is a third category of advertisement for which 'express consent' is required from the planning authority. Procedurally this is akin to the determination of an application

TABLE 3.2 Classes of advertisements that may be displayed with deemed consent

Class	*Description*
Class 1	Advertisements of government departments, councils, statutory undertakers including public transport undertakers.
Class 2	Miscellaneous advertisements relating to the premises on which they are displayed, for example, plaques for professional services, signs for institutions, accommodation facilities, public houses.
Class 3	Miscellaneous temporary advertisements relating, for example, to the sale or letting of land or premises, livestock sales, building work, local events, agricultural methods, travelling circus or fair.
Class 4	Illuminated advertisements on business premises.
Class 5	Advertisements, other than illuminated advertisements, on business premises.
Class 6	An advertisement on a forecourt of business premises.
Class 7	Flag advertisements, for example, in relation to the construction of houses.
Class 8	Advertisements on hoardings related to building operations.
Class 9	Directional advertisements for potential buyers and others to a residential development site.
Class 10	Advertisements inside buildings.
Class 11	Advertisements for Police Service of Northern Ireland-approved neighbourhood watch schemes.
Class 12	Advertisements on sites used for the preceding ten years for the display of advertisements without express consent.
Class 13	Advertisements relating to a pending Parliamentary, European Parliamentary, Northern Ireland Assembly or district council election – to be removed within 14 days after the close of the poll.

Source: Adapted from Schedule 3, the Planning (Control of Advertisements) Regulations (Northern Ireland) 2015.

for planning permission, as discussed in Chapter 2, and involves consultation by the planning authority should there be issues around road safety. Conditions can be applied on any express consent regulating the display of the advertisements, the use of the site to which the application relates and removal or discontinuance at the end of a specified period. Where an application for express consent is refused or granted subject to conditions an appeal can be lodged with the Planning Appeals Commission. Additionally, an appeal can be pursued against the failure of the council to take a planning decision on an application for express consent. Typically, applications will be lodged for banner-type advertisements, multi-sheet display panels, shroud advertisements, shop signs that go beyond the provisions of deemed consent and flat electronic signs.

All three categories of advertisement must be displayed in accordance with a common set of conditions. These are set out in Schedule 1 of the Planning (Control of Advertisements) Regulations (Northern Ireland) 2015 (p. 12):

> (1) Any advertisements displayed, and any site used for the display of advertisements, shall be maintained in a clean and tidy condition to the reasonable satisfaction of the council. (2) Any structure or hoarding erected or used principally for the purpose of displaying advertisements shall be maintained in a safe condition. (3) Where an advertisement is required to be removed under these Regulations the removal shall be carried out to the reasonable satisfaction of the council. (4) No advertisement may be displayed without the permission of the owner of the site or any other person with an

> interest in the site entitled to grant permission. (5) No advertisement may be displayed on or so as to obscure, or hinder the ready interpretation of, any road traffic sign, railway signal or aid to navigation by water or air, or so as otherwise to render hazardous the use of any road, railway, waterway (including coastal waters) or aerodrome (civil or military).

However, unlike in England, the granting of express consent is not time bounded; planning authorities there are permitted to specify the appropriate period by way of a condition and, where no period is specified, the default period is set at five years. In Northern Ireland, as previously stated, the corresponding regulations simply allow for the imposing of a condition requiring the removal of any advertisement or the discontinuance of any use of land authorised by the consent at the end of a specified period (Article 11.-(3)c). Additionally, the regulations provide for the discontinuance of a deemed consent in order to remedy a substantial injury to the amenity of a locality or a danger to public safety (Article 7.-(1)), and they allow for the revocation or modification of consent for the display of an advertisement, but with a potential follow-on claim for compensation (Article 16 and Article 17).

Conservation Areas

The designation of Conservation Areas is a recognition of the particular amenity quality of the built environment heritage in parts of our cities, towns and villages. In Britain the introduction of Conservation Areas was provided for by the Civic Amenities Act 1967 and, since then, each of the jurisdictions within the United Kingdom has embraced the deployment of this planning instrument. Numbers range from nearly 10,000 in England to over 600 in Scotland, over 500 in Wales and 60 in Northern Ireland (Figure 3.1). Their character can vary quite considerably, reflecting changing emphases in planning policy (for example, from preservation to regeneration) and responsiveness to local cultural preferences (for example, vernacular building styles). But common to all is their intrinsic value derived from a special plan form, architectural or historic interest which is considered worthy of protection and enhancement.

Legislative and policy evolution

The issuing of the Planning (Northern Ireland) Order 1972 introduced the concept of Conservation Areas into Northern Ireland, making their designation the responsibility of the Department of the Environment in consultation with the Historic Buildings Council and the relevant local authority. Section 104 of Planning Act (Northern Ireland) 2011 empowers councils and the Department for Communities to designate a Conservation Area following consultation with the Historic Buildings Council or such other persons or bodies as may be prescribed; in the interim the list of formal consultees has been widened to include the Department for Infrastructure because of public works implications. It is, perhaps, noteworthy that the former Department of the Environment has stated that it anticipated a few proposals to designate new Conservation Areas in Northern Ireland (Explanatory Memorandum to the Planning (Conservation Areas) (Consultation) Regulations (Northern Ireland) 2015, S.R. 2015 No. 177). However, given that the designation of Conservation Areas in Northern Ireland can be traced back to the 1970s, the issue being raised by District Councils during the preparation of their new Local Development Plans is whether they continue to retain the

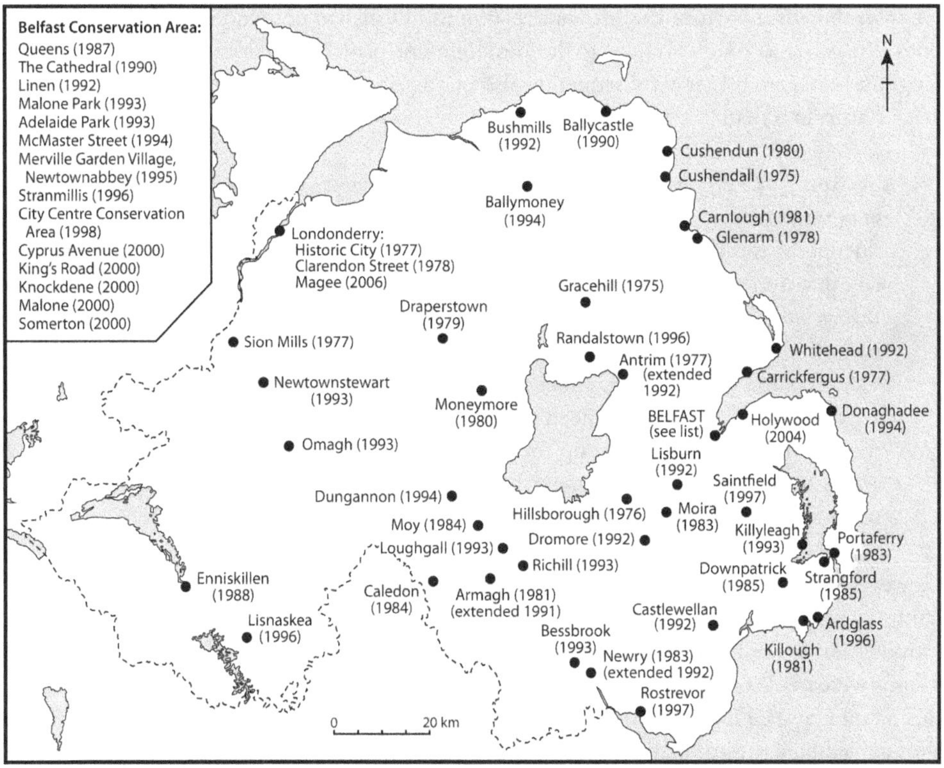

FIGURE 3.1 Conservation Areas in Northern Ireland, 2022 (with dates of designation). *Source*: Updated and adapted from Mourne Heritage Trust and the Ulster Architectural Heritage Society (2002).

character and appearance that justified this special status. More than likely, existing designations will be revisited to determine if changes are necessary. However, Conservation Area designations can only be varied or cancelled by the authority which made the designation under Section 104 (4) of the 2011 Planning Act, and thus the situation is that councils are currently unable to vary or cancel a department designated Conservation Area (Department for Infrastructure 2022:65).

When a designation-related decision is taken, sufficient particulars to identify the area affected must then be published in at least one newspaper circulating in the council's district, by that council or, as may be the case, by the Department for Infrastructure. Additionally, it has been standard practice to publish for each designated area a Conservation Area guide that provides illustrated information on the historical context, a description or character appraisal of the area, the designation boundary in map format and guidelines for development proposals. These booklets constitute supplementary planning guidance, and their content is a material consideration in development management decision making. The weight to be given to the Conservation Area booklet as supplementary planning guidance was examined *In the Matter of an Application by Barry Gilligan, Valerie Allen, Anthony Cox, Maeve Fee, Henry Glover and Malone Park Residents' Association for Judicial Review and In the Matter of a Decision by the Planning Service of the Department of the Environment for Northern Ireland Dated 16 January 2002* (2002 NIQB 53) wherein the affidavit from the principal planning officer had contended that the document does not carry as much weight as the development plan or a Planning Policy Statement. Coghlin J disagreed, a view subsequently also endorsed by the Court of Appeal (2003 NICA 10), in his ruling that:

> Section 7 of PPS1, which contains Policy BH12, relates to Conservation Areas in general, whereas it is quite clear that the Malone Park/Adelaide Park Conservation Area document was generated with the requirements of one specific area in mind. Thus, if any question of priority arises, it seems to me that it would be reasonable to anticipate that it would be that of the latter over the former. Such an approach would appear to be entirely consistent with the wording of paragraph 7.12 of PPS6 which deals with Conservation Area design guides and provides that: 'The Department will therefore attach great weight to the need for proposals for new development to accord with the specific guidance drawn up for each particular Conservation Area'. Accordingly, it seems to me that the Department has misunderstood and misinterpreted its policy in relation to this consideration.

However, as noted by Hendry (1977), the initial legislation did not provide special powers to control development beyond the automatic protection of Listed Buildings. It was only with the introduction of the Planning (Amendment) (Northern Ireland) Order 1978 that specific provisions were made for the control of demolition in Conservation Areas; Article 4 (2) required that a building shall not be demolished without the consent of the Department. This concern about avoiding harm to the built environment integrity of a Conservation Area has been carried forward by the Planning Act (Northern Ireland) 2011 with Section 105 stipulating that 'Conservation Area Consent' be obtained prior to any demolition, including the structural alteration of a building where the alteration comprises demolishing a part of the building. The Planning (Conservation Areas) (Demolition) Regulations (Northern Ireland) 2015 set out the operational procedures for obtaining Conservation Area Consent for the demolition in a Conservation Area of a building which is neither a Listed Building nor an ecclesiastical building nor a building guarded, protected or scheduled under the Historic Monuments and Archaeological Objects (Northern Ireland) Order 1995. An additional power under Article 4 of the Planning (General Permitted Development) Order (Northern Ireland) 2015 allows for the removal of certain permitted development rights in any area considered expedient to do so by the Department for Infrastructure or council. *Planning Policy Statement 6: Planning, Archaeology and the Built Environment* (The Planning Service 1999) suggests that this provision may have a role in protecting important features of interest within a Conservation Area; their removal opens up the scope for enforcement action if an Article 4 direction is in place. Additional measures to protect visual amenity include the requirement of planning permission for the siting within a Conservation Area of a satellite dish on a chimney, wall or roof slope that faces onto and is visible from a road.

A further and potentially important dimension to planning legislation in this sphere is the incorporation of financial provisions into statute following on again from the Planning (Amendment) (Northern Ireland) Order 1978; Article 5 (1) made it possible to offer grants or loans to defray, in part or in whole, any expenditure relating to the promotion, preservation or enhancement of the character or appearance of a Conservation Area or any part of it. Nonetheless, as recorded again by Hendry (1993: 419) applicants for Conservation Area grants in that early period had a low success rate that was partly explained by the relatively small amount of money available in any year. Moreover, conditions imposed on the use of materials (often illustrated within the Conservation Area guides) prompted many property owners to ignore or reject grant aid which covered only 20 percent of approved works. Section 106 of the Planning Act (Northern Ireland) 2011 has also incorporated Conservation Area financial provisions albeit within the contemporary context of public sector spending austerity:

> The Department may, in relation to any Conservation Area, make grants or loans for the purpose of defraying in whole or in part any expenditure incurred or to be incurred in or in connection with, or with a view to the promotion of, the preservation or enhancement of the character or appearance of the area or any part thereof.

Other financial provisions have sought to enhance the built environment of Conservation Areas, for example, the Community Regeneration and Special Improvement Programme of the International Fund for Ireland that ran during the 1990s and targeted small towns and villages generally. Additionally, the long-running Townscape Heritage Initiative, funded by Heritage Lottery, has helped stimulate a local partnership approach to the repair of building fabric, the restoration of authentic details, and utilising vacant floor space in historic buildings within Conservation Areas. And launched in 2020, the Village Catalyst project brings together a partnership comprising the Department for Communities (Historic Environment Division), the Architectural Heritage Fund, the National Heritage Lottery Fund, the Department of Agriculture, Environment and Rural Affairs and local community social enterprise groups. It aims to restore disused historic buildings in villages, increase opportunities for a range of local social engagement activities by improving access to key services in rural areas through the provision of enhanced facilities and, ultimately, act as a catalyst for further village regeneration. Projects supported include those within Conservation Areas (for example, the conversion of an 18th-century post office into a social enterprise training hotel in Gracehill) and other village settings (for example, the development of a wellbeing and co-working hub in a Listed Building in Ederney).

Demolition within Conservation Areas

It is against this backdrop of legislative and policy evolution that the discussion can now turn to consider more fully the thorny issue of demolition within Conservation Areas. Planning Policy Statement 6: 35 sets out the policy baseline position in this regard:

> The Department will normally only permit the demolition of an unlisted building in a Conservation Area where the building makes no material contribution to the character or appearance of the area. Where Conservation Area Consent for demolition is granted this will normally be conditional on prior agreement for the redevelopment of the site and appropriate arrangements for recording the building before its demolition.

The amplification of this policy states that a presumption in favour of retaining any building which makes a positive contribution to the character or appearance of a Conservation Area will operate and that, in determining proposals for demolition of unlisted buildings, account will be taken of the part played in the architectural or historic interest of the area by the building for which demolition is proposed and, in particular, of the wider effects of demolition on the building's surroundings and on the Conservation Area as a whole. The relationship between these policy considerations and development control decision making has been thrown into sharp focus *In the matter of an application by the Ulster Architectural Heritage Society for judicial review* (2014 NIQB 21) whereby the respondent had granted consent to demolish a four-storey 19th-century former warehouse in a city centre Conservation Area in Belfast (Photograph 3.3).

In a lengthy consideration of the evidence, Treacy, J gave attention to whether planning policy was lawfully and fairly applied to the decision:

> In exercising the supervisory function of the Judicial Review Court the court is only empowered to review whether the Department applied its policy lawfully and fairly, I am not empowered to interfere with the expert planning judgement of the Department. To this end, questions of the weight to be attached to the material considerations which weighed in the balance of the final decisions are within the sole province of the Department. However, where there is a presumption in the policy, the Department are obliged to have regard to that presumptive weight and cannot in their decision undermine that presumption. Presumptions in policy are there to protect the overriding objectives of the policy itself and to undermine them would be to nullify the policy ... The duty of the department is to ensure land use in the public interest and it has developed policies which guide planning decisions. One of the 'public interests' defined in planning policy is the preservation of the built heritage. Conservation Areas are designated as such because taking the buildings together, they form an area

PHOTOGRAPH 3.3 Belfast city centre Conservation Area building. *Source*: Photograph by authors, 2016.

> of particular architectural or historical interest. There is a presumption in the policy that any building which adds to this interest factor should be retained. When weighing up an application for development, and particularly for demolition consent, in a Conservation Area, the department must have regard to the presumption in favour of retaining. Against this presumption, it must weigh all other material considerations. The Department enjoys discretion as to the weight to be given to these other considerations. In the group decision of 25 May 2011 the group explicitly give 'determining' weight to the fact that the proposals for retention and repair were not economically viable. That is, the fact that the repair options submitted by the owner are insufficient to create a profit for the developer is considered sufficient to outweigh the presumption in favour of retaining the building. The problem with this decision is that the Department is not obliged to change the status quo. The instant application is not the end of the line for the building. It is not the only option. If this application were refused, the building would continue in use as it is at the moment for the time being. There would be no necessary detriment to the area. Nothing bad will immediately happen if the building continued in its present use save the effect of time on the fabric of the building which it is estimated will eventually render the building unusable in 30 years in the absence of alternative proposals. To my mind where the Department is obliged to have regard to a presumption, and there are two options available to it, one which respects the presumption, and one which does not, the existence of the presumption-respecting option is a very material consideration and one which must be considered and given its full presumptive weight. Key documents leading to the ultimate decision overlooked the existence of this presumption, and none of the documents gave any or proper consideration to the presumption-respecting option that was open to the department. For this reason I hold that the Department failed to properly take into account a very relevant consideration.

The judgment in this case clearly prefaces the thrust of the analysis of judicial review, set out in Chapter 7 of this book, regarding the stated duty in all development management decisions to properly take into account relevant considerations. But the case also very usefully pinpoints the tension that can arise in mediating between public and private interests in a planning system that is committed to proactively facilitating development. The consideration of amenity looms large in Conservation Areas and, in that regard, it is helpful that the *Strategic Planning Policy Statement* (Department of the Environment 2015a: 40) has revisited the policy guidance set out in Planning Policy Statement 6 and gives a much more precautionary intonation to decision making in this arena:

> In managing development within a designated Conservation Area the guiding principle is to afford special regard to the desirability of enhancing its character or appearance where an opportunity to do so exists, or to preserve its character or appearance where an opportunity to enhance does not arise. Accordingly, there will be a general presumption against the grant of planning permission for development or Conservation Area Consent for demolition of unlisted buildings, where proposals would conflict with this principle. This general presumption should only be relaxed in exceptional circumstances where it is considered to be outweighed by other material considerations grounded in the public interest.

Simply put and in line with provisions of the Planning (Northern Ireland) Act 2011, there is now a higher legislative test for redevelopment proposals within a Conservation Area that extends to both preservation and enhancement, rather than a mere no harm test that was commonly applied under the now-repealed Planning (Northern Ireland) Order 1991 (Department for Infrastructure 2017: 18).

On those occasions when demolition is permitted within a Conservation Area, the Planning Act (Northern Ireland) 2011, in extending the special controls on Listed Buildings to non-listed buildings in a Conservation Area (Section 105(6)), allows for a condition to be imposed that the building shall not be demolished before a contract for the carrying out of works of redevelopment of the site has been made, and planning consent has been granted for the redevelopment for which the contract provides (Section 91(6)). It is, however, not a legal requirement that the proposals for demolition and redevelopment be considered together (Planning Appeals Commission 2002), desirable though that may be.

Areas of Townscape and Village Character

Areas of Townscape and Village Character are defined in the Addendum to Planning Policy Statement 6 (The Planning Service 2005: 6) as areas that 'exhibit a distinct character normally based on their historic built form or layout' and which in the main 'derives from the cumulative impact of the area's buildings, their setting, landscape and other locally important features'. While Conservation Areas enjoy designation at any time under primary legislation, Areas of Townscape and Village Character must rely on development plans to facilitate their designation process. Part 33 of the Schedule to The Planning (General Permitted Development) Order (Northern Ireland) 2015 expressly states that the latter designations relate to a departmental development plan or a local development plan prepared under Part 2 of the Planning Act (Northern Ireland) 2011 or a draft of such a plan. Areas of Townscape Character initially appeared as a policy statement in the *Belfast Urban Area Plan 2001* and were described at that time as 'areas of special architectural and historic character' and for which development control statements would be prepared, as required, 'to protect and enhance their particular character' (Department of the Environment 1990: 42). The subsequent *draft Belfast Metropolitan Area Plan* (The Planning Service 2004) carried forward this provision, while more recent preparation work for the *Belfast City Council Local Development Plan 2020–2035*, for example, lists 57 Areas of Townscape Character. The designation has been rolled out across Northern Ireland and, by 2019, a total of 177 Areas of Townscape and Village Character had been identified in pre-2015 development plans (Department for Communities 2020).

The issue has emerged, however, as to whether they might be regarded as second-tier Conservation Areas and, indeed, the ad hoc Northern Ireland Planning Commission established under the auspices of the National Trust had recommended in 2004 that the then-existing Areas of Townscape Character should 'urgently' be designated as Conservation Areas and that 'the designation 'Areas of Townscape Character' should be abolished once existing areas have been redesignated as Conservation Areas' (5). This proposal has been strongly refuted by Galway and McEldowney (2006) in an exhaustive empirical analysis of the comparative quality of Conservation Areas and Areas of Townscape and Village Character, with the former deemed 'special places' and the latter as 'the remarkable commonplace'. Accordingly, it is argued that implementation of the suggested abolition of Areas of Townscape and Village Character would inadvertently undermine the credibility of Conservation Area status, reduce the number of areas capable of receiving some protective designation and remove a significant opportunity for local discretion in the protection of local environments (ibid.: 417).

The demolition or partial demolition of unlisted buildings is prohibited within Areas of Townscape and Village Character designated in a departmental development plan, a Local Development Plan or a draft plan unless consent has been obtained by way of planning permission. The 2015 *Strategic Planning Policy Statement for Northern Ireland* notes that the demolition of an unlisted building in an Area of Townscape or Village Character should be permitted only where the building makes no material contribution to the distinctive character of the area and subject to appropriate arrangements for the redevelopment of the site. Exceptions under the Planning (Demolition – Description of Buildings) Direction 2015 comprise:

- buildings that do not exceed 115 cubic metres;
- the whole or any part of a gate, wall, fence or other means of enclosure less than one metre high which is adjacent to a road or open public space or in other cases less than two metres high.

Additionally planning policy gives particular attention to the control of advertisements in an Area of Townscape or Village Character to ensure that the scale, size, proportions, dominance, siting, materials and illumination do not adversely impact visual amenity or contribute to clutter. New development must not detract from or fail to maintain the distinctive character of these designated areas. Finally, as noted in the *Addendum to Planning Policy Statement 6: Areas of Townscape Character*, supplementary planning guidance and local policies will normally be incorporated into the relevant development plan and will be taken into account as a material consideration in assessing development proposals.

The application of these measures can be briefly considered by way of illustration in the case of Newcastle, where two Areas of Townscape Character were initially proposed in the *Ards and Down Area Plan 2015*, published in draft form in December 2002. These comprised sea-facing locations that focused on the Rock and the Harbour to the south of the town. Following the public local inquiry into the draft plan, the Planning Service of the Department of the Environment accepted in its March 2009 Adoption Statement a recommendation from the Planning Appeals Commission that these two Areas of Townscape Character be amalgamated and extended to include properties along King Street. In discussing that element of the Area of Townscape Character known as 'The Rock' the adopted plan draws attention to the prevailing backdrop of the Victorian seaside terraced boarding houses with their stuccoed, two and three storeys and bay windows appearance. More particularly, the plan states:

> This townscape has a significance and architectural style similar to elsewhere in Newcastle and other seaside resorts of the same era. It is therefore likely to experience similar development pressures prejudicial to its present character and distinctiveness. These could variously include redevelopment, multiple-occupancy, apartments, changes of use, fragmentation of sites and minor but cumulative alteration. To protect the special character of the area there will be a clear presumption in favour of conservation and re-use of existing buildings. Proposals should value and respect the physical characteristics and artefacts of existing development and pattern of land use, be subordinate in terms of massing, form, style and function, and be accurate in matters of detail, material and finishes.
>
> *(The Planning Service 2009b: 86)*

The significance of this statement is thrown into sharp focus by the unauthorised demolition of a boarded-up house at 141 Central Promenade, Newcastle, in January 2009 (see Belfast Telegraph 2009). The former three-storey Victorian building (Photograph 3.4) formed part of a sea-facing terrace, and its demolition created a prominent visual gap in the streetscape structure of the town. The property had been situated within an Area of Townscape Character in the then *draft Ards and Down Area Plan 2015* and thus enjoyed a presumption of designated protection unless demolition was approved by way of a planning application. Subsequently in August 2016 a planning application (LA07/2016/1143/F) was lodged for full planning permission for, inter alia, a new three-storey building with vehicle underpass on the gap site, which, in turn, would allow for backland apartment development at the gap site and on an adjacent plot of land at 143 Central Promenade. Planning permission was granted on 31 August 2017. An amended proposal (LA07/2020/0200/F) was then submitted in February 2020 and which in its preparation had canvassed the demolition of the property adjacent to the gap site. This was deemed unacceptable by the planning authority as was the proposal for a pastiche facade design. The analysis of the case officer (see: https://epicdocs.planningni.gov.uk/ViewDocument.aspx?guid=21b442ce-de6c-45ab-aec9-ecfb9ee359df) draws attention to the importance of protecting the integrity of the Area of Townscape Character (ATC):

> additional regard must be had to the impact of demolition and redevelopment proposals within an ATC. Demolition of a building in an ATC is therefore a material planning consideration. The impact of their demolition cannot be assessed in isolation and divorced from the merits of the redevelopment scheme. The agent was made aware of Planning's view from the outset that the building to be demolished was deemed to be

PHOTOGRAPH 3.4 Property at 141 (centre) and 143 (left) Central Promenade, Newcastle within The Harbour, The Rock and King Street Area of Townscape Character.
Source: Photograph by Neil Galway 2003.

> worthy of retention and contributed positively to the character of the ATC and that the redevelopment scheme was not appropriate to maintain or enhance the overall character and appearance or respect the built form of the ATC. The supporting information provided by the agent was considered to fall short of the policy requirements set out in PPS 6. Following negotiations with the agent a pastiche design approach which would see the existing façade of the building being replicated in a new build form was also deemed to be unacceptable. The scheme was subsequently amended to retain the existing façade and tie in the remaining redevelopment scheme with this. While Planning would prefer the retention of the building in its entirety, given that it is the façade that contributes variety and interest to the streetscape it was deemed that this was an appropriate compromise.

Planning permission was granted on 19 January 2021, inter alia, for street facade retention and partial demolition of the existing building at 143 Central Promenade adjacent to the gap site. The important point here is that the designated Area of Townscape Character was central to leveraging a negotiated outcome, which, when implemented, will generate beneficial use of the lands as a whole and contribute positively to the restoration of the visual amenity of this part of Newcastle.

Listed Buildings

Listed Buildings are dealt with in Part Four, Chapter 1 of the Planning Act (Northern Ireland) 2011 with Section 80 requiring 'the Department' (currently the Department for Communities), in consultation with the appropriate local authority and the Historic Buildings Council, to compile a list of buildings of special architectural or historic interest; any building included in the list is referred to as a 'Listed Building'. Section 85(1) makes it an offence to execute (or cause to be executed) any works for demolishing a Listed Building, or for altering or extending it in any manner which would affect its character as a building of special architectural or historic interest, unless the works are authorised. Section 85(2) provides that these works are authorised only if consent ('Listed Building Consent') has been granted for them and they are carried out in accordance with that consent. Applications for Listed Building Consent are dealt with by councils, although under Section 88(1) of the 2011 Act the Department for Infrastructure (as the regional planning authority) may give directions requiring applications for Listed Building Consent to be referred to it instead of being dealt with by councils. Section 85 shall not apply, however, to works for demolition, alteration or extension of a listed ecclesiastical building being used for ecclesiastical purposes. Church development remains subject to planning permission and if a church is a Listed Building it does not, moreover, enjoy permitted development rights to carry out small works without planning permission. If a building is no longer used as a church, the 'ecclesiastical exemption' does not apply.

The listing scheme in Northern Ireland

Field-based research to compile Northern Ireland's first Listed Buildings record began in 1969, with the statutory listing of buildings commencing in 1974 (see Northern Ireland

Audit Office 2011). A key influence and contributor to the listing process during that formative period was the Ulster Architectural Heritage Society (see Latimer 2015), whose prolific engagement in this field has produced an extensive portfolio of gazetteers, lists and themed publications to inform conservation policy and practice. A notable success from this early campaigning activity was the much-needed introduction of legislation for listing within the Planning (NI) Order 1972. In 2021 there were over 8,900 Listed Buildings in Northern Ireland. Listing includes a building and its curtilage with attached/unattached objects or structures which form part of the land and which has done so since before 1 October 1973. A range of built environment features can be listed which go well beyond conventional buildings to include street furniture and bridges. This includes Royal Mail post boxes, red public telephone kiosks and tiled street signs produced by Belfast Corporation during the early to mid-1900s. Listed Building status also applies to interior finishes such as wall panelling, plasterwork and fireplaces.

The principal statutory criteria for listing are special architectural interest and/or special historic interest, with planning policy guidance elaborating on both meaning and application (Department for Communities – Historic Environment Division 2019a). The architectural interest can embrace factors such as style, proportion, ornamentation, plan form, structure, innovatory qualities, alterations, interiors, setting and group value; historic interest can encompass age, rarity, authenticity, historic importance, authorship, historical associations and the social, cultural or economic importance of the building. This interest can be applicable at international, Northern Ireland and local scales, but essentially the key test permeating these criteria is the identification of 'special' attributes. Buildings listed by the Department are divided into four categories: Grade A, Grade B+, Grade B1 and Grade B2 (Table 3.3). These are purely administrative categories in contrast to the statutory gradings elsewhere in

TABLE 3.3 The Northern Ireland listing categories

Category	*Description*
Grade A	Buildings of greatest importance to Northern Ireland including both outstanding architectural set-pieces and the least altered examples of each representative style, period and type.
Grade B+	High-quality buildings that because of exceptional features, interiors or environmental qualities are clearly above the general standard set by grade B1 buildings. Also buildings which might have merited Grade A status but for detracting features, such as an incomplete design, lower quality additions or alterations.
Grade B1	Good examples of a particular period or style. A degree of alteration or imperfection of design may be acceptable. Generally B1 is chosen for buildings that qualify for listing by virtue of a relatively wide selection of attributes. Usually these will include interior features or where one or more features are of exceptional quality and/or interest.
Grade B2	Special buildings which meet the test of the legislation. A degree of alteration or imperfection of design may be acceptable. B2 is chosen for buildings that qualify for listing by virtue of only a few 11 attributes. An example would be a building sited within a Conservation Area where the quality of its architectural appearance or interior raises it appreciably above the general standard of buildings within the Conservation Area.

Source: Department for Communities – Historic Environment Division 2019a.

the United Kingdom. Detailed information on over 9,000 buildings, most of which have been listed, can be accessed on the Northern Ireland Buildings Database (see https://www.communities-ni.gov.uk/services/buildings-database). Section 80 of the Planning Act (Northern Ireland) 2011 provides for the Department for Communities to de-list any building on the basis of a significant loss, perhaps, of historic fabric and, following consultation with the Historic Buildings Council and the relevant local authority, to serve a notice on the owner and occupier to the effect that the building has ceased to be listed, with an explanation for the decision.

Listed Building Consent

Listed Building Consent is required before carrying out any work for demolition, alteration or extension of a Listed Building, which would affect its character, externally or internally, as a building of special architectural or historical interest. It is an offence to carry out such works without this consent. Listed Building Consent is not a planning approval and, thus, it may also be necessary to apply for planning permission for external and substantial works. Where a development proposal involves works of alteration to a Listed Building, both the Listed Building Consent application and the planning application should be submitted together to the planning authority, to enable simultaneous consideration. This is important, as the impact of the proposed works on the significance of a Listed Building will often have direct impact on the appropriateness of the development proposal. The statutory provisions in regard to Listed Building Consent are detailed in the Planning (Listed Buildings) Regulations (Northern Ireland) 2015 and require that a design and access statement shall accompany the application. This will set out how the works take account of the special architectural or historic importance of the building, the particular physical features of the building that justify its designation as a Listed Building and the building's setting. There is a statutory requirement that planning authorities must consult with the Department for Communities and take account of any representations that are received from its Historic Environment Division as the statutory consultee (see Department for Communities – Historic Environment Division 2019b). Additionally, under Section 89 of the Planning Act (Northern Ireland) 2011there is a duty on a council to notify the Department for Infrastructure that it intends to grant Listed Building Consent. This makes the granting of listing building consent by a council subject to time-bounded departmental consideration. Should consent be granted, works must commence within a period of five years from the date of the grant. If consent is refused, or granted subject to conditions, or a decision is not forthcoming within a period of eight weeks from the council (or the Department for Infrastructure in the case of called in / regionally significant development proposals), an appeal may be lodged with the Planning Appeals Commission.

The range of matters aired at appeal in regard to Listed Building Consent include, for example, shop signs (Planning Appeals Commission 2012/A0167), retention of existing white Unplasticised Polyvinyl Chloride (UPVC) double-glazed windows (Planning Appeals Commission 2012/A0135), retractable awnings and window canopies (Planning Appeals Commission 2008/A0113), bay window construction (Planning Appeals Commission 2004/A245) and removal of a statue (Planning Appeals Commission 2004/A205). All these appeals were dismissed. Proposals for the demolition of a Listed Building do arise and in that regard one of the most interesting cases to come before the Planning Appeals Commission concerned a modernist style house at Strand Road, Portstewart, built in 1959, extended in 1968 and listed in 1994 (Photograph 3.5). This proposal comprised both demolition and replacement and thus two appeals were determined simultaneously, one in regard to Listed Building Consent to demolish a Listed Building (Planning Appeals Commission 2004/A369) and a full planning application for a replacement dwelling (Planning Appeals commission 2004/A370). In his report to the Planning Appeals Commission recommending dismissal, the presiding Commissioner opined:

> As one of a small number of listed modernist houses its importance is high. This is not diminished by the subjective opinion that other house designs exist by the same architect that may be considered better. The Listed Building contains originality and is recognisable as a piece of modernist Ulster Architecture. The replacement proposal, in my opinion, is informed to some extent by the Listed Building, e.g. the proposed length of the building, its horizontality and the angle to the stonework has an affinity to the existing butterfly roof. The Listed Building has withstood the passage of time and for the appellant to compare his proposal with the Listed Building is in itself is a measure of the Listed Building's design qualities. Every era, including the modern era, has left a legacy of buildings that provide a sense of place and constitute material evidence of how people lived in the past and a source of reference for contemporary architecture. The appellant accepts that the defects in the building can be rectified and the objectors submitted evidence that a sensitive approach would resolve the difficulties with privacy in the existing layout of the house. The merits of the replacement proposal have been well articulated by the appellant. It reflects the spirit of contemporary design and could

PHOTOGRAPH 3.5 Listed Building at 108 Strand Road, Portstewart. *Source*: Photograph by David Houston, 2015; NIEA evaluation available at www.doeni.gov.uk/niea/content-databases-buildlist.htm, accessed 21 October 2015.

> be judged to be an example of the current era. However, the existing building at No108 Strand Road is listed and is not only a good example of the Modern International Style but also a building of its time, built in the mid 20th Century. The architectural, chronological and historical significance of the Listed Building cannot be replaced by the proposals before me.

The corporate decision of the Planning Appeals Commission was to refuse Listed Building Consent.

Preserving the setting of a Listed Building

Section 91 (2) of the Planning Act (Northern Ireland) 2011 places a statutory duty on decision makers to have special regard to the desirability of preserving not only a Listed Building but also its 'setting'. This is more fully articulated within the 2015 *Strategic Planning Policy Statement for Northern Ireland*:

> Listed Buildings of special architectural or historic interest are key elements of our built heritage and are often important for their intrinsic value and for their contribution to the character and quality of settlements and the countryside. It is important therefore that development proposals impacting upon such buildings and their settings are assessed, paying due regard to these considerations, as well as the rarity of the type of structure and any features of special architectural or historic interest which it possesses. Development involving a change of use and / or works of extension / alteration may be permitted, particularly where this will secure the ongoing viability and upkeep of the building. It is important that such development respects the essential character and architectural or historic interest of the building and its setting, and that features of special interest remain intact and unimpaired. Proposals should be based on a clear understanding of the importance of the building/place/heritage asset, and should support the best viable use that is compatible with the fabric, setting and character of the building.
> *(ibid.: 39)*

Extant planning guidance in *Planning Policy Statement 6: Planning, Archaeology and the Built Environment* (The Planning Service 1999) states that development will not normally be permitted:

> which would adversely affect the setting of a Listed Building. Development proposals will normally only be considered appropriate where all the following criteria are met: (a) the detailed design respects the Listed Building in terms of scale, height, massing and alignment; (b) the works proposed make use of traditional or sympathetic building materials and techniques which respect those found on the building; and (c) the nature of the use proposed respects the character of the setting of the building.

The important point here is that the visual impact of a development proposal has to be considered in relation to Listed Buildings which may be adjacent, close to or even well removed from the application site. In that regard, detailed guidance from the Department for Communities – Historic Environment Division (2018: 8) makes the crucial observation that setting and curtilage may vary and that a more expansive appreciation of heritage context is necessary:

> Curtilage may be a component of the settings of certain heritage assets, particularly Listed Buildings. Curtilage is not legally defined in planning policy, but the extent of curtilage can be described as the consideration of ownership (past and present) around heritage assets. Setting often extends beyond property boundaries, fences, walls, hedges or other 'curtilage' structures.

Additional amenity considerations relating to noise, nuisance or general disturbance may also be relevant. The question consequently arises as to what constitutes the setting of a Listed Building? In the case Planning Appeals Commission 2014/A0140, the Commissioner advanced the following pithy explication:

> While the term 'setting' is not defined in the policy, it is generally viewed as a broad concept with its precise extent varying from site to site in response to particular conditions and circumstances, including historical factors. The area within which the building is visible can comprise its setting as can views from the Listed Building if they provide an understanding or appreciation of its historical background.

In this planning appeal, the Northern Ireland Environment Agency successfully argued against the proposed introduction of a flat roofed three-storey stair tower in the central or link part of the appeal site as historically there would not have been a high building in this location. The Agency argued that the proposal would, therefore, unacceptably alter the order and rhythm of the established architectural arrangement due to its overall height and massing and that its scale and form would adversely affect the view from the rear of the Listed Building. Its analysis was that the proposal would jar with the language of the street, including the Listed Building, in terms of both the proposed materials and the elevation treatment.

Development proposals which at first sight may seem rather innocuous can also be deemed to adversely impact the setting of Listed Buildings. In the case Planning Appeals Commission 2011/A0028 regarding the addition of a balcony above the entrance door to a house adjacent to a three-storey terrace of Listed Buildings in Kerr Street, Portrush (Photograph 3.6), the Commissioner in her decision concluded:

> The proposed balcony would be visually jarring. It would draw the eye and degrade the appreciation of the adjacent Listed Buildings, their setting and the visually harmonious grouping ... Whilst acknowledging that a balcony of this size would add to the appellant's amenity, this particular proposal is unacceptable.

Perhaps the final word on this matter from the Historic Environment Division of the Department for Communities (2018:7) is a summing up of what 'setting' means. It advises that the term 'applies to the physical space that is part of – and contributes to – the significance and distinctive character of a heritage asset, and through which the asset may be seen, experienced, understood and enjoyed'). In short, what is being described here fits well with an understanding of the concept of amenity.

PHOTOGRAPH 3.6 Three-storey Listed Buildings in Kerr Street, Portrush. *Source*: David Houston, 2014.

Preserving buildings at risk

The listing of a building does not automatically guarantee its preservation and in situations where owners are unwilling to maintain properties with special architectural and historic value, a great danger exists that important components of Northern Ireland's built heritage will eventually be lost. The research and advocacy of the Ulster Architectural Heritage Society has been vital in drawing attention to the need for engaged public policy that can include the availability of restoration funding to owners, community-based building preservation trusts and other charitable organisations. Considerable information is now available on the status of buildings at risk, and there is an emergent body of before and after exemplars that can illustrate good practice in this sphere (see Ulster Architectural Heritage Society 2008). Particular attention has been given to raising the profile of traditional unprotected buildings in the countryside which not only celebrate the vitality of this vernacular architecture but also demonstrate the scope to sensitively extend and alter domestic scale properties (see Latimer 2005; Mourne Heritage Trust 2004). Legislation, nonetheless, still remains key to action and a number of provisions are in place.

Section 81 of the Planning Act (Northern Ireland) 2011 provides for a council to serve a Building Preservation Notice on the owner and occupier of a non-listed building in its district if it appears to the council that the building is of special architectural or historic interest and is in danger of demolition or alteration to its character. The immediate effect is that the building enjoys the same level of protection for a period of six months as a Listed Building. The Building Preservation Notice requests that the Department for Communities considers the merits of listing the building, which the Department may subsequently confirm or reject. Councils are currently liable for compensation where a Building Preservation Notice ceases to have effect without the building having been listed by the Department for Communities (Department for Infrastructure 2022: 64). This type of instrument existed prior to the current legislation, and over the period 2003–2011, following the passage of the Planning (Amendment) (Northern Ireland) Order 2003, a total of thirty-seven Building Preservation Notices were issued by the Department of the Environment (Public Accounts Committee 2011).

Section 161 of the Planning Act (Northern Ireland) 2011 makes provision for urgent works perceived as necessary for the preservation of a Listed Building (or a building in a Conservation Area). Under that legislation the Department for Communities retains the power to issue an Urgent Works Notice, but as District Councils now also hold this power and are the local planning authority, the Department would issue such notices only in exceptional circumstances. The expense of these works for affording temporary support or shelter for the building can be recovered by the council or Department. The process commences with a warning letter requesting negotiated compliance but, if ignored, will then move to a more formal procedure. This requires giving the owner of the building notice in writing (an Urgent Works Notice) of the details of the work to be carried out and the costs of the work. The owner, within twenty-eight days of the service of this notice, may then appeal to the Planning Appeals Commission on the following grounds:

- that some or all of the works were unnecessary for the preservation of the building;
- that temporary arrangements affording support and shelter have continued for an unreasonable length of time; and
- that the amount specified in the notice is unreasonable or that its recovery would cause the owner hardship.

The Planning Appeals Commission must then give reasoned notice of its determination and state the amount recoverable with these outlays to be regarded as a civil debt recoverable summarily.

Urgent Works Notices have been employed in Northern Ireland under Article 80 of the Planning (Northern Ireland) Order 1991 and in the period following a Heritage Crime Summit, convened in August 2011 by Minister for the Environment Alex Atwood, the Northern Ireland Environment Agency was provided with additional funding to enable direct action when negotiated solutions had failed. Several cases were pursued at that time. In March 2012, for example, an Urgent Works Notice was issued in regard to a pre-1830 Georgian farmhouse in the environs of Portrush (Photograph 3.7), which had been listed in 1977; the aim was to keep the building watertight and secure. After being served by the Urgent Works Notice, the owner undertook works at his own expense, including re-roofing the property with a natural slate, to secure the building and make it wind and weather-proof.

Finally, Section 202 of the Planning Act (Northern Ireland) 2011 empowers the Department for Communities to compulsorily acquire a Listed Building where reasonable steps are not being taken to properly preserve it. Any lands adjoining the Listed Building can additionally be acquired for purposes of preserving the building or its amenities, giving access to the building or assisting with its management. However, the commencement of proceedings must be signalled at least two months previously by the serving of a repairs notice on the owner specifying what is reasonably required to be carried out to preserve the building. The repair notice can be withdrawn at any time by the Department.

PHOTOGRAPH 3.7 Corrstown House Listed Building, Portrush. *Source*: Photograph by David Houston, 2015; NIEA evaluation available at www.doeni.gov.uk/niea/content-databases-buildlist.htm, accessed 21 October 2015.

A landmark case in Northern Ireland, which exemplifies the consequences of direct action by the former Department of the Environment to secure building preservation, is Sion Stables (Photograph 3.8). This Grade B+ listed stables block is situated within Sion Mills Conservation Area, designated in 1977, and makes a significant contribution to the local streetscape. While the owner had been given encouragement by the Department of the Environment over a lengthy period from 1984 to undertake restoration work, the lack of a successful outcome prompted the serving in May 2003 of a repairs notice under Article 109 of the Planning (NI) Order 1991. In September 2003, the Department issued a notice of intention to vest. However, in the face of subsequent unauthorised dismantling works, the Department then moved to issue an urgent works notice in June 2008, followed immediately by the publication of a notice making the Vesting Order and a follow-on injunction in the High Court restraining access to the site by the respondent. (In a subsequent appeal to the Planning Appeals Commission (2008/E020) regarding the recovery of costs, the Commissioner noted in passing that there would have been no need for an urgent works notice had the Department followed up its 2003 repairs notice by compulsorily acquiring the building. However, it was the unauthorised works that the appellant had carried out to the building in May and June 2008 that had precipitated the need for the urgent works notice along with a temporary stop notice.) In its determination in November 2008 in an appeal against the costs set out in the urgent works notice, the Planning Appeals Commission confirmed the amount recoverable from the appellant as £50,000 (excluding value-added tax) and deemed that this would not cause hardship. Thereafter, the Stables block was passed to HEARTH (an established building preservation trust), which, in partnership with Sion Mills Buildings Preservation Trust, completed a restoration project in 2014 that now offers a visitor education facility, museum, craft shop and restaurant.

PHOTOGRAPH 3.8 Sion Stables, Sion Mills. *Source*: Photograph by authors, 2016.

Scheduled Monuments

The Northern Ireland Buildings Database includes not only Listed Buildings but also details of buildings and structures protected by 'scheduling' (for example, the industrial archaeology bridge cranes and the rails on which they run in the dry dock at the former Harland and Wolff shipyard in Belfast; Photograph 3.9). This process comes under the Historic Monuments and Archaeological Objects (Northern Ireland) Order 1995 whereby the Department for Communities has a duty to compile and maintain a schedule of monuments. Monuments on the schedule have statutory protection, and inclusion is at the Department's discretion albeit drawing upon consultation with the Historic Buildings Council. Occupied dwellings and churches still in ecclesiastical use cannot be scheduled. Interestingly the criteria for scheduling partly mirror those used for Listed Buildings and include period, rarity, survival and group value. The schedule contains almost 2,000 Scheduled Historic Monuments in Northern Ireland that have been compiled over the period since the introduction of the Ancient Monuments (Northern Ireland) Act 1926 (Northern Ireland Environment Agency 2015). Once a monument has been scheduled, 'Scheduled Monument Consent' is required from the Department for Communities – Historic Environment Division before any works are carried out which would have the effect of demolishing, destroying, damaging, removing, repairing, altering, adding to, disturbing, flooding or tipping on the monument and its surrounding red-lined zone of legal protection. Consent can be granted only for detailed proposals and, unlike planning permission, there is no provision for the granting of outline consent. Where both Scheduled Monument Consent and planning permission are required, Scheduled Monument Consent must be determined first.

Scheduled Monuments offer considerable scope for wider stakeholder involvement in their management, promotion and protection and which is well illustrated by the case of Derry City Walls, dating from the 17th century and comprising one of the most complete

PHOTOGRAPH 3.9 Harland and Wolff bridge crane, Belfast. *Source*: Authors, 2012; NIEA comments available at www.doeni.gov.uk/niea/content-databasesbuildlist.htm, accessed 21 October 2015.

walled town enclosures within Ireland and the United Kingdom. The walls comprise the largest upstanding monument in state guardianship in Northern Ireland over the period from 1955, and they are a Scheduled Historic Monument under the provisions of the Historic Monuments and Archaeological Objects (Northern Ireland) Order 1995. The current curatorship of the monument is set out in a suite of documents comprising the *Derry City Walls Conservation Plan* (Department of the Environment 2015c) and the *Derry Walls Management Plan* (Department for Communities 2021) overseen by an interorganisational Derry Walls Management Group established in 2009 and a dedicated day-to-day City Walls' manager post created in 2019. Derry City Walls have tourism flagship significance attracting almost 450,000 visitors in 2019. Accordingly, the planning interventions go well beyond conservation and regulatory control to include improvements to historical interpretation, signage, physical access, tree and public realm maintenance, security and events management.

Landscape amenity

The landscape amenity of Northern Ireland has long attracted the attention of those with an interest in protecting its special environmental qualities. Sterrett (2003:120) notes, for example, that as far back as 1947, the then Planning Advisory Board had praised the unique quality of the Ulster countryside with its 'restfulness and unspoiled charm' but that in some instances there was 'irreparable damage by thoughtless and careless development'. It called for legislative action to protect certain areas of particular beauty and interest along with the designation of National Parks as appropriate. However, 'its proposals were greeted with almost total silence by the Government of the day' (Greer and Jess 1987:113) and only with the passage of the Amenity Lands Act (Northern Ireland) 1965 was this legislative base provided. Subsequently, proposals for National Parks were advanced in the late 1960s for parts of Fermanagh, south Armagh and the Sperrins, but, in the face of political opposition, these were not followed through. Instead, the designation of eight Areas of Outstanding Natural Beauty (AONB) was confirmed in the belief that strict development control would ensure a degree of landscape protection until National Parks were finally designated (Buchanan 1982). The enactment of the Nature Conservation and Amenity Lands (Northern Ireland) Order 1985 allowed for the redesignation of AONB (Figure 3.2) and conjoined this with landscape management capabilities for environmental conservation and recreation access that could go well beyond the association of designation with planning control. In short, neither Areas of Outstanding Natural Beauty nor National Parks may be designated under planning legislation.

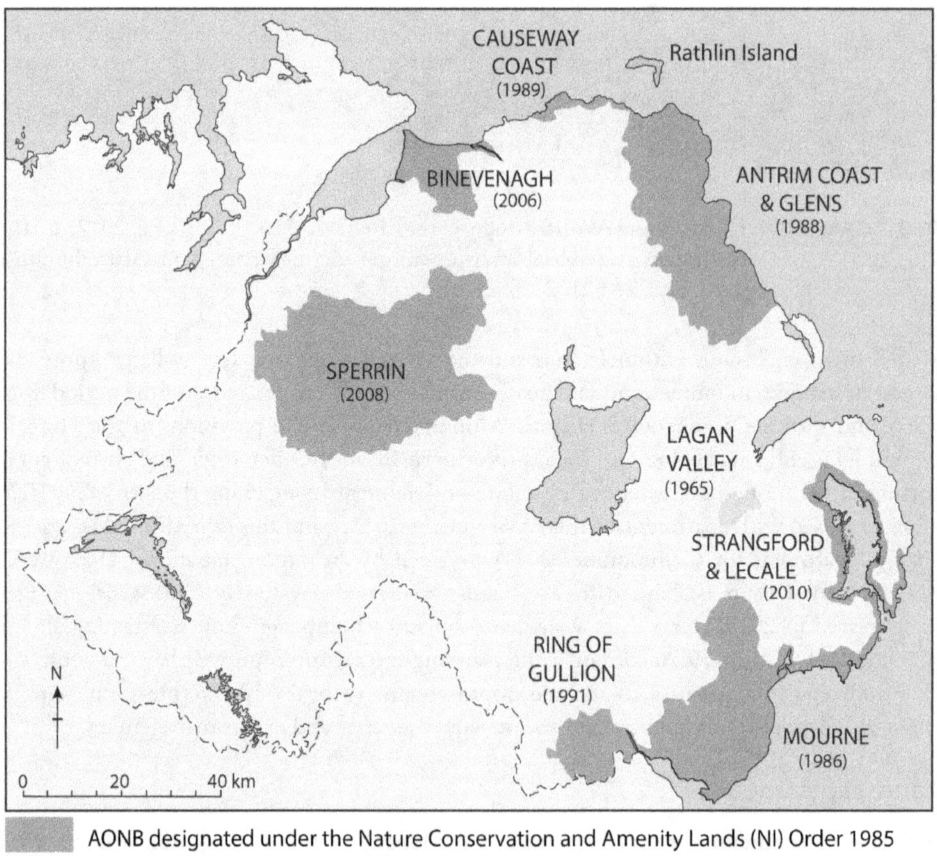

AONB designated under the Nature Conservation and Amenity Lands (NI) Order 1985

AONB designated under the Amenity Lands (NI) Act 1965

FIGURE 3.2 Areas of Outstanding Natural Beauty in Northern Ireland

Nonetheless, there is regional planning guidance on development management within the prized landscapes of Northern Ireland. The *Regional Development Strategy 2035* (Department for Regional Development 2010) identifies 'Strategic Natural Resource' areas which overlap in the main with designated and previously proposed AONB. It calls for the protection, conservation and enhancement of special landscape areas from inappropriate development and also reiterates giving consideration to the establishment of one or more National Parks in order to conserve and enhance the natural, built and cultural heritage of areas of outstanding landscape value while promoting the social and economic development of the communities they support (ibid:53). The *Strategic Planning Policy Statement for Northern Ireland* (Department of the Environment 2015a: 83) adds the following:

> Development proposals in AONB must be sensitive to the distinctive special character of the area and the quality of their landscape, heritage and wildlife, and be in the accordance with relevant plan policies. In assessing proposals, including cumulative impacts in such areas, account will also be taken of the Landscape Character Assessments and any other relevant guidance including AONB Management Plans and local design guides.

Essentially the way forward now lies with District Councils making policy provisions within their Local Development Plans on how best to protect landscape amenity. This will certainly acknowledge the existence of already designated AONB but will complement this by adopting a locally sensitive nomenclature to achieve landscape amenity objectives. So, for example, the Newry, Mourne and Down District Council (2018) *Local Development Plan 2030: Preferred Options Paper* mentions its three AONB covering some 55 percent of the district. It gives particular attention to the issue of 'Sensitive Upland Landscapes' and advances a preferred option of extending existing 'Special Countryside Areas' within these uplands in order to protect their high scenic quality 'from development in all but exceptional circumstances' (169). The deeply controversial designation of a National Park for the Mournes is not alluded to (for further discussion, see: Bell and Stockdale 2015). Fermanagh and Omagh District Council (2018) in its *Local Development Plan 2030: Draft Plan Strategy* explicates policy to 'protect against any development which could be likely to adversely affect the scenic quality of the AONB and to protect against the piecemeal erosion of the distinctive character of the Sperrins AONB' (148). It adds to this by the designation of 'Special Countryside Areas' where 'inappropriate or unnecessary development could erode their unique quality and amenity value' (149), along with proposed 'Areas of High Scenic Value' within which 'proposals for development will only be permitted where it is demonstrated that they would not adversely affect or change either the quality or character of the landscape or the settings of the loughs' (150).

Tree Preservation Orders

Trees can make a significant contribution to the visual amenity of the built environment and countryside landscape by giving texture, colour, enclosure and screening and supporting biodiversity. Accordingly, it is not surprising that strong emotions can be expressed when there is significant felling of mature trees and woodlands (see Belfast Telegraph 2012). In 2021, for example, the proposed felling of trees at the Exploris Aquarium in Portaferry to facilitate an extension of car parking spawned the creation of a 'Stop the Chop' campaign group that generated considerable media attention. The proceedings of a Northern Ireland Assembly

Environment Committee meeting in 2002 to review the provisions of the then Planning (Amendment) Bill encapsulate the depth of public concern:

> Mrs Nelis: I understand the difficulties of protecting trees given prior planning permission, but how do you propose to do it? We all have experience of developers clearing sites and cutting down trees, resulting in disruption to and protests from people. How can we protect the environment?
>
> The Chairperson: Many trees have been destroyed, and there is much dissatisfaction about developers clearing away good mature trees for financial gain and nothing else. No consideration is shown for the environment. Unfortunately, trees are destroyed and taken away before the Department arrives. That happens often, and usually at around 4 o'clock in the morning. That is the 'sneaky beaky' way in which these boys do it, and it must be stopped.
>
> *(Northern Ireland Assembly 2002a)*

The subsequent Planning (Amendment) (Northern Ireland) Order 2003 revised the earlier 1991 legislation in this sphere by introducing a new power for the Department to enforce the duty to replace trees subject to a Tree Preservation Order (TPO), setting out specific grounds of appeal against enforcement notices in relation to trees, giving power to the Department to enter onto land to replant trees subject to a TPO, and to recover any costs incurred as a civil debt and, finally, a new duty on an owner to replace trees that are removed in a Conservation Area. Such provisions, however, fall short of what was being pressed for by Assembly members at that time and in its defence the Department pointed to operational challenges regarding wholesale tree protection:

> Mr Small: We have concerns about blanket protection to prevent the removal of any tree without prior approval from the Department. That would have serious operational difficulties, as well as implications for house owners who simply wanted to trim or cut trees in their gardens. Our intention is to be more proactive in the use of Tree Preservation Orders and in how we use the powers given to us by them. We are conscious of the recent occasions on which developers have gone into sites, cleared them and created development sites. The Minister is aware of the problem and is concerned about it. We propose to move more quickly with Tree Preservation Orders and to work more proactively to identify where they might offer better protection.
>
> *(Northern Ireland Assembly 2002a)*

The current legislative procedures attached to Tree Preservation Orders are set out in The Planning (Trees) Regulations (Northern Ireland) 2015. The planning system relies on the instrument of TPOs to help regulate change and within Section 121 of the Planning Act (Northern Ireland) 2011, both councils and the Department have a duty to ensure that adequate provision is made for the preservation or planting of trees when granting planning permission for development; TPOs can be used to give effect to conditions imposed on planning permission. More generally, TPOs may be used to protect trees, groups of trees or woodlands which add to the character and appearance of an area, which have special value in terms of amenity, history or rarity, and which may or may not be under threat. The Department for Infrastructure and councils may make TPOs in line with these considerations. Within Conservation Areas, trees are protected as if they had TPOs imposed. On the other hand, Areas of Townscape Character designated by the development plan process do not currently provide statutory protection of trees. If a tree is protected by a TPO, the council's consent is required before it may be felled or pruned in any way. Under Schedule 4, Part 2 of The Planning (Trees) Regulations (Northern Ireland) 2015, where an application is made to a council for consent to cut down, uproot top or lop trees under a TPO and if that consent is refused or granted subject to conditions, then the applicant may appeal to the Planning Appeals Commission. Guidance issued by the Department of the Environment (2011a) and carried forward by the Department for Infrastructure in October 2019 until such times as councils have an adopted Plan Strategy, indicates that consent is not required for the removal of dead or dangerous trees. Replanting of trees of an appropriate size and species in the same location as soon as is reasonable can be required. Moreover, the owner must ensure that he has proof that the tree is dead or dangerous and it is recommended that the planning authority is made aware of the proposed works prior to them being carried out. Anyone who is unsure as to the condition of a tree is advised to obtain the advice of a qualified arboriculturist (ibid.:6). The Department for Infrastructure has a power to amend or revoke a TPO in consultation with a council, but should a council wish to amend or revoke a TPO it must make a request that the Department for Infrastructure does so on its behalf. This requirement will be subject to review (Department for Infrastructure 2022: 65).

To move speedily, in a situation where, for example, a tree or woodland may be under threat, the legislation provides for the making of a provisional TPO to take effect immediately on the date specified (Section 123). The destruction of trees subject to a TPO is an offence punishable by a fine on conviction. There is a duty by the landowner to replace any removed tree subject to a TPO with a replacement tree as soon as possible and, in the case of woodlands, it is deemed sufficient to replace the trees removed, uprooted or destroyed by planting the same number of trees on or near the affected land or on such other land as may be agreed between the council and landowner, and in such places as may be designated by the council (Section 125(3)). The legislation would seem, accordingly, to introduce a degree of locational flexibility to woodland replanting requirements. Previously in 2002, in an interim response by the Woodland Trust to the former Planning (Amendment) Bill, it was argued that this degree of flexibility was less than desirable:

> This appears to mean that if an area of woodland protected by a TPO was cleared, there would be no absolute requirement to replant on the cleared area. If this area remained unplanted, it would then presumably be unprotected from development, since a TPO in its current form can only be made to protect the *trees* which make up a woodland, rather than the land or other plants which make up a woodland area. The incentive would thus remain for a developer to clear protected trees from a site and risk a fine, with the possibility remaining of building on the cleared land ... For the reasons outlined above, we would strongly suggest that all TPO-protected trees be afforded the same level of protection and the same replacement requirements (ie replanting *at the same place*), and that the exceptions for trees forming part of a woodland ... be removed.
> *(Northern Ireland Assembly 2002b)*

There are few reported cases determined by the Planning Appeals Commission on this matter, but an interesting decision relates to the challenge to a tree replacement condition attached to a consent by Belfast City Council in 2015, on grounds of health and safety, for the felling of three trees covered by a TPO (2015/A0123). The TPO had been imposed on the land in 1996 and in 2005 planning permission had been issued for a housing scheme subject to a condition requiring the preservation of the trees covered by the TPO. The completed development, however, encroached as approved into the root protection area of the three protected trees in question, impacting negatively on their stability. The Planning Appeals Commission ruled that the Department, in granting planning permission for the housing development which encroached so significantly into the root protection area, had undermined the purpose of the TPO. Because the approved development had rendered the protected trees unsafe, the presiding Commissioner considered that it would be unfair to penalise the appellant by requiring the planting of three replacement trees. Accordingly, it was determined that the replacement condition attached to the consent by Belfast City Council to fell the trees should be dispensed with and the appeal was granted.

Conclusion

This chapter has explored the regulatory context for advertisements, Conservation Areas, Areas of Townscape and Village Character, Listed Buildings, Scheduled Monuments, landscape amenity, trees and woodlands and underlines the central role that amenity protection plays in the Northern Ireland planning system. The key attribute here is retaining and enhancing visual quality in the public interest and, in that regard, it is clear that planning practice is required to mediate between the perceived advantages for landowners regarding land use change and that more abstract but potent aspiration which gives planning intervention its quintessential legitimacy. In this regard, the special categories of development management examined in this chapter are vital for the continued making and reshaping of quality places in Northern Ireland.

Legislation

Amenity Lands Act (Northern Ireland) 1965
Ancient Monuments (Northern Ireland) Act 1926
Civic Amenities Act 1967
Historic Monuments and Archaeological Objects (Northern Ireland) Order 1995
Planning Act (Northern Ireland) 2011
Planning (Amendment) (Northern Ireland) Order 1978
The Nature Conservation and Amenity Lands (Northern Ireland) Order 1985
The Planning (Amendment) (Northern Ireland) Order 2003
The Planning (Conservation Areas) (Consultation) Regulations (Northern Ireland) 2015
The Planning (Conservation Areas) (Demolition) Regulations (Northern Ireland) 2015
The Planning (Control of Advertisements) (Amendment) Regulations (Northern Ireland) 2009
The Planning (Control of Advertisements) Regulations (Northern Ireland) 2015
The Planning (Demolition – Description of Buildings) Direction 2015
The Planning (General Permitted Development) Order (Northern Ireland) 2015
The Planning (Listed Buildings) Regulations (Northern Ireland) 2015
The Planning (Northern Ireland) Order 1972
The Planning (Northern Ireland) Order 1991
The Planning (Trees) Regulations (Northern Ireland) 2015

Cases cited

BP Amoco PLC and John Kelly Ltd and Glenshane Tourist Services Limited (2001 NICA 3)

Calderdale Metropolitan Borough Council v Windy Bank Dairy Farm Ltd and another (2010 EWHC 2929 Admin)

In the Matter of an Application by Barry Gilligan, Valerie Allen, Anthony Cox, Maeve Fee, Henry Glover and Malone Park Residents' Association for Judicial Review and In the Matter of a Decision by the Planning Service of the Department of the Environment for Northern Ireland Dated 16 January 2002 (2002 NIQB 53)

In the Matter of an Application by Barry Gilligan and Others for Judicial Review (2003 NICA 10)

In the Matter of an Application by the Ulster Architectural Heritage Society for Judicial Review (2014 NIQB 21)

Planning Appeals Commission (2002) *On an appeal to the Commission under Article 51 of the Planning (Northern Ireland) Order 1991 by R&A Developments against the non- determination of an application for Conservation Area Consent for the demolition and removal of buildings at 27–39 Manor Street and 1–5 Saltworks Street and Old Pier Inn, Railway Street, Donaghadee.* (2001/A114)

Planning Appeals Commission (2004) *Appeal by Mrs D Horsman against the refusal of Listed Building Consent for the removal of a statue known as Francis Jack, the second Earl of Kilmorey, Mourne Park House, Mourne Park, Kilkeel, Townland of Ballyrogan, Co Down.* (2004/A215)

Planning Appeals Commission (2004) *Appeals by Mr T Turkington against the non- determination of applications for Listed Building Consent to demolish a Listed Building (Appeal ref. 2004/A369) and a full planning application (Appeal Ref. 2004/A370) for a replacement dwelling at 108 Strand Road, Portstewart.* (2012/A369 and 2012/A370)

Planning Appeals Commission (2004) *Appeal by Mr S Robinson against the non- determination of applications for full planning permission (2004/A244) and Listed Building Consent (2004/A245) for a dwelling (amendment to the design and construction of a bay window) at 1a Clanbrassil Terrace, Clanbrassil Road, Holywood.* (2004/A245)

Planning Appeals Commission (2008) *Appeal by Birchsilver Ltd, T/A Café Vaudeville against the refusal of Listed Building Consent for 2 retractable awnings and timber housing boxes at ground floor level and 7 individual window canopies at first floor level, 27–39 Arthur Street, Belfast.* (2008/A0113)

Planning Appeals Commission (2008) *Appeal by Reverend John D F Withy in respect of an urgent works notice relating to the stable block at Zion House, 120 Melmount Road, Sion Mills, County Tyrone.* (2008/E020)

Planning Appeals Commission (2011) *Appeal by Mr & Mrs Storey against the refusal of full planning permission for a traditionally styled balcony to front elevation, 35 Kerr Street, Portrush.* (2011/A0028)

Planning Appeals Commission (2012) *Appeal by Mr Eddie Clarke against the refusal of Listed Building Consent for conversion of adjoining buildings to 3 tourist units and restoration of a listed pig crew, 40 Point Road, Killough.* (2012/A0237)

Planning Appeals Commission (2012) *Appeal by Herbel Restaurants Ltd against the refusal of Listed Building Consent for 2 No. shops signs and 1 No. projecting sign, Donegall Place & 1–5 Fountain Lane, Belfast.* (2012/A0167)

Planning Appeals Commission (2012) *Appeal by Mr Martin McGovern against the refusal of Listed Building Consent for white uPVC double glazed windows, 26 Darling Street, Enniskillen.* (2012/A0135)

Planning Appeals Commission (2014) *Appeal by Mr C Haughian against the refusal of full planning permission for 4 self contained apartments, 12 Great George's Street, Warrenpoint, Newry.* (2014/A0140)

Planning Appeals Commission (2015) *Appeal by Odyssey Trust Company against the refusal of consent for the retention of two banner type advertisements and fixings at Odyssey Arena, 2 Queen's Quay, Belfast.* (2015/A0017)

Planning Appeals Commission (2015) *Conditional grant of consent dated 4/8/15 to fell 3 trees.* (2015/A0123)

Planning Appeals Commission (2019) *Appeal by Clear Channel NI Ltd against the refusal of consent to display an advertisement comprising the replacement of existing 2x48 sheet standard display boards with 2x48 sheet standard light boxes at 2-10 Bridge End, Belfast.* (2019/A0067)

References

Advertising Association. 2021. *Strong Rebound from Pandemic Losses Forecast for UK Ad Market but Full Recovery Will Stretch into 2022.* Available at https://adassoc.org.uk/?s=value+to+economy.

Belfast Telegraph. 2009. *Family Told to Pay for Bulldozer Damage.* 20 February.

Belfast Telegraph. 2012. *121 Alleged Breaches of Tree Protection Orders … but Just One Prosecution.* 28 March.

Bell, J. and A. Stockdale. 2015. 'Comparing the Political Imperative to Designate New National Parks in a Devolved Scotland and Northern Ireland', in F. Gaffikin, S. McKay, M. Murray, B. Murtagh and K. Sterrett (eds) *A Sense of Place: Multidisciplinary Essays in Honour of Malachy McEldowney.* Belfast: Queen's University Belfast, pp. 239–54.

Blackhall, J.C. 2005. *Planning Law and Practice.* London: Cavendish Publishing Limited.

Buchanan, R.H. 1982. 'Landscape', in J.G. Cruickshank and D.N. Wilcock (eds) *Northern Ireland Environment and Natural Resources.* Belfast: Queen's University Belfast, pp. 265–89.

Commission on Flags, Identity, Culture and Tradition. 2021. *Final Report.* Belfast: The Executive Office (available at https://www.executiveoffice-ni.gov.uk/sites/default/files/publications/execoffice/commission-on-fict-final-report.pdf).

Cullingworth, B., V. Nadin, T. Hart, S. Davoudi, J. Pendlebury, G. Vigar, D. Webb and T. Townshend. 2014. *Town and Country Planning in the UK.* Abingdon: Routledge.

Department for Communities - Historic Environment Division. 2018. *Guidance on Setting and the Historic Environment.* Belfast: Department for Communities.

Department for Communities – Historic Environment Division. 2019a. *Criteria for the Scheduling of Historic Monuments and the Listing of Buildings of Special Architectural or Historic Interest, with Associated Procedures.* Belfast: Department for Communities.

Department for Communities – Historic Environment Division. 2019b. *Services and Standards Framework.* Belfast: Department for Communities.

Department for Communities. 2021. *The Derry City Walls Management Plan.* Belfast: Department for Communities.

Department for Communities. 2020. *NI Heritage Statistics as Recorded on 31 March 2019.* Belfast: Department for Communities.

Department for Infrastructure. 2017. *Historic Environment: Development Management Practice Note 5.* Belfast: Department for Infrastructure.

Department for Infrastructure. 2022. *Review of the Implementation of the Planning Act (Northern Ireland) Act 2011: Report.* Belfast: Department for Infrastructure.

Department for Regional Development. 2010. *Regional Development Strategy 2035.* Belfast: Department for Regional Development.

Department of the Environment. 1990. *Belfast Urban Area Plan 2001.* Belfast: HMSO.

Department of the Environment. 2011. *Tree Preservation Orders - A Guide to Protected Trees.* Belfast: Department of the Environment.

Department of the Environment. 2015a. *A Strategic Planning Policy Statement for Northern Ireland, Planning for Sustainable Development*. Belfast: Department of the Environment.

Department of the Environment. 2015b. *Practice Note 07: Control of Advertisements*. Belfast: Department of the Environment.

Department of the Environment. 2015c. *Derry City Walls Conservation Plan*. Belfast: Department of the Environment.

Economist. 2011. 'Billboard Boom: The Future of Out-of-Home Advertising is Rosy and Digital', *The Economist*, 20th April.

Fermanagh and Omagh District Council. 2018. *Local Development Plan 2030: Draft Plan Strategy*. Enniskillen: Fermanagh and Omagh District Council.

Galway, N. and M. McEldowney. 2006. 'Place and Special Places: Innovations in Conservation Practice in Northern Ireland', *Planning Theory and Practice* 7(4): 397–420.

Gomez, J.E.A. 2013. 'The Billboardization of Metro Manila', *International Journal of Urban and Regional Research* 37(1): 186–214.

Greer, J. and P. Jess. 1987. 'Town and Country Planning', in R.H. Buchanan and B.M. Walker (eds) *Province, City and People*. Antrim: Greystone Books, pp. 101–24.

Hendry, J. 1977. 'Conservation in Northern Ireland', *Town Planning Review* 48(4): 373–88.

Hendry, J. 1993. 'Conservation Areas in Northern Ireland: An Alternative Approach', *Town Planning Review* 64(4): 415–34.

Historic Environment Division. 2021. *Conservation Principles: Guidance for the Sustainable Management of the Historic Environment in Northern Ireland*. Belfast: Department for Communities.

Lane, W.R., K.W. King and T. Reichert. 2011. *Kleppner's Advertising Procedure*. Upper Saddle River, NJ: Pearson.

Latimer, K. 2005. 'Preserving Character in the Towns and Villages of Northern Ireland', in M. McEldowney, M. Murray, B. Murtagh and K. Sterrett (eds) *Planning in Ireland and Beyond: Multidisciplinary Essays in Honour of John V. Greer*. Belfast: Queen's University Belfast, pp. 157–68.

Latimer, K. 2015. 'Caring for the Past or Past Caring? Challenges for the Voluntary Sector in Conservation', in F. Gaffikin, S. McKay, M. Murray, B. Murtagh and K. Sterrett (eds) *A Sense of Place: Multidisciplinary Essays in Honour of Malachy McEldowney*. Belfast: Queen's University Belfast, pp. 275–84.

Lewis, M. 1996. 'Brand Packaging', in D. Cowley (ed) *Understanding Brands*. London: Kogan Page Ltd, pp. 135–64.

Mourne Heritage Trust. 2004. *Traditional Buildings in Ireland: Home Owners Handbook. Featuring the Mourne Homesteads Experience*. Newcastle: Mourne Heritage Trust.

Mourne Heritage Trust and the Ulster Architectural Heritage Society. 2002. *Jewels in the Crown or Albatrosses?* The Proceedings of a Conference on Reinventing Northern Ireland's Conservation Areas, Jointly Organised by Mourne Heritage Trust and the Ulster Architectural Heritage Society, Newcastle, 7 November 2001.

Mynors, C. 2009. *The Control of Outdoor Advertising and Graffiti*. Crayford: Shaw and Sons Ltd.

Newry, Mourne and Down District Council. 2018. *Local Development Plan 2030: Preferred Options Paper*. Newry: Newry, Mourne and Down District Council.

Northern Ireland Assembly. 2002a. *Committee for the Environment Minutes of Proceedings, Minutes of Evidence, Written Evidence, Submissions and Associated Commentary (Up to and Including the Committee's Meeting on 10 October 2002)*. Report 03/02. Belfast: Northern Ireland Assembly.

Northern Ireland Assembly. 2002b. *Written Submissions to the Committee*. Report 03/02. Appendix 2, Number 337. Belfast: Northern Ireland Assembly.

Northern Ireland Audit Office. 2011. *Safeguarding Northern Ireland's Listed Buildings*. Belfast: Northern Ireland Audit Office.

Northern Ireland Environment Agency. 2015. *Historic Monuments of Northern Ireland Scheduled Historic Monuments 1st April 2015*. Belfast: Northern Ireland Environment Agency.

Northern Ireland Human Rights Commission. 2011. *Flag Flying: Briefing Paper on Human Rights Compliance and Commission Policy*. Belfast: Northern Ireland Human Rights Commission.

Planning Portal for England and Wales. 2021. *Glossary*. Available at https://www.planningportal.co.uk/directory_record/101/amenity.

PML Group NI. 2015. *Northern Ireland Planning Guide, 2015*. Belfast: PML Group NI.

Police Service of Northern Ireland, et al. 2005. *Joint Protocol in Relation to the Display of Flags in Public Areas*. Belfast: Police Service of Northern Ireland.

Public Accounts Committee. 2011. *Report on Safeguarding Northern Ireland's Listed Buildings: Together with the Minutes of Proceedings of the Committee Relating to the Report and the Minutes of Evidence*. Belfast: Northern Ireland Assembly.

Sterrett, K. 2003. 'The Countryside Aesthetic and House Design in Northern Ireland', in J. Greer and M. Murray (eds) *Rural Planning and Development in Northern Ireland*. Dublin: Institute of Public Administration, pp. 117–44.

The National Trust. 2004. *A Sense of Place: Planning for the Future in Northern Ireland. Summary and Recommendations of the Northern Ireland Planning Commission*. Saintfield: The National Trust.

The Planning Service. 1998. *Planning Policy Statement 1: General Principles*. Belfast: The Planning Service.

The Planning Service. 1999. *Planning Policy Statement 6: Planning, Archaeology and the Built Environment*. Belfast: The Planning Service.

The Planning Service. 2004. *Belfast Metropolitan Area Plan 2015 (BMAP) Draft Plan November 2004*. Belfast: The Planning Service.

The Planning Service. 2005. *Addendum to Planning Policy Statement 6: Areas of Townscape Character*. Belfast: The Planning Service.

The Planning Service. 2006. *Planning Policy Statement 17: Control of Outdoor Advertisements – Summary of Consultation Responses*. Belfast: The Planning Service.

The Planning Service. 2009a. *Proposed Amendments to the Planning (Control of Advertisements) Regulations (Northern Ireland) 1992 – Analysis of Consultation Responses*. Belfast: The Planning Service.

The Planning Service. 2009b. *Ards and Down Area Plan 2015*. Belfast: The Planning Service.

Ulster Architectural Heritage Society. 2008. *Built Heritage at Risk – Northern Ireland (Volume 8)*. Belfast: Ulster Architectural Heritage Society.

4

CONTINUITY AND CHANGE IN DEVELOPMENT PLANS

Stephen McKay

Introduction

Although most people's experience of planning tends to be via development management, a key component of the system is the development plan. Indeed, it is arguably the principal mechanism for the delivery of the planning function. The land use plan, covering all or part of the territorial area of a local authority, has long been an integral part of the statutory town and country planning framework in each of the jurisdictions of the United Kingdom. Plans can translate policy into land-based allocations, coordinate public and private sector investment and indicate how development is to be regulated. In effect, they provide a rationale for demonstrating that public sector decisions are not arbitrary and they constitute a tool for efficient management (Healey 1993). Key considerations include international, national and regional policies with social, economic and environmental implications. Indeed, the importance of coherent spatial guidance that crosses scale was established in the case of *Blythe Valley District Council v Persimmon Homes (North East) Ltd.* (2008 ECWA Civ 861) where part of a development plan was quashed by the Court of Appeal because of a disconnect with national policy. Development plans in the United Kingdom are not prescriptive in the sense that development which conforms to the plan is lawful, whilst development that does not is deemed unlawful. Rather, they are quasi-legal policy documents that are not fully legally binding. Consequently, they are indicative in the sense that they do not confer development rights (i.e., allocate planning permission) but are material in the determination of planning applications.

This chapter provides insight into the nature and content of development plans in Northern Ireland. In this context, it will consider the evolving legislative, institutional and policy framework and give particular attention to the changes which have occurred since the devolution of statutory planning responsibilities to local government in 2015. The chapter examines the rubric of new-generation development plans (Local Development Plans) and their preparation procedures from inception to adoption.

DOI: 10.4324/9781003218319-4

Contextual background

The launching of a development plan is seldom a media event, but back in March 1972 the prime-time BBC Northern Ireland news bulletin, *Scene Around Six*, featured the publication of the *Mid Down Area Planning Statement.* A planner from Down County Council is interviewed in Hillsborough and explains the ambitions of the plan in terms of new housing, social facilities and recreation open space, road improvements and statutory Conservation Area designations. The public are to be offered a 'say' in the content of the final plan and are invited to a series of two-day exhibitions and a 'talk' in each of the main settlements within the plan area. The colour footage deploys a combination of green fields, streets with traffic and historic townscapes to project a stage on which development changes will occur over the following fifteen years. The value of this short, three minutes retrospect (BBC Rewind – A Portal to the Past) lies in its explication of a fundamental rationale for planning intervention – the imagination of powerful long- and short-term possibilities. Development plans, as set out in the de rigueur reading for planners at that time, *A Manual on Form and Content* (Ministry of Housing and Local Government, Welsh Office 1970), can be summarised as follows:

> the aims of the plan should be designed to satisfy social and economic aspirations as far as possible through the creation of an efficient physical structure, and a good environment; the policies should be set out to provide a framework for the continuous process of making planning decisions, and the proposals should be expressed as firm statements of intended action. The development plan is therefore essential to the working of the planning system. (5)

During the intervening years to the present the procedures and content attached to plan-making have evolved considerably in line with new regulatory and organisational imperatives, but the spirit and purpose of development plans, as outlined in the 'red book' development plans manual, have largely remained constant.

The Planning (Northern Ireland) Order 1972, at the time of its ratification, provided a much-needed code of new planning law for Northern Ireland, bringing it broadly into line with that which applied in the rest of the United Kingdom (House of Lords 1972). The Town and Country Planning Service was established in Northern Ireland in 1973 when the responsibilities of the local planning authorities briefly passed to the Ministry of Development and then to the Department of Housing, Local Government and Planning before being incorporated into the Department of the Environment for Northern Ireland. The Department's statutory planning functions were outlined in The Planning (Northern Ireland) Order 1991 and associated subordinate legislation. The Town and Country Planning Service subsequently became an Executive Agency (The Planning Service) of the Department of the Environment for Northern Ireland on 1 April 1996 and, following the devolution of authority to the Northern Ireland Executive in 1998, the Department of the Environment for Northern Ireland was renamed the Department of the Environment. The Planning Service remained an Executive Agency within the Department of the Environment until 2015, at which time most planning functions were restored to local government. In 2016 the Department of the Environment was dissolved and high-level planning responsibilities were situated in a newly established Department for Infrastructure, which previously, in the main, had been the Department for Regional Development. Across this lengthy period from 1972 land use planning in Northern Ireland has operated through development plans, development control and enforcement. These activities have been informed by regional planning frameworks, development plans, planning policy and planning guidance.

Regional planning frameworks

In line with the bedding-down of the post-1972 planning system, the Department of the Environment for Northern Ireland published in 1977 the *Regional Physical Development Strategy for Northern Ireland 1975–1995*. Its principal objective was to create 'a spatial pattern of population distribution and investment … thought to be the most likely to give the best results for the economic development and social wellbeing of Northern Ireland as whole' (cited by Goodyear and Eastwood 1979: 325). Its concerns revolved around tackling inner-city decline in Belfast and rural deprivation across the region by concentrating public sector investment in a suite of twenty-three growth centres (in reality, the main town of each district council area outside the Belfast Urban Area). The strategy was specifically intended to provide a framework for the follow-on preparation of development plans. The 1998 Belfast/Good Friday Agreement promised a renewed commitment to regional planning by stating that the British Government would make progress on:

> a new regional development strategy for Northern Ireland, for consideration in due course by the Assembly, tackling the problems of a divided society and social cohesion in urban, rural and border areas, protecting and enhancing the environment, producing new approaches to transport issues, strengthening the physical infrastructure of the region, developing the advantages and resources of rural areas and rejuvenating major urban centres.

The outcome was the publication in 2001 of *Shaping Our Future: Regional Development Strategy for Northern Ireland 2025* by the Department for Regional Development. It was subsequently updated in 2012 to cover the period out to 2035. Each document comprises a long-range vision for the physical development of Northern Ireland and has provided a framework within which to locate subsequent development plans by giving guidance on housing growth indicators, along with policies for infrastructure and economic development. In operational terms and within the context of the 2001 document, development plans prepared by the Department of the Environment were initially obliged 'to be consistent with the regional development strategy' (3). The Planning (Amendment) (Northern Ireland) Order 2003 substituted this development plan consistency requirement with a softer 'general conformity with' the regional development strategy. However, with responsibility for the preparation of development plans having now transferred to local councils, the requirement in the updated 2012 document is that 'these must take account of the Regional Development Strategy' (13).

Development plans

A significant function of the former Planning Service was to prepare development plans for each of the twenty-six local government districts in consultation with the district council and the local community. Each development plan was concerned with the location and quantity of development that should be permitted within each district. Three types of development plan could be produced:

- Area Plans: covering a large area such as a whole district council (see, for example, *Fermanagh Area Plan 2007*) or a group of council areas (see, for example, *Banbridge, Newry and Mourne Area Plan 2015*);
- Local Plans: covering an area within a district council such as a town centre (for example, *Ballymoney Town Centre Local Plan 1991–2002*), a single settlement (see for example, *Carryduff Local Area Plan 1988–1993*) or a special amenity area (see, for example, *Lagan Valley Regional Park Local Plan 2005*); and
- Subject Plans: covering a particular issue, such as small settlement development (see, for example, *Limavady District Hamlet Subject Plan 1989–1999*) or houses in multiple occupancy (see, for example, *Houses in Multiple Occupation (HMOs) Subject Plan for Belfast City Council Area 2015*).

The development plans include a written statement as well as maps, diagrams and illustrations. As a transitional measure, plans prepared by the Planning Service of the former Department of the Environment remain operational until district councils adopt new Local Development Plans. Given the considerable age of many of these plans, the weight to be given to their content is likely to be reduced with greater weight being given in decision making to other material considerations, including more recent policy and guidance. In such circumstances, the weight attributed to material considerations remains a matter for the decision-taker (Northern Ireland Audit Office 2022: 24).

Planning policy

In 1993 the Department of the Environment published its first compendium of planning policies, some 117 in all, within *A Planning Strategy for Rural Northern Ireland*. This document sought to reconcile the diverse and conflicting aspirations of different interest groups, abandoned a rigid hierarchy of settlements, promoted the potential of broad-based rural development and promised generous development limits around smaller settlements (Greer and Murray 2003: 11). Over the period to 2021 most of these policies have been superseded by Planning Policy Statements (PPS), which apply to the whole of Northern Ireland and are subject-specific. They deal, for example, with archaeology and built heritage, waste management, transportation and land use, open space, sport and outdoor recreation, telecommunications, natural heritage, quality residential environments, access, movement and parking. Only a very few of the policy provisions in *A Planning Strategy for Rural Northern Ireland* remain operational, for example, in regard to minerals, and it is intended that these remaining policies and extant PPS will continue as retained planning policies under transitional arrangements in the lead up to the adoption of the Plan Strategy component of Local Development Plans. Thereafter they will cease to have effect in those district council areas with adopted Plan Strategies and ultimately will be cancelled when all Plan Strategies across Northern Ireland are adopted. Additionally, in 2015 the Department of the Environment published *The Strategic Planning Policy Statement for Northern Ireland – Planning for Sustainable Development (SPPS)*, which consolidates some twenty separate policy publications into one document and sets out strategic subject planning policy for a wide range of planning matters. It also provides the core planning principles to underpin the delivery of the two-tier planning system and sets the strategic direction for district councils to bring forward detailed operational policies within future Local Development Plans. The document is judged to be in general conformity with the Northern Ireland Regional Development Strategy 2035, and its provisions must be taken into account by district councils when preparing their Local Development Plans.

Planning guidance

A number of other documents have been produced over time to demonstrate how policies operate in practice. Planning guidance is set out, for example, in Development Control Advice Notes (DCAN). DCAN provide more detail about the criteria and technical standards that may be relevant for certain types of development, for example, amusement centres, taxi offices, residential and nursing homes, public houses, restaurants and hot food bars. Amenity considerations loom large. When a district council adopts its Plan Strategy, existing guidance contained in DCAN will cease to have effect in the district of that council and shall not be material from that date, whether any planning application has been received before or after that date. Additionally, a wide range of documents provide supplementary planning guidance related to the design of buildings and places, economic development, heritage conservation, renewable energy, transportation and trees. Those that are regional in focus will be retained unless and until replaced by the Department for Infrastructure. Another suite of important supplementary planning guidance, prepared by the former Department of the Environment, comprises Conservation Area booklets that detail the spatial extent of designations, examine townscape character and offer customised design solutions and development criteria for the built environment.

Local Development Plans in Northern Ireland

The legislation dealing with the preparation of statutory Local Development Plans (LDP) is contained in Part 2 of the Planning Act (Northern Ireland) 2011, whilst the Planning (Local Development Plan) Regulations (Northern Ireland) 2015 and the Planning (Local Development Plan) (Amendment) Regulations (Northern Ireland) 2016 specify the role of the council in the plan preparation process, the Department for Infrastructure (hereinafter referred to as the Department) and any independent examiner; the sequencing which must take place from inception to adoption of the plan; the form and content of the development plan document; and the functions relating to annual monitoring and review. Under this legislation there are differences from the previous provisions which were used in the preparation of development plans by the Department of the Environment. One of the most significant changes is found under s. 6 (2) of the 2011 Act, which specifies that Local Development Plans will be comprised of two documents to be prepared and adopted in sequence:

(a) the Plan Strategy (PS); and
(b) the Local Policies Plan (LPP).

The purpose of the PS is to provide the strategic policy framework for the plan area as a whole across a range of topics whilst taking account of the *Regional Development Strategy 2035* and any policy issued by the Department such as *A Strategic Planning Policy Statement for Northern Ireland 2015* (SPPS) (Department of the Environment 2015a). As noted above, transitional arrangements have been put in place under s. 12 (4) of the 2011 Act until such time as a council adopts its PS. Specifically, local planning authorities will continue to apply retained planning policy in tandem with the SPPS and relevant supplementary and best practice guidance. Subsequent to adoption, existing policy retained under these transitional arrangements shall cease to have effect in the district of that council and shall not be material from that date, whether a planning application has been received before or after that date (ibid. 2015a). In the event of conflict between the SPPS and any policy retained under the transitional arrangements, the provisions of the SPPS take precedence. Additionally, S. 6(3) of the 2011 Act provides clarification where the situation arises that a policy contained in a Local Development Plan conflicts with another policy in that plan. In such cases, the conflict must be resolved in favour of the policy which is contained in the last development plan document to be adopted or, if appropriate, approved. S. 6(4) of the 2011 Act provides for the implementation of a plan-led system whereby, in making any determination under the 2011 Act, regard is to be had to the Local Development Plan and the determination must be made in accordance with the plan unless material considerations indicate otherwise. Specifically, in the determination of any planning application for operational development or use the LDP will be the primary consideration. However, as Lord Guest commented in *Simpson v Edinburgh Corporation* (1961 SLT 17) this did not mean, in his opinion, a requirement to "slavishly adhere" to the plan (cited by Moore et al. 2014). His Lordship went on to state the following:

> In the view and nature of a development plan … I should have been surprised to find an injunction on the planning authority to follow it implicitly, and I do not find anything in the Act to suggest this was intended … The planning authority are to consider all material planning considerations, of which the development plan is one.

With this in mind attention turns now to the LDP preparation process.

The Local Development Plan preparation process

The LDP preparation process comprises four stages. Stage 1 is denoted by a series of activities which include the preparation of a Statement of Community Involvement, the preparation of a programme management Timetable and a Preferred Options Paper. Stage 2 involves the preparation of a Plan Strategy, the Independent Examination of the Plan Strategy and adoption of the Plan Strategy. Stage 3 involves the preparation of a Local Policies Plan, the Independent Examination of the Local Policies Plan and adoption of the Local Policies Plan. There is also Stage 4, whereby there are mandatory requirements to produce an Annual Monitoring Report and to conduct a Review of the Local Development Plan. Each of these stages is now considered in detail, with the narrative drawing upon the content of published departmental practice notes.

Stage 1

Statement of Community Involvement

At the outset of the process the council must prepare a Statement of Community Involvement (SCI). The purpose of the SCI is to outline how a council proposes to engage the community and stakeholders in exercising its planning functions. The SCI should set out a council's policy regarding the involvement of persons who appear to a council to have an interest in the matters relating to development in its area (Department of the Environment 2015b). S. 3 of the 2011 Act states that the council and the Department must attempt to agree the terms of the SCI. It goes on to specify that, where the council and the Department cannot agree the terms of the SCI, the Department may direct that the statement must be in the terms specified in the direction. The council must comply with the direction. In this context, the 2011 Act states that the Department may prescribe:

(a) the procedure in respect of the preparation of the Statement of Community Involvement;
(b) the form and content of the statement;
(c) publicity about the statement;
(d) making the statement available for inspection by the public; and
(e) the manner in which: (i) representations may be made in relation to any matter to be included in the statement; and (ii) those representations are to be considered;
(f) circumstances in which the requirements of the statement need not be complied with.

Specifically, the SCI should explain how both the community and stakeholders will be involved in the preparation of the Local Development Plan whilst specifying the steps that a council will take to facilitate community involvement. It should provide transparency, putting signposts in place so that everyone understands how, where, why, when and with whom public participation will take place in the plan preparation process. Crucially, the SCI should set out the standards to be met by a council in terms of community involvement, building upon the minimum requirements set out in both the Planning (Local Development Plan) Regulations (NI) 2015 and the Planning (Statement of Community Involvement) Regulations (NI) 2015.

Timetabling

S. 7 of the 2011 Act states that a council must prepare and keep under review a Timetable for the preparation and adoption of the council's Local Development Plan. Whilst it must attempt to reach agreement on the terms of the Timetable with the department, where agreement cannot be reached the department may direct that the Timetable must be in the terms specified in the direction (s. 7 (3)). It is incumbent upon the council to comply with the direction (Department of the Environment 2015c). In this context, the department may prescribe the procedure in respect of the preparation of the Timetable; the form and content of the Timetable; the time at which any step in the preparation of the Timetable must be taken; publicity about the Timetable; making the Timetable available for inspection by the public and; circumstances in which the requirements of the Timetable need not be complied with.

Regulations 5 to 8 of the Planning (Local Development Plan) Regulations (Northern Ireland) 2015 set out the requirements for the preparation, form, content, agreement and availability of the Timetable. Regulation 5 specifies that the council must, in the preparation of the Timetable, consult the Planning Appeals Commission and those consultation bodies that the council considers appropriate. The underpinning rationale for this is that it facilitates effective programme management by allocating sufficient time and resources and prioritising work where necessary in order to avoid unforeseen delays in the preparation process. Regulation 6 explains that the Timetable must include indicative dates for each stage of the preparation of the Local Development Plan, including indicative dates for:

(i) publication of the Preferred Options Paper;
(ii) publication of the plan strategy and publication of the Local Policies Plan; and
(iii) adoption of the Plan Strategy and adoption of the Local Policies Plan.

The Timetable also provides for carrying out an appraisal of the sustainability of the Plan Strategy, an appraisal of the sustainability of the Local Policies Plan, and the preparation of a report on the findings of these appraisals. Under Regulation 7 the Timetable must be approved by resolution of the council prior to submission to the department, and then be submitted to the Department for Infrastructure for its agreement. When this has occurred the department must respond within four weeks of receipt, unless it has, before the expiry of that period, notified the council that it requires more time to consider the Timetable. If, before the end of the period specified, the department has failed to respond to the receipt by it of the Timetable, it will be deemed to be agreed at the end of that period. Under Regulation 8, once the Timetable is agreed or deemed to be agreed, the council must make a copy of the agreed Timetable available for inspection during normal office hours at its principal offices and such other places within its district as it considers appropriate. Furthermore, it must give notice by local advertisement that the Timetable is available for inspection, the place and times at which it can be inspected and publish the Timetable on its website.

The Preferred Options Paper

The Preferred Options Paper (POP) marks the end of Stage 1 of the LDP plan preparation process. The POP is the first of three public consultation documents that a council will issue during the LDP process (the others being at later stages: the draft Plan Strategy and the draft Local Policies Plan). The purpose of the POP is to stimulate focused debate on key issues of strategic significance which are likely to influence the direction of future development within a plan area over the course of the lifespan of the plan. It outlines the vision and strategic objectives of the LDP and identifies options that can respond to key issues, including the council's preferred options and their underpinning rationale. As noted above, Regulation 6 (2) of the LDP Regulations requires the Timetable to include an indicative date for the publication of the POP. Although a SCI and Timetable must be agreed with the Department prior to the preparation of the Plan Strategy and Local Policies Plan, a council should undertake the majority of preparatory work for its POP in advance of, or in parallel with, the SCI and Timetable approval process in order to maximise time savings and to expedite the publication of the POP. A council should publish its POP approximately four months following the agreement with the Department of its SCI and Timetable. The requirements for the preparation of the POP are located in Part 3 of the Local Development Plan Regulations. In this context, Regulation 9 states that before a council complies with Regulation 11 (public consultation), it must, for the purpose of generating alternative strategies and options, engage the consultation bodies and, in preparing the Preferred Options Paper, the council must take into account any representation received from those bodies.

As the POP will influence the preparation of the Plan Strategy and Local Policies Plan, the options and alternatives contained therein should focus on how the Plan Strategy or Local Policies Plan will implement the strategic vision and objectives whilst taking account of the *Northern Ireland Regional Development Strategy 2035*, prevailing planning policy and any other policy and advice issued by the Department. A council must also consider other information, such as its Community Plan and any other local strategies in order to ensure that the subsequent LDP is locally distinctive and as comprehensive as possible to facilitate an integrated and coordinated approach to the planning and development of the area. The options considered should be realistic and deliverable. Therefore, in some cases, only one reasonable option may be available. Under these circumstances, the alternative would be to consider the scenario without the implementation of the plan, i.e., 'do nothing' option. If a council considers that there are no alternative options available, then this should be explained and justified (Department of the Environment 2015d).

Regulation 10 of the LDP Regulations specifies the provisions relating to the availability of the POP. Before a council prepares a development plan document it must do the following:

(a) make the following documents available for inspection during normal office hours at the places referred to in paragraph (b) below: (i) a copy of the Preferred Options Paper; (ii) such supporting documents as in the opinion of the council are relevant to that paper; (iii) a document containing a statement indicating the period specified by the council under Regulation 11(2) as the period within which representations on the Preferred Options Paper may be made; and (iv) notice of the address to which representations are to be sent.
(b) the places referred to in paragraph (a) above are: (i) the council's principal offices; and (ii) such other places within the council district as the council considers appropriate;
(c) send to the consultation bodies the information set out at paragraph (a) above:
(d) give notice by local advertisement of the following: (i) the title of the Local Development Plan; (ii) a statement of the fact that the Preferred Options Paper is available for inspection and the places and times at which it can be inspected; (iii) a brief description of the content and purpose of the Preferred Options Paper; and (iv) details of how further information on the Preferred Options Paper may be obtained; and (e) publish on its website the following: (i) the Preferred Options Paper; (ii) such supporting documents as in the opinion of the council are relevant to the Preferred Options Paper; and (iii) the document mentioned in paragraph (a)(iii) and notice mentioned in paragraph (a)(iv) above.

The final stage in the preparation of the POP relates to the actual public consultation process. This is dealt with under Regulation 11 of the LDP Regulations. Under these provisions any person may make representations about a council's Preferred Options Paper and, as such, these must be made within a period which the council specifies and sent to the address specified. The consultation period must be a period of not less than eight weeks or more than twelve weeks starting on the day on which the council complies with the regulation, and a council must take account of any representations made before it prepares a development plan document.

Stage 2

The Plan Strategy

The purpose of the Plan Strategy is not only to set out a council's objectives in relation to the development and use of land in the council area but also the strategic policies for the implementation of those objectives and other relevant matters. The Plan Strategy is based on both the information gathered and the public and stakeholder responses to the POP. As integral components of this process, a council must carry out a range of assessments including a Sustainability Appraisal (SA) incorporating Strategic Environmental Assessment (SEA), Equality Impact Assessment (EQIA), Habitats Regulations Assessment (HRA) and Rural Proofing before publishing the draft Plan Strategy in tandem with all of the above assessments for public consultation. Thereafter, the Plan Strategy must be subject to an Independent Examination before adoption (Department of the Environment 2015e).

On previous occasions, protracted judicial review proceedings have scrutinised the Strategic Environmental Assessment process linked to the preparation of development plans. In *(1) An Application by Seaport Investments Limited (2) An Application By Magherafelt District Council, F P McCann (Developments) Limited, Younger Homes Limited, Herron Bros Limited, G Small Contracts And Creagh Concrete Products Limited* (2007 NIQB 62) challenges were mounted against the *draft Northern Area Plan 2016* (dNAP) and the *draft Magherafelt Area Plan 2015* (dMAP) and comprised three key strands:

a) Ineffective transposition of the European Directive 2001/42/EC (SEA Directive), the purpose of which was to ensure that environmental assessment is integrated into plans and programmes by the Environmental Assessment of Plans and Programmes Regulations (Northern Ireland) 2004 Regulations (SEA Regulations) in relation to the designation of Department of the Environment as the 'Consultation Body' for the purposes of environmental assessment; the absence of appropriate timeframes in the SEA Regulations.
b) The manner in which the process of environmental assessment was carried out, in particular, the sequencing of its conduct and the preparation of the Draft Plan;
c) The content of the published environmental reports.

A conjoined judgment dealing with both cases was handed down by the High Court in Northern Ireland on 7 September 2007. The ruling from Weatherup J (2007 NIQB 62) can be summarised as follows:

(a) the designation of the Department of the Environment for Northern Ireland ("the Department") as a consultation body under Regulation 4 of the SEA Regulations does not properly transpose Article 6(3) of the SEA Directive, in that the Department cannot be both the responsible authority and the consultation body;
(b) the absence of appropriate time frames in Regulation 12 of the SEA Regulations does not properly transpose Article 6(2) of the SEA Directive;
(c) the environmental reports prepared for dNAP and dMAP are not in substantial compliance with Schedule 2 to the SEA Regulations (and Article 5 and Annex 1 of the SEA Directive);
(d) the sequencing of the environmental reports and the draft plans in each instance was not in compliance with Regulations 11 and 12 of the SEA Regulations and Articles 4 and 6 of the SEA Directive.

The Department of the Environment then reviewed with its legal advisors the consequences of the judgment. In light of all the circumstances, the Department took the view that, taking into account the judgment and its implications, it was not feasible to comply fully with the SEA Directive and the SEA Regulations in the preparation of either the dNAP or the dMAP. Accordingly, the Minister for the Environment made Non-Feasibility Determinations for each plan on 6 November 2007. A remedies hearing was held on 8 November 2007, and judgment was handed down on 13 November 2007. Weatherup J. refused to quash either of the draft Area Plans or their associated environmental reports. He indicated that in reaching this decision he took into account the Non-Feasibility Determinations issued by the Department which were relevant in both cases, subject to the outcome of any future judicial challenge. In regard to remedies on grounds (a) and (b) above, the Applicants sought to have the SEA Regulations quashed by the court. Weatherup J refused to quash the regulations and instead made the following declarations – the designation of the Department of the Environment (DOE) as 'the consultation body' under Regulation 4 of the SEA Regulations does not properly transpose articles 5.4 and 6.3 of the SEA Directive, where DOE is also the responsible body that promotes the plan or programme; the absence of appropriate timeframes in Regulation 12 of the SEA Regulations does not properly transpose the requirements of article 6.2 of the SEA Directive.

But this was not the end of the matter. The Department then appealed to the Court of Appeal against the findings by the High Court that the 2004 Regulations did not adequately transpose the requirements of the SEA Directive. The Court of Appeal subsequently referred a number of questions to the European Court of Justice (ECJ) (*Seaport 1*) regarding the correct interpretation of the SEA Directive. On 11 December 2006 leave was granted to Murdock Group Ltd to bring a judicial review challenge against the publication of the *draft Banbridge, Newry and Mourne Area Plan 2015* (dBNMAP). The grounds of challenge were similar to those in the challenges against dNAP and dMAP. It was agreed to adjourn the hearing of that case until judgment had been issued on *Seaport 1*. However, in light of the High Court judgment in the dNAP and dMAP cases and the grant of leave to the Murdock Group, the Department thereafter considered the status of dBNMAP together with the adequacy of the Strategic Environmental Assessment it had prepared. In line with its consideration of dNAP and dMAP, the Department again formed the view that it was not feasible to comply fully with the SEA Directive and the SEA Regulations in the preparation of dBNMAP. Accordingly, the Minister of the Environment made a Non-Feasibility Determination in respect of dBNMAP on 27 February 2008.

The judicial challenge to dBNMAP was withdrawn at the High Court on 9 June 2008. But on 27 June 2008 Seaport (NI) Ltd was granted leave to apply for judicial review of the Department's Non-Feasibility Determination issued for dNAP (*Seaport 2*). The Non-Feasibility Determination in respect of dMAP was not challenged. On 23 April 2010 the *Seaport 2* challenge was withdrawn. However, the ECJ continued to consider the matters raised in *Seaport 1* as this referral arose from an appeal by the Department. On 20 October 2011, the ECJ handed down its ruling, which concluded that there had been no failure on the part of the United Kingdom correctly to transpose the SEA Directive into the law of Northern Ireland (Case C-474/10). Moreover, the ECJ judgment provided guidance on the circumstances in which a body that was responsible for the preparation of a plan or programme governed by the Directive might also fulfil the role of a consultation authority. From the point of view of development plan preparation the important point to note here is that a period of more than four years was occupied with litigation around procedural protocols. In preparing Local Development Plans district councils and the Department for Infrastructure need to be cognisant of potential Timetable delays.

The provisions for the preparation of the Plan Strategy are set out in s.8 of the 2011 Act (as amended by the Local Government Act (Northern Ireland) 2014). The Plan Strategy must set out the council's objectives in relation to the development and use of land in its district; its strategic policies for the implementation of those objectives; and such other matters as may be prescribed. It is mandatory that a Plan Strategy must be prepared in accordance with the agreed Timetable and the relevant council's SCI. In the course of preparing a Plan, a council must take account of the Regional Development Srategy 2035 (Department for Regional Development, 2010); its current Community Plan; any policy or advice contained in guidance issued by the Department; such other matters as the Department may prescribe or, in a particular case, direct; and may have regard to such other information and considerations as appear to the council to be relevant (Department of the Environment 2015e). Furthermore, under s. 8 of the 2011 Act, the council must also carry out an appraisal of the sustainability of the Plan Strategy and prepare a report of the findings of the appraisal. A Plan Strategy has standing only if it is adopted by resolution of the council; or approved by the Department in accordance with s. 16(6) of the 2011 Act.

Part 4 of the LDP Regulations contain the provisions for the form and content of the Plan Strategy and indeed these apply to any development plan document. In this context, Regulation 12 specifies that such a document must contain a title which must give the name of the council district for which the development plan document is prepared and indicate whether it is a Plan Strategy or a Local Policies Plan, a sub-title which must indicate the date of the adoption of the development plan document. It must also contain a reasoned justification of the policies contained in it. Those parts of a development plan document which comprise the policies of the development plan document and those parts which comprise the reasoned justification must be readily distinguishable. Regulation 13 requires that a development plan document must contain a map or maps describing the policies and proposals set out in the development plan document so far as practicable to illustrate such policies or proposals spatially. Such a proposals map must be sufficiently detailed to enable the location of proposals for the development and use of land to be identified. Importantly, in the case of any contradiction between the written statement and any other document forming part of a Local Development Plan, the written statement shall prevail.

Part 5 of the LDP Regulations sets out the procedural requirements for the Plan Strategy. Accordingly, prior to submitting a development plan document to the Department under s.10 of the 2011 Act the council must make the following documents available for inspection during normal office hours at the council's offices or other places within the council district as the council considers appropriate:

(i) a copy of the development plan document;
(ii) a copy of the Sustainability Appraisal report under s. 8(6)(b) or 9(7)(b), as the case may be;
(iii) such supporting documents as in the opinion of the council are relevant to the preparation of the Local Development Plan;
(iv) a document containing a statement indicating the period within which representations on the development plan document may be made; and
(v) notice of the address to which representations can be sent.

The council must also send to the consultation bodies a copy of the documents, give notice in the *Belfast Gazette* and by local advertisement of the title of the development plan document, and a statement of the fact that the development plan document is available for inspection and the places and times at which it can be inspected. It must also publish on its website the development plan document, any supporting documents that in the opinion of the council are relevant, and the statement indicating the period within which representations on the development plan document may be made.

With regard to public consultation on a development plan document, Regulation 16 provides that any person may make representations about a development plan document and that these must be made within a period of eight weeks starting on the day the council complies with the regulations. Similarly, the representation must be sent to the address specified pursuant to the regulations. Thereafter, as soon as reasonably practicable after the expiry of the specified period, the council must make the following documents available for inspection at the places referred to above: a copy of the representations, a document containing a statement indicating the time within which counter-representations may be made and notice of the addresses to which counter-representations are available for inspection. Similarly, it must publish the representations on its website; give notice in the *Belfast Gazette* and by local advertisement of the fact that representations are available for inspection and the places and times at which they can be inspected; notify the consultation bodies of the fact that representations are available for inspection and the places and times at which they can be inspected; and notify any person who has made (and not withdrawn) a representation in accordance with Regulation 16(2) of those matters; any such representations must be: (a) made within a period of eight weeks starting on the day the council complies with Regulation 15(d), and (b) sent to the address specified pursuant to Regulation 15(a)(v).

Regulation 18 states that any person may make representations about a site-specific policy representation (in Regulations 17, 19 and this regulation referred to as 'counter-representations'). Such counter-representations must be made within a period of eight weeks, starting on the day the council complies with Regulation 17(1)(a), and (b) sent to the address specified in Regulation 17(1)(a)(iii). Importantly, counter-representations must not propose any changes to the development plan document. With regard to the availability of representations on site-specific policy representations, as soon as reasonably practicable after the expiry of the period referred to in Regulation 18(2)(a) the council must make a copy of the counter-representations available for inspection during normal office hours at the places specified, compliant with the regulations.

Prior to the submission of documents for Independent Examination under s. 10 of the 2011 Act, as per Regulation 20, a council must consider all representations made under Regulation 16 and, as the case may be, Regulation 18. The documents to be submitted are as follows:

(a) the report under s. 8(6)(b) or, as the case may be, s. 9(7)(b);
(b) the Statement of Community Involvement;
(c) evidence that the council has complied with its Statement of Community Involvement;
(d) copies of the notices referred to in Regulation 10(a)(iv), 15(a)(iv), and 17(1)(a)(iii);
(e) the Timetable;
(f) a statement setting out (i) a summary of the main issues raised in representations made in accordance with Regulation 11(2), and (ii) how those main issues have been taken into account in the preparation of the development plan document;

(g) a statement setting out (i) if representations were made in accordance with Regulation 16(2) or 18(2), the number of representations made and a summary of the main issues raised in those representations, or (ii) that no such representations were made;
(h) copies of any representations made in accordance with Regulation 16(2) or 18(2), and (i) such supporting documents as in the opinion of the council are relevant to the preparation of the development plan document.

The council must also send a copy of the draft Plan Strategy to the Department. When these requirements have been complied with the Plan Strategy document is then ready to undergo Independent Examination.

Independent Examination of the Draft Plan Strategy

S.10 of the 2011 Act sets out the requirements for Independent Examination of the Local Development Plan document for the initial draft Plan Strategy and the subsequent draft Local Policies Plan. In this context, in compliance with the legislation, the council is required to submit every development plan document together with other documentation as is prescribed to the Department for Independent Examination. These include the Plan Strategy or Local Policies Plan, Sustainability Appraisal report, the council's SCI, evidence that the council has complied with its SCI, copies of notices of the address to which representations are to be sent in relation to the public consultation on the Preferred Options Paper, notice of the address to which representations are to be sent in relation to the public consultation on the Local Development Plan document and notice of the address to which counter-representations are to be sent in relation to representations made during the public consultation process. Also included are the Timetable, a statement setting out a summary of the main issues raised in representations made on the council's Preferred Options Paper and how those main issues have been taken into account in the preparation of the Local Development Plan document, a statement setting out if representations/counter-representations were made in accordance with the LDP Regulations, the number of representations made and a summary of the main issues raised in those representations, or that no representations were made. In addition, copies of any representations/counter-representations and such supporting documents that, in the opinion of the council, are relevant to the preparation of the development plan document must be included.

The Department must then cause an Independent Examination to be carried out by the Planning Appeals Commission or a person appointed by the Department, unless the Department directs the council to withdraw or modify the document. The function of the Independent Examination is to ensure that the Local Development Plan meets the requirements of the 2011 Act with regard to the preparation of Timetable, Plan Strategy and the Local Policies Plan. Furthermore, it must confirm that it stands up to the test of Soundness in terms of content, conformity and the process by which it is produced. This requires considering the alignment of the development plan document with central government regional plans, policy and guidance, and the coherence, consistency and effectiveness of the content of the development plan document. (The Soundness tests are set out in full in Department for Infrastructure (2017) Development Plan Practice Note 06: Soundness: Version 2). The Independent Examination process provides that any person who makes representations seeking to change a development plan document must be given the opportunity to appear before and be heard by the person carrying out the examination and, subsequently, the person appointed to carry out the examination must make recommendations and give reasons for the recommendations. The independent examiner must then send the report to the Department for its consideration in terms of making a direction to the council regarding the adoption of the development plan document.

Adoption of the Plan Strategy

In accordance with Regulation 24(1), the independent examiner must send the advisory report to the Department for its consideration in terms of making a direction to the council regarding the adoption of the development plan document. Subsequently, the Department must consider the recommendations made by the independent examiner and direct the council to adopt the development plan document as originally prepared; adopt the development plan document with such modifications as may be specified in the direction; or withdraw the development plan document. The Department must give reasons for a direction given, and the council must comply with this. Similarly, under subs. (1) a development plan document is adopted by a resolution of the council.

Stage 3

The Local Policies Plan

It is important that any Local Policies Plan is consistent with a council's Plan Strategy and take account of the *Northern Ireland Regional Development Strategy 2035* and any policy issued by the Department (Department of the Environment 2015f). The Local Policies Plan is required to establish the council's policies in relation to the development and use of land in its district and such other matters as may be prescribed. It must be prepared in accordance with the Timetable and the council's Statement of Community Involvement. Furthermore, in the course of preparing a Local Policies Plan the council must take account of its current Community Plan; any policy or advice contained in guidance issued by the Department; such other matters as the Department may prescribe or, in a particular case, direct; and may have regard to such other information and considerations as appear to the council to be relevant.

Part 4 of the LDP Regulations sets out the requirements for the form and content of the Local Policies Plan and the details of the additional matters to be taken into account in preparing the Local Policies Plan. It is important to highlight that a Local Policies Plan document must contain a reasoned justification of the policies contained in it. A council's local policies, including its site-specific proposals, designations and land use zonings, should show how the council's vision, objectives and strategic policy as set out in its Plan Strategy are to be achieved. They should be tailored to the local area and set out a council's approach towards the development and use of land in its district. In particular, topic-based local policies and proposals should reflect the regional strategic objectives and policies contained in the SPPS and logically flow from the content of its Plan Strategy. Important themes are, for example, housing in settlements, economic development, natural and built heritage, development in the countryside, town centres and retailing, open space, recreation and transportation. Local policies may draw upon other local strategies and/or plans of a council, particularly those which may have implications for the development and use of the land, for example, community and regeneration strategies, and seek to provide an integrated approach to the future development of a council area. Local policies and site-specific proposals, for example, zonings and designations, should be founded on a reliable evidence base, for example, assessments of housing and employment need. The Local Policies Plan should identify and define settlement development limits in accordance with the settlement hierarchy set out in the Plan Strategy, whilst taking into account the *Northern Ireland Regional Development Strategy* Spatial Framework Guidance, any policy or advice issued by the Department and the council's current Community Plan. It should allocate and zone land for housing and for economic growth and regeneration. It should seek to conserve and sustain and enhance the area's environmental qualities, local distinctiveness and sites of environmental importance in terms of landscape character and diversity, wildlife and habitats, townscape and archaeology. Sites of local nature conservation importance, areas of townscape or village character and local landscape policy areas should be identified. Local transportation proposals, in accordance with the Department's approach to ensuring a sustainable transport future (see Department for Regional Development 2011), and other local infrastructural and utility needs should be mentioned. The Local Policies Plan should promote the development of sustainable local tourism, open space, recreational and other rural and urban community facilities that will positively contribute to the amenity and wellbeing of the population. And, finally, the Local Policies Plan should facilitate the promotion of equality of opportunity and good relations between persons of different religious belief, political opinion or racial group.

A council should also aim to ensure that its Local Policies Plan is both realistic and deliverable, taking into account the resources available and any potential constraints which may arise during the plan period. Furthermore, to allow for unforeseen circumstances arising, for example, from the withdrawal of funding or the cancellation/postponement of infrastructure proposals, a council should aim to incorporate a degree of flexibility within its Local Policies Plan. The council's Community Plan is important here as it sets out proposals for meeting local needs and services and can help to justify the implementation of certain policies and proposals contained in the Local Policies Plan.

In the same way as the Plan Strategy, the preparation of the Local Policies Plan is also subject to a number of other statutory requirements, such as a Sustainability Appraisal incorporating the requirements of Strategic Environmental Assessment, an Equality Impact Assessment and a Habitats Regulations Assessment. S. 9(7) of the 2011 Act sets out the requirement to carry out a Sustainability Appraisal for the Local Policies Plan. This incorporates an assessment of environmental effects and should be an integral part of the plan-making process by providing a robust evidence base for the Local Policies Plan. This will play an important role in demonstrating the soundness of the development plan document. A council should also undertake an appraisal of reasonable alternatives and any likely significant effects of the draft Local Policies Plan against sustainability assessment objectives as agreed with the consultation body. Following the appraisal of reasonable alternatives and any likely significant effects arising from the draft Local Policies Plan, a council must prepare a sustainability assessment report. The purpose of the report is to present the findings of the appraisal and show how reasonable alternatives and any likely significant effects from the implementation of the draft Local Policies Plan have been taken into account in the decision-making process. The report should aim to provide transparency by documenting the sustainability assessment and plan preparation process and highlighting any changes to policies and proposals along with mitigation measures as a result of the sustainability assessment of the draft Local Policies Plan. It should provide an audit trail of policy thinking and development. In addition to the Sustainability Appraisal, a council must also undertake a Habitats Regulations Assessment and an Equality Impact Assessment. The overall aims of the Habitats and Birds Directives are to maintain or restore the conservation status of habitats and species of policy interest. Special Areas of Conservation and Special Protection Areas, designated to afford protection to habitats and species, are cited in the Habitats and Birds Directives. These designations form a suite of sites which are collectively known as the Natura 2000 Network. A Habitats Regulations Assessment is required by The Conservation (Natural Habitats, etc.) Regulations (Northern Ireland) 1995. It aims to assess possible adverse effects on Natura 2000 sites (Special Areas of Conservation, Special Protection Areas) and Ramsar Sites as a result of the implementation of policies and proposals contained in a Local Policies Plan. The process of Habitats Regulations Assessment may be undertaken in parallel with the Sustainability Appraisal and a council should liaise with the Shared Environmental Service to ensure that it meets all legislative requirements. A council is also required to ensure that its Local Policies Plan is prepared in accordance with the Section 75 statutory obligations of the Northern Ireland Act 1998. Consequently, a council must undertake an Equality Impact Assessment to determine the potential impacts upon Section 75 groups as a result of the policies and proposals contained in the development plan document.

With regard to availability and public consultation on the Local Policies Plan, Regulation 15 of the LDP Regulations sets out the requirements for the availability of development plan documents. A council must:

(a) make the following documents available for inspection during normal office hours at the places referred to in paragraph (b) namely (i) a copy of the development plan document; (ii) such supporting documents as in the opinion of the council area relevant to the preparation of the Local Development Plan; (iii) a document containing a statement indicating the period within which representations on the development plan document may be made; (iv) notice of the address to which representations can be sent;
(b) the places referred to in paragraph (a) are: (i) the council's principal office; and (ii) such other places within the council district as the council considers appropriate;
(c) send to the consultation bodies: (i) the information set out in paragraph (a); (ii) a copy of the development plan document; and (iii) a copy of the sustainability appraisal report;
(d) give notice in the *Belfast Gazette* and by local advertisement of the following: (i) the title of the development plan document; and (ii) a statement of the fact that the development plan document is available for inspection and the places and times at which it can be inspected;
(e) publish on its website: (i) the development plan document; (ii) such of the supporting documents as in the opinion of the council are relevant; (iii) notice of the address to which representations can be sent.

Regulation 16 of the LDP Regulations sets out the requirements for public consultation on a development plan document. It states: (1) Any person may make representations about a development plan document. (2) Any such representations must be: (a) made within a period of eight weeks starting on the day the council complies with Regulation 15(d); and (b) sent to the address specified pursuant to Regulation 15(a) (iv). A council must make available for inspection during normal working hours a copy of the draft Local Policies Plan and associated documents at its principal office and such other places as the council considers appropriate; make a statement that representations can be made on the draft Local Policies Plan and the address to which the representations can be sent; send this information to the consultation bodies which were engaged in the preparation of the draft Local Policies Plan; and give notice in the *Belfast Gazette* and by local advertisement and on its website setting out that the draft Local Policies Plan has been prepared, the places and times at which it can be inspected, a brief description of the content and purpose of the draft Local Policies Plan and details of how further information may be obtained. To help reduce the number of unsubstantiated and speculative representations, a council should set out requirements for the submission of representations in order to facilitate the management and consideration of all representations and counter-representations submitted on the development plan document(s). A council may wish to provide a standard form to assist in this process and help focus on the information required to either object to or support the policies and proposals contained in the development plan document. In particular, representations should provide evidence to demonstrate why the draft Local Policies Plan is unsound and/or how any proposed changes would make the draft Local Policies Plan more sound. Furthermore, representations should demonstrate how any proposed changes meet the sustainability objectives of the draft Local Policies Plan and should be submitted in full within the specified public consultation period. This requirement will help to enhance the quality and efficiency of the decision-making process as it will enable all information to be considered by all involved parties and help to ensure a fairer, more comprehensive and thorough debate of issues at Independent Examination.

Stage 4

Annual Monitoring Report and Five-Year Review of the Local Development Plan

A council must undertake comprehensive and reliable monitoring in order to establish how the objectives of the Local Development Plan are being achieved and determine whether any changes are required to ensure the continued delivery of the Local Policies Plan. A council's consideration of its Community Plan, as well as monitoring of the Local Policies Plan, forms part of the tests of soundness. Under the provisions of s. 24 of the 2011 Act every council must make an annual report to the Department for Infrastructure. In this context, the report must contain information regarding the extent to which the objectives set out in the Local Development Plan are being achieved. Furthermore, the annual report must be made at such time as may be prescribed, be in such form as may be prescribed and contain such other matter as may be prescribed. There is also statutory provision for the review of Local Development Plans. Under s. 13 of the 2011 Act a council must carry out a review of its Local Development Plan at such times as the Department may prescribe and which in the *Strategic Planning Policy Statement for Northern Ireland* is set at every five years. The council must report to the Department on the findings of its review and a review must be in such form as may be prescribed, and it must be published in accordance with such requirements as may be prescribed. Furthermore, under s.14, the council may at any time prepare a revision of its Plan Strategy; or its Local Policies Plan. The council must prepare a revision of its Plan Strategy or its Local Policies Plan at such times and in such manner as may be prescribed if, following a review under s. 13, it thinks that the Plan Strategy or the Local Policies Plan should be revised or if the Department directs the council to do so.

Conclusion

The jurisdiction of Northern Ireland has now formally adopted a plan-led system, and thus development plans have become increasingly important tools for decision making. The scale and degree of complexity involved in preparing a new suite of plans comprising a Plan Strategy followed by a Local Policies Plan cannot be underestimated. Nevertheless, with planning authorities being mindful of the growing threat of judicial review, it is more important than ever that due diligence is paid not only to *doing the right thing* but *doing the thing right* in terms of following the basic steps set out in the legislation and plan preparation practice notes.

Legislation

Northern Ireland Act 1998
Planning Act (Northern Ireland) 2011
The Conservation (Natural Habitats, etc.) Regulations (Northern Ireland) 1995
The Planning (Amendment) (Northern Ireland) Order 2003
The Planning (Local Development Plan) (Amendment) Regulations (Northern Ireland) 2016
The Planning (Local Development Plan) Regulations (Northern Ireland) 2015
The Planning (Northern Ireland) Order 1972
The Planning (Northern Ireland) Order 1991
The Planning (Statement of Community Involvement) Regulations (Northern Ireland) 2015

Cases cited

Blythe Valley District Council v Persimmon Homes (North East) Ltd. (2008 ECWA Civ 861)
In the Matter of (1) An Application by Seaport Investments Limited (2) An Application by Magherafelt District Council, F P McCann (Developments) Limited, Younger Homes Limited, Herron Bros Limited, G Small Contracts and Creagh Concrete Products Limited. (2007 NIQB 62).
Judgment of the Court (Fourth Chamber) of 20 October 2011 (reference for a preliminary ruling from the Court of Appeal in Northern Ireland (United Kingdom) – Department of the Environment for Northern Ireland v Seaport (NI) Ltd, Magherafelt District Council, F P McCann (Developments) Limited, Younger Homes Limited, Herron Bros Limited, G Small Contracts and Creagh Concrete Products Limited. (Case C-474/10).
Simpson v Edinburgh Corporation. (1961 SLT 17).

References

Department of the Environment. 2015a. *Strategic Planning Policy Statement for Northern Ireland: Planning for Sustainable Development.* Belfast: Department of the Environment.

Department of the Environment. 2015b. *Development Plan Practice Note 02: Statement of Community Involvement,* Version 2. Belfast: Planning Guidance Team.

Department of the Environment. 2015c. *Development Plan Practice Note 03: Timetable.* Belfast: Planning Guidance Team.

Department of the Environment. 2015d. *Development Plan Practice Note 05: Preferred Options Paper.* Belfast: Planning Guidance Team.

Department of the Environment. 2015e. *Development Plan Practice Note 07: The Plan Strategy.* Belfast: Planning Guidance Team.

Department of the Environment. 2015f. *Development Plan Practice Note 08: The Local Policies Plan*. Belfast: Planning Guidance Team.

Department for Infrastructure. 2017. *Development Plan Practice Note 06: Soundness: Version 2*. Belfast: Planning Guidance Team.

Department for Regional Development. 2010. *Regional Development Strategy 2035: Building a Better Future*. Belfast: Department for Regional Development.

Department for Regional Development. 2011. *Ensuring a Sustainable Transport Future: A New Approach to Regional Transportation*. Belfast: Department for Regional Development.

Goodyear, P.M. and D.A. Eastwood. 1979. 'Regional Development and Spatial Equity in Northern Ireland', *Economic and Social Review* 10(4): 321–40.

Greer, J. and M. Murray. 2003. 'Unravelling the Rural in Northern Ireland', in J. Greer and M. Murray (eds), *Rural Planning and Development in Northern Ireland*. Dublin: Institute of Public Administration, pp. 3–35.

Healey, P. 1993. 'The Communicative Work of Development Plans', *Environment and Planning B: Urban Analytics and City Science* 20(1): 83–104.

House of Lords. 1972. Available from https://api.parliament.uk/historic-hansard/lords/1972/oct/25/planning-northern-ireland-order-1972 (accessed 16 November 2021).

Ministry of Housing and Local Government, Welsh Office. 1970. *Development Plans: A Manual on Form and Content*. London: Her Majesty's Stationery Office.

Moore, V., M. Purdue and A. Bowes. 2014. *A Practical Approach to Planning Law*. Oxford: Oxford University Press.

Northern Ireland Audit Office. 2022. *Planning in Northern Ireland*. Belfast: Northern Ireland Audit Office.

5

EVOLUTION AND APPLICATION OF ENFORCEMENT

Jane McKay and Stephen McKay

Introduction

The Northern Ireland system of development management[1] is based on Section 24 (1) of the Planning Act (Northern Ireland) 2011, hereinafter referred to as the 2011 Act, which provides that planning permission is required for the carrying out of any development of land. In this context, 'development' is defined, under Section 23 (1), as the carrying out of building, engineering, mining or other operations in, on, over or under land, or the making of any material change in the use of any buildings or other land as 23 (subject to subsections (2) to (6)). To establish a comprehensive understanding of the fundamental principles of mainstream planning practice it is essential to realise the distinction between 'operations' and 'use'. The true meaning of the concepts can best be grasped through case law. Lord Parker (*Cheshire CC v Woodward* 1962 2 Q.B. 126) stated than an 'operation' is some act which changes the physical characteristics of the land, or of what is under it, or of the air above it, 'use' refers to the purpose to which the land and buildings are devoted. Lord Denning (*Parkes v Secretary of State* 1978 1 WLR 1308) made the comment 'operations' comprise activities that result in some physical alteration to the land which have some degree of permanence in relation to the land itself, whereas 'use' comprises activities which are done in, alongside or on the land but do not interfere with the actual characteristics of the land.

Taken together, the effect of Section 23 (1) and Section 24 (1) is that every minor operation or activity would require planning permission, for example, the ploughing of farmland. Since such a degree of control would be unnecessary and indeed impractical to operate, the legislation provides for reasonable limits to be placed on the extent of control in two ways. First, it excludes certain operations and uses from the definition of development (Section 23 (3)), and, second, it empowers the planning authority to make Development Orders which can exclude certain classes of development from the need to obtain planning permission (Section 24 (3)).

Under certain circumstances, such as those specified in the Planning (General Permitted Development) Order (Northern Ireland) 2015, planning permission is granted for specified classes of development, thereby avoiding the need for a planning application. Such

DOI: 10.4324/9781003218319-5

developments are normally of a minor nature or represent changes of use from more to less significant forms of operation. Any other class of development is likely to require an application for planning permission. It is deemed to be the responsibility of would-be developers to take steps to ascertain whether such an application is required. For this purpose, the 2011 Act provides that developers may approach the planning authority for a determination. To ensure that unacceptable development does not take place, it is essential to have a deterrent mechanism. Development management, therefore, necessarily involves measures for planning enforcement referred to under Part 5 of the 2011 Act.

It is not normally a criminal offence to carry out development without first getting planning permission.[2] If, however, the development is deemed to be unacceptable, it is the responsibility of the offender to take remedial action. Failure to comply with such requirements may result in the initiation of enforcement procedures. The purpose of planning enforcement is, therefore, to protect the integrity of the planning system by enabling planning authorities to remedy any harm to amenity or 'other interest of acknowledged importance' that may result from unauthorised development (DOE (NI) 1999). The nature of the system means that the decisions on whether to take enforcement action, and, if so, what action is best suited to particular circumstances, are matters for the planning authority's discretion. Indeed, as Millichap (1998) stated, the three planning tasks, labelled the planning trinity, are all intimately related – none is more important than the other if the ultimate goal is the effective use of the planning legislation's powers. Forward planning, development management and enforcement are all mutually dependent and mutually reinforcing. To deploy any of these mechanisms ineffectively is to undermine the legitimacy of the entire planning system.

Unfortunately, it is apparent that this has often been the case with enforcement and it is no surprise that, through time, it has come to be referred to as the Cinderella of planning (Department of the Environment 1975; Ministry of Housing, Communities and Local Government 2020), evidenced by the fact that it has frequently been referred to as the weakest link in the planning system (Lewis 2003; Cullingworth et al. 2015). Cameron Blackhall (2017) highlights a number of factors which have been responsible for this undesirable reputation, including the difficulties in collating substantive evidence to sustain formal enforcement action, the technicality and complexity of the system and time limitation, the latter of which has become particularly contentious in the jurisdiction of Northern Ireland (see below).

The main purpose of this chapter is to analyse the event sequence which has taken place in the development of the planning enforcement system from inception to the present. The rationale is to facilitate an understanding of why the enforcement machinery has been crafted and how it is implemented in practice. Three critical avenues of enquiry are explored. First, the emergence of and changes which have taken place in enforcement legislation in England and Wales, which have heavily influenced developments in Northern Ireland, are examined. This focuses on key changes in the law relating to enforcement regulations from the introduction of planning legislation in the Housing Town Planning, Etc. Act 1909 through to the Town and Country Planning Act 1947 and to the significant developments enacted by the Town and Country Planning Act 1990 and the Planning and Compensation Act 1991. This provides a platform for the scrutiny of the system which exists in Northern Ireland whereby the focus of attention shifts to an analysis of measures introduced under the Planning (Northern Ireland) Order 1972, consolidated by the Planning (Northern Ireland) Order 1991 and the existing legislative toolkit which came into force under the provisions of the Planning

Act (Northern Ireland) 2011. The latter is the essential basis for the current planning enforcement legal procedures within this jurisdiction.

The emergence of enforcement: from 1909 to 1947

The enforcement of planning control has been fundamental to the legitimisation of, what has been identified by Emmett (1994) as, the modern planning system introduced by the Town and Country Planning Act 1947. Evidence, however, would clearly indicate that it is essential to examine the earlier developments in planning when the discipline was endeavouring to influence the development of modern cities and remove the maladies of the industrial revolution.

Planning control was introduced into England and Wales under the Housing Town Planning, Etc. Act 1909. This built upon the public health measures of the earlier legislative schemes (Public Health Act 1875) which sought to address the problems of urban squalor and overcrowding in the 19th century (Cullingworth et al. 2015). Attempts under the public health legislation helped to produce more sanitary housing, but the grim regularity of monotonous housing along straight streets caused a backlash that led to the establishment of amenity considerations in planning (Cherry 1974: 64-65). The legislation of 1909, with its emphasis on amenity, ushered in the modern era of planning. One notable feature of this new system was the enforcement regime, which provided local authorities with significant powers to regulate development in the public interest.

Section 54 of the 1909 Act empowered local authorities to make a 'town planning scheme ... as respects any land which is in the course of development or appears to be used for building purposes, with the general object of securing proper sanitary conditions, amenity and convenience in connection with the laying out and use of the land and of any neighbouring lands' (Heap 1996). Whereas the Public Health Act 1875 provided the framework for controlling the construction of individual buildings, the 1909 legislation provided the opportunity to regulate building development as a whole. Land no longer had to be considered as isolated plots which could be developed independently. Under section 54 each plot had to be developed with regard to 'amenity and convenience in connection with the laying out and use of the land and any neighbouring lands'. This is a fundamental principle which continues to be embroidered in all land use plans whereby neighbouring developments must be mutually compatible. Consequently, it is not acceptable to have potentially hazardous industrial development located within a predominantly residential area. With specific reference to planning enforcement, Section 7 of the 1909 legislation provided that buildings built in contravention of the 'planning scheme' could be 'removed, pulled down or altered' at the offender's expense (Wood 1949: 45). The 1909 legislation thus followed a draconian enforcement code that was instituted in the public health legislation to deal with the unsanitary conditions found in the sprawling urban areas of Victorian cities.

The Housing, Town Planning, &c. Act 1919 section 45 gave consideration to interim development whereby during the preparation of a scheme, development of land could proceed while the developer was in the knowledge that compensation would be paid if the development was detrimentally affected by the scheme. A special consent procedure was adopted when the Town Planning (General Interim Development) Order 1922 instituted a special regime whereby local authorities could give an applicant permission to proceed with development. This legislation had the ability to impose such requirements or conditions as the local authority might think appropriate on such permissions.

By 1925 it was recognised that housing and town planning had become distinctive entities. The two could no longer be considered effectively together, and, consequently, planning legislation was consolidated into the Town Planning Act of that year. The next major watershed in planning law occurred with the Town and Country Planning Act 1932, which extended the scope for planning action by enabling local authorities to make planning schemes for both urban and rural land. In addition, the 1932 legislation put the procedures for seeking interim permission on a much clearer statutory footing. The precise mechanics were later set out in the Town and Country Planning (General Interim Development) Order of 1933. This piece of legislation covered consultation outlines permission and permitted development provision (Heap 1996). Overall, however, from 1933 until the end of the Second World War, the powers of local authorities to enforce planning control were so severely restricted under the legislation enacted that it proved difficult to take effective action against those who were in breach of the regulations (Grant 1982). Planning schemes were only a partial control over development, they were inflexible and the prospect of having to pay compensation to those who sustained loss as a result of the making of a planning scheme deterred many authorities from taking action under the legislation. Indeed, the failure by planning authorities to prepare planning schemes can be seen as an early example of undermining the 'planning trinity'.

Major land use problems began to emerge during the 1930s, providing clear evidence that planning powers were insufficient to deal with the developing situation. In response, a series of what have been recognised as landmark reports were produced by the Barlow Commission[3] and the Scott[4] and Uthwatt[5] Committees. The main findings of these studies were that planning control should cover the whole country and that there should be a national policy for industrial location and population distribution. It was also concluded that there should be a central planning authority concerned with urban redevelopment; reduction of congestion; achieving industrial balance within and between regions; examining the potential of garden suburbs, satellite towns and trading estates; research and information about industry, resources and amenities and correlation of local planning schemes in the national interest (McLoughlin 1969). It was not, however, until after the Second World War that the necessary legislation was enacted which gave effect to the recommendations contained in the reports – legislation which sought to ensure that planning controls could be enforced throughout the state.

The development of enforcement from 1947

The Town and Country Planning Act 1947 established the basis of the modern planning system in England and Wales. Many of the provisions relating to the enforcement of planning control which exist today were introduced as part of that legislation. Under the 1947 Act, the local planning authority could serve an Enforcement Notice where a breach of planning control had been committed and it was expedient to do so, having regard to the development plan and other material considerations (Brand 1988). Enforcement action was entirely a matter at the discretion of the planning authority, and there was no criminal liability for an actual breach of planning control. The time limit for enforcement action under the 1947 Act was four years from the date of the breach (except in the case of minerals development where the four-year period ran from the time of discovery). Where an Enforcement Notice was served under the 1947 Act the offender had recourse to one of three options, which included complying with the requirements of the Enforcement Notice; making an application to the planning authority for retrospective planning permission, or submitting an appeal to the magistrates' court. If the Enforcement Notice took effect, at this point a criminal offence was deemed to have taken place where the Notice required discontinuance of a use of land. In the event of the Notice requiring steps other than the discontinuance of a use of land, for example, the carrying out of building works, the planning authority could enter the land, carry out the steps themselves and recover the costs from the owner.

The mechanisms introduced by the 1947 Act, according to Hendry (1989), did provide significant gains for the planning process. However, the appeal process introduced to protect the rights of landowners resulted in the magistrates' court being given a significant role to play. Furthermore, an appeal could be made on the basis that the breach specified in the Notice had not actually occurred. If this could be proved then the case against the offender would fail. It also meant that the system became highly technical and, consequently, on many occasions for this very reason, ineffective. Evidence to support this line of thinking comes from Carnwath (1989), who indicated that the technical nature of the law, coupled with a judicial attitude which was highly protective of property rights, saw the creation of a body of case law which emphasised the need for strict adherence to the letter of the Act in the drafting and service of the Enforcement Notice. In this respect, arguably the leading case is that of *East Riding County Council v Park Estate (Bridlington) Ltd* (1957), where Lord Simonds accepted that the Act was highly technical and as it encroached on private rights the court must insist on strict and rigid adherence to formalities (Grant 1982). Carnwath (1989) indicated that such a technical approach led to many Notices failing as nullities. Particular difficulties were experienced in controlling the caravan sites which sprang up in the 1950s in response to the housing shortage and, as a consequence, the enforcement provisions of the Town and Country Planning Act 1947 were reformed by the Caravan Sites and Control of Development Act 1960. The latter introduced a right of appeal to the Minister against an Enforcement Notice, who was given the power on appeal to correct any 'informality, defect or error' in the Notice that was not considered material. The 1960 Act also introduced criminal liability for failure to comply with the requirements of an Enforcement Notice.

The Caravan Sites and Control of Development Act 1960 went some way towards alleviating the difficulties of a rigid interpretation of the requirements of the legislation. Indeed, case law evidence shows that this was recognised by the courts. In the Court of Appeal, for example, in the case of *Miller-Mead v Minister of Housing and Local Government* (1963 2 QB 196), the court held that:

> the subject, who is being told he is doing something contrary to planning permission and that he must remedy it, is entitled to say that he must find out from the four corners of the document exactly what he is required to do or to abstain from doing.

The judgment continues:

> Supposing then … that the owner or occupier … could not tell with reasonable certainty what steps he had to take to remedy the alleged breaches. The Notice would be bad on its face and a nullity.

Upjohn L J stated a simple test for the validity of an Enforcement Notice was to ask the question:

> does the Notice tell (the recipient) fairly what he has done wrong and what he must do to remedy it?

Importantly, the court considered the power of the Secretary of State to amend an Enforcement Notice and it was indeed held that he/she could amend a Notice which was otherwise invalid but not one which was upon its face a nullity. Lord Denning MR said:

> He can correct errors so long as having regard to the merits of the case the correction can be made without injustice.

The Town and Country Planning Act 1962 subsumed the provisions of the Caravan Sites and Control of Development Act 1960 but did not contain any significant provisions relating to breaches of planning control. Consideration was, however, given in the Town and Country Planning Act 1968, where the four-year time limit for taking enforcement action against breaches of planning control comprising a material change of use (other than to a use as a single dwelling house) was replaced by the removal of any immunity period for uses that were begun after the end of 1963. The four-year time limit remained for operational development and breaches of related conditions.

The Town and Country Planning Act 1968 made two significant changes to planning enforcement legislation. In the first instance, it amended the 'four-year rule' by excluding from its scope most developments which are classified as constituting a material change of use of buildings or other lands. This meant that such development was vulnerable to enforcement action at any time, though provisions were included to allow the issuing of established use certificates in respect of an act of development comprising a material change of use which had already become immune from enforcement under the previous 'four-year rule' (Brand 1988). Second, it introduced Stop Notices, which can be used by planning authorities to counter the effect of making an appeal against an Enforcement Notice (that is to suspend its operation) and can require the alleged breach of planning control to cease three days after the service of the Stop Notice.

Between 1968 and 1990 planning enforcement legislation was the subject of changes which were more in the nature of systematic progression rather than a radical overhaul. No changes at all were made by the Town and Country Planning Act 1971, but subsequent legislative amendments extended the scope of the Stop Notice procedure to a wider range of development and set a twelve-month time limit for its use from the date of service. Furthermore, the changes provided planning authorities with strong powers to obtain relevant information which significantly weakened the hand of the recipient. Alterations to the enforcement mechanism under the Local Government and Planning (Amendment) Act 1981 further clarified the provisions by incorporating recommendations of the Dobry Report (1975) into the planning system. Dobry (1975) indicated that planning enforcement was 'probably … the weakest link' in the planning system and identified a number of specific areas of criticism, including technicality and complexity, lack of urgency, inflexibility, lack of teeth and policy ambivalence. The Dobry Report (1975) recognised that a selective approach to enforcement was inevitable if resources were to be used effectively. The margins between lawful and unlawful development were not considered to be sufficiently clear-cut to form the basis for definition as a criminal offence, and the magistrates' court was not considered to be a suitable forum in which to argue the technicalities of planning law. It was therefore felt that unauthorised development should not become an immediate criminal offence. The basic conclusion was that there was no sensible alternative to working within the framework of planning law. All subsequent guidance issued by the central government stressed the discretionary nature of the power for any planning authority to undertake statutory enforcement procedures. Circulars 22/80 (D.O.E. 1980) and 4/87 (D.O.E. 1987) made it clear that the service of an Enforcement Notice should be regarded as a last resort where significant harm is being caused with the threat of costs being awarded against an authority that takes 'needless' action.

Although the 1990 Town and Country Planning Act consolidated the previous legislation and remains the primary legislation for planning enforcement, a number of radical changes have been implemented as a result of the Planning and Compensation Act 1991. The changes introduced under the Act were based on the recommendations of the Carnwath Report (1989) on the review of the legislative provisions relating to the enforcement of planning control. The objective of the Carnwath review was to assess the scope and effectiveness of the legislative provisions relating to the enforcement of planning control (Bourne 1992). The final report identified three key elements which had to be addressed if the goal of effective enforcement was to be achieved. In the first instance, it was deemed to be essential to bring offending activities within planning control. Second, effective methods were deemed necessary to ensure that undesirable effects are remedied or at the very least mitigating measures are used, and third, that punishment or deterrence mechanisms must be effective and be seen to be effective (Carnwath 1989). It was also recommended that in achieving these, the system must be seen to operate fairly and, as far as possible, it should aim for speed, clarity and cost-effectiveness. The vast majority of the recommendations made by Carnwath were adopted in the Planning and Compensation Act 1991, which were seen by many observers as putting right some of the apparent weaknesses in the prevailing planning legislation (Berry 1991). These now form key tenets of planning enforcement not only in England and Wales but also in Northern Ireland. With this in mind, attention now turns to an examination of the enforcement toolkit which is available to planning authorities in the jurisdiction of Northern Ireland.

Planning enforcement legislation and practice in Northern Ireland

Section 24 of the Planning Act (Northern Ireland) 2011, hereinafter referred to as the 2011 Act, provides that planning permission is required for the carrying out of any development of land. In this context, 'development' is defined, under section 23, as the carrying out of building, engineering, mining or other operations in, on, over or under land, or the making of any material change in the use of any buildings or other land.[6]

In Northern Ireland, councils have the power to issue a Submission Notice under section 43 (1) of the 2011 Act which requires that a planning application be submitted where it appears that development has been carried out without the necessary planning permission. Such a Notice must be served within five years of the date upon which the development was begun (see below under time limitation). It is an offence not to comply with this Notice in the time specified and, importantly, the issue of a Submission Notice does not constitute enforcement action as per the 2011 Act. In this context, a key function of the Notice is to ensure that the recipient complies with the requirement to obtain planning permission retrospectively whilst enabling a council to impose planning conditions to control the development (where appropriate)[7], as well as ensuring that the offender pays the appropriate fee. Part 5 of the 2011 Act contains the machinery for the enforcement of planning control, the key provisions of which are now considered.[8]

In the first instance, section 132 deals with the issue of time limitation and, whereas in Great Britain the periods for operational development and material change of use are four and ten years, respectively, for operational development and material change of use,[9] the 2011 Act specifies that where there has been a breach of planning control consisting in the carrying out without planning permission of building, engineering, mining or other operations in, on, over or under land, no enforcement action may be taken after the end of the period of five years beginning with the date on which the operations were substantially completed. Similarly, where there has been a breach of planning control consisting in the change of use of any building to use as a single dwelling house, no enforcement action may be taken after the end of the period of five years, beginning with the date of the breach. In the case of any other breach of planning control, no enforcement action may be taken after the end of the period of five years, beginning with the date of the breach. The context of the disconnect between the situation in Northern Ireland and Great Britain is of interest, given the rationale for the emergence of the need for a distinction to be made between operational development and material change of use.

As part of the programme for the devolved planning system of the new local government in Northern Ireland, fresh planning legislation was required. In this context, an extensive consultation process took place whereby questions were asked regarding the content of the new Planning Act. Specifically, what was wrong with the existing system and how might this be remedied were queried (Department of the Environment 2009). Subsequently, a consultation paper, coupled with responses by the Department of the Environment to comments received, was published to take forward legislative change. Though the issue of four- and ten-year rules was raised, no suggestion was made to introduce change. It was, therefore, no surprise that when the Planning Bill 2010, the precursor to the 2011 Act, was published, section 131 (1) dealt with time limitation and provided that for operational development the time limitation period for immunity would, as previously, be four years. Similarly, section 131 (3) indicated that, as previously, immunity for a material change of use would be ten years. What followed, however, has been cause for concern. When the Planning Act (Northern Ireland) 2011 was subsequently published, section 132 (3) specified that in the case of any material change of use 'no enforcement action may be taken after the end of the five years beginning with the date of the breach' (page 84). In effect, the four-year rule for operational development and the ten-year rule for material change of use had been changed to a uniform five-year rule for both operational development and material change of use.

It is a matter of concern that, in the absence of any supporting evidence, this change has been implemented, particularly when post-1947 legislation has continuously viewed differentiation between the two classes of development as a necessity. Indeed, all evidence presented since has been to the contrary, with most pointing to a need to expand the ten-year limitation period. Research conducted in England by Richard Humphreys QC (2011) has, for example, suggested that the ten-year rule should be doubled to twenty years. The Law Society, however, whilst at variance with this opinion, explained how the difficulties presented by the ten-year rule could be remedied. It suggested that the provisions of the Localism Act 2011 may go some way towards remedying concealment (Law Society 2011). Section 124 of the Act allows enforcement action to be taken within six months, beginning with the date on which evidence of the apparent breach of planning control came to the authority's knowledge. To understand how this development originated it is important to consider two judgments: first, *Fidler v First Secretary of State Reigate & Anor* (2004 EWCA Civ 1295) and, second, *Welwyn Hatfield BC v. Secretary of State for Communities and Local Government & Beesley* (2011 2 WLR 905).

In both *Fidler* and *Beesley* the defendants had concealed residential buildings, waited until the immunity period expired and then applied for Certificates of Lawful Use. In the case of *Fidler*, the defendant built his house beneath a haystack covered by a tarpaulin without planning permission. Having been refused a Certificate of Lawful Use, the court ordered that the building be demolished because the straw bales and tarpaulin were, in the eyes of the law, necessary parts of the building operation and the building was not substantially complete until they had been removed. Importantly, however, the judge ruled that concealment did not in itself provide a legitimate basis for the enforcement to succeed, as hiding something does not take away the lawful rights that accrue due to the passage of time. This was considered by many to be an unsatisfactory ruling. In the *Beesley* case, a farm building for agricultural purposes was given consent but was then used purely for residential purposes for a period in excess of four years. Subsequently, the Beesleys applied for a Certificate of Lawful Use for residential purposes. Again, the application was refused and the case went all the way to the Supreme Court, which ruled that not to take action against such concealed development would be unthinkable:

> It would damage public confidence in planning legislation and any law abiding citizen would be astonished to suppose that the (owner's) dishonest scheme, once discovered, would not be enforced against, but rather crowned with success, thereby undermining the maxim *volenti non fit injuria* that 'one cannot profit from one's own wrongdoing'.
>
> (Welwyn Hatfield BC v. Secretary of State for Communities and Local Government & Beesley *(2011 2 WLR 905))*

It is significant to note that whilst the remedy implemented was essentially identified back in 1947, when it was similarly applied to mineral development, its wider value was not appreciated for sixty-four years. Importantly, the evidence from the courts, compounded with the provisions of the Localism Act 2011, suggests not only Northern Ireland's failure to learn from good practice but irrationality in the adoption of an approach which precipitates some of the problems that have haunted the system for decades.

The Planning Contravention Notice

The Planning Contravention Notice (PCN), which requires the recipient to provide information on activities on land, is provided for under sections 133-134 of the 2011 Act. Specifically, where it *appears* to a council that there may have been a breach of planning control in respect of any land in its district, it may serve a PCN on any person who is the owner or occupier of the land or has any other estate in it; or is carrying out operations on the land or is using it for any purpose. A PCN may require the person on whom it is served to give such information as to any operations being carried out on the land, any use of the land and any other activities being carried out on the land; and any matter relating to the conditions or limitations subject to which any planning permission in respect of the land has been granted, as may be specified in the Notice. It may also require the person on whom it is served, so far as that person is able, to state whether or not the land is being used for any purpose specified in the Notice or any operations or activities specified in the Notice are being or have been carried out on the land; to state when any use, operations or activities began; to give the name and postal address of any person known to that person to use or have used the land for any purpose or to be carrying out, or have carried out, any operations or activities on the land; to give any information that person holds as to any planning permission for any use or operations or any reason for planning permission not being required for any use or operations; to state the nature of that person's estate (if any) in the land and the name and postal address of any other person known to the first mentioned person to have an estate in the land. Furthermore, a PCN may give notice of a time and place at which any offer which the person on whom the Notice is served may wish to apply for planning permission, to refrain from carrying out any operations or activities or to undertake remedial works; and any representations which that person may wish to make about the Notice, will be considered by the council, and the council must give that person an opportunity to make in person any such offer or representations at that time and place.

The PCN must also inform the person on whom it is served of the likely consequences of failing to respond to the Notice and, in particular, that enforcement action may be taken. It must also explain the effect of section 185(5)(b) of the 2011 Act whereby no compensation is payable in the case of a claimant who was required to provide information under section 133 in respect of any loss or damage suffered by the claimant which could have been avoided if he or she had provided the information or had otherwise cooperated with the council or, as the case may be, the Department for Infrastructure (hereinafter referred to as the Department) when responding to the Notice. Furthermore, the 2011 Act states that any requirement of a PCN must be complied with by giving information in writing to the council and, under section 134, if, at any time after the end of the period of 21 days beginning with the day on which a PCN has been served on any person, that person has not complied with any requirement of the Notice without proving reasonable excuse, that person shall be guilty of an offence and liable on summary conviction to a fine not exceeding level 5 on the standard scale. Similarly, if the recipient of a PCN makes any statement purporting to comply with a requirement of a PCN which that person knows to be false or misleading in a material particular; or recklessly makes such a statement which is false or misleading in a material particular, that person shall be guilty of an offence and liable on summary conviction to a fine not exceeding level 5 on the standard scale.

It is not possible to appeal to the Planning Appeals Commission (PAC) against the serving of a PCN, however, as set out in *R v Teignbridge DC, ex parte Teignmouth Quay Co. Ltd* (1995 2 PLR), prior to serving a PCN, it must be *apparent* to the planning authority that a breach of planning control has taken place and not because it thinks there *might* have been a breach. In the above case, there was nothing to suspect that the development was in breach of the regulatory framework but a PCN was served. It was suggested that this might have been with a view

> to appeasing local residents who had objected to the development; and satisfying the Teignmouth Harbour Environmental Co-ordinating Committee, which had doubts about the legality of the development.
>
> *(RICS, accessed 3.8.21)*

In that context, the Teignmouth Quay Co. successfully sought a court order to quash the PCN where it was ruled that the planning authority

> must reach its own conclusion as to whether there may have been a breach of planning control, and refusal by a landowner to co-operate with inquiries by a council is not sufficient, on its own, to satisfy the minimal statutory requirements without which the Contravention Notice may not be served.
>
> *(RICS, accessed 3.8.21)*

In summary, the PCN sends out a deterrent message to the recipient which may be adequate to discontinue a breach of planning control whilst, simultaneously, enabling the planning authority to prepare the ground for the issuing of an Enforcement Notice where it is expedient to do so (Cameron Blackhall 2017).

The Breach of Condition Notice

The provision of the Breach of Condition Notice (BCN) under section 152 of the 2011 Act is crafted to remedy a preexisting problem whereby the recourse was to employ other weaponry against a breach of a planning condition, thereby incurring the difficulties associated with collating substantive evidence to sustain formal enforcement action. Indeed, it has been suggested that the often-protracted complexity of the process led to an increased lack of compliance as the threat of any deterrent sanction was *de minimis*.

> In 1989, evidence was given to the review of enforcement procedure conducted by Robert Carnwath QC that the Enforcement Notice procedure was insufficiently flexible to secure the enforcement of conditions. The view was expressed that '...enforcement action was rarely an efficient means of dealing with breaches of conditions relating to the period of construction of a project on such things as noise or working hours, since a Stop Notice would be too drastic and in most cases enforcement is too slow and unsure'. It was thought there was a need for a summary remedy which would enable conditions to be enforced without enabling the merits to be reopened through the full panoply of an enforcement appeal.
>
> *(Moore and Purdue 2014: 397)*

In deciding whether to serve a BCN, it is the responsibility of a council not only to ensure that there has been an apparent breach of condition but also to ensure that the planning condition in question is precise, reasonable, necessary, enforceable, relevant to planning and relevant to the development to be permitted. Specifically, that it is legally valid. Under the provisions of section 152 the council may, if any of the conditions are not complied with, serve a BCN on any person who is carrying out or has carried out the development; or any person having control of the land, requiring that person to secure compliance with such of the conditions as are specified in the Notice. A BCN must specify the steps which the council considers ought to be taken, or the activities which the council considers ought to cease, to secure compliance with the conditions specified in the Notice. It must also specify a period allowed for compliance with the Notice, which must not be less than twenty-eight days beginning with the date of service, or a period as extended by a further Notice served by the council on the person responsible. If at any time after the end of the period allowed for compliance with the Notice any of the conditions specified in the Notice is not complied with, and the steps specified in the Notice have not been taken or, as the case may be, the activities specified in the Notice have not ceased, the person responsible is in breach of the Notice. If the person responsible is in breach of the Notice they are guilty of an offence. In this context, it is a defence for a person charged with an offence to prove that he/she took all reasonable measures to secure compliance with the conditions specified in the Notice; or that the recipient no longer had control of the land. If found guilty of an offence the recipient of the BCN is liable, on summary conviction, to a fine not exceeding level 3 on the standard scale.

A recent case in Northern Ireland throws an interesting light on dealing with alleged breaches of planning conditions. *In the Matter of an Application by William Donnelly for Leave to Apply for Judicial Review* (2020 NIQB 35), the applicant sought to challenge the failure of Fermanagh and Omagh District Council to take enforcement action in respect of alleged breaches of condition in the planning permission granted to mining works near his home. More specifically, a planning condition restricted works within an area of blanket bog, but boreholes were drilled in 2015 in breach of that planning permission. Nonetheless, the council determined that it would not proceed with enforcement having taken into account the passage of time, the absence of any evidence of physical damage as a result of the boreholes, the opinions of other relevant statutory agencies and the absence of any satisfactory remedy. In his consideration Humphreys J opined:

> the law affords a margin of appreciation or latitude to the decision maker. I am satisfied in this case that the Council took into account all material considerations and that the decision not to pursue enforcement action was a proportionate one in all the circumstances … The statutory enforcement regime does not concern itself with punishment, at least in the first instance. As section 140 of the 2011 Act sets out, an Enforcement Notice should both set out the breach of planning control identified and the steps which the Council requires to be taken or the activities which it requires to cease. Given that the unlawful drilling took place over 4 years ago, and has not occurred since, and the lack of any physical damage to be remedied, it is difficult to see what purpose an Enforcement Notice would serve. As such, I have concluded that the Council was entitled to arrive at the decision which it did, namely that it was not expedient in all the circumstances to issue an Enforcement Notice. The claim that such a decision was irrational in the *Wednesbury* sense was plainly unarguable. This is not one of the rare cases in which the Court should intervene in relation to the failure to take enforcement action.

The Enforcement Notice

Section 131 of the 2011 Act states that the carrying out of development without the planning permission required, or failing to comply with any condition or limitation subject to which planning permission has been granted, constitutes a breach of planning control whilst the issue of an Enforcement Notice, or the service of a Breach of Condition Notice, constitutes taking enforcement action.

Development carried out without planning permission is, therefore, unlawful, but the provisions of the 2011 Act do not make it a criminal offence even if planning permission has been applied for, refused and dismissed on appeal. However, once an Enforcement Notice has been issued, comes into an effect and has not been complied with, an offence has been committed. In relation to Enforcement Notices, there are two separate offences as set out below.

(1) *The 'Owner' offence.* Where the owner of the land to which an Enforcement Notice relates is in breach of the Notice, he or she shall be guilty of an offence (section 147 (2) of the 2011 Act). 6.22 Where any person has been charged with a section 147 (2) offence, it is a defence for that person to show that he or she did everything they could be expected to do to secure compliance with the Notice (section 147 (3) of the 2011 Act).
(2) *The 'Person in Control' offence.* Section 147 (4) of the 2011 Act states that any person who has control of, or an estate in the land to which the Notice relates (other than the owner) must not carry on any activity which is required to cease or cause to permit that activity to carry on. If he or she does so, he or she is in breach of the Notice and guilty of an offence (section 147 (5) of the 2011 Act) (Department for Infrastructure 2016).

Sections 138 and 139 of the 2011 Act provide the council or the Department respectively to serve an Enforcement Notice. The power to serve is discretionary where it appears that there has been a breach of planning control, and it is considered expedient to do so having regard to the provisions of the development plan and to any other material considerations. The council or the Department does not have to satisfy itself beyond doubt that a breach has occurred or that there are no possible grounds for appeal; it is for the appellant to establish such grounds (*Ferris v SSE* (1988 JPL 777)). However, the matters which are the subject of an action must have taken place; a prospective breach is not sufficient (*R v Rochester-Upon-Medway CC, ex-party Hobday* (1990 JPL 17).

Contents and effect of an Enforcement Notice

An Enforcement Notice must state the matters which appear to the council or, as the case may be, the Department, to constitute the breach of planning control, and must specify the steps which the council or, as the case may be, the Department requires to be taken, or the activities which the council or the Department requires to cease, in order to achieve, wholly or partly, any of the following purposes: remedying the breach by making any development comply with the terms (including conditions and limitations) of any planning permission which has been granted in respect of the land, by discontinuing any use of the land, or by restoring the land to its condition before the breach took place; or remedying any injury to amenity which has been caused by the breach. An Enforcement Notice may, for example, require the alteration or removal of any buildings or works; the carrying out of any building or other operations; any activity on the land not to be carried on except to the extent specified in the Notice; or the contour of a deposit of refuse or waste materials on land to be modified by altering the gradient or gradients of its sides. Where an Enforcement Notice is issued in respect of a breach of planning control consisting of the demolition of a building, the Notice may require the construction of a replacement building which is as similar as possible to the demolished building. In this context, a replacement building must comply with any requirement imposed by any statutory provision applicable to the construction of buildings and may differ from the demolished building in any respect that, if the demolished building had been altered in that respect, would not have constituted a breach of planning control.

An Enforcement Notice must specify the date on which it is to take effect. The Notice must also specify the period at the end of which any steps are required to have been taken or any activities are required to have ceased and may specify different periods for different steps or activities. The Enforcement Notice must also specify such additional matters as may be prescribed, and regulations may require every copy of an Enforcement Notice served under section 138 or 139 to be accompanied by an explanatory note giving prescribed information as to the right of appeal under section 143.

Section 140 of the 2011 Act goes on to state that where an Enforcement Notice in respect of any breach of planning control could have required any buildings or works to be removed or any activity to cease, but does not do so, and all the requirements of the Notice have been complied with, then, so far as the Notice did not so require, planning permission shall be treated as having been granted[10] by virtue of section 55 of the 2011 Act in respect of development consisting of the construction of the buildings or works or, as the case may be, the carrying out of the activities. Furthermore, where an Enforcement Notice requires the construction of a replacement building, and all the requirements of the Notice with respect to that construction have been complied with, planning permission shall be treated as having been granted by virtue of section 55 in respect of development consisting of that construction.

With regard to the contents of the Notice, it is important that the allegation is clearly specified. Specifically, an Enforcement Notice must tell the recipient in plain terms what he has done wrong and what he needs to do to put it right (*Miller-Mead v Minister of Housing and Local Government* (1963 22 QB 196)).

Where an Enforcement Notice relates to a material change of use, as set out in *Westminister CC v SSE and Aboro* (1983 JPL 602), it is not essential to recite the previous use; rather, it is sufficient to allege that there is a material change to the new use. In situations where there is an alleged breach of condition, for purposes of clarity, it is important to provide sufficient information on the planning permission and condition to which the breach relates. Whilst it is important to provide an accurate plan, the absence of this per se does not make the Notice a nullity but may constitute a defect which it may or may not be possible to correct without injustice. In this context, the case of *Coventry Scaffolding Company (London) Ltd v Barker* (1987 JPL 127) is of particular interest. Here, the Enforcement Notice was not deemed to be a nullity when it did not specify the street number of the premises where the alleged breach took place. Furthermore, as in *R v SSE ex parte Ahern (London) Ltd* (1989 JPL 757), issues emerging regarding the allegation may be capable of correction. Indeed, it is apparent that there is the expansive use of this power to correct even fundamental errors (PAC 2003). Furthermore, it is important to note that, as in *Kaur v SSE (*1990 JPL 814), requesting that a situation should be remedied by undertaking work 'in accordance with the scheme to be submitted to and approved by the Planning Authority' invalidates enforcement.

The purpose of the steps to be taken is to make the development comply with the terms of any planning permission granted or, wholly or partly, to remedy the breach of control or any injury to amenity emerging from it (Harwood 2020). Importantly, a breach of planning control does not have to be remedied in total when something less will suffice. The steps to remedy a breach must always be both clear and unambiguous, and, as in *Bath CC v SSE* (1983 JPL 937), an Enforcement Notice should not impose more onerous steps than are required to restore the land to its condition before the breach took place and where a use has induced unlawful intensification it may be suitable to specify that this should be reduced to an acceptable level (PAC 2003).

Dates and time intervals

A series of key dates and time interval requirements must be complied with in preparing and serving an Enforcement Notice. In the first instance, the date on which the Notice is issued is critical as appropriate serving can only take place not more than twenty-eight days after issue. In the same context service of the Enforcement Notice must take place at least twenty-eight days before the date on which it comes into effect in order to allow the recipient a reasonable time period in which to lodge an appeal to the PAC. In the event of an appeal not being lodged, the Notice comes into effect on the date specified and the period for compliance with the steps required to remedy the breach of control commences. The time periods contained in the Notice for compliance with the specified steps to remedy should not be unreasonable. As set out in *R (Lynes and Lynes) v. West Berkshire DCl* (2003 JPL 1137), any Notice which fails to contain a period or periods for compliance or where compliance is required 'immediately or forthwith' is a nullity.

Under sections 141 and 142 of the 2011 Act the council or the Department, respectively, may withdraw an Enforcement Notice issued by it; or waive or relax any requirement of such a Notice and, in particular, may extend any period specified in accordance with section 140 (9).[11] These powers may be exercised whether or not the Notice has taken effect. Importantly, the council or, as may be the case, the Department must, immediately after exercising the powers, give notice of the exercise to every person who has been served with a copy of the Enforcement Notice or would, if the Notice were reissued, be served with a copy of it. Furthermore, the withdrawal of an Enforcement Notice does not affect the power of the council to issue a further Enforcement Notice.

Appeals against an Enforcement Notice

Under the provisions of section 143 of the 2011 Act a person having an estate in the land to which an Enforcement Notice relates or a person who on the date on which the Enforcement Notice is issued occupies the land to which it relates by virtue of a licence and continues to occupy the land as aforesaid when the appeal is brought, may, at any time before the date specified in the Notice as the date on which it is to take effect, appeal to the PAC against the Notice, whether or not a copy of it has been served on that person. Under section 143 (3) an appeal may be brought on any of the following grounds:

(a) that, in respect of any breach of planning control which may be constituted by the matters stated in the Notice, planning permission ought to be granted or, as the case may be, the condition or limitation concerned ought to be discharged;
(b) that those matters have not occurred;
(c) that those matters (if they occurred) do not constitute a breach of planning control;
(d) that, at the date when the Notice was issued, no enforcement action could be taken in respect of any breach of planning control which may be constituted by those matters;
(e) that copies of the Enforcement Notice were not served as required by section 138 or, as the case may be, section 139 of the 2011 Act;
(f) that the steps required by the Notice to be taken, or the activities required by the Notice to cease, exceed what is necessary to remedy any breach of planning control which may be constituted by those matters or, as the case may be, to remedy any injury to amenity which has been caused by any such breach;

(g) that any period specified in the Notice in accordance with section 140 (9) falls short of what should reasonably be allowed.

An appeal under section 143 of the 2011 Act must be made by serving written notice of the appeal on the PAC before the date specified in the Enforcement Notice as the date on which it is to take effect, and such notice must indicate the grounds of the appeal and state the facts on which it is based. Before determining an appeal under section 143, the PAC must, if either the appellant or the council or, as the case may be, Department so desires, afford to each of them an opportunity of appearing before and being heard by the Commission.[12] Where an appeal is brought under section 143, the Enforcement Notice shall be of no effect pending the final determination or the withdrawal of the appeal.

The grounds of appeal: appeal under Section 143 (3) (a)

With regard to an appeal brought under section 143 (3) (a), specifically the grounds that, in respect of any breach of planning control that may be constituted by the matters stated in the Notice, planning permission ought to be granted or, as the case may be, the condition or limitation concerned ought to be discharged; any assessment as to whether planning permission should be granted must not take account of the fact that development has been carried out without permission. Importantly, existing development may be a material consideration as set out in *R -v- Westminster DC exp Pothecary* (1998 JPL 335). This considers the 'fall back' position. Specifically, the likely outcome if the Notice is upheld is an important material consideration to the merits of any deemed application. Similarly, as set out in *Burge v SSE* (1988 JPL 487), permitted development rights are a material consideration to the planning merits. On appeal to the PAC, as set out in *Hobday v SSE* (1990 JPL 923), whilst the presiding Commissioner is not bound to seek out remedies which might be appropriate but which are not raised by the parties, planning conditions which are proposed by the parties as compromise remedies must be taken into account (PAC 2003). Furthermore, natural justice demands that no party should be taken by surprise by conditions attached to any permission issued. In situations where the breach of planning control is non-compliance with a condition, particular care must be taken as the aim of the appeal is the discharge of the condition or the substitution of something less onerous. Although it is possible for a Commissioner to substitute a more onerous condition, strong justification is required, as the overarching rationale is that the severity of the requirements of the Notice as a whole cannot be increased (PAC 2003).

The grounds of appeal: appeal under Section 143 (3) (b)

With regards to an appeal brought on ground (b) under section 143 (3) that the matters stated in the Notice have not occurred; this is simply a matter of what was happening on the site or premises on the date of issue of the Notice.

The grounds of appeal: appeal under Section 143 (3) (c)

In the case of an appeal on ground (c) of section 143 (3), that those matters (if they occurred) do not constitute a breach of planning control, the relevant date is the date the Notice was issued (see above), not the date of the appeal or the date of the decision. Whilst the definition of development is set out in section 23 of the 2011 Act, significantly, the same is not true for the definition of a material change of use which depends on matters relating to fact and degree, extent and intensity. The provisions of section 23 (32) (e) which specifies 'in the case of buildings or other lands which are used for the purpose of any class specified in an order made by the Department for the purpose of this section, the use of the buildings or other land or, subject to the provisions of the order, of any part thereof for any other purpose of the same class' only apply once the authorised use has been implemented (*Kwik Save Discount Group Ltd. v Secretary of State for Wales* (1981 JPL 198). There is no requirement for planning consent for a change within the same use class, even if the characteristics are significantly different (*Carpet Decor Limited v SSE* (1981 JPL 806)), but the converse does not necessarily apply (PAC 2003); the Planning (Use Classes) Order Northern Ireland 2015 is permissive only and not restrictive, and it is a question of fact and degree whether uses within different classes are, in fact, materially different. Change of use is usually indicated by a change from one class to another or to a *sui generis* use. However, this still remains a matter of fact and degree. In this context, it is helpful to have a framework to help establish whether or not a COU has occurred. In this context, there is a helpful common law framework containing an overarching question – what is the planning unit? – along with three subsidiary questions – what is the primary use of that unit? Are there any lawful ancillary uses? What is the status in law of the existing use? Whilst neither the courts nor the statutes tell us what the planning unit is, as a general rule, the planning unit can be taken as the whole of an area under the same ownership or occupancy unless, within that whole, there exists another unit or units which have a separate function and/or which are physically distinct areas occupied for a substantially different or unrelated purpose. Leading cases in this respect are *Church Commissioners for England v Secretary of State for the Environment* (1995 2 PLR 99) and *Burdle and Williams v New Forest DC and SSE* (1972 1WLR 1207). These cases set out three broad tests, which are as follows:

1. *What is the primary use of that unit?* The purpose of this is twofold: to establish the existing use against which a new or proposed use can be compared and to provide a baseline against which to consider other ancillary uses which may exist. The established principle of ancillary (or incidental) uses recognises that new activities may be started in a building or within its curtilage without any need for permission, provided they remain ancillary to the main use.
2. *Are there any lawful ancillary uses?* In this case, there are three key issues. First, are they related to the planning unit; second, are they related to the primary use; and third, are they subsidiary or incidental to the primary use? In this context whenever an ancillary use intensifies to the extent where it becomes a primary use per se, then it may be apparent that a material change of use has occurred. In determining whether an activity is located outside the definition of development contained in section 23 (32) (c) of the 2011 Act the assessment should ask if it is incidental to the normal use. The provision of a factory canteen for use by the workers in the factory during the operational hours of the factory might well be within the scale regarded as ancillary or incidental to the primary use and as such within the exclusions. However, if it remains open when the factory is closed and invites members of the public, it is likely a material change of use will result. Similarly, as set out in *Wallington v SSW* (1991 JPL 942), the keeping of dogs in large numbers will amount to a material change of use if they are outside what normally occurs within the dwelling house and its curtilage. In this case the landowner

appealed against an Enforcement Notice, saying that though she kept a large number of dogs (forty-four), this was for domestic pleasure purposes and was only incidental to the use as a private domestic dwelling. The appeal, however, failed. It was held that it is sensible to have regard to what would be the normal use of a dwelling house (i.e., a generic approach) in order to decide whether, as a matter of fact and degree, the activities should reasonably be regarded as incidental to the enjoyment of the particular dwelling house as a dwelling house (PAC 2003). A use involving such a large number of dogs was not merely incidental.

3. *What is the status in law of the existing use?* Again, there are three issues to be considered. First, does the use benefit from planning permission; second, is the use immune from enforcement action; and third, is the use a nil use (*Hughes v Secretary of State for the Environment, Transport and the Regions and Another* (2000 1 PLR 76)?[13] This brings into play the concept of abandonment, which involves the cessation of a use in such a way and in such a time as to give the impression to a reasonable onlooker (applying an objective rather than a subjective test) that it was not to be resumed. The expressed intentions and the actions of the appellant are relevant. Once planning permission has been granted it inures for the benefit of the land and cannot be abandoned while it is still capable of implementation (*Pioneer Aggregates v SSE and Another* 1983 JPL 733) (PAC 2003). However, where there are two permissions which are clearly inconsistent, only one can be implemented. A use cannot survive the destruction of buildings and installations necessary for it to be carried on. If permission is granted for a new building to be put to a new use, the previous use rights on the open site will be expunged, but a replacement building within a planning unit can still be put to the use which continues in respect of the unit as a whole.

Once it is concluded that development has taken place, the next step is to decide whether planning permission is required. This calls for detailed consideration of the terms of any existing planning permission, the possibility of reversion to the previous lawful uses on the expiry of temporary permission and the provisions of the Planning (General Permitted Development) Order (Northern Ireland) 2015.[14]

In condition cases it may be argued that the condition which was the subject of the Notice has been complied with or was invalid from the start or fails the accepted tests for conditions. If a condition is invalid, no breach of planning control can be founded on it and the appeal under ground (c) should succeed. The condition in question must also be still in force, for example, a condition requiring that a dwelling be completed strictly in accordance with the submitted plans would not be effective against alterations made with the benefit of permitted development rights which arise once the dwelling is completed and occupied. Only a permission which expressly prohibited such works and specifically excluded permitted development rights would be effective against subsequent alterations which would otherwise be permitted development. In dealing with Enforcement Notices concerning breaches of condition, note the case *Handoll and Suddick-v-Messrs Warner Goodman and Streat and others DC* (1995 JPL 930), which states that if the planning permission granted did not authorise the building constructed, compliance with the conditions of that permission cannot be enforced (PAC 1998). As an alternative to an Enforcement Notice, a Breach of Condition Notice may be served. If an appeal succeeds under ground (c), the remaining grounds do not fall to be considered since there has been no breach of planning control.

The grounds of appeal: appeal under Section 143 (3) (d)

In the case of an appeal on ground (d) of section 143 (3), that at the date when the Notice was issued, no enforcement action could be taken in respect of any breach of planning control which may be constituted by those matters; the five-year rule applies to all operational development and material changes of use under the provisions of section 132 of the 2011 Act. However, it is important to be cognisant of the principle of 'second bite'. The 'second bite' provision allows a planning authority to take enforcement action in certain circumstances, notwithstanding that the limitation periods for enforcement action have passed. The planning authority can take further enforcement action in respect of any breach of planning control if, during the period when enforcement action can be taken ends with that action being taken, the planning authority has taken or purported to take enforcement action in respect of that breach.

The principle is demonstrated in the case of *Jarmain v Secretary of State for the Environment, Transport and the Regions* (2000 EWCA Civ 126), which indicates that the courts will not tolerate a return to 'arid technicalities' (Carnwath 1989). In this case, the appellant had obtained temporary planning permission in 1983 for a mobile home, which, after a number of renewals, expired in 1995. When a further attempt to renew the approval failed, it was discovered by the planning authority that the accommodation had been changed by the appellant from a mobile home to a permanent single-storey dwelling. An Enforcement Notice was issued by the planning authority in March 1996 referring to the condition on the most recent planning permission requiring the use to be discontinued and the site restored to its former condition. The Notice did not refer to the new development and instead stated: It appears that the condition has not been complied with because the mobile home remains on the land following the expiry of the said planning condition.

An appeal was lodged against the Enforcement Notice, and in April 1997 four years after the mobile had been transformed into a single-storey dwelling, an application was made for a lawful development certificate.[15] The local planning authority withdrew the Enforcement Notice and issued a further one that did allege the carrying out of development without planning permission. The appellant appealed against this Notice on the following ground: At the date when the Notice was issued, no enforcement action could be taken in respect of any breach of control which may be constituted by those matters.

This was rejected both by the inspector and by a deputy judge when the decision was challenged in the High Court. Before the Court of Appeal, everything turned upon the proper approach to section 171B (4) (b) of the Town and Country Planning Act (1990) (inserted by the 1991 Act). This allows further enforcement action in respect of any breach of planning control if, during the period of four years ending with that action being taken, the local planning authority has taken or purported to take enforcement action in respect of that breach. The court recognised that a purist approach to the wording of the section would require the breach that founded the first Notice to be identical to the breach that founded the second in terms of description as well as substance. Both the inspector and the deputy judge had both taken a pragmatic approach to the interpretation of the section, placing greater emphasis on the physical reality than the technical description of the alleged breach (Edwards and Martin 2000). Brooke LJ in the Court of Appeal said that the latter approach was correct and to allow the appeal would be to allow a return to 'the world of pettifoggery and arid technicalities' that had attracted such strong judicial disapprobation earlier.[16]

The grounds of appeal: appeal under Section 143 (3) (e)

In the case of an appeal on ground (e) of section 143 (3), that copies of the Enforcement Notice were not served as required by section 138 or, as the case may be, section 139; copies of the Notice must be served on owners, occupiers or any person having any estate in the land, which is materially affected by the Notice. The term owner is defined in the legislation, but occupier is not and the status of an occupier should be ascertained. Occupiers may be leasees, licensees by virtue of an oral or written licence or mere trespassers who have no right of appeal. Relevant factors are the degree of control, duration of occupation and the nature of the occupation. Caravan dwellers who have occupied a site for some time are occupiers (*Stevens v London Borough of Bromley* (1972 23 PCR 142) as are residents of bedsitting accommodation but market-stall holders who pay a weekly licence fee are not.

The grounds of appeal: appeal under Section 143 (3) (f)

In the case of an appeal on ground (f) of section 143 (3) that the steps required by the Notice to be taken, or the activities required by the Notice to cease, exceed what is necessary to remedy any breach of planning control which may be constituted by those matters or, as the case may be, to remedy any injury to amenity which has been caused by any such breach; the legislation allows the council or, where appropriate, the Department to specify the steps to achieve, wholly or partly, remedying the breach of planning control or remedying any injury to amenity. In the case of a building erected without planning permission, the appropriate requirement is to demolish it and remove the materials. If the breach is an unauthorised material change of use, then the normal requirement will be to discontinue the use and remove any equipment or materials brought onto the site to sustain it. In condition cases the aim should be to make the development comply with any conditional planning permission granted. In occupancy condition cases, the appropriate requirement is simply to comply with the condition leaving the appellant with the choice as to exactly how to comply.

Importantly, it is possible for an Enforcement Notice relating to an unlawful material change of use to include requirements to remove operational development carried out solely to facilitate the unauthorised development. In *Somak Travel v SSE* (1987 55 P&CR 250; 1987 JPL 630), a residential unit had undergone an unauthorised change of use to an office. Integral to this was the installation of a staircase linking to the shop on the floor below. In this case, even though the staircase was internal works and not development in its own right, the Enforcement Notice against the change of use to the office could lawfully require its removal (Harwood).[17]

The requirements must not purport to prevent an appellant from doing something which he/she is entitled to do without planning permission. This is the principle as enunciated in *Mansi v Elstree RDC* (1964 16 PCR 153), set out in *South Ribble Borough Council v SSE* (1990 JPL 808). An example is that of farm gate sales of home-grown produce. Such sales, when conducted from part of the planning unit on which the produce is grown, are considered to be ancillary to the agricultural use of the farm as a whole, however great the volume (*Allen v SSE* 1990 JPL 340). The Mansi principle extends to degrees of use which were lawful or have become immune from enforcement (*Denham Developments v SSE* 1974 JPL 347). The requirements should be stated precisely, and vague language should be avoided (PAC 2003). Such a defect could render a Notice a nullity (see below). The overriding principle is that the Notice should tell an appellant fairly what he has done wrong and what he must do to remedy it (*Miller-Mead v Minister of Housing and Local Government* 1963 22 QB 196).

The grounds of appeal: appeal under Section 143 (3) (g)

In the case of an appeal on ground (g) of section 143 (3) that any period specified in the Notice in accordance with section 140 (9) falls short of what should reasonably be allowed; although the Notice may have been issued sometime before the appeal decision, the appellant is entitled to assume success and is entitled to a reasonable period for compliance after the Notice takes effect. In considering what is reasonable, account should be taken of the physical steps required, possible difficulties in finding other premises and personal and financial hardship. A view is generally taken that periods for compliance exceeding twelve months are excessive though they have been permitted in exceptional circumstances. A longer period not only blunts the urgency but may call into question the expediency of issuing the Notice in the first place. Arguably, it may be appropriate to consider temporary permission, but it is very difficult to justify giving planning permission for development that would not be acceptable had it not already been carried out. It may be preferable to leave further time extensions to the discretion of the council, or where appropriate, the Department, rather than give temporary permission for an unacceptable development. The period for compliance runs from the date of the appeal decision, the effectiveness of the Notice having been suspended pending determination of the appeal.

Nullity and invalidity

An Enforcement Notice may be challenged on the grounds of whether it is a nullity or invalid. These grounds are significantly different, and it is important to distinguish between the two. It was held in *Millar-Mead v Minister of Housing and Local Government* (1963 22B 196); a Notice which is a nullity is 'so much waste paper' and it can in consequence be ignored completely by the recipient. The pertinent question to ask is: 'when will a Notice be a nullity as opposed to merely being invalid?' Upjohn LJ identified the following situations where a Notice would be a nullity in *Miller-Mead v Minister of Housing and Local Government* (1963 22 QB 196):

- where on its face the Notice fails to contain the matters required by the legislation,
- where the Notice is uncertain so that the recipient does not know what he has done wrong or what he is required to do.

A Notice may also be a nullity on the grounds of being *ultra vires*, for example, where the decision to issue it was made for improper reasons or because the planning authority did not believe that there had been a breach of control at the time of resolution as held in the case of *R v Rochester-upon-Medway BC, ex p Medway BC, ex p. Hobday* (1990 JPL 17). An Enforcement Notice is invalid where it has been issued outside the period permitted under the 2011 Act, or because it alleges a breach of planning control when none has in fact taken place. In this situation, the Notice cannot be ignored by the recipient, and appropriate action must be taken if the recipient is to avoid compliance with the Notice. In *Bracken v East Hertfordshire District Council* (2000 CO/686/2000) a builder had continued to store building materials and waste on agricultural land at the rear of his house without planning permission. East Hertfordshire District Council had previously served Bracken with an Enforcement Notice, alleging that he had changed the land from an agricultural use to the storage of building materials and waste. He was convicted of the offence in the magistrates court. Bracken contested the validity of the Notice, claiming that the items did not constitute building materials and that the storing of them did not amount to a material change of use. He said that the fencing and tiles were present when he bought the house in 1976. He also said that the wheelbarrows had domestic as well as work uses and that a digger was used for repairing the driveway and clearing horse manure. Lord Justice Roch, sitting with Lord Justice Harrison, said that a council planning officer had demonstrated through the use of photographs that there was 'ample evidence' on which the magistrates were entitled to make their findings. The Notice was valid on the basis that the appellant had been in breach of the Enforcement Notice on at least three occasions during the period stated. The evidence was overwhelming, and the case against the appellant was beyond reasonable doubt.

The Fixed Penalty Notice

A Fixed Penalty Notice may be issued by a council under sections 153, 154 and 155 of the 2011 Act as an alternative to prosecution in certain circumstances. The purpose of this additional enforcement measure is to assist councils and add rigour to the planning enforcement function (Department for Infrastructure 2016).[18] The Fixed Penalty Notice offers a recipient the opportunity to discharge liability for prosecution in respect of a breach of an Enforcement Notice or Breach of Condition Notice by paying the council a penalty of an amount specified in the Notice within twenty-eight days. This period commences on the day the Notice is served. Each council is responsible for developing a planning enforcement strategy which should encompass the application of Fixed Penalty Notice legislation. The underpinning rationale is that a Fixed Penalty Notice should be served where it is considered effective in addressing the breach and proportionate to the nature of the breach. In the event whereby an authorised officer of a council has reason to believe a person has committed an offence under section 147 or 152 of the 2011 Act within a council's district, a council may issue a Notice offering that person the opportunity of discharging any liability of conviction for that offence by payment of a fixed penalty to the council.

In the event where a Notice is issued under section 153 or 154 of the 2011 Act a council must ensure, first, that no proceedings are instituted for the offence before the expiration of twenty-eight days following the date of the Notice; and, second, that the recipient of the Notice shall not be convicted of that offence if the Fixed Penalty Notice is paid before the expiration of that period. In terms of the level of fine for a Fixed Penalty Notice, the thresholds are set out in the Planning (Amount of Fixed Penalty) Regulations (Northern Ireland) 2015. In this context, Article 2 (a) states that the amount payable in respect of a breach of an Enforcement Notice is £2,000, whilst Article 2 (b) states that the amount payable in respect of a Breach of Condition Notice is £300.

With regard to the contents of a Fixed Penalty Notice section 153 (3), section 153 (4) and section 154 (3) and (4) of the 2011 Act require that a Fixed Penalty Notice must specify or state the steps specified within the Enforcement Notice or Breach of Condition Notice which have not been taken; the activity specified within the Enforcement Notice or Breach of Condition Notice which has not ceased; the period during which proceedings will not be taken for the offence; the amount of the fixed penalty; and the person to whom and the address at which the fixed penalty may be paid.

With regard to payment of a Fixed Penalty Notice, sections 153 (6) and 154 (6) of the 2011 Act state that payment of the fixed penalty may be made by pre-paying and posting a letter containing the amount of the penalty (in cash or otherwise) to the appropriate authority. Where a letter is issued in accordance with subsection section 153 (6) or 154 (6) of the 2011 Act a council should regard payment as having been made at the time at which that letter would be delivered in the ordinary course of postal deliveries. The amount payable is to be reduced by 25 percent if payment is made within the first fourteen days of the period specified in section 153 (2), section 153 (10) and 154 (10) of the 2011 Act. Finally, there is no right of appeal against a Fixed Penalty Notice. However, a recipient can make representations to the council that the breach to which the Notice relates had, in fact, been corrected, and they should not therefore be required to pay the fixed penalty.

The Stop Notice

The result of an appeal against an Enforcement Notice means that the requirements of the Notice are suspended until the outcome of the appeal is known. The consequence is that the development can continue unabated without penalty. In certain situations, it may be the case that the abeyance of the Notice is of such concern that the planning authority considers it imperative to take urgent action to address the impacts of the alleged breach of planning control during the specified period for compliance contained in the Enforcement Notice. It can therefore serve a Stop Notice. Under section 150 of the 2011 Act where a council considers it expedient that any relevant activity should cease before the expiry of the period for compliance with an Enforcement Notice, it may, when it serves the copy of the Enforcement Notice or afterwards serve a Stop Notice referring to, and having annexed to it a copy of, the Enforcement Notice and prohibiting the carrying out of that activity on the land to which the Enforcement Notice relates, or any part of that land specified in the Stop Notice.[19] The Stop Notice cannot therefore be served on its own but must not only refer to but also have a copy of the Enforcement Notice attached. The Stop Notice cannot be served where the Enforcement Notice has taken effect nor can it prohibit anyone from continuing to use any building, caravan or other structure situated upon the land as their permanent residence whether as owner, occupier, tenant, patient, guest or otherwise. The date specified in a Stop Notice must be the date when the Notice is served, unless the council considers that there are special reasons for specifying a later date, but the date specified in the Notice must, in any case, be a date not later than twenty-eight days from the date when the Notice is first served on any person. The Stop Notice shall cease to have effect on the following grounds:

(a) the Enforcement Notice referred to in it is withdrawn or quashed; or
(b) the period for compliance with the Enforcement Notice expires; or
(c) notice of withdrawal of the Stop Notice is first served; or
(d) if or to the extent that the activities prohibited by it cease, on a variation of the Enforcement Notice referred to in it, to be relevant activities.

A Stop Notice may be served by the council on any person who appears to it to have an estate in the land or to be engaged in any activity prohibited by the Notice; and where a Stop

Notice has been served in respect of any land, the council may display a 'Site Notice' stating that a Stop Notice has been served; and that any person contravening the Stop Notice may be prosecuted, giving the date when the Stop Notice takes effect and indicating its requirements. The council may at any time withdraw a Stop Notice (without prejudice to its power to serve another) by serving notice to that effect on persons served with the Stop Notice; and if a Site Notice was displayed in respect of the Stop Notice, displaying a Notice of the withdrawal in place of the Site Notice. Anyone who contravenes a Stop Notice after a Site Notice has been displayed or the Stop Notice has been served on that person is guilty of an offence and liable on summary conviction, to a fine not exceeding £100,000, on conviction on indictment, to a limitless fine. In determining the amount of any fine to be imposed on a person convicted of an offence under this section, the court must in particular have regard to any financial benefit which has accrued or appears likely to accrue to that person in consequence of the offence. In proceedings for an offence under this section, it is a defence for the accused to prove that the Stop Notice was not served on the accused and that the accused did not know, and could not reasonably have been expected to know, of its existence.

Under section 185 of the 2011 Act, a person who, when a Stop Notice under section 150 or 151 is first served, has an estate in or occupies the land to which the Stop Notice relates may be entitled to be compensated by the council in respect of any loss or damage directly attributable to the prohibition contained in the Notice on the following grounds:

(a) the Enforcement Notice is quashed on grounds other than those mentioned in section 143 (3) (a);
(b) the Enforcement Notice is varied, otherwise than on the grounds mentioned in section 143 (3) (a), so that any activity the carrying out of which is prohibited by the Stop Notice ceases to be a relevant activity;
(c) the Enforcement Notice is withdrawn by the council or the Department otherwise than in consequence of the grant of planning permission for the development to which the Notice relates;
(d) the Stop Notice is withdrawn.

Compensation is not payable in respect of the prohibition in a Stop Notice of any activity which, at any time when the Notice is in force, constitutes or contributes to a breach of planning control; or in the case of a claimant who was required to provide information under section 133 or 240 in respect of any loss or damage suffered by the claimant which could have been avoided if he or she had provided the information or had otherwise cooperated with the council or, as the case may be, the Department when responding to the Notice.

As noted above, a Stop Notice may be issued under section 151 of the Planning Act (Northern Ireland) 2011, where the Department 'considers it expedient that any relevant activity should cease before the expiry of the period for compliance with an Enforcement Notice'. This provision became the subject of a judicial review in an *Application by Friends of the Earth* (2017 NICA 41), which concerned the failure of the Minister with responsibility for the then Department of the Environment to issue a Stop Notice to the Shaftesbury Estate of Lough Neagh Limited as owners of the bed of Lough Neagh and a number of businesses involved in sand extraction from Lough Neagh. It was reported in the Court of Appeal judgment that up to 1.5 million tons of sand are extracted from the lough each year, that sand extraction has been proceeding for eighty years without planning permission and that the lough is a Special Protection Area, a Ramsar site and an Area of Special Scientific Interest. A letter dated 25 September 2014 from the Department to the Shaftesbury Estate and the sand traders stated that the dredging of sand was unauthorised and that the activity should cease. Operations did not cease, and on 27 May 2015, the Department issued to the Shaftesbury Estate and the sand traders an Enforcement Notice that required the cessation of the working of minerals within one day of the Notice taking effect on 30 June 2015, unless an appeal was lodged beforehand. The Shaftesbury Estate and the sand traders appealed against the Enforcement Notice to the PAC the effect of which was that the Enforcement Notice had no effect pending the final determination or the withdrawal of the appeal. In July 2015 the Minister considered whether to issue a Stop Notice but deferred a decision pending an environmental report from the Northern Ireland Environment Agency. With officials becoming concerned about the ongoing extraction of sand from Lough Neagh pending the completion of appeal proceedings before the PAC, it was recommended in a submission to the Minister on 9 November 2015 that a Stop Notice should be issued compelling the cessation of operations and which was affirmed in a subsequent submission to the Minister on 19 November 2015. The Minister issued his decision on 20 November 2015 stating:

> A Stop Notice should not be issued at this time. In my view it would not be a proportionate response in a situation where there is no evidence that the dredging, which has been going on since long before the site's designations, is having any impact on the environmental features of the lough. It would not be in the wider public interest to risk such potential economic harm until there is some understanding of the environmental situation as outlined above. However, I will review the situation as soon as the relevant information is available.

In its judgment the Court of Appeal concluded that the matter of issuing a Stop Notice 'remains one of discretion for the decision maker', but that in this case it should be underpinned by applying the precautionary principle:

> We return to the decision under challenge and the statement of the Minister in the decision letter that there is 'no evidence that the dredging … is having any impact on the environmental features of the lough'. This is the wrong approach. It is acknowledged by the Department that these operations are likely to have a significant effect on the environment. It is not known what that effect will be. The precautionary principle applies. It operates on the basis that there should be no planning permission until it is established that there is no unacceptable impact on the environment. The Minister's decision proceeds on the basis that there is an absence of evidence of an unacceptable impact on the environment. The proper approach is to proceed on the basis that there is an absence of evidence that the operations are not having an unacceptable impact on the environment … What has been disregarded in the letter of decision, where it deals with the Stop Notice, is that these operations are considered likely to have significant impact, that the nature and extent of that impact has not been established, that prior to the grant of permission is the requirement to establish that there will be no significant impact and that it is imperative that the precautionary principle be applied. What must be put in the balance is the absence of evidence that there is no harm. To approach the matter with a requirement for evidence of harm is the negation of the precautionary principle. The

> decision will be quashed...The Court refers the matter back to the Department so that a determination may be made as soon as possible on whether or not to issue a Stop Notice.

The Temporary Stop Notice

A Temporary Stop Notice can be served under section 135 of the 2011 Act if the council considers that there has been a breach of planning control and it is expedient that the activity is stopped immediately.[20] A Temporary Stop Notice is unlike a Stop Notice in that it can be served independently from an Enforcement Notice and provides the council with a twenty-eight-day window to decide whether further enforcement action is appropriate and what action should be taken, without the breach intensifying by being allowed to continue. Stop Notices cannot prohibit a person from continuing to use any building, caravan or other structure situated on land to which the Temporary Stop Notice relates as that person's permanent residence. A Temporary Stop Notice must be in writing and must specify the relevant activity which the council considers amounts to the breach. It must prohibit the carrying on of the activity (or part of the activity as specified in the Notice). Furthermore, it must set out the council's reasons for issuing the Notice. The Temporary Stop Notice may be served on a person who the council *thinks*, first, is carrying on the activity, second, is an occupier of the land or, third, has an estate in the land.

In accordance with section 135 (5) of the 2011 Act, a copy of the Temporary Stop Notice must be displayed on the land along with a statement of the effect of the Notice and of the effect of contravening the Notice and thereby committing an offence. A Temporary Stop Notice has effect from the time a copy of it is displayed on the land and ceases to have effect at the end of the twenty-eight-day period, starting on the day the copy Notice is displayed or at the end of a shorter period specified in the Notice. Furthermore, the council has the power to withdraw a Temporary Stop Notice under section 135 (7) of the 2011 Act.

Any person who contravenes, or causes or permits the contravention of the provisions of a Temporary Stop Notice after it has been served on them or a Site Notice displayed, is guilty of an offence under section 137 (1) and (2) of the 2011 Act. Under section 137 (6) of the 2011 Act anyone guilty of contravening the Stop Notice is liable to a fine of up to £100,000 on summary conviction and on indictment to an unlimited fine. Under section 137 (5) of the 2011 Act, it is a defence for the accused to prove that the Temporary Stop Notice was not served on them and they did not know, and could not have been reasonably expected to know, of its existence. As is the case with a Stop Notice compensation may be payable in certain circumstances set out in section 187 (1). Specifically, the council may be liable to pay compensation to any person who had an estate in the land at the time the Temporary Stop Notice was served. Compensation may be paid if a Notice is issued and one or more of the following applies. First, the activity which is specified in the Notice is authorised by planning permission or a development order; second, a certificate of lawfulness of existing use or development in respect of the activity is issued or granted; or, third, withdrawal of the Temporary Stop Notice. The loss or damage sustained by a claimant includes any sum payable in respect of a breach of contract by the taking of action necessary to comply with the Temporary Stop Notice.

An injunction

The purpose of an injunction is to provide a legal and equitable remedy in the form of a special court order that compels a party to do or refrain from specific acts. The doctrine that reflects this is the requirement that an injunction can be given only when there is no adequate remedy at law. Injunctions are intended to make whole again someone whose rights have been violated. Nevertheless, when deciding whether to grant an injunction, courts also take into account the interests of non-parties (that is, the public interest) (OTS 2017). Under the provisions of section 156 of the 2011 Act, there is compendious injunction provision for planning, listed buildings, conservation areas and trees (Harwood 2020). Where the council considers it necessary or expedient for any actual or apprehended breach of planning control, any actual or apprehended contravention of sections 85 (1) or (5), 126 or 127; or, any actual or apprehended contravention of the hazardous substances control to be restrained by injunction, it may apply to the court for an injunction.

In this context, a ruling in Belfast is particularly interesting. Specifically, with regard to the building of an unauthorised wall around the premises of the Consul General of the People's Republic of China, which is located within a Conservation Area. Specifically, Belfast City Council ('the Council') applied *ex parte* for an interim injunction against Madame Zhang Meifang, the Consul General of the People's Republic of China, Belfast, seeking to restrain her or her servants and agents from carrying out certain development work at 75–77 Malone Road, Belfast BT9 6SH ('the premises') without the benefit of planning permission. However, the High Court in June refused to order a halt to the works after ruling that the Chinese Consul General, Madame Zhang Meifang, is protected by diplomatic immunity. In this case, the ruling (2020 NICh 12) was set out by McBride J as follows:

> Although I have found that the respondent enjoys immunity in this case, immunity is a procedure which protects someone who is carrying out what may otherwise be an unlawful act. I consider, on the basis of the papers before me that there is at least an arguable case that the acts complained of are unlawful acts because they are being done in breach of planning law. Gillian Ormiston in an email to the respondent's solicitors dated 28 February 2020 advised as follows: 'should planning permission be required to do the works intended – which seems to be the case (and have been advised that the address is within a Conservation area) – the Consulate MUST apply and gain the relevant planning permissions … Without prejudice to their privileges and immunities, it is the duty of all persons enjoying such privileges and immunities to respect the laws and regulations of the receiving State. You and your client should please bear in mind that there will likely be consequences if any unauthorised building/refurbishment takes place at the property. I very strongly recommend that you work with Belfast City Council Planning Authority to arrive at a satisfactory conclusion. For our part we are prepared to work with all parties to achieve such an outcome'. Mr Scoffield described the 'consequences' referred to in this email as the 'nuclear option' whereby the Secretary of State for Foreign and Commonwealth Affairs can withdraw the grant of permission to use the premises as consular premises and if he does so the Chinese State's right to occupy the premises will terminate. That would be a very serious consequence. Therefore, it is my view that it is all the more important at this stage for the respondent to engage in discussions with the Council so that those serious consequences do not occur. I hope that my ruling in this case, that Madame Zhang Meifang has immunity, will assist in having discussions re-commenced. It seems to me from the letter sent by the respondent's solicitors dated 18th June 2020 that if the respondent had been given an assurance by the Council that the respondent had immunity, the respondent would

> have continued to engage in discussions and would have applied for planning permission. It is my hope that the ruling I have given now gives that assurance and because of that the Consul General will hold to her agreement to continue to have meaningful discussions with the Council to re-apply for planning permission and in the interim period to stop, on a voluntary basis, the works that are being carried out.

At the time of writing (2021) a planning application has been submitted by the Consulate General of the People's Republic of China to Belfast City Council and is in the process of being determined.

Conclusion

The integrity of the planning system is dependent upon effective enforcement. Public respect can diminish rapidly if unauthorised development, which is unacceptable on planning grounds, is allowed to proceed without intervention before serious harm to amenity results. It is, therefore, essential that a solution is found before the situation reaches an intractable stage or, indeed, the problem becomes impossible to remedy. The onus is, therefore, on the planning authority to defend the public interest yet remain reasonable in its approach, thereby instilling confidence in the system and deterring would-be offenders from contravening regulations. Effective enforcement of the law, whilst simultaneously respecting the right of the individual, is never going to be easy. In sum, the enforcement puzzle is real and complex and, as can be seen above, the enforcement armoury continues to evolve, albeit sometimes with contentious results. It is apparent that there is no single solution but by taking steps to enhance the capacity to enforce and the commitment of offenders to comply, balancing the merits of the stick with those of the carrot, our rudimentary knowledge of how to resolve the complexities of the planning enforcement system can only grow.

Legislation

Caravan Sites and Control of Development Act 1960
Housing, Town Planning, Etc., Act 1909
Housing, Town Planning, &c. Act 1919
Human Rights Act 1998
Land Development Values (Compensation) Act (Northern Ireland) 1965
Local Government and Planning (Amendment) Act 1981
Localism Act 2011
Planning Act (Northern Ireland) 2011
Planning and Compensation Act 1991
Public Health Act 1875
The Planning (General Permitted Development) Order (Northern Ireland) 2015
The Planning (Northern Ireland) Order 1972
The Planning (Northern Ireland) Order 1991
Town and Country Planning Act 1947
Town and Country Planning Act 1962
Town and Country Planning Act 1968
Town and Country Planning Act 1971

Town and Country Planning Act 1990
Town and Country Planning (General Interim Development) Order 1933
Town Planning (General Interim Development) Order, 1922

Cases cited

Allen v SSE (1990 JPL 340)
Barling v SSE (1980 JPL 594)
Bath CC v SSE (1983 JPL 937)
In the Matter of Belfast City Council and Madame Zhang Meifang and The Consul General of the People's Republic of China, Belfast (2020 NICh 12)
Bracken v East Hertfordshire District Council (2000 CO/686/2000)
Burge v SSE (1988 JPL 487)
Burdle and Williams v New Forest DC and SSE (1972 1WLR 1207)
Carpet Decor Limited v SSE (1981 JPL 806)
Cheshire CC v Woodward (1962 2 Q.B. 126)
Church Commissioners for England v Secretary of State for the Environment (1995 2 PLR 99)
Coventry Scaffolding Company (London) Ltd v Barker (1987 JPL 127)
Denham Developments v SSE (1974 JPL 347) *Ferris v SSE* (1988 JPL 777)
Fidler v First Secretary of State Reigate & Anor (2004 EWCA Civ 1295)
Handoll and Suddick-v-Messrs Warner Goodman and Streat and others DC (1995 JPL 930)
Henry Boot Homes v Bassetlaw DC (2003 JPL 1030)
Hobday v SSE (1990 JPL 923)
Hughes v Secretary of State for the Environment, Transport and the Regions and Another (2000 1 PLR 76)
In an Application by Friends of the Earth (2017 NICA 41)
In the Matter of an Application by William Donnelly for Leave to Apply for Judicial Review (2020 NIQB 35)
Jarmain v Secretary of State for the Environment Transport and the Regions (2000 EWCA Civ 126)
Kaur v SSE (1990 JPL 814)
Kwik *Save Discount Group Ltd. v Secretary of State for Wales* (1981 JPL 198)
Lipson v SSE (1976 JPL 33PCR 95)
Mansi v Elstree RDC (1964 16 PCR 153)
Miller-Mead v Minister of Housing and Local Government (1963, 22 QB 196)
Murfitt v SSE (1980 JPL 598)
Nourish v Adamson (1998 JPL 859)
Parkes v Secretary of State (1978 1 WLR 1308)
Pioneer Aggregates v SSE and Another (1983 JPL 733)
R v East Sussex ex p. Reprotech (2002 JPL 821)
R (Lynes and Lynes) v. West Berkshire DCl (2003 JPL 1137)
R V Rochester-Upon-Medway CC, ex-party Hobday (1990 JPL 17)
R v SSE ex parte Ahern (London) Ltd (1989 JPL 757)
R v Teignbridge DC, ex parte Teignmouth Quay Co. Ltd (1995 2PLR 1)
R v Westminster DC exp Pothecary (1998 JPL 335)
Somak Travel v SSE (1987 55 P&CR 250; 1987 JPL 630)
South Ribble Borough Council v SSE (1990 JPL 808)
Stevens v London Borough of Bromley (1972 23 PCR 142)

Wallington v SSW (1991 JPL 942)
Welwyn Hatfield BC v. Secretary of State for Communities and Local Government & Beesley (2011 2 WLR 905)
Westminister CC v SSE and Aboro (1983 JPL 602)

Notes

1 Formerly referred to as development control.
2 See below for a fuller discussion.
3 Report of the Royal Commission on the Distribution of the Industrial Population (Cmd 6153, 1940)
4 Report of the Committee on Land Utilisation in Rural Areas (Cmd 6378, 1942)
5 Interim and Final Reports of the Expert Committee on Compensation and Betterment (Cmd 6291, 1941; and Cmd 6386, 1942)
6 Subject to subsections (2) to (6) of the 2011 Act.
7 Whereas the unauthorised development may be unacceptable per se, the situation can potentially be remedied via the imposition of planning conditions.
8 There are additional powers of enforcement relating to listed buildings, hazardous substances, advertisements and replacement of trees.
9 With the exception of material change of use to a single dwelling house where the time limitation period is four years.
10 Whereby planning permission which may be granted includes planning permission for development carried out before the date of the application.
11 The period at the end of which any steps are required to have been taken or any activities are required to have ceased.
12 Sections 41 (Notice, etc., of applications for planning permission)and 45(2) (whereby a development order may provide that a council or the Department must not determine an application for planning permission before the end of such period as may be specified by the development order) of the 2011 Act shall apply, with any necessary modifications, in relation to an appeal to the PAC under this section as they apply to an application for planning permission to the council or the Department.
13 It is established in *Barling v SSE* (1980 JPL 594) that ancillary or incidental use rights are extinguished on cessation of the primary use.
14 And also the Planning (General Permitted Development) (Amendment) Order (Northern Ireland) 2020.
15 The time limitation period in England is four years as opposed to five years in the jurisdiction of Northern Ireland.
16 The case should not be seen as opening the door to highly technical challenges to 'second bite' notices. The court in this case reached the decision it did with some reluctance and it is likely that in future cases there will be similar reluctance to retreat from the robust attitude of the court in *Jarmain*. There will have to be a real difference in the scope of a second bite notice and the earlier notice to justify the court interfering (Thomson Reuters 2021).
17 See also *Murfitt v SSE* 1980 JPL 598.
18 For fuller discussion see Department for Infrastructure 2016.
19 Under section 151 the Department may serve a Stop Notice but not without consulting the appropriate council. In this context a notice served by the Department under has the same effect as if it had been served by a council and, accordingly, the provisions of section 150 apply, with any necessary modifications, to the service of a Stop Notice by the Department as they apply to the service of a stop notice by a council.
20 For fuller discussion see Department for Infrastructure 2016.

References

Berry, J. 1991. 'The Planning and Compensation Act 1991: Planning Provisions', *The Property Journal (Northern Ireland)* 12(3): 1–11.
Bourne, F. 1992. *Enforcement of Planning Control*. Second Edition. London: Sweet and Maxwell.

Brand, C.M. 1988. *Enforcement of Planning Control.* London: Longman.

Cameron Blackhall, J. 2017. *Planning Law and Practice.* Third Edition. London: Routledge-Cavendish.

Carnwath, R. 1989. *Enforcing Planning Control.* London: HMSO.

Cherry, G.E. 1974. *The Evolution of British Town Planning.* London: Hill.

Cullingworth, B., V. Nadin, T. Hart, S. Davoudi, J. Pendlebury, G. Vigar, D. Webb and T. Townsend. 2015. *Town and Country Planning in the UK.* Fifteenth Edition. Oxon: Routledge.

Department of the Environment. 1975. *Investigation into Planning Enforcement, The Dobry Report.* London: HMSO.

Department of the Environment. 1980. *Enforcing Planning Control.* Circular 22/80. London: HMSO.

Department of the Environment. 1987. *Enforcing Planning Control.* Circular 4/87. London: HMSO.

Department of the Environment. 2009. *Reform of the Planning System in Northern Ireland: Your Chance to Influence Change.* Belfast: Department of the Environment.

Department for Infrastructure. 2016. *Enforcement Practice Note 4: Enforcement Procedures.* Belfast: Planning Guidance Team.

Edwards, M., and J. Martin. 2000. 'No More Pettifoggery', *Estates Gazette* 43: 179.

Grant, M. 1982. *Urban Planning Law.* London: Sweet and Maxwell.

Harwood, R. 2020. *Planning Enforcement.* Third Edition. Haywards Heath: Bloomsbury Professional.

Heap, D. 1996. *An Outline of Planning Law.* Eleventh Edition. London: Sweet and Maxwell.

Hendry, J. 1989. *The Development of Planning in Northern Ireland.* Belfast: Queen's University Belfast.

Humphreys, R. 2011. '20 Years of the Ten Year Period for Enforcement: Time for Reform', JPL 5: 522–23.

Law Society. 2011. *The Ten Year Period for Enforcement: Is It Really Time for Reform?* London: Law Society.

Lewis, M. 2003. 'Expediency in Enforcement – Planning's "Weakest Link?"', *Journal of Planning and Environment Law*: 55: 1109–12.

Millichap, D. 1998. *Planning Ghosts – Past, Present and Future*, London: Linklaters.

Ministry for Communities, Housing and Local Government. 2020. *Planning for the Future.* London: OGL.

Moore, V. and M. Purdue. 2014. *A Practical Approach to Planning Law.* Thirteenth Edition. New York: Oxford University Press.

McLoughlin, J.B. 1969. *Urban and Regional Planning: A Systems Approach.* London: Faber & Faber.

OTS.2017.https://www.otssolicitors.co.uk/news/your-guide-to-injunctions-and-emergency-remedies/.

PAC. 2003. *Appeals against Enforcement Notices.* Unpublished. Belfast: Planning Appeals Commission.

RICS. 2021. https://www.isurv.com/directory_record/4951/r_v_teignbridge_dc_ex_parte_teignmouth_quay_co_ltd.

Thomson Reuters. 2021. *Validity of 'Second Bite' Planning Enforcement Notices.* https://uk.practicallaw.thomsonreuters.com/8-107-1714?__lrTS=20180123013745784&transitionType=Default&contextData =%28sc.Default%29.

Wood, W. 1949. *Planning and the Law.* London: Marshall.

6

DEVELOPER CONTRIBUTIONS AND 'PLANNING AGREEMENTS'

Thomas Murphy, John McCord and Michael McCord

Introduction

The mechanisms used to secure social or community benefits from increases in land value through the planning system are longstanding topics of planning law and policy debate. Development, however beneficial, will almost inevitably result in some form of public cost consequences, which, if they are to be avoided or moderated, will require expenditure or improved infrastructure. A new housing development may, for example, mean increased traffic that may require upgraded roads or additional public transport provision; it may result in environmental impacts that will need to be offset by, for example, flood defences, replacement habitats or enhanced waste management provision. Development may also result in an additional burden on local services, such as schools and hospitals. It has long been expected that developers need to contribute towards meeting the consequent costs of their development – why (the argument goes) should the public purse pick up the public costs of development and the developer simply pocket the profits? Furthermore, as developers benefit from an uplift in land value[1] as a result of the award of planning permission by public authorities, it may thereby be considered justified that a contribution is made, not simply towards costs but also towards the positive advancement of planning objectives such as the provision of open space, recreational facilities or affordable housing, either through the provision of such within the development itself (at the sacrifice of space that could have been used more profitably) or through a commuted sum for provision elsewhere.[2]

Developer contributions towards the costs that flow directly from development have been a commonplace and (relatively) uncontroversial feature of planning law on both sides of the Irish Sea for many years. In contrast, contributions towards less obvious costs and towards the promotion of particular objectives have become more commonplace but have the potential to be controversial: would this amount to the extraction of 'planning gain' from developers and is it legitimate to use the planning system in this way? From the perspective of the planning authority, it may be considered that it would be failing in its duty not to maximise the public benefit that can be achieved from the award of planning permission. To developers, this may sound like a development tax and taxes are supposed to be imposed by Parliament rather

DOI: 10.4324/9781003218319-6

than by local government. Furthermore, they may consider that they have no real choice but to pay, even though the viability of projects may consequently become troublesome. These perspectives may, in turn, be less obvious to the general public, who, particularly in respect of controversial developments, may simply take the view that developers are 'buying' planning permission and that public representatives are 'selling' it. Negotiated agreement 'beyond the public gaze' has attracted an abundance of metaphors, leading many academics and jurists to coin it as: 'buying and selling' planning permission; the 'auction' of planning decisions; the 'marketisation' of the planning system; and 'buying' planning gain, to state but a few.

Across the United Kingdom, considerable attention has been directed towards developer contributions as a mechanism to acquire planning gain from private development through the planning process. In Northern Ireland developer contributions operate under a different legislative and policy environment compared with Great Britain. Whilst a burgeoning corpus of scholarship has considered at length the operation of developer contribution instruments in other parts of the United Kingdom, this topic has received relatively little attention within the Northern Ireland setting. Accordingly, the positioning of this chapter relates to developer contributions within the contemporary regulatory and planning reform context. The chapter will critically explore the legal and policy matrix in which developer contributions currently operate by way of planning agreements.

Developer contributions in Great Britain

Charting the partial historiography of British land use legislation identifies that a range of largely unsuccessful measures to redistribute or recover a portion of 'windfall' capital from the development of land through direct betterment capture have made their way onto the statute books. The pioneering Housing, Town Planning, Etc., Act 1909 reflected the birth of town planning and gave local authorities in England and Wales the power, when adopting land use schemes, to claim half the uplift value of a property by virtue of the scheme.[3] Despite its laudable intentions, this was a relatively timid and unworkable measure due to exigencies in exacting the levy.[4] An ephemeral attempt to introduce national betterment levies followed in the form of the landmark Town and Country Planning Act 1947, which saw the nationalisation of development rights and implemented a development charge. However, the 1947 Act did not apply to Northern Ireland.[5]

The impugned issue of the relationship between land use control, developer activity and betterment remained the subject of an assortment of taxation mechanisms. This was reflected in an array of development- and infrastructure-based direct and indirect instruments that acted 'as *de facto* if not *de jure* betterment taxation',[6] including stamp duty land tax, capital gains tax, inheritance tax, direct development by governmental bodies and localised land value capture arising from development rights, including developer obligations and planning gain. A form of taxation was witnessed in the unique Betterment Levy introduced in 1967, which applied at a rate of 40 percent to increases in land value attributable to changes of a limited type in land use and development value, although this did not fall on general increases in land value.[7] Repeal of the Betterment Levy in 1971[8] was accompanied by the extension of the tax on capital gains to development value, as observed through the unpopular Development Gains Tax in 1973 and the Development Land Tax in 1976, which was abolished in 1985.[9] Ultimately, however, these measures failed due to a cocktail of political, administrative and economic reasons, resulting in a policy shift from betterment taxation to planning charges.[10]

Concomitant with the respective land taxation regimes was the enactment of the Town and Country Planning Act 1971, which introduced legal agreements within England and Wales as a mechanism for land value capture by allowing for planning obligations to be secured through a system of negotiated development-specific agreements.[11] The advent of this legislation empowered a local authority to enter into a contractually binding and legally enforceable mechanism with developers in order to regulate development activity. Although the lineage of similar types of agreement can be traced to the 1947 Act, and indeed earlier legislation, planning obligations were not codified until this later legislation and its related guidance. They were rarely used until the 1980s, when Thatcherite free-market principles increasingly encouraged the use of developer contributions to support off-site infrastructure and other community benefits. Through the subsequent enactment of the Town and Country Planning Act 1990 these became known as 'Section 106 agreements'. A similar approach was followed in Scotland, where the legal basis for planning obligations is through Section 75 of the Town and Country Planning (Scotland) Act 1997, as amended – referred to as 'Section 75 agreements'.[12] The equivalent legislative base in Northern Ireland was under Article 40 of the Planning (Northern Ireland) Order 1991 (as amended) or the so-called Article 40 developer contribution agreement. Through legislative amendments and enactments, the nature of planning obligations in England, Wales and Scotland expanded to include infrastructure connected with the proposed development and the provision of affordable housing. In Northern Ireland, planning agreements did not extend to the provision of affordable and social housing and this was not remedied in the Planning Act (Northern Ireland) 2011. We return to that matter later in this chapter.

Since the 1990s negotiated planning obligations/agreements have, therefore, evolved into the principal mechanism of the planning system to extract land value uplift at the local level and prescribe for, compensate or mitigate the impact of development activity.[13] Practice guidance on the use of agreements and model modes of such agreements were published in an attempt to provide clarity and consistency in their operation in order to preserve the integrity of the planning system and address forceful concerns that planning agreements were facilitating the 'buying' of development schemes. Following various reviews that addressed, inter alia, the type of issues mentioned above, a major initiative for land value capture in England and Wales and, more recently in Scotland, but not in Northern Ireland, has been the introduction of the Community Infrastructure Levy (CIL) under the Planning Act 2008, the Community Infrastructure Levy Regulations 2010 (as amended) and the Planning (Scotland) Act 2019.[14] In *R. (on the application of Fox Strategic Land and Property Ltd) v Chorley BC* [2014] EWHC 1179 (Admin) 1, the nature of CIL under Part 11 of the Planning Act 2008 was considered by the court, where Lindblom J observed that:

> The Community Infrastructure Levy (CIL) enables a local planning authority to raise funds from the development of land in its area to pay for infrastructure of various kinds... Some landowners and developers will regard it as an unnecessary and unwelome tax on development. Authorities, however, will see it as an effective and fair means of ensuring that new infrastructure is provided for communities that need it.[15]

CIL came into force in 2010 and represented a legislative ambition to constrain the ability of local authorities to negotiate customised agreements related to development activity. In its original form, CIL was intended to be a simplified, transparent and efficient mechanism to replace the discretional planning obligations due to growing dissatisfaction with the operation of Section 106 agreements, related planning delays and administrative complexity.

CIL permits 'charging'[16] authorities (in essence, local planning authorities) to publish a charging schedule[17] that sets rates of financial contribution required from developers for specified types of development, with a 'chargeable amount' payable by the developer for each eligible development based on a formula laid down in the regulations.[18] Thus, CIL allows local authorities to generate revenue from developers to fund infrastructure alongside a scaled-back system of Section 106 developer contributions. As with Section 106, CIL is a local levy and in choosing to implement the CIL regime, the local authority applies a non-negotiable flat fee to new developments, calculated with reference to the subordinate CIL regulations. The sums recoverable under the local CIL tariff are fixed for non-site-specific infrastructure. There is no discretion in striking a balance across the delivery of development plans, development viability and necessary infrastructure.[19] In contrast, Section 106 agreements do retain discretion in the reduction of contributions relating to circumstances where the uplift in land value may be diminutive and, therefore, they do remain necessary to retain scaled-back planning obligations that can facilitate some development activity, including affordable housing. Thus, the implementation of CIL in England and Wales[20] reflects a hybrid planning system of flat-rate fees and complementary negotiated processes where CIL is 'additional' to core public funding of infrastructure and is not a direct replacement for Section 106 agreements.

There are obvious attractions to such an approach but, at the same time, it has not been wholly popular or without issue. Whilst it was anticipated that CIL would provide greater certainty and fairness in contributions to the community, in practice the supposed benefits of the scheme have been somewhat diluted, if not eroded, by challenges relating to blunt development typologies, exemptions, lack of sensitivity to development viability and delays relating to site-specific negotiations.[21] Critically, CIL has not shifted the individuated and negotiated process to a rule-based system. Rather, it has added a new dimension for the local authority to implement. This means that each mechanism can operate in parallel and the outcome may well be that developers pay CIL to help fund uncoupled infrastructure provision and also pay Section 106 contributions to mitigate the impact of their projects, including the provision of affordable housing.[22] This has resulted in considerable criticism arising from a complicated toolbox of regulations, with research indicating that the uptake of the CIL framework has been slow and piecemeal.[23] Consequently, the negotiated agreement remains the main category of developer contributions across Great Britain. In a recent White Paper for England, *Planning for the Future*, the Ministry of Housing, Communities and Local Government has proposed further fundamental reform.[24] As noted above, a particular issue is that CIL constitutes an additional method of securing developer contributions, while planning obligations still remain important, particularly in securing contributions related to affordable housing (which lies outside the scope of the existing CIL). The 2020 White Paper advocates abolition of both CIL and planning obligations and their replacement by the creation of a new 'Infrastructure Levy', which has the aim of raising more revenue than the existing system and eliminating negotiation-based delays.

In summary, planning gain instruments within Great Britain have evolved from a foundational tax on the uplift in land value under 'betterment' to a negotiated process through planning agreements and extending to contemporary tariff- or levy-based approaches through CIL.[25] Thus, in practice, presently three common mechanisms may allow a public authority to obtain contributions from developers: (1) Planning conditions (but which generally exclude the matter of financial payments), (2) Planning obligations/agreements and (3) the Community Infrastructure Levy.

Developer contributions in Northern Ireland

Historically, the planning regime under the Planning (Northern Ireland) Order 1991, and its predecessors, was a unitary system that saw the Department of Environment being responsible for planning and district councils holding the role and status of consultee. However, reform under the Local Government Act (Northern Ireland) 2014 took effect on 1 April 2015 and brought about a major overhaul of the planning system that had been in existence for some four decades. The new regime was driven by government policy designed to make councils more efficient and cost effective, innovative and empowered to discharge a broader range of public functions.[26] Under the Planning Act (Northern Ireland) 2011 (and subordinate legislation), each council is the local planning authority for its district council area and is responsible for devising its own Local Development Plan (LDP) and determining most planning applications. This recognises that planning policies can vary across different areas and that the benefits of local planning can be better captured through an enhanced understanding of the needs and aspirations of local communities rather than through a centralised agency that was often in conflict, or out of touch, with local government accountability.

The important point in the context of this chapter is that, through the reform and transfer of planning powers, the delivery of planning agreements has also changed. Previously, planning agreements were the responsibility of the departmental Planning Service and were relatively narrow in scope. In terms of developer contributions, the 1998 *Planning Policy Statement 1: General Principles* (now cancelled) provided that developers 'may be required' to make 'some degree of contribution' in circumstances where a development requires provision of, or improvement to, infrastructural works above those specified in development plans; requires earlier implementation; and is dependent on off-site work.[27] Wedded to the devolved policy agenda and anticipation that local authorities could be more autonomous, heterogenous and fiscally responsive, planning agreements now serve as the principal mechanism to deliver developer contributions mainly entered into by district councils as provided for through Section 76 of the 2011 Act.[28] The new planning system is steered by the 2015 *Strategic Planning Policy Statement for Northern Ireland* (SPPS)[29], which establishes the overarching strategic direction and core planning principles from which councils should develop their own planning policies in order to further sustainable development.[30] In the context of developer contributions, the SPPS is more expansive than the previous Planning Policy Statement guidance and states that contributions may be required in an increased number of circumstances.[31] In addition, the SPPS comments that in some circumstances community benefits, such as payments to the community, in-kind benefits and shared ownership arrangements may be 'voluntarily' offered to communities that are likely to be affected by a development.[32] However, such voluntary community benefits cannot be considered material considerations in decision making and are distinct from developer contributions and planning conditions. Specific procedural and good practice guidance is provided in the 2017 *Development Management Practice Note 21: Section 76 Planning Agreements* (DMPN 21), which outlines the legislative requirements relating to the use of planning agreements.[33] Other local guidance has followed suit and it is notable that Belfast City Council has published a *Developer Contributions Framework 2020* document that sets out the Council's approach to securing developer contributions as part of the planning application process.[34]

Critical reflections

Until relatively recently the position in both Northern Ireland and Great Britain was that developer contributions were secured primarily via planning agreements (as they are known in Northern Ireland) or planning obligations (as they are known in Great Britain) with some (very restricted) use of planning conditions in this context.[35] Four issues require brief consideration: the nature of planning agreements, planning agreements as material considerations in deciding planning applications, the validity of a planning agreement and challenges to planning agreements.

The nature of planning agreements

As noted above, in Northern Ireland the principal mechanism available to planning authorities to secure developer contributions is the planning agreement under Section 76 of the Planning Act (Northern Ireland) 2011.[36] Therefore, planning agreements are not new to the Northern Ireland lexicon[37], and, in substantive terms, Section 76 is very similar to Section 106 of the Town and Country Planning Act 1990, applicable in England and Wales, and Section 75 of the Town and Country Planning (Scotland) Act 1997 (as amended) in Scotland. The provision of Section 76 enables the relevant planning authority to enter into planning agreements for the purpose of facilitating, regulating or restricting the development or use of land and which may include contributions to be paid by a developer to the relevant authority, or any Northern Ireland department, to offset the impact of the proposed development, in conjunction with granting planning permission. Planning agreements are invariably a private agreement negotiated between the developer and local authority on an application-by-application basis. This means, in accordance with the Section 76 regime, that such agreements are discretional and can operate with flexibility and are customised to the specific circumstances of the proposed development scheme to take account of the prevailing market conditions and required obligations on the part of the developer. However, a by-product of this approach is that planning agreements may be complex, veiled in secrecy, lacking public scrutiny, time-consuming, costly and potentially subject to legal dispute and judicial review.[38]

In this context, agreements can take various guises but usually entail some form of contribution to the community in order to promote the acceptability of the planning application. According to the Strategic Planning Policy Statement (SPPS) for Northern Ireland developer contributions can be used:

- where a proposed development requires the provision or improvement of infrastructural works over and above those programmed in an LDP;
- where earlier than planned implementation of a programmed scheme is required;
- where a proposed development is dependent upon the carrying out of works outside the site; and
- where archaeological investigation or mitigation is required.

An obvious difference between Northern Ireland and Great Britain is that in the latter these arrangements are known as 'planning obligations'.[39] Under Section 106(1) of the Town and

Country Planning Act 1990 anyone "interested in land" in the area of a local planning authority may, 'by agreement or otherwise' enter into a planning obligation. Section 75(1) of the Town and Country Planning (Scotland) Act 1997, in a similar vein, states that a person in respect of land in the district of a planning authority may '(a) by agreement with that authority, or (b) unilaterally' enter into a planning obligation. The term 'planning obligation' is a deliberate recognition that while such arrangements will often in practice be agreed with the relevant planning authority, agreement is not necessary under the legislation. Where an obligation is entered into unilaterally (or otherwise) by the developer, it may well be a material consideration in deciding a planning application even though the planning authority has not sought, desired or agreed to it.

Section 76 of the Planning Act (Northern Ireland) 2011 (with language clearly different from what pertains in Great Britain) contains no words to apparently indicate the possibility of a developer entering into a unilateral undertaking. Section 76(1) provides:

> 76.— Planning agreements
> (1) Any person who has an estate in land may enter into an agreement with the relevant authority (referred to in this Section and Sections 77 and 78 as '*a planning agreement*'), enforceable to the extent mentioned in subsection (4)—
>
> (a) facilitating or restricting the development or use of the land in any specified way;
> (b) requiring specified operations or activities to be carried out in, on, under or over the land;
> (c) requiring the land to be used in any specified way;
> (d) requiring a sum or sums to be paid to the authority on a specified date or dates or periodically; or
> (e) requiring a sum or sums to be paid to a Northern Ireland department on a specified date or dates or periodically.

That interpretation is consistent with *Development Management Practice Note 21: Section 76 Planning Agreements* issued by the Department for Infrastructure, which states:

> Section 76 of the 2011 Act enables the relevant authority, that is either a council or, as the case may be, the Department to enter into a planning agreement with any person who has an estate in land.[40]
>
> Planning agreements should always be between willing parties. Applicants do not have to agree to a proposed planning agreement. However, if the barriers to the grant of planning permission have not been addressed this may lead to a refusal of planning permission or non-determination of the application. An appeal may be made against the non-determination or refusal of planning permission.[41]

Similarly, planning policy guidance on the predecessor to Section 76 (Article 40 of the Planning (Northern Ireland) Order 1991) is written very much from the perspective of the power of the planning authority to enter into agreements (rather than the right of a developer to enter into one).[42] Nevertheless, in the *Developer Contributions Framework 2020* published by Belfast City Council it is asserted that:

Planning agreements may take the following forms:

- Unilateral agreement – the applicant is the sole party to the agreement.
- Bi-lateral agreement – the applicant and council are signatories to the agreement.
- Multi-lateral agreement – the applicant and council are signatories as well as other parties such as the landowner (if different from the applicant) or government departments."[43]

In a similar vein, the Department of the Environment, *Best Practice Guidance to Planning Policy Statement 23 'Enabling Development for the Conservation of Significant Places'* published in April 2014 (and based on English Heritage guidance) stated:

> 7.2.2 Planning agreement: An agreement entered into by a person with an interest in land, either by way of agreement with the planning authority or unilaterally, that conforms to the requirements of Article 40 of the Planning (Northern Ireland) Order 1991.

Unilateral obligations (akin to those in Great Britain) were considered by the Northern Ireland Assembly Committee for the Environment as the Planning Bill (which became the Planning Act (Northern Ireland) 2011) passed through the Assembly and some potential advantages of such an approach were suggested:

> If developers could draw up unilateral obligations, they would be volunteering to do something to facilitate their developments. If developers submitted unilateral obligations to the DOE for applications, or the Commission for appeals, the long delays in negotiating and drafting agreements could be avoided, which could speed up the process of delivering appeal decisions."[44]

However, this tends to suggest that there was no belief that Article 40 of the Planning (Northern Ireland) Order 1991 had allowed for unilateral obligations. Section 76 is not identical to Article 40: in particular, Article 40 stated: 'The Department may enter into an agreement with any person who has an estate in land' whereas Section 76 states that 'Any person who has an estate in land may enter into an agreement with the relevant authority.' While it might be argued that this represents a change of focus, with an emphasis on the capacity of the developer rather than the power of the planning authority, at no point does Section 76 refer to such an arrangement as being anything other than an 'agreement': there is no mention of the words 'obligation' or 'unilateral' and there seems to be no material divergence in this respect between the wording of Article 40 and Section 76.[45] Perhaps the interpretation articulated by Belfast City Council is correct. However, a 'unilateral agreement' does not really sound much like an agreement at all and, if the possibility of unilateral arrangements had been contemplated, the legislators could surely have borrowed some more appropriate terminology.

If a 'unilateral agreement' is not possible under Section 76 it would seem that this divergence from the Great Britain legislation gives Northern Ireland planning authorities a greater position of power: they cannot have an obligation to which they are not in favour effectively imposed upon them as something they would have to consider in the context of deciding a planning application (a decision that they could potentially have to justify in any appeal). Consequently, if a planning application would be acceptable in planning terms only if an agreement was concluded to address particular matters and the planning authority declined to enter into such an agreement, then this would effectively ensure rejection of the application. It is noteworthy that a unilateral undertaking under Section 106 cannot bind the local authority as it is not party to the agreement.

Planning agreements as material considerations in deciding planning applications

Lord Hoffman in *Tesco Stores Ltd v Secretary of State for the Environment*[46] stated:

> to describe a planning decision as a bargain and sale is a vivid metaphor. But I venture to suggest that such a metaphor (and I could myself have used the more emotive term 'auction' rather than 'competition' to describe the process of decision-making process in the Plymouth case) is an uncertain guide to the legality of a grant or refusal of planning permission… in a case of competition such as the Plymouth case, in which it is contemplated that the grant of permission to one developer will be a reason for refusing it to another, it may be perfectly rational to choose the proposal which offers the greatest public benefit in terms of both the development itself and related external benefits.[47]

Section 45(1) of the Planning Act (Northern Ireland) 2011 provides that in determining a planning application the planning authority 'must have regard to the Local Development Plan, so far as material to the application, and to any other material considerations.' Whether any given factor is, or is not, a material consideration in the decision whether to award planning permission is a matter of law for the courts to determine.[48] Conversely, the weight to be attached to the various material considerations that relate to a particular decision is a matter of planning judgement with which the courts cannot interfere unless *Wednesbury* unreasonable.[49] The starting point for a discussion of material considerations in the context of planning agreements/obligations requires examination of the decisions of the House of Lords in *Newbury District Council v Secretary of State for the Environment*[50] and *Tesco Stores Ltd v Secretary of State for the Environment*.[51] In the former, the House established three principles for the validity of conditions attached to planning permission (which strictly limit planning gain by the imposition of conditions) and, in the latter, it was concluded that only two of these applied in deciding whether a planning obligation constituted a material consideration in the context of the decision whether to award planning permission.

Newbury was specifically concerned with the validity of a planning condition and when considered in the context of 'material considerations' the House of Lords concluded that a valid planning condition must:

a) Be imposed for a planning purpose and not for any ulterior motive;
b) Fairly and reasonably relate to the development in question; and
c) Not be so unreasonable that no reasonable authority could have imposed them (not be *Wednesbury* unreasonable).

While, in the instant case these principles were applied to determine the validity of a planning condition, their Lordships were clear that the first two principles (or rather their underpinning basis) derived from the broader wording of Section 29(1) of the Town and Country

Planning Act 1971 while the third, the application of the *Wednesbury* reasonableness test, was a generally applicable requirement of administrative law. Lord Scarman, for example, stated that Section 29 (1) 'expressly mentions the first two tests.'[52] In *Tesco*, Lord Hoffman said:

> as Lord Scarman pointed out in Newbury [1981] A.C. 578, 619A, the first two tests are a judicial paraphrase of the planning authority's statutory duty in Section 70(2) of the Act of 1990 to have regard to the provisions of the development plan and 'any other material considerations.' This duty applies to as much to the decision to grant planning permission… as to the decision to impose conditions.[53]

In the (slightly) earlier case of *R v Plymouth City Council, Ex p Plymouth and South Devon Co-operative Society Ltd*[54], a Court of Appeal featuring Hoffman LJ had concluded that all three Newbury tests applied in determining whether a planning obligation was a material consideration. Two supermarket chains proposed significant planning obligations. Sainsbury's offered approximately £3.6 million of contributions to be applied in respect of, inter alia, a tourist information centre, an art gallery display facility, a birdwatching hide, an £800,000 contribution towards a park and ride facility and up to £1m for infrastructure works to make another site suitable for industrial use. Tesco similarly offered the development of crèche facilities for working mothers, a wildlife habitat, a water sculpture and land for a park and ride facility. However, the Court of Appeal found that (unlike the position with conditions) such benefits did not have to be *necessary* to the decision as to whether (or not) to grant planning permission in order to be material considerations. It was sufficient that they were planning benefits and, in this case, they did reasonably relate to the particular development as the park and ride facility would, for example, counteract an increase in traffic caused by the development. However, in *Tesco,*[55] Lord Hoffman (as he had by then become) concurred, despite his observations as quoted above, with the leading judgment of Lord Keith that went a step further and concluded that of the three *Newbury* principles, only the first and third apply in the context of deciding the materiality of a planning obligation/agreement: a planning obligation can be a material consideration in deciding a planning application even if it does not reasonably relate to the development in question (so long as it had *some* connection) – it simply has to relate to a planning matter and must not be perverse. In *Tesco* the County Council considered that a link road to relieve town centre traffic congestion was necessary for Witney, Oxfordshire, and Tesco offered £6.6m to fully pay for it. In terms of physical proximity with the proposed link road, there was a tenuous relationship with Tesco's application for planning permission for a superstore (it would potentially contribute to more congestion) but the Secretary of State concluded that the offered obligation was not reasonably related in scale to the increase in traffic likely to be produced by the superstore. The House of Lords concluded this was a material consideration but that on the facts the Secretary of State in this case actually had (contrary to the contention of Tesco) rightly taken it into account, albeit affording it (as he was entitled to do) little weight. In summary, the *Tesco* case established that in order for a planning agreement/obligation to be material to the award of planning permission only the first and third tests laid down for conditions in *Newbury* had to be met: the agreement/obligation must have a planning purpose and must not be *Wednesbury* unreasonable but need not fairly and reasonably relate to the development in question. Nevertheless, in order to be a material consideration an agreement/obligation will still have to have *some* connection. The decision in *Tesco*, however, flies in the face of planning policy. Throughout all parts of the

United Kingdom it is clear that planning authorities should enter into planning agreements/obligations only where all three *Newbury* principles are satisfied and, in fact, the planning policy requirements extend beyond the *Newbury* principles.

In England the longstanding policy position on planning obligations has been placed on a legal footing (and *Tesco* has, in essence, been statutorily reversed) by the Planning Act 2008. In England the legal position is now governed by Part 11 of the Community Infrastructure Levy Regulations 2010 (as amended):

> 122.— (1) This regulation applies where a relevant determination is made which results in planning permission being granted for development.
> (2) A planning obligation may only constitute a reason for granting planning permission for the development if the obligation is—
>
> (a) necessary to make the development acceptable in planning terms;
> (b) directly related to the development; and
> (c) fairly and reasonably related in scale and kind to the development.

No similar legal provision exists in Northern Ireland under the Planning Act (Northern Ireland) 2011. However, the 2017 *Development Management Practice Note: Section 76 Planning Agreements* does rehearse these three guiding principles (paragraph 5.2) and advises that 'a planning agreement is a material consideration in considering an application and, therefore, should be completed before permission can be issued' (paragraph 8.2). It also notes that an 'agreement does not in itself confer planning permission nor does it determine the outcome of a related application' (paragraph 5.1).

The validity of a planning agreement

A planning authority in Northern Ireland might be tempted to devise a 'pooled fund' approach using Section 76. As Lord Hodge observed in the Scottish case of *Aberdeen City and Shire Strategic Development Planning Authority v Elsick Development Company Ltd* (2017 UKSC 66):

> there appears to be much that can be said in favour of such a scheme. It enables a planning authority to facilitate development within its area. Inclusion of such a scheme in a development plan allows a public debate during the statutory process of the approval of the plan. The scheme allows developers in the area to assess the viability of their proposed developments knowing the extent of their liability to the Fund before they spend large sums pursuing their applications.

Validity was the core issue in question in *Elsick* and the judgment of Lord Hodge (with whom the other members of the Supreme Court unanimously agreed) now provides an authoritative discussion of the key principles that will apply to planning agreements under Section 76 of the Planning Act (Northern Ireland) 2011. In *Elsick* the planning authority published a proposed strategic development plan and draft supplementary planning guidance which, in essence[56], required developers to enter into planning obligations to make financial contributions into a pooled fund. The purpose of the fund was the provision of infrastructure necessary to deal with the cumulative impact of development in the Aberdeen Housing Market Area. Contributions to the fund, in the case of residential developers, were set at a fixed sum per residential unit, were not tied to the impact of a particular development and were additional to contributions necessary to mitigate impacts specific to any particular development. Elsick Development Ltd, which proposed to develop approximately 4,000 houses together with commercial, retail and community facilities, objected to this policy and only entered into a planning obligation to contribute to the fund subject with the express caveat that no contributions would be due if the supplementary planning guidance was to be found invalid. Specifically, Elsick objected on the basis that they were required to make contributions that would be applied to infrastructure interventions at locations where their development had only a trivial impact (if any). The Supreme Court agreed, finding that the implementation of this policy was not within the powers of the planning authority under the Town and Country Planning (Scotland) Act 1997, Section 75 of which governs planning obligations in Scotland and thus the local authority was not legally entitled to require contributions to the fund.

In his lead judgment, Lord Hodge considered the scope of Section 75 and said that although planning authorities had wide discretion about both the circumstances in which they were permitted to seek planning obligations and the nature of those obligations, the 'express words' of that Section required there to be a relationship between the obligation and the land to be burdened by that obligation.[57] Thus, he stated:

> In my view, it was not within the power of the planning authority to require a developer to enter into such an obligation which would be irrelevant to its application for permission as a precondition of the grant of that permission.[58]

In keeping with the established body of case law pertaining to the scope of planning obligations, the ruling of the Supreme Court was largely predictable and reasserted forceful emphasis on the orthodox planning custom and convention that linked requirements with the individual scheme. Essentially, the impact of the development on the infrastructure for which the contributions were sought was trivial and too remote to be relevant to the development in question and thus form the basis of a planning obligation. While acknowledging that there appeared to be 'much that can be said in favour' of the proposed transport funding scheme,[59] Lord Hodge concluded that it was simply not permitted by the 1997 Act. The lower courts had already found that the connection between the Elsick development and some of the proposed transport projects was 'at best trivial',[60] and so was not 'imposed for a purpose in relation to the development and use of the burdened site as Section 75 requires.[61] Thus, he concluded that:

> If planning authorities in Scotland wish to establish a local development land levy in order to facilitate development, legislation is needed to empower them to do so.[62]

Given this outcome, the Scottish government then decided to remove this general restriction on planning obligations by introducing a Scottish infrastructure levy as part of the Planning (Scotland) Act 2019. Without equivalent enabling legislation there may well be legal limits to a Northern Ireland local authority going down a comparable pathway to what was being tested in Aberdeen.

Challenges to planning agreements

Challenges relating to planning agreements will typically come in one of two forms. By far the most common source of challenge will be a person who is aggrieved with a development to which an agreement relates (rather than the agreement per se). A good example is the High Court case of *Conlon v Belfast City Council*[63] in which a residents' group challenged a decision of Belfast City Council to grant planning permission for an office development. The challenge was based on a number of grounds, including a legally impermissible consideration by the Council to take into account the developer's willingness to contribute £230,000 to the cost of public infrastructure works. It was contended that the developer contribution was a 'voluntary contribution' that lies outside the remit of a Section 76 agreement and cannot be considered in decision-making. While the Applicant was ultimately successful on another ground of challenge, the argument relating to the planning agreement was rejected by the High Court. The court rehearsed the Northern Ireland planning policy provisions and concluded that:

> A developer's willingness to execute a Section 75 agreement incorporating a financial obligation does not fall foul of DPMN 21 … An offer of a community benefit 'on a voluntary basis' stands in stark contrast to a contractually binding mechanism enforceable in accordance with the enforcement provisions of Section 76.

On the other hand, if a developer has a problem, it is likely to be with the agreement per se, though it might still be surprising to find that a developer, who has entered into an agreement and (typically) obtained the related planning permission, would subsequently challenge the validity of that 'agreement' with a view to declining to perform aspects of it.[64] In line with the notion of a planning agreement being, in essence, a contract between two parties, the statutory framework provides very limited opportunity for any challenge. A significant advantage for a planning authority is that agreements (unlike conditions attached to the grant of planning permission) are not subject to appeal (per se) to the Planning Appeals Commission.[65] Once agreed and formally constituted a planning agreement can be modified or discharged only on the very limited basis provided by Sections 77 and 78 of the Planning Act (Northern Ireland) 2011 and supplemented by the Planning (Modification and Discharge of Planning Agreement) Regulations (Northern Ireland) 2015. Firstly, Section 77(1)(a) provides that modification or discharge can be achieved by agreement of the parties. In the absence of agreement, any person bound by a planning agreement may apply to the relevant (planning) authority with which the agreement was entered into (and which has, presumably, declined to agree modification or discharge) seeking modification or discharge. However, unless the agreement expressly provides to the contrary, no such application may be made until the expiry of a period of five years beginning with the date on which the agreement is entered into. In the (not inconceivable) event that such an application is unsuccessful an appeal may then be made to the Planning Appeals Commission, which may determine:

a) that the planning agreement shall continue to have effect without modification;
b) if the agreement no longer serves a useful purpose, that it shall be discharged; or
c) if the agreement continues to serve a useful purpose, but would serve that purpose equally well if it had effect subject to the modifications specified in the application, that it shall have effect subject to those modifications.

The determination of the Planning Appeals Commission is final, but the basis upon which it is made, or what threshold an appellant would need to demonstrate, is unclear.

Developer contributions and affordable and social housing in Northern Ireland

A significant issue is that, compared with analogous provision in Great Britain, the pre- and post-Northern Ireland reform statutory language and attendant policy mechanisms do not obviously provide for developer contributions for social and affordable housing provision. Therefore, Northern Ireland remains the only region in the United Kingdom that does not have a contribution scheme to provide affordable housing in mixed tenure developments delivered through the planning system. This position is in stark contrast with the position in England, Wales and Scotland, where increasingly the matter of developer contributions is being defined by the need for affordable and social housing provision.

Legislative reforms in England and Wales through Section 106 of the Town and Country Planning Act 1990, as amended, have been used to ensure that developers applying for planning permission for developments are required to make a contribution to the provision of affordable and social housing. Such a contribution to the provision of affordable and social housing consists of one-off, in-kind and financial contributions, one-off payment and phased payments, maintenance payments and pooled contributions. The practical outworking of the use of developer contributions under Section 106 is implemented primarily through the *National Planning Policy Framework* in England and *Planning Policy Wales*. In this context, planning obligations/agreements can be used as a device to implement policy in this sphere and, in particular, a developer may agree to allocate a proportion of a site to affordable housing or, alternatively, to contribute a commuted sum in lieu (for the development of such housing elsewhere). Clearly, the contribution of a proportion of a site to affordable housing will sufficiently relate to the land (and the development) to meet the requirements set out by Lord Hodge in *Elsick*. Similarly, contributing a commuted sum in lieu should also suffice as it is, in essence, paying for affordable housing that would otherwise have to have been on the particular site. If payment of a commuted sum individually negotiated between the developer and the planning authority in this respect is legitimate, then it is not much of a stretch to say that a pooled fund for affordable housing based on a standard charging schedule set by the planning authority for all housing developments would be likewise. That practice would thus also seem to be sufficiently related to the land (and the development) as the alternative would be the construction of affordable housing on the land itself. It is also not impossible that an approach of this nature might be challenged in a similar manner to *Elsick* (but probably not for the same reason). Of course, a likely ground of complaint from developers would be viability (depending on the nature of the charging schedule), although it is rather unclear if this would actually be a sustainable ground for legal challenge.

Lack of affordable housing is a significant problem in Northern Ireland, a point clearly identified in the content of Programme for Government documents[66] and the growing volume of research relating to latent housing need. Planning agreements in Northern Ireland could (in theory) support the delivery of developer contributions for the much-needed provision of affordable and social housing. However, developer contributions have remained, until relatively recently, strangely absent from the legislative and planning policy matrix in this sector. Indeed, in view of the growing ethos and practice of using developer contributions across the United Kingdom as a significant vehicle to support the affordable and social housing agenda, proposals for developer contributions for affordable and social housing were proposed in Northern Ireland as an addendum to *PPS12 Housing and Settlements* and included integrated developments.[67] The then Minister for Social Development announced:

> That contribution, which is a normal part of the planning process elsewhere on these islands, will require developers to provide social and affordable housing as a proportion of all new proposed housing developments.[68]

In 2012, the Minister for Social Development launched the strategic plan *Facing the Future: A Housing Strategy for Northern Ireland*[69]. A key action in the strategy was the 'consideration' of developer contributions to support housing sustainability, although this was qualified by the recognition that the introduction of developer contributions needs to take account of wider economic circumstances and that the conditions were not right to introduce the scheme at that time.[70] Critically, however, the strategy commented that Northern Ireland should be ready to introduce a scheme when circumstances improve. Following this, a joint consultation was issued in 2014 by the Department for Social Development and the Department of the Environment on the possible introduction of a Developer Contributions Scheme for social and affordable housing in Northern Ireland. It is fair to say that the outcome of the consultation was somewhat mixed. The preponderance of consultee views fell against the proposals and highlighted the need for additional economic research. This research was subsequently commissioned and reported general opposition by the development industry to the introduction of a scheme of any sort at that point in time.[71] Issues raised comprised the fragile housing market, slow value growth, major development land access constraints, scale of local development companies and market dichotomy and variation outside Belfast. This position was largely borne out by market intelligence and a viability analysis, which indicated that:

> a developer contribution scheme without any public subsidy would not be viable in most of Northern Ireland but, in Belfast, some level of contribution should be achievable and provide a reasonable return to both developer and landowner.[72]

Accordingly, it was concluded that the introduction of a scheme with a common standard of affordable housing across the region is not realistic and, that for most of the region a developer contribution scheme would not work. In the Belfast Metropolitan Area, however, a scheme could be considered but would best be left to local decision-makers working through the Local Development Plan process – where targets and delivery mechanisms could be fully debated. Thus, the research recommended, inter alia, that regional policy should include a requirement that Local Development Plans identify a local target for a developer contribution scheme that takes into account need and viability targets, and that delivery mechanisms for a developer contribution scheme should be set locally as part of the Local Development Plan preparation process. It was further identified that whilst a target of nil contributions may be justified across most locales of Northern Ireland, nonetheless, taking account of viability and flexible responses to schemes, a limited contribution to affordable housing should be possible in some areas without impacting on the recovery of the housing market.[73] It was also recommended that thresholds for developer contributions should form part of a locally defined policy and that regional guidance could prescribe a standard threshold that can be varied with reference to local market evidence and development plans.[74]

The policy landscape for affordable and social housing subsequently moved on with the publication of the 2015 *Strategic Planning Policy Statement for Northern Ireland*. It details that the development plan process will:

> be the primary vehicle to facilitate any identified need by zoning land or by indicating, through key site requirements, where a proportion of a site may be required for social/affordable housing. This will not preclude other sites coming forward through the development management process.[75]

However, the SPPS does not provide guidance for a developer contribution scheme; instead, referencing the 2014 public consultation document from the Department for Social Development: *Developer Contributions for Affordable Housing*, and the parallel Department of the Environment publication: *PPS 22 (Consultation) Draft: Affordable Housing*, which, as noted above, had sought views on a policy for securing developer contributions towards affordable housing in Northern Ireland. The latter stated that the introduction of developer contributions:

> will promote a mix of housing particularly in terms of tenure, price and household composition including families with children, single person households and older people.[76]

Furthermore, in terms of developer contributions, as a point of general observation it is interesting to note that despite the clear policy direction, affordable and social housing is not mentioned in the detail or language of the more recent Belfast City Council *Developer Contributions Framework 2020*.[77] Arguably, this inferentially suggests that affordable and social housing is not considered as a developer contributions issue '*yet*' by Belfast City Council or indeed beyond given the lack of any other local policy frameworks on this issue. That said, the 2020 framework does provide guidance for undertaking viability assessments (applicable to developer contributions), and there is detail relating to developer appraisal, albeit this stops a little short of on- or off-site contributions for affordable and social housing. Furthermore, this area of prospective policy may also be constrained by the lack of a definition of 'viability assessment' and varied opinions around the definition of 'affordable housing'.[78] Perhaps it is time to reconsider the implementation of a site-specific approach in Northern Ireland with mechanisms for the delivery of affordable and social housing through the planning process being put in place. In this context, it is important that the negotiation of site-specific developer contributions through planning agreements is not undermined by a power imbalance between planners and developers. It may be necessary to consider tariff standardisation to ensure, so far as practical, that high value sites are under-taxed and that low value sites are not rendered unviable.

Conclusion

It is clear that negotiated developer contribution agreements remain a key mechanism used by planning authorities to extract planning gain from the development process and have been the main instrument for land value capture by local government across the United Kingdom. At the same time, there is policy tension between public-value capture, the planning system, development activity and the growing 'fuzziness' between what should be private or public sector investment financed by the private sector.[79] This protracted debate focuses on the efficacy of policy instruments, the legitimacy of planning agreements, standardisation and transparency in the planning process within differentiated and shifting policy regimes at different levels of governance. For this reason, developer contributions continue to attract considerable attention in the context of shifting and different legislative and policy reforms. This includes the nature of the planning agreement, the provision of social and affordable housing through agreements, community benefits and the infrastructure levy mechanism. Clearly, Northern Ireland has not followed the same track and while the law relating to planning agreements in Northern Ireland is very similar to that relating to planning obligations in England and Wales, it must be noted that it is not identical and has not been reformed in the same way. Indeed, the legal evolution of developer contributions in Northern Ireland has ceased at the negotiated legal agreement and thus tension between the Section 76 regime for discretionary and negotiated planning agreements and the voluntary or prescribed use of CIL does not yet exist. Nonetheless, discussion of developer contribution reform through the implementation of a regional or sub-regional CIL-style of policy has been mooted and it would be fair to speculate that any such legislative provisions for Northern Ireland would need to reflect on the Great Britain experience and, particularly, the interaction between CIL and planning agreements.[80] As local authorities in Northern Ireland find new ways to finance development through the planning process, legislation will need to be brought forward to provide for a possible direction of travel towards an infrastructure levy and, arguably, pooled funds. Developer contributions can also help provide for affordable and social housing. Whilst market conditions may not fully support this approach, perhaps explaining the lack of political appetite to advance this important issue, a reasoned analysis suggests that market fundamentals should not detract from providing this option within the legislation and planning policy frameworks of Northern Ireland.

Notes

1 In *R v South Northamptonshire District Council, ex parte Crest Homes plc* [1993] JPL 200 Section 106 contributions were directly calculated on the basis of uplift – each developer was required to contribute 20 per cent of enhanced land value for residential developments and 17 per cent for industrial/commercial developments. The Court of Appeal found this approach to be lawful on the particular facts of the case. In a recent (2020) White Paper: *Planning for the Future*, the Ministry of Housing, Communities and Local Government has recommended reform of developer contributions in England and the introduction of a system based directly and explicitly on uplift value.

2 For a discussion of the nature of contributions see: Alex Lord, Richard Dunning, Bertie Dockerill, Gemma Burgess, Adrian Carro, Tony Crook, Craig Watkins and Christine Whitehead. 2018. *The Incidence, Value and Delivery of Planning Obligations and Community Infrastructure Levy in England in 2016-17* (London: Ministry of Housing, Communities and Local Government).

3 This Act did not extend to Scotland or Ireland.

4 Pete Wyatt, 2017. 'Experiences of running negotiable and non-negotiable developer contributions side-by-side', *Planning Practice and Research* 32:2, 152–70, at p. 153.

5 Section 120(4) of the Town and Country Planning Act 1947

6 See: Pat McAllister, Edward Shephard and Peter Wyatt. 2018. 'Policy shifts, developer contributions and land value capture in London 2005–2017', *Land Use Policy* 78, 316–26, at 316.

7 See: Land Commission Act 1967,

8 Land Commission (Dissolution) Act 1971

9 See: Colin Jones, James Morgan and Mark Stephens. 2018. *An Assessment of Historic Attempts to Capture Land Value Uplift in the UK* (Scottish Land Commission). See also: Pete Wyatt. 2017. 'Experiences of running negotiable and non-negotiable developer contributions side-by-side', *Planning Practice and Research* 32: 2, 152–70, at p. 152.
10 See: Colin Jones, James Morgan and Mark Stephens. 2018. *An Assessment of Historic Attempts to Capture Land Value Uplift in the UK.* (Scottish Land Commission).
11 Town and Country Planning Act 1971, Section 52.
12 In Scotland these were not the only form of planning obligation. Section 69 Agreements were often used in relation to roads.
13 Pete Wyatt. 2017. 'Experiences of running negotiable and non-negotiable developer contributions side-by-side', *Planning Practice and Research* 32: 2, 152–70.
14 For England, see: SI 2010/948 as amended by the CIL (Amendment) Regulations 2011, SI, 2011/987, SI 2012/2975, SI 2013/982, SI2014/385, SI 2015/836. For Scotland, see: Sections 54 to 58 and schedule 1 of the 2019 Act which sets out powers to introduce an infrastructure levy, and to issue associated regulations and guidance.
15 At Para 18.
16 See: Section 206 of the Planning Act 2008
17 See: Section 211 of the Planning Act 2008 and Regulation 14 of the CIL Regulations 2010. See also: *R. (on the application of Fox Strategic Land and Property Ltd) v Chorley BC* [2014] EWHC 1179 (Admin).
18 Although it should be noted that there are a number of exemptions to the charge.
19 CIL charges are set by the local authority, based on the size and type of the new development. It is payable on most developments over 100 square metres or where a new dwelling is created. The local authority can set different rates for different geographical zones in their area and for different intended uses of development. This is a local decision based on economic viability and the infrastructure needed.
20 In respect of Wales, Part 11 functions transferred (W.) (24.5.2018) by The Welsh Ministers (Transfer of Functions) Order 2018 (S.I. 2018/644), arts. 1(1), 44(1) (with art. 44(2)).
21 Alex Lord, Richard Dunning, Bertie Dockerill, Gemma Burgess, Adrian Carro, Tony Crook, Craig Watkins, and Christine Whitehead. 2018. *The Incidence, Value and Delivery of Planning Obligations and Community Infrastructure Levy in England in 2016/17* (London: Ministry for Housing Communities and Local Government).
22 Pete Wyatt. 2017. 'Experiences of running negotiable and non-negotiable developer contributions side-by-side', *Planning Practice and Research* 32: 2, 152–70, at p. 156.
23 Richard Dunning, Alex Lord, Berna Keskin and Malachy Buck. 2019. 'Is there a relationship between planning culture and the value of planning gain? Evidence from England', *Town Planning Review* 90, 453–71. at pp. 466–67. See also: Pat McAllister, Edward Shephard and Peter Wyatt. 2018. 'Policy shifts, developer contributions and land value capture in London 2005–2017', *Land Use Policy* 78, 316–26.
24 Ministry of Housing, Communities and Local Government. 2020. *White Paper: Planning for the Future.* (London: Ministry of Housing, Communities and Local Government).
25 Anthony Crook, John Henneberry and Christine Whitehead. 2016. *Planning Gain: Providing Infrastructure & Affordable Housing* (Chichester: Wiley Blackwell).
26 See: Explanatory Notes to the Planning Act (Northern Ireland) 2011. Available at: https://www.legislation.gov.uk/nia/2011/25/notes/contents
27 Department for the Environment. 1998. *Planning Policy Statement 1: General Principles*, at p. 24. Note that this is no longer retained planning policy.
28 (Department for Infrastructure. 2017).
29 Department of Environment. 2015. *Strategic Planning Policy Statement: Planning for Sustainable Development.* Available at: https://www.planningni.gov.uk/SPPS. Note that the SPPS has a statutory basis under Part 1 of the Planning Act (Northern Ireland) 2011 which requires the Department to formulate and co-ordinate policy for securing the orderly and consistent development of land and the planning of that development.
30 It is also worth noting that Section 25 of the Northern Ireland (Miscellaneous Provisions) Act 2006 requires the council to act in the way that is best calculated to contribute to the achievement of sustainable development. This is embedded as the principal objective of the planning system – to secure the orderly and consistent development of land whilst furthering sustainable

development and improving well-being. See: Part 1, Section 1 of the Planning Act (Northern Ireland) 2011).

31 See: Para. 5.69 *Strategic Planning Policy Statement*, at p. 35.

32 Para. 5.71 *Strategic Planning Policy Statement*, at p. 36

33 Department of Infrastructure. 2017. *Development Management Practice Note 21: Section 76 Planning Agreements*. Available at: https://www.infrastructure-ni.gov.uk/sites/default/files/publications/infrastructure/dmpn-21-section-76-planning-agreements-v1-jan-2017_0.pdf

34 Belfast City Council. 2020. *Developer Contributions Framework 2020*, at Paras 1.2 and 1.4. Available at: https://www.belfastcity.gov.uk/documents/developer-contributions-framework

35 For an overview of developer contributions across the United Kingdom, see: Graham Winter, Louise Smith, Suzie Cave and Alan Rehfisch. 2016. 'Comparison of the Planning Systems in the Four UK Countries', *House of Commons Library Research Paper* 713-2015, 20 January, pp. 34-36.

36 Additionally (and similarly) Article 122 of the Roads (NI) Order 1993 provides the relevant department with the power to enter into an agreement in the specific context of highway infrastructure.

37 A 'Planning agreement' is a legally binding agreement between the relevant parties, normally between the council, landowner, applicant and any other parties directly involved in fulfilling the agreement.

38 See: Ilse Helbrecht and Francesca Weber-Newth. 2018. 'Recovering the politics of planning: Developer contributions and the contemporary housing question', *City* 22: 1, 116 –29. See also: Anthony Crook, Henneberry, J. and Christine Whitehead. 2016. *Planning Gain: Providing Infrastructure and Affordable Housing* (Chichester: Wiley Blackwell).

39 Though under earlier legislation, Section 52 of the Town and Country Planning Act 1971 and Section 50 of the Town and Country Planning (Scotland) Act 1972, the predecessors of these arrangements were known as 'agreements'. In practice, obligations under Section 106 of the TCPA 1990 frequently are agreed and are often known as 'agreements'.

40 Para. 2.1. Issued January 2017.

41 Para. 8.3. Department for Infrastructure. 2017. *Development Management Practice Note Section 76 Planning Agreements*. Available at: https://www.infrastructure-ni.gov.uk/sites/default/files/publications/infrastructure/dmpn-21-section-76-planning-agreements-v1-jan-2017_0.pdf.

42 PPG1: https://www.planningni.gov.uk/index/policy/planning_statements_and_supplementary_planning_guidance/pps01/pps01_development_control/pps01_planning_agreements.htm.

43 See Para. 2.8 Belfast City Council. 2020. *Developer Contributions Framework 2020*. Available at: https://www.belfastcity.gov.uk/documents/developer-contributions-framework

44 Mrs Maire Campbell, Chief Commissioner of the PAC, oral evidence to the Northern Ireland Assembly, Committee for the Environment, Planning Bill, 26 January 2011, page 42.

45 The explanatory notes to Section 76 state: 'This Section enables any person who has an estate in land to enter into a planning agreement with either a council or the Department (whichever is the relevant authority)'.

46 [1995] 1 WLR 759.

47 At 782.

48 See: *Belfast City Council v The Planning Appeals Commission* [2018] NIQB 17; *Tesco Stores v Dundee City Council* [2012] UKSC 13. More generally, see: *In Re McFarland* [2004] UKHL 17 at [24], per Lord Steyn.

49 *Tesco*, per Lord KeithPp 764, 770.

50 [1981] AC 578.

51 [1995] 1 WLR 759.

52 [1981] A.C. 578, 619. See also the comments of Lord Lane at 627.

53 *Tesco* 779

54 (1994) 67 P & CR 78.

55 [1995] 1 WLR 759.

56 Technically speaking this was a voluntary scheme but not, as Lord Hodge observed at paragraph 57 of his judgment "in any real sense".

57 *Elsick*, at Para 58.

58 At Para 63.

59 At Para 59.

60 At Para 60.

61 At Para 61.

62 At Para 64.
63 [2018] NIQB 49.
64 See for example the sentiments of Lord MacLean (Court of Session, Outer House) in *McIntosh v Aberdeenshire Council*, 1998 S.C.L.R. 435, 440-441.
65 *Where planning permission is not granted because (inter alia) of failure to successfully negotiate a planning agreement with the planning authority then an appeal might be made (against refusal of planning permission) but it is questionable whether a developer would want to go down this road. For a discussion in the context of the English legislation see David Brock. 2002. 'Planning Obligations – Ideas for Reform' Journal of Planning Law* 12, 16–18.
66 See: The *draft NI Programme for Government 2016-21* where Housing is a priority action. Available at: https://assets.publishing.service.gov.uk/government/uploads/system/uploads/attachment_data/file/856998/2020-01-08_a_new_decade__a_new_approach.pdf. In addition, the *New Decade New Approach* document of January 2020 details a commitment by the NI Executive to examine the issue of housing affordability and that the Executive will also enhance investment and agree a target for new social and affordable home starts.
67 Per Minister for Social Development to Northern Ireland Assembly, 26 February 2008.
68 Speech by Margaret Richie, Minister for Social Development to the Northern Ireland Assembly, 26 February 2008.
69 Department for Social Development 2012. *Facing the Future: The Housing Strategy for Northern Ireland 2012-2017*. Available at: https://www.communities-ni.gov.uk/publications/facing-future-housing-strategy-northern-ireland.
70 Ibid., at p. 4.
71 Department for Social Development. 2015. *Developer Contributions for Affordable Housing in Northern Ireland Report of Study*. (Edinburgh: Three Dragons Heriot-Watt University, at para. 263, p. 84.
72 Ibid.
73 Ibid., at p. 90.
74 Ibid.
75 Department of Environment. 2015. *Strategic Planning Policy Statement: Planning for Sustainable Development*, at Para 6.143.
76 See: Department of Environment. 2014. *Draft PPS 22 'Affordable Housing.*
77 Except for the Appendices. 'Social housing' appears only in Appendix B Glossary. 'Affordable' housing appears in Appendix A: Viability Appraisal Checklist and Appendix B Glossary.
78 The SPPS sets out definition of affordable and social housing respectively. However, there has been recent debate relating to such definition and, in June 2019, the Department for Communities launched a consultation seeking views on its proposals to change the 'Definition of Affordable Housing'. The purpose of the consultation was to agree a definition of affordable housing that considers the needs of a wider range of groups, some of whom are not currently finding their needs adequately met by the market.
79 Anthony Crook and Christine Whitehead. 2002. 'Social housing and planning gain: is this an appropriate way of providing affordable housing?', *Environment and Planning A: Economy and Space* 34, 1259–79, at p. 1266. See also: Anthony Crook and Christine Whitehead. 2019. 'Capturing development value, principles and practice: why is it so difficult?', *Town Planning Review* 90, 359–81.
80 For a recent discussion of planning gain and CIL see: Tola Amodu. 2020. 'Revisiting the rules. The pervasiveness of discretion in the context of planning gains: the case of the community infrastructure levy', *Public Law* Oct, 643-60.

7

JUDICIAL REVIEW IN THE PLANNING ARENA

Stephen McKay and Michael Murray

Introduction

Judicial review is a procedure that allows the use and non-use of powers and duties of public bodies and officials to be challenged in court. Litigation can arise from perceived unlawful decisions not authorised by legislation, irrationality, procedural impropriety and breaches of citizens' rights and freedoms. The scope of judicial review is wide-ranging across multiple spheres of public policy and, in the context of this book, it has become a significant feature of the UK planning systems. The increasing recourse to judicial review as a way by which to challenge ministerial and local authority decision making is evidenced by the establishment in 2014 of a specialist Planning Court in England and Wales within the Administrative Court in the Queen's Bench Division of the High Court of Justice, to deal with planning matters that include applications and appeals relating to planning permission, enforcement decisions, compulsory purchase orders and highways and other rights of way. From March 2014 to the end of June 2020, it is estimated that a total of 377 judgments have been handed down by the Planning Court with an additional 105 rulings by the Court of Appeal and 11 by the Supreme Court (see: https://simonicity.com/2020/08/15/introducing-the-planning-court-case-explorer/). In Northern Ireland the judicial review procedure has gained strong traction with cases being brought forward by individuals, commercial interests, local resident associations, special interest groups, local authorities and government agencies and departments. Indeed, the current frequency and high profile of judicial review in this jurisdiction can be illustrated by the publication in 1991 of a seminal paper (Hadfield 1991) that highlighted some twenty-two cases as demonstrating the width of matters covered by judicial review; none were planning related. In contrast, by 2009 Northern Ireland witnessed the publication of its first authoritative compendium of planning-related judicial review cases (Orbinson et al. 2009). Over the interim, judicial review has continued its supervisory role in the planning system. It is firmly established as a key element of contemporary planning practice and thus merits a measure of discussion in this book.

It is important to note that judicial review is not an appeal against an unfavourable planning decision, although for affected third parties, for instance, this may be perceived as the

DOI: 10.4324/9781003218319-7

only procedural option given the absence of third-party planning appeals in Northern Ireland. Simply put, the key question to be addressed in judicial review is whether a public decision in a planning case has been taken fairly and lawfully as set out by Treacy J in *Newry Chamber of Commerce and Trade's Application* (2015 NIQB 65):

> The Court will not conduct an appeal against the planner's judgement: it will not substitute its judgement on the weight to be attached to the relevant factors in place of the planner's judgement on that question. It will however review the legality of the planning process on the basis of the well understood principles of public law where a case is made out that the planner has made an error of reviewable kind.

Accordingly, the jurisdiction of the court in planning matters, as Lowry LCJ stated in *R v Secretary of State for Northern Ireland ex parte Blair* (1985 Queen's Bench Division (Crown Side)), is not appellate but supervisory. It has been likened to a legal audit of the decision-making process (see *Mooreland and Owenvarragh Residents' Associations' Application* (2014 NIQB 130)). The duty of the court as Carswell J has said, citing Lord Donaldson MR, is to act as referee and ensure that the rules have been observed (see *R v Secretary of State ex parte Genova Ltd and another* (1992 Queen's Bench Division (Crown Side)). Judicial review links clearly to the accountability of public bodies within a mature democracy and, in that regard, the scope for legal redress has been cogently expressed by Gillen J *In the Matter of an Application by Sheridan Millennium Limited for Leave to Apply for Judicial Review* (2007 NIQB 27):

> it is critical to identify the decision and the nature of the attack on it. Unless there is a public law element in the decision and unless the obligation involves breaches of duties or obligations owed as a matter of public law, the decision will not be reviewable. However this should not mask the purity of the principle that a public body in exercising a statutory function cannot escape being subject to judicial review if it abuses the power vested in it.

The chapter is structured into three main sections. First, we introduce the judicial review procedure and examine matters relating to the pre-action protocol; leave; promptitude and time limit; the requirement of standing; litigants in person; duty of candour and discovery; remedies; and costs. Second, we consider the grounds for judicial review related to illegality, irrationality and procedural impropriety and illustrate the application of these criteria with reference to lack of proper inquiry, consultation shortfall, error of approach and actual bias/predetermination or apparent bias by drawing on selected cases. Third, by way of conclusion we consider the role of an independent judiciary, attempts by the state to circumscribe the remit of judicial review in regard to planning and development and emergent directions for practice within the wider UK context. The brief narrative must inevitably draw on a limited number of judgments handed down by the Northern Ireland courts, but as an aid to future research in this field, a more complete list of authorities is included as Appendix 1 to this book.

The judicial review procedure

Legislation regarding judicial review is set out in the Judicature (Northern Ireland) Act 1978 with additional statutory provisions set out in Order 53 of the Rules of the Court of Judicature

(NI) 1980, amended in 2017. More complete guidance in regard to judicial review proceedings in Northern Ireland is set out in Judicial Review Practice Direction 3/2018 issued by the Lord Chief Justice's Office, Belfast. This deals with the range and content of papers that should be submitted to the Judicial Review Office in the Royal Courts of Justice in Belfast. Mention is made of the need to specify the decision under challenge, the relief being sought and the grounds being relied upon, together with the preparation of affidavits and exhibits. The procedure is denoted by three key components:

- the pre-action protocol, conceived as an early dispute resolution mechanism between the parties in order to avoid litigation and which, if unsuccessful, can then move to an applicant entering the judicial review court;
- an initial application for leave (permission by the court) to apply for judicial review which may be granted on the papers or following a leave hearing;
- a subsequent substantive hearing, if leave is granted.

Should leave be refused, that decision may be appealed as, for example, *In the Matter of an Application by John Hill for Leave to Apply for Judicial Review* (2007 NICA 1) where the Court of Appeal granted leave on the ground that the Department of the Environment had not had proper regard to planning policy in the determination of the Notice Party's planning application. This section of the chapter explores a number of matters that arise in regard to judicial review practice.

Pre-action protocol

Before initiating a judicial review action, a prospective applicant should seek to resolve the issue to hand by informal discussions, complaint procedures or other potential remedies, such as recourse to Ombudsman services. Essentially this recognises that judicial review is a measure of last resort following the exhausting of alternative remedies. The pre-action protocol also engages with that search for resolution and comprises a formal letter to the respondent setting out, inter alia, the matter being challenged, including the date and details of the impugned decision, act or omission, the details of the action that the respondent is expected to take, perhaps a request for further information and, finally, a proposed reply date. Judicial Review Practice Direction 3/2018 provides helpful information on shaping these communications from the prospective applicant and the respondent and suggests that, as a general rule: (a) the putative applicant's pre-action protocol letter will be transmitted not later than seven weeks following the date of the impugned decision, act or measure and (b) the agency or agencies to whom the applicant's pre-action protocol letter is directed shall respond within the ensuing period of three weeks.

The importance of this rubric of resolution behaviour is underscored *In the matter of an Application by Coulters Hill Residents Ltd for Leave to Apply for Judicial Review v Planning Appeals Commission* (2020 NIQB 1), wherein the applicant sought to challenge the decision of the Planning Appeals Commission to allow, in part, an appeal against an enforcement notice issued by Ards and North Down Borough Council regarding alleged breaches of planning control related to quarrying activities. In his ruling that dismissed an application for leave to apply for judicial review, McCloskey, J commented:

> The inter-related principles of judicial review as a remedy of last resort and the exhaustion of alternative remedies also arise. The well-established principles of legal certainty

> and the finality of litigation (*interest rei publicae ut sit finis litium*) explain in part why this is so. It is not the function or duty of our legal system to devise a mechanism whereby the applicant, in fulfilment of its stated objects … can maintain or extend or renew its objections to the activities on the subject site … Notwithstanding it is clear that, as a minimum, the applicant has at its disposal the facility of assembling all evidence now available and, based thereon, inviting the Council to issue a fresh enforcement notice in specified terms. This course has not been pursued. In this respect the written communication from the applicant to the court, dated 28 November 2019, is of special note:
>
>> 'The Applicant has served a PAP (Pre-application protocol) letter on the [Council] on 14 November [2019] … . [and] … . is preparing to submit a new application for leave to review the Council for lack of enforcement'.
>
> This court, in furtherance of the overriding objective, would caution the applicant to ensure that before issuing any judicial review challenge against the Council of the kind mooted it must first present its case for enforcement action in comprehensive terms, supported by the totality of all available material evidence, as indicated above.

Leave

The initial application for leave to apply for judicial review is primarily concerned with testing whether there exists an arguable case for granting the relief being sought. An application for leave to apply for judicial review must be made by lodging an *ex parte* docket (a summary note of intention to proceed by the applicant, along with what is being applied for and the formal documents being relied upon), an Order 53 Statement, affidavits and exhibits. The Order 53 Statement provides information on the applicant, the respondent, the impugned decision, act or omission, the relief being sought and the grounds upon which the challenge is based. On occasion the court may direct amendments to the Order 53 Statement in order to crystallise the grounds of challenge. *In the Matter of an Application by Dean Blackwood, Representing River Faughan Anglers Limited for Judicial Review v Derry City and Strabane District Council* (2018 NIQB 87) McCloskey J described the grounds of challenge in their initial incarnation as 'somewhat diffuse and prolix' but which via initial amendments of the Order 53 Statement were able to be reformulated into six grounds which provided the basis for granting leave to apply for judicial review. This preliminary stage is also designed to:

> prevent the time of the Court being wasted by busybodies with misguided or trivial complaints of administrative error, and to remove the uncertainty in which public officers and authorities might be left as to whether they could safely proceed with administrative action while proceedings for judicial review of it were actually pending even though misconceived. (*R v Inland Revenue Commissioners, ex parte National Federation of Self-Employed and Small Businesses Ltd.* (1982 AC 617))

As noted by Anthony (2014: 72) the requirement of arguable case has been described as a modest hurdle that may vary depending on the material available at the leave stage and on the view of the judge determining the application. The significance of having an arguable case is illustrated *In the Matter of William Young for Judicial Review* (2012 NIQB 15) whereby the applicant sought leave to challenge the decision of the Department of the Environment granting planning permission for a house on farmlands near Greyabbey, County Down. The site is located within the countryside and thus the proposal was the subject of detailed policy tests set out in *Planning Policy Statement 21: Sustainable Development in the Countryside*. The most relevant policy in this case is identified as CTY10 Dwellings on Farms: 'Planning permission will be granted for a dwelling house on a farm where all of the following criteria can be met: (a) the farm business is currently active and has been established for at least 6 years'. The applicant argued that there had been no active farm business on the subject site and that the criteria in CTY10 were not met. Following his examination of the detailed case notes of the respondent, Treacy J concluded to the contrary:

> The unanimous judgement, therefore, of all the officials who dealt with the impugned planning application is that it satisfied the requirements of CTY10 namely that the land had been maintained in good agricultural and environmental condition. The applicant's objection to the contrary were fully considered and rejected and I see no public law basis for interfering with the assessment of those entrusted with this task. In reality the applicant's judicial review is an impermissable attempt to challenge the merits of the planning judgement of the proposed respondent … Accordingly the applicant has not made out any arguable case for leave and the application must be dismissed.

What also comes through from this judgment is the much-stated remit of the court in planning cases to confine itself to the lawfulness of public decision making and not to evaluate *de novo* the factors informing planning judgement. But in those circumstances where leave is granted we are again reminded by Treacy J in *Crusaders Football Club's Application* (2013 NIQB 68) that:

> the grant of leave says nothing at all as to how the case will ultimately be resolved following the provision and consideration of all relevant evidence and full submissions on all the substantive issues that arise. The grant of leave merely signifies that the case is arguable or worthy of further consideration.

Moreover, the court may grant leave on only some of the grounds of the challenge stated in the Order 53 Statement.

Promptitude and time limit

It was long-established practice that an application for leave to apply for judicial review must be made promptly and within a period of three months from the date on which the grounds for the application first arose. The requirement of promptitude for making an application for leave to apply for judicial review was removed by The Rules of the Court of Judicature (Northern Ireland) (Amendment) 2017 that came into operation on 8 January 2018. The time limit is now three months. Previously, a court could refuse to grant leave if the application had not been brought promptly or within the judicial review time limit.

This matter of promptitude attracted considerable judicial comment. In *Corbo Properties' Application* (2012 NIQB 107), Horner J suggested that there may be cases in which a substantial delay may not be fatal, citing, for example, a planning permission, which, if implemented, would have such detrimental consequences that, despite the delay, the public interest was such that relief had to be granted. In *Blackburn's (Louisa) Application (Leave Stage)* (2010 NIQB 71) concerning the vesting of properties in the Village area of Belfast, Treacy J ruled that delay caused by the initial refusal of legal aid was not an adequate excuse given the large number of other affected house owners. *In The Matter of an Application by Francis McHenry for Leave to Apply for Judicial Review* (2007 NIQB 22) regarding a decision of the Planning Appeals Commission, Gillen J concluded that while the application had not been brought promptly, albeit within the three-month period, some measure of delay was justified in order to acquire legal advice before proceeding further. In *Tesco Stores Limited's Application* (2010 NIQB 15), Weatherup J has more fully stated:

> The application for judicial review commenced on 9 March 2009 which was just within the three month period. The obligation to commence proceedings for judicial review is to apply promptly and in any event within three months, with the court having power to extend time for good reason. The obligation in the first place is to apply promptly. An applicant must explain any delay in applying promptly … The application could not be said to have been made promptly … In commercial matters and in relation to issues concerning planning permission there is a need for particular vigilance and for expedition in the response to decisions considered to be adverse to one's interests. If there has been delay the applicant must satisfy the court that there is good reason to extend time. Establishing good reason also requires that an explanation be provided of the reasons for delay … The nature and character of the challenge and of the grounds of challenge and the prospects of success are considerations in the exercise of the discretion to extend time for good reason.

The question that can arise for an applicant is when the clock starts to run? In *Tesco Stores Limited* the court was satisfied that time began to run against the applicant when planning permission was granted and not when a previous notice of opinion to approve had been issued. The principle rehearsed here recognises the completion of the statutory decision-making process as the start date for the formal initiation of any challenge to the granting of planning permission, although the courts have advised that an engagement with the planning process before any decision is made can be a material consideration. In *Musgrave Retail Partners (NI) Ltd's Application (Leave Stage)* (2012 NIQB 109), Maguire J declared:

> While every case will depend on its own facts, and there is no rule of thumb as to what constitutes a timely application, great expedition will normally require that the application be made well within the outer time limit for judicial review applications of three months. This will especially be so if the challenger is an entity which is well aware of a trade competitor's application for planning permission and has objected to it and monitored its progress. In the present case it is clear that the applicant falls clearly within the category just mentioned. It objected to the various planning applications made by the developer in respect of the building of a supermarket at Coalisland including the two applications for planning permission at issue in these proceedings. It engaged planning experts to monitor the progress of the developer's applications and it was in correspondence with the Planning Service about them. Many of the points giving rise to the judicial review had been rehearsed by the applicant with the Department prior

> and subsequent to the grant of the planning permissions. In short, the applicant was an entity familiar with the main issues now raised and had plenty of time in advance of the actual date of grant to have begun preparation for a legal challenge. Notwithstanding the above, as it turned out, no challenge was mounted for nearly three months ... In these circumstances the court concludes without difficulty that the application for judicial review in this case was not initiated promptly.

Turning to good excuse as an explanation for any delay in launching this judicial review application, Maguire J declined to exercise his discretion in favour of the applicant:

> Given the nature of the challenge in this case, the resources of the applicant and the applicant's state of knowledge of the permissions being challenged and its access to legal and other advice, this is not a case where one could write off delay simply on the ground that mounting a judicial review is a difficult or complex matter.

However, a further point highlighted by this case is that the promptitude requirement was related to domestic law grounds of challenge and did not apply to EU grounds of challenge in a leave application. Leave in *Musgrave Retail Partners* was granted in respect of the applicants' EU grounds only. As explained by Maguire J:

> The reasoning behind the distinction between domestic law and EU grounds of challenge is one which derives from the operation of the certainty and effectiveness principles contained in EU law. The criterion of promptitude is viewed as incompatible with these as in effect it is viewed as erecting a discretionary time limit. While this approach originates in authorities in the European Court of Justice dealing with public procurement ... the reasoning has now been applied to the test of promptitude in the context of judicial reviews in the planning context where EU grounds of challenge are at issue. The Court considers that the distinction is now well established and that accordingly it should accept it.

As observed above, the now amended requirement for promptitude had previously generated considerable judicial comment. Its removal dates back to June 2015 when the Department of Justice initiated a consultation process on a proposal that there should be no requirement to bring judicial review proceedings promptly in any case and that all proceedings should, rather, be required to be brought within three months of the date of the reviewed decision. The Department took the view that this change would provide a greater degree of certainty to the judicial review process and ensure continued compliance with obligations arising from EU law (Department of Justice 2015a: 2–3). In its analysis of the consultation submissions, the Department of Justice (2015b) noted that all respondents agreed that it was appropriate to remove the requirement to bring a judicial review promptly, that the power to extend the time limit in appropriate cases should be retained and that it would be inappropriate to remove the promptitude requirement for applications made on EU grounds (and under the Aarhus Convention) but retain it for applications made on domestic grounds. Moreover, most respondents that expressed a view did not consider that the proposal to remove the promptitude requirement would have an adverse impact on any particular category of judicial review. However, the Bar Council did signal concern that the review of planning decisions could be adversely affected in that change could result in delays that could financially impact on developers and other interested third parties.

Notwithstanding the removal of the promptitude requirement, the judicial review court still retains the discretion to extend the period beyond three months within which an application shall be made where it is considered that there is good reason. This issue arose *In Gibson's (Christine) Application* (2016 NIQB 82), whereby the applicant sought leave to apply for judicial review of planning permission granted by Newry Mourne and Down District Council for the construction of ferry terminal facilities in Greencastle, County Down, to allow for the operation of a vehicle ferry across Carlingford Lough to Greenore. Additionally, the granting of a marine licence by the Department of Agriculture, Environment and Rural Affairs (Marine and Fisheries Division) was challenged. In his ruling that refused leave to apply for judicial review, Colton J noted that the application had been lodged almost eleven months after the planning consent decision, that the applicant had been fully aware of the decision when it was made and that the granting of leave will result in significant prejudice to the ferry company, which enjoyed the benefit of the planning permission and the marine licence. The Managing Director of the ferry company in his affidavit had stated:

> I had understood that an important purpose of the 3 month period was to provide sufficient certainty to allow those who benefit from planning permission to proceed with their development. We would certainly not have proceeded with the vast majority of the public expenditure if proceedings had been lodged within the 3 months of the grant of the 2015 permission.

The judgment of the court gave great weight to this material consideration and concluded that permission to proceed to judicial review would be prejudicial to the interests of the ferry company and that the time period should not be extended. In so doing, Colton J noted the approach adopted in England and Wales where Sales, LJ in *Gerber v Wiltshire Council and Others* (2016 EWCA Civ 84) explicated key principles regarding the implementing of a planning permission in the face of a challenge in the courts:

> Once the planning permission is granted the owner of the land to which it relates is entitled to rely upon it and there is a substantial risk that he will begin investing effort and money to do so without waiting any lengthy period before he does. Also, planning permission will have been granted because the grantor is satisfied that it is in the overall public interest for the development to occur, without any further delay. The basic rules regarding notification of applications for planning permission set out in the 2010 order are designed to afford potential objectors a fair opportunity to learn about and object to an application for planning permission before it is granted. The court's approach in relation to an application to extend time for judicial review has to strike a fair balance between the interests of the objector, the interests of the developer and the public interest.

A subsequent ruling by the Court of Appeal (2017 NICA 77) concluded that Colton J 'was entitled to take into account the prejudice to the third party rights and detriment to good administration which would be occasioned by allowing this late application'. The decision made by Colton J was affirmed and the appeal for leave to apply for judicial review was dismissed.

As a final note on this matter, it is worth mentioning the relationship between commencing judicial review proceedings within time and progressing the litigation expeditiously. *In the Matter of an Application by Rural Integrity (Lisburn 03) Limited for Leave to Apply for Judicial Review and in the Matter of a Decision of Lisburn and Castlereagh City Council* (2021 NIQB 32) Scoffield J ruled:

> one must also recognise that, once judicial review proceedings have been lodged, the applicant itself is no longer in sole control of the speed with which the proceedings are progressed. That is partly because significant responsibility for managing the case moves to the Court in the exercise of its case management powers; but also because progress will be dependent (at least to some degree) on issues such as court resources, judicial availability and that of the parties' representatives, and steps to be taken by other parties such as proposed respondents or notice parties. That is not to say that there will not be cases where a judicial review application should be dismissed for want of prosecution; nor to downplay the importance of expedition and finality in the public law sphere. It is simply a recognition that, although the applicant bears a responsibility to act reasonably and seek to drive the litigation on, matters are not entirely in the applicant's own hands.

The requirement of standing

There is a requirement on applicants for leave to apply for judicial review to demonstrate *locus standi*, the threshold test being sufficient interest in the matter to which the application relates. If standing is not demonstrated it follows that an application will be dismissed. The early determination of this consideration, if challenged by a respondent, is important not least in saving the costs associated with a subsequent substantive hearing should leave be granted. In the search for guidance regarding the interpretation of standing, useful insight is provided by the Northern Ireland Court of Appeal *In the Matter of an Application by D for Judicial Review* (2003 NICA 14). This related to a challenge by the Department for Regional Development to a High Court declaration that it had acted unlawfully in deciding not to prosecute an Orange Lodge that had erected an Orange arch in Glengormley in 2001. Carswell LCJ, in delivering the judgment of the Court of Appeal, explored the facets of standing:

> There has been much discussion of the topic of standing in textbooks and legal periodicals and examples abound in the reported cases, yet it is difficult to pin down any authoritative statement of the principles to be applied by a Court in determining the question. It appears to be incontestable that the Courts have tended in recent years to take a more liberal attitude to matters of standing. We would tentatively suggest that the following propositions may now be generally valid:
>
> a) Standing is a relative concept, to be deployed according to the potency of the public interest content of the case.
> b) Accordingly, the greater the amount of public importance that is involved in the issue brought before the Court, the more ready it may be to hold that the applicant has the necessary standing.

c) The modern cases show that the focus of the Courts is more upon the existence of a default or abuse on the part of a public authority than the involvement of a personal right or interest on the part of the applicant.
d) The absence of another responsible challenger is frequently a significant factor, so that a matter of public interest or concern is not left unexamined.

Much emphasis is placed above on the public importance of the matters being raised by way of judicial review. Nonetheless, we are informed of the often narrowness of this concept by the judgment of Gillen J *In the Matter of an Application by James Stewart Moore for Judicial Review* (2007 NIQB 23) when he restates the comments of Dyson J in *R v Lord Chancellor, ex parte Child Poverty Action Group* (1999 1 WLR 347):

> I should start by explaining what I understand to be meant by a public interest challenge. The essential characteristics of a public law challenge are that it raises public law issues which are of general importance, where the applicant has no private interest in the outcome of the case. It is obvious that many, indeed most judicial review challenges, do not fall into the category of public interest challenges so defined. This is because, even if they do raise issues of general importance, they are cases in which the applicant is seeking to protect some private interest of his or her own.

It is within this context that judicial review cases on planning matters in Northern Ireland can point to a wide sweep of applicants with standing. These include government departments (for example, *Department of the Environments Application* (2014 NIQB 4)); local authorities (for example, *In the Matter of an Application by Omagh District Council for Judicial Review* (2007 NIQB 61)); public agencies (for example, *In the Matter of an Application by William Young for Judicial Review* (2007 NICA 32)); membership organisations (for example, *National Trust's Application* (2013 NIQB 60)); pressure groups (for example, *Application by Friends of the Earth for Judicial Review* (2006 NIQB 48)); ad hoc single issue groups (for example, *Alternative A5 Alliance's Application for Judicial Review* (2013 NIQB 30)); aggrieved institutions (for example, *In the Matter of an Application by Queen's University of Belfast for Judicial Review* (1997 Queen's Bench – Crown Side)); aggrieved businesses (for example, *Central Craigavon's Application* (2010 NIQB 102)); and aggrieved individuals (for example, *In the Matter of an Application by Brian Quinn and Michael Quinn* (2013 NIQB 24)).

The lack of standing to apply for judicial review remains open, however, as a ground on which to submit that leave should be refused. This issue surfaced *In Doyle's (Ellen) Application* (2014 NIQB 82), whereby the applicant sought to challenge a decision of the Planning Appeals Commission to grant planning permission to the University of Ulster for a mixed-use regeneration scheme that included a multi-storey car park in Belfast. The evidence indicated that the plaintiff did not respond to a public advertisement regarding the application for planning permission, did not respond to a public advertisement regarding the university's planning appeal against the refusal of planning permission and that, consequently, the applicant had not participated at any stage of the process. In his ruling, Treacy J addressed the issue of standing:

> 'I accept the submission of the [Planning Appeals Commission] that where, as here, members of the public are provided with a reasonable opportunity to participate in a

> quasi-judicial process, a person who does not so participate cannot ordinarily be said to have sufficient interest in the outcome of that process'.

Being unaware of the impugned decision at the time it was made, in the context of public advertisements and considerable publicity prior to and following the planning appeal decision, prompted further comment by Treacy J:

> Being unaware of the advertisements cannot be a sufficient basis to confer standing. It would undermine the clear statutory purpose underpinning Art21 and Art32(6) that the advertisements pursuant to those sections are intended to be sufficient to provide interested parties with a reasonable opportunity of participating in the statutory planning process. Further, it would introduce uncertainty since a person not involved in the process could, as here, emerge late in the day to mount a challenge including seeking to rely on points not taken by any of the participants in the appeal even though better placed challengers who actually participated in the process have not sought judicial review. For the above reasons I conclude that the applicant does not have sufficient interest to bring this application.

Litigants in person

In introducing the publication of a guide for litigants in person involved in judicial review planning and environment cases (Deb, A. et al. 2019), the Lord Chief Justice of Northern Ireland, The Right Honourable Sir Declan Morgan, commented:

> The right to a fair trial is enshrined in Article 6 of the European Convention on Human Rights. It is an overarching right, not qualified by whether or not a party is legally represented. There are two important strands to Article 6 rights in the context of those who are unrepresented: firstly, is the litigant in person able to participate effectively in the proceedings to a level where he or she is able to influence them so that the court can ensure procedural and substantive justice? Secondly is there a fair balance between the parties in the opportunities given to them to present their case in a manner which does not disadvantage them with respect to the other side? Effective participation and equality of arms are the bedrock of a personal litigant's right to a fair trial.
>
> *(p. ii)*

Litigants have appeared in person in planning and environment cases in Northern Ireland, on occasion accompanied by a McKenzie Friend (a person who supports and advises a personal litigant in court but who does not speak on their behalf unless invited to do so by the court). *In the Matter of an Application by Mr Chris Murphy to Apply for Judicial Review; In the Matter of a Decision by the Department for Infrastructure on 17 August 2016 to Proceed with the Castledawson to Toome Dualling Scheme* (2017 NIQB 35) the applicant appeared as a personal litigant and in the final words of the judgment that dismissed the application, Keegan J, concluded:

> I offer a final word to the applicant. I want to record that he conducted his case impeccably with the assistance of his wife. I know that he will be disappointed by this decision

> but I commend him for the care and attention he has applied to his case and for raising environmental awareness in relation to this important issue.

What comes through here is the high standard of case preparation and presentation required by personal litigants. At the same time, the court exercises a duty of due consideration to all participants and which has been reinforced by Gillen LJ in delivering the judgment of the Court of Appeal *In the Matter of an Application by Christine Gibson for Leave to Apply for Judicial Review and In the Matter of a Decision of Newry, Mourne and Down District Council Made on 29 June 2015 and 25 February 2016 and In the Matter of a Decision by the Department of Agriculture, Environment and Rural Affairs (Marine and Fisheries Division) made on 3 June 2016* (2017 NICA 77) when it was stated:

> we take this opportunity to restate the now oft repeated principle that personal litigants cannot have an unfair advantage against represented parties. They cannot seek to rely on inexperience or lack of proper appreciation of what the law requires. Courts must strive to bring structure and focus to their arguments with firm case management at the earliest stage available to ensure that there is clarity to the issues at stake and that the personal litigant is confined to relevant issues. There are great advantages to such an early management system. In addition to assisting the judge in determining whether or not leave should be granted, it may help to focus the real issues between the parties and provide an opportunity for early ground rules for the conduct of the hearing.

The important point embedded here is that judicial review is not concerned with the merits of a decision and will not intervene unless a public law wrong has been established on the evidence and not by way of conjecture. For litigants in person in planning cases, and in a situation where there is no third-party planning appeal remedy within Northern Ireland, this important distinction may be difficult to grasp. It is against this backdrop that seminal research by McKeever et al. (2018), concerned with those going more widely through the Northern Ireland court system as self-litigants without legal representation, is to be welcomed. In securing that effective participation and equality of arms championed above, the published report makes not less than thirty-nine recommendations that embrace changes to culture, administration, support to self-litigants and engagement with the legal profession.

Duty of candour and discovery

Parties have a duty to lay before the court all material documents and the late introduction of a document into the proceedings can cause an adjournment to enable an affidavit to be sworn exhibiting it (see *In the Matter of an Application by Kinnegar Residents' Action Group, Park Road and District Residents' Association, Old Stranmillis Residents' Association, Belfast Holyland Regeneration Association and Cultra Residents' Association for Judicial Review* (2007 NIQB 90)). The sharing of documentation is embedded in the principle of the duty of candour that speaks to openness and transparency and which has been endorsed in judicial comment as a practice to be commended. *In the Matter of an Application by Paul McNamara for Leave to Apply for Judicial Review v Lisburn and Castlereagh City Council* (2018 NIQB 22) that challenged the granting of planning permission for an infill dwelling across the road from the applicant's dwelling, McCloskey J, in his review of the written evidence, opines:

> I draw attention to another issue of good litigation practice. The Council's solicitor has exhibited to her affidavit the manuscript notes made by her at the public meeting of the Planning Committee on 28 June 2017. This is a commendable illustration of every judicial review respondent's duty of candour to the court. The 'Attendance Note' of the solicitor, who attended the relevant meeting is in printed form and, hence, readily comprehensible. As the solicitor who swore this affidavit recognises, this court 'has been requesting copies of such materials in similar cases …'. To this I add that the production of such materials will almost invariably be required by the duty of candour.

On occasion, an applicant for judicial review may formally seek the disclosure of additional documents deemed relevant to the case. However, there is no automatic right to discovery within this arena and the courts have been guided by principles that limit its scope. *In the matter of Applications by (1) AM Developments UK Limited (2)Belfast City Council and Belfast Chamber of Trade and Commerce (3) Bow Street Mall Limited and Lisburn Chamber of Commerce (4) Central Craigavon Limited for Judicial Review and In the Matter of a Decision of the RT Hon Lord Rooker, Minister with Responsibility for the Department of the Environment* (2006 NIQB 9) the plaintiffs were requesting the disclosure of documents that were argued as being necessary for the purpose of challenging the reasoning and procedural process followed by the Minister in arriving at an impugned decision to approve an out of town shopping development at Sprucefield and where the anchor store would be occupied by John Lewis Plc. Girvan J summarised the governing principles relating to discovery as follows:

> It is essential to examine carefully the issues which arise in any particular application for judicial review to ascertain whether discovery is necessary for the resolution of some issue arising in the application. Unless there is some *prima facie* case for suggesting that the evidence relied on by the deciding authority is in some respect incorrect or inadequate it is improper to allow discovery of documents, the only purpose of which would be to act as a challenge to the accuracy of the affidavit evidence. The court will consider that discovery is not necessary for disposing fairly of the case where the applicant is unable to positively identify material which suggests that the respondent has acted improperly in making the impugned decisions.

In this case, the court did not accede to the application for discovery, with Girvan J commenting that 'the nature and width of the claim for discovery smacks clearly of a fishing expedition'. However, discovery can bring to light additional substantive grounds for judicial review. In *Quinn's (Brian) Application* (2013 NIQB 24) an Amended Order 53 Statement was submitted for the court hearing following discovery that enlarged on the initial grounds on which relief was being sought. Nonetheless, in those instances when discovery is granted, the courts have advised caution in regard to further applications for additional documents should the released documentation not serve the purpose for which it is intended. Gillen J *In the Matter of an Application by Sheridan Millennium Limited for Judicial Review and In the Matter of a Decision by the Department for Social Development and in the Matter of a Decision by Laganside Corporation* (2007 NIQB 92) concluded at the end of his judgment granting discovery of a due diligence assessment of the Victoria Square development in Belfast: 'Accordingly any further application for documents arising from the scheme at Victoria Square will command the anxious scrutiny of the Court to ensure the exercise is not becoming unhelpful, oppressive, time consuming or flooding the Court with needless paper'.

Remedies

The Northern Ireland judicial review practice note (Northern Ireland Courts and Tribunal Service 2013) indicates that it is a requirement to signal the relief being sought when issuing proceedings. When planning-related decisions are being challenged the courts may do the following:

- issue an order of *certiorari* to quash the decision at issue, which will have the effect of rendering the grant of planning permission a nullity and meaning that the planning authority will have to make a fresh decision that may or may not be different. In *R v Secretary of State for Northern Ireland ex parte Johnston* (1984 Queen's Bench Division – Crown Side) Kelly, LJ quashed the decision of the Planning Appeals Commission on the ground that the appointed member had refused an adjournment and deprived the applicant of presenting his case. *In Boswell's Application* (2009 NIQB 95) a decision of the Planning Appeals Commission dismissing the applicant's appeal against the refusal of planning permission by the Department of the Environment in June 2006 for the retention of a serviced site for a single traveller family on land within the green belt of Portadown was quashed. The court was not satisfied that the Commissioner had conducted an evaluation of the suitability of alternative accommodation and in that regard had failed to take into account a material consideration. A new hearing was subsequently convened by the Planning Appeals Commission which fully explored all material considerations, the appeal was allowed and full planning permission was granted subject to conditions in March 2011 (Commission Reference: 2006/A1986);
- issue a declaration that the decision was wrong in law with the court giving expression as to what the legal position is. In an *Application by Friends of the Earth for Judicial Review* (2006 NIQB 48), Weatherup J made a declaration that the Department of the Environment through the Water Service had failed to take into account a relevant consideration in regard to new connections to waste water systems, that relevant consideration being whether the system is compliant with the requirements of the Urban Waste Water Treatment Regulations (Northern Ireland) 1995;
- issue an order of *mandamus* to require the decision maker to perform a duty or reach a different decision. Within the world of contemporary Northern Ireland planning cases this remedy tends to be little used but can be fully appreciated *In the Matter of an Application by UK Waste Management Limited for Judicial Review* (1999 CARF2798) where the Court of Appeal determined that the Department of the Environment did not have power to defer consideration of a planning application for an extension to the company's landfill operation at Mallusk and thus issued an order of *mandamus* directing the Department of the Environment to determine the company's planning application;
- issue an order of prohibition that forbids a respondent from acting in a certain way. *In Re North Down Borough Council's Application* (1986 Queen's Bench Division) Carswell J in dealing with a planning appeal for a deemed refusal of planning permission for change of use of the former Tonic Cinema in Bangor and the refusal of an adjournment by the presiding Commissioner, made orders prohibiting the presiding Commissioner from

> furnishing the Planning Appeals Commission with a report of the hearing held by him, and the Planning Appeals Commission from receiving or taking into account any such report from that presiding Commissioner.

A further remedy is to seek an injunction, which is an additional curb on behaviour but, as noted by Orbinson et al. (2009), resorting to this remedy is rare in planning cases. In, for example, *57 Developments Limited and The Department of the Environment for Northern Ireland* (2004 NICh 3), the defendant had issued a counterclaim restraining by injunction the plaintiff from commencing or continuing individual site works, general site works and placement of caravans at Ballyhalbert Caravan Park, County Down. While Weatherup J acknowledged the desirability that the issues between the parties be resolved, he nonetheless concluded that: 'there is no immediacy or other factor arising in the present proceedings that necessitates the grant of an injunction in the present circumstances. The defendant has not considered it necessary to take immediate enforcement action through the course of this dispute'.

A combination of remedies may be requested and granted, partially granted or refused and in that regard the court exercises judicial discretion on the way forward. In *Corbo Properties' Application* (2012 NIQB 107), Horner J opined that: 'the principle of legality suggests that an unlawful decision should be struck down and relief granted but in certain exceptional circumstances it may be appropriate for the judge to exercise his discretion and not grant relief. The exercise of that discretion, which depends on the particular facts and circumstances of each case, can revolve around detriment to good administration, the demonstrated harm to persons resisting the quashing, or the avoidance of adverse public consequences. It can result in the awarding of declaratory relief rather than making a quashing order'. This outcome surfaced *In the Matter of an Application by Omagh District Council for Judicial Review* (2007 NIQB 61) in which Gillen J had concluded that draft Planning Policy Statement 14 dealing with the regulation of development in the countryside was *ultra vires* the powers of the Department for Regional Development and that the related decision of the Minister to introduce that policy statement was itself *ultra vires*. In a preliminary judgment delivered on 7 September 2007, the court made an order of *certiorari* quashing the decision of the Minister and a declaration that Planning Policy Statement 14 was unlawful and *ultra vires*. The matter was subsequently revisited in a remedies hearing with final judgment delivered on 25 October 2007. The exercise of judicial discretion regarding public interest is captured well in the comments of Gillen J:

> draft PPS 14 was a draft land use policy which had been produced in the aftermath of a four-year gestation period during which the time of public officials, the involvement of the public at large and the expenditure of large sums of public money have been part of the process. Thus there has been for example, a full public consultation exercise, a report on the consultation exercise, the publication of an Issues Paper to invite comment from a wide range of interested parties, organisations and the public, detailed submissions emanating from local councils, an equality impact assessment and an environmental assessment. No legal challenge to any of these measures was mounted as they progressed over the last four years by this Applicant or any other party. Whilst the maxim *fiat justitia ruat coelum* must normally apply, I have been persuaded that any order by this court which might serve to immediately nullify or render useless those earlier steps prior to the ministerial announcement and the introduction of PPS 14 might amount to the

> adverse public consequences which the authorities have cautioned against. Arguably those steps and the expenditure of public money thereby engendered might not necessarily fall foul of a quashing order. Nonetheless to ensure that some degree of measured reflection is introduced before any precipitous action is taken in regard to these earlier steps, I consider that declaratory relief is preferable to an order of certiorari. Moreover I have come to the conclusion that a declaration to the effect that the decision of the Minister with responsibility for regional development made on 16 March 2006 whereby he purported to introduce a new planning policy statement PPS 14 was *ultra vires*, coupled with a declaration that the planning policy statement PPS14 was also *ultra vires* are sufficient remedies for the Applicant in this case. This will clearly state the law as I find it and open the door to such other consequential remedies to give effect to these declarations as are deemed appropriate in individual circumstances without necessarily contributing to the problems of the potential waste of public time and money to which I have earlier adverted.

This judgment prompted the reissuing of Planning Policy Statement 14 by the Department of the Environment within whose legislative compass the responsibility lawfully rested.

Costs

The determination of costs in regard to judicial review is a matter for the court to deal with. The general, though not inflexible, rule is that applicants and respondents will be responsible for their own costs at the leave stage. In this vein, McCloskey J has noted *In the matter of an Application by Coulters Hill Residents Ltd for Leave to Apply for Judicial Review v Planning Appeals Commission* (2020 NIQB 1) the general principle that an applicant should not be liable for the costs of a proposed respondent in the event of a decision refusing leave to apply for judicial review. However, the general rule at a subsequent substantive hearing is that the losing party will pay a proportion or all of the other party's costs as well as their own. The capacity to cover financial outlays can vary, not surprisingly, between big business and private individuals and, in this regard, it may be possible for individuals to apply for and receive legal aid to meet their costs and that normally precludes a successful respondent from recovering its costs. Judicial review proceedings can be expensive and thus a further option that may be invoked is for the court, upon application by a plaintiff, to issue a protective costs order that can limit any exposure to financial risk.

The case *In the Matter of an Application by Ciara Patricia Thompson for Judicial Review* (2010 NIQB 38) is helpful by way of explication. The applicant who lived in a housing development in Bryansford was granted leave to apply for a judicial review of a planning permission granted for a nearby housing scheme. She was a supporter of the Tollymore Community Forum, which, in turn, had supported local residents in objecting to the planning application giving rise to the impugned decision. The applicant had a low income and was financially eligible for legal aid but this was refused, as noted by Morgan LCJ:

> Her application for legal aid funding was refused on the basis of Regulation 5 (11) of the Legal Aid General Regulations (Northern Ireland) 1965 which provides that where an application is made by or on behalf of a person in connection with a cause or matter in which numerous persons have the same interest, the appropriate committee shall

> refuse the application if they are satisfied that it would be reasonable and proper for the other persons having the same interest in the matter as the applicant to defray so much of the costs as would be payable from the fund in respect of the proceedings if a certificate were issued.

The court ordered, however, that because the case did raise one issue of general public importance in the locality in which the applicant resides, albeit on the borderline, and that exposure to unlimited costs would make it inevitable that the applicant could not proceed, that any award of costs against the applicant in respect of the hearing at first instance should not exceed £10,000. A subsequent judicial review application by the plaintiff to review and alter this protective costs order was dismissed by Treacy J (2010 NIQB 120).

On occasion, the court can make an order requiring an applicant to make security for costs. This matter has surfaced among a clutch of environmental actions brought to the judicial review court over the period March 2018 to September 2019 with leave to apply for judicial review granted on 7 June 2018 in one of those cases, namely *Rural Integrity (Lisburn 01) v Planning Appeals Commission*. As noted by McCloskey J *In the matter of 32 Applications by Rural Integrity (Lisburn 01)Limited and Related Companies for Leave to apply for Judicial review* (2020 NIQB 25) the granting of leave in *Lisburn 01* was followed by an application by the Planning Appeals Commission for an order compelling the applicant company to make security for costs while the court was also invited by the applicant to determine an application for a protective costs order. By its order dated 9 March 2019 the court ruled that in the event that the applicant company having to pay costs to the Planning Appeals Commission, the amount recoverable would not exceed £10,000 and that the applicant was to make security for these costs in the same amount by 19 March 2019. The reasoned judgment of the court for this decision was set out on 10 April 2019 (2019 NIQB 40):

> The acutely one sided and unbalanced nature of the accommodation which the Applicant … is seeking from the court is unmistakable. It has three central components: his contention that the court should not order security for costs in any amount against the Applicant, his quest to secure a protective costs order for the Applicant restricting its potential costs exposure to £100 maximum and his intention to seek to recover some £6,000 from the Planning Appeals Commission (PAC) in the event of the judicial review succeeding. This arrangement, if sanctioned by the court, would result in the PAC being unable to recover any costs if the challenge fails, in a context where its estimated costs are at least £20,000 plus VAT or, alternatively, having to pay some £6,000 costs in the event of the Applicant's challenge succeeding. It falls to the court to strike a balance which is harmonious with the applicable legal rules and principles and the principle of proportionality.

The application for judicial review was subsequently dismissed by an Order from McCloskey J by reason of failure to provide security in the sum of £10,000 and costs were awarded against the applicant by reason of the failure to comply with the Security for Costs Order (see 2020 NICA 12).

Finally, it is important to note that the United Kingdom is legally bound after Brexit to stand by the provisions of the Aarhus Convention, which allow for access to justice in environmental matters. The relevant statutory rule in Northern Ireland is The Costs Protection

(Aarhus Convention) (Amendment) Regulations (Northern Ireland) 2017, which amends previous regulations made in 2013 and which allows for the limit on the amount of costs that can be recovered from an applicant in a Convention case to be lowered in order to avoid prohibitive expense to the applicant. Additionally, the regulations allow the limit on the amount of costs that can be recovered from the respondent in these cases to be increased in order to prevent prohibitive expense to the applicant. The regulations specify that the court shall order that any costs recoverable from an applicant shall not exceed £5,000 where the applicant is an individual and £10,000 where the applicant is a legal person or an individual applying in the name of a legal entity or unincorporated association, with discretion to decrease these amounts. The court shall further order that the costs recoverable from a respondent shall not exceed £35,000 with discretion to increase the amount. The exercise of that discretion is based on a determination of prohibitive expense and which is defined in the regulations by having regard to the financial means of the applicant or where the likely costs for an applicant are considered unreasonable having regard to the situation of the parties; whether the applicant has a reasonable prospect of success; the importance of what is at stake for the applicant; the importance of what is at stake for the environment; the complexity of the relevant law and procedure; and whether the case is frivolous.

The grounds for judicial review

A successful judicial review requires that the plaintiff demonstrates the existence of one or more grounds upon which the challenge is constructed. In that regard the much-cited comments of Lord Diplock in the case *Council of Civil Service Unions and Others and Minister for the Civil Service* (1985 AC 374) give helpful guidance:

> Judicial review has I think developed to a stage today when without reiterating any analysis of the steps by which the development has come about, one can conveniently classify under three heads the grounds upon which administrative action is subject to control by judicial review. The first ground I would call illegality, the second irrationality and the third procedural impropriety. That is not to say that further development on a case by case basis may not in course of time add further grounds ... By illegality as a ground for judicial review I mean that the decision-maker must understand correctly the law that regulates his decision-making power and must give effect to it ... By irrationality I mean what can by now be succinctly referred to as *Wednesbury* unreasonableness ... It applies to a decision which is so outrageous in its defiance of logic or of accepted moral standards that no sensible person who had applied his mind to the question to be decided could have arrived at it ... I have described the third head as procedural impropriety rather than failure to observe basic rules of natural justice or failure to act with procedural fairness towards the person who will be affected by the decision. This is because susceptibility to judicial review under this head covers also failure by an administrative tribunal to observe procedural rules that are expressly laid down in the legislative instrument by which its jurisdiction is conferred, even where such failure does not involve any denial of natural justice.

The unpacking of these three main grounds has developed in line with case law and at a generic level a battery of more specific and geometrically varying elements can be identified upon which a judicial review application may be framed. Table 7.1 offers a stylised synthesis of grounds for judicial review, albeit that the categorisation is far from rigid. Lord Irvine LC in the case *Boddington v British Transport Police* (1998 UKHL 13) commented on this general fluidity as follows:

> Categorisation of types of challenge assists in an orderly exposition of the principles underlying our developing public law. But these are not water tight compartments because the various grounds for judicial review run together. The exercise of a power for an improper purpose may involve taking irrelevant considerations into account, or ignoring relevant considerations; and either may lead to an irrational result. The failure to grant a person affected by a decision a hearing, in breach of principles of procedural fairness, may result in a failure to take into account relevant considerations.

Additionally, these listed grounds are case sensitive and may or may not be successfully argued. In this vein, judicial comment has been aired in regard to the awarding of costs, as to whether the costs involved in proceedings have been unnecessarily increased by the number of issues raised by an applicant against a backdrop when the overwhelming majority of an applicant's grounds of challenge are rejected (see, *Alternative A5Alliance's Application for Judicial Review* (2013 NIQB 30)). Thus, while applications for judicial review may invariably present multiple

TABLE 7.1 Grounds for judicial review

Illegality	*Irrationality*	*Procedural impropriety*
• *ultra vires* • mistake of fact • error of approach • relevant considerations not taken into account • irrelevant considerations taken into account • action or non-action contrary to domestic or European Court of Human Rights statutory provisions • action or non-action pre-Brexit are contrary to EU statutory provisions, and post-Brexit are contrary to transposed EU regulations • lack of proper enquiry • policy misunderstanding / incorrect application of policy • inflexible application of policy that fetters discretion • unjustified departure from policy	• unreasonableness • unexplained inconsistency • unfair balancing / unsound evidential grounding of considerations in decision making • bad faith / improper motive / abuse of power	• procedural unfairness / not sharing information or not granting adjournment for effective case management • actual bias / predetermination or apparent bias • frustrating of legitimate expectations • duty to consult shortfall • inadequate reasons

Source: Informed in part by Orbinson et al. 2009: 29–90, and by Patterson et al. 2015: 3–71.

grounds for determination, for the purpose of this part of the chapter it is helpful to consider key cases that explicate some more frequent considerations comprising lack of proper enquiry, consultation shortfall, error of approach and actual bias/predetermination or apparent bias.

Lack of proper enquiry

Planning decision making requires that all material considerations are taken into account and in that regard a key ground for judicial review is that an inadequate or no enquiry into relevant matters has taken place. The considerations can be wide-ranging: in *Roberts (Gregory) Application* (2015 NIQB 20) the grant of planning permission to the Cats Protection League for a facility in a rural location near Hillsborough was quashed; it was reported that there was no enquiry by the respondent into whether alternative sites were or could be available instead of the approved rural location and in any reconsideration O'Hara J opined that the availability or otherwise of alternative sites is likely to be a significant issue. In *River Faughan Anglers Limited's Application* (2014 NIQB 34), the plaintiff argued unsuccessfully that the Department of the Environment had erred in making its determination by failing to base its decision on sufficient information or enquiry about whether the proposals would be likely to have significant environmental effects. In *Quinn's (Brian) Application* (2013 NIQB 24) regarding the refusal of a planning appeal against the refusal of planning permission for the development of a wind farm within the Sperrins Area of Outstanding Natural Beauty, it was argued, inter alia, that the Commissioner had failed to make enquiries in relation to the issues of renewable energy targets and economic considerations; that decision by the Planning Appeals Commission was quashed on the basis of a materially flawed assessment of benefits. Inadequate enquiry as a ground for judicial review also extends to the highest level of decision making as illustrated by a successful challenge against ministerial involvement in a proposed development for a John Lewis Plc store and an accompanying twenty-nine units at Sprucefield on the fringes of the Belfast urban area. This case gives a very full explication of the need for proper enquiry, with Girvan J stating *In the Matter of an Application for Judicial Review by (1) Bow Street Mall Limited (2) Lisburn Chamber of Commerce (3) Multi-Development UK Limited (4) Central Craigavon Limited (5) Belfast City Council (6) Belfast Chamber of Trade and Commerce and In the Matter of a Planning Decision made by the Department of the Environment* (2006 NIQB 28):

> The Minister appears to proceed it on the basis that since the Planning Service viewpoint was that it was unnecessary to have an inquiry in the context of their proposal to refuse the application he should proceed in the same way in the context of his decision which was to accede to the application. It would have been helpful for the Minister to have had his mind directed to the issues that needed consideration if he was minded to reject the advice to refuse. As it was, he was not so directed. There was, however, a logical difference between the desirability of a public inquiry in the context of a proposed refusal and in the context of a proposed grant of the permission. If planning permission was going to be refused the applicant had a right to proceed to the Commission where the Department's viewpoint could be challenged. This provided a form of public forum for investigation of the issues. If the Department was going to go ahead and grant permission then a hearing before the Planning Appeals Commission would be less likely and there would be probably no public forum at which the issues could be

> reviewed. …The ultimate answer to the question whether the decision not to direct a public inquiry was flawed appears to me to be related to the question whether the decision to issue the Notice of Opinion and the grant of outline planning permission were flawed on the ground of lack of proper enquiry. … the issue of the need for the size and layout of the overall development were accepted at face value. Kerr LJ pointed out that planning authorities must be careful not to give way too readily to assertions of financial constraints as a ground for relaxing policies formulated in the public interest. Commercial entities intent on maximising profits and seeking to persuade planners to accept their proposals will seek to employ all arguments to their economic advantage and to present their case in its most compelling form. From the developer's point of view there is nothing wrong with such a approach. Planners, however, must approach their functions with due regard to the wider interests at stake and approach the ipse dixit contentions of planning applicants in a questioning frame of mind. If after proper consideration of the issues properly explored the decision is in favour of planning permission the decision maker could not be challenged as having acted irrationally or perversely. In the present circumstances, in the absence of any real exploration of the issues relating to the alleged financial imperatives of an overall development of this magnitude, the conclusion that it was an appropriate development was one which no reasonable planning authority properly directing itself could have reached having regard to the Departmental conclusions that the vitality and viability of other centres could be significantly affected by a development of this size.
>
> Viewing the matter in that way the conclusion by the Minister that a public inquiry was unnecessary because it would not have brought out anything new was *Wednesbury* unreasonable. A central aspect of the application had gone unexplored and unquestioned. Looking at the matter in another way, by wrongly assuming that the Department had all the information and data it required to make a rational decision the Minister had left out of account a relevant consideration, namely that the information for the Department was insufficient to reach a proper decision. Accordingly, the Minister's decision not to hold a public inquiry was flawed. It would, of course, have been open to the decision-maker to carry out investigations and inquiries into the unexplored issues short of directing a public inquiry. Had the Department done so and obtained information which satisfied it of the inevitable need for the connection between the 29 units and the JLP store and, if the Department rationally considered that information adequate, then the need for a public inquiry might not have arisen. However, the Department did not take adequate steps to inform itself on a central issue either through the mechanism of a public inquiry or by investigation short of such inquiry.

Consultation shortfall

The basic requirements for consultation as part of the planning decision-making process have been set out in *R v Devon County Council, Ex parte Baker* (1995 1 All ER 73) whereby consultation must take place at a time when proposals are still at a formative stage; that the proposer must give sufficient reasons for any proposal in order to allow intelligent consideration and response; that adequate time must be given for such consideration and response; and that the product of any consultation must be conscientiously taken into account in finalising any statutory proposals. Further principles regarding consultation shortfall are set out by Horner J

In the Matter of an Application by Patrck Lynch for Judicial Review (2020 NICA 32) in which the written judgment adopts a commentary from Auburn et al. (2013):

> As in all cases where a court finds that a decision is unlawful, where such unlawfulness arises as a result of a breach of a duty to consult, the court has a discretion as to what remedy, if any, to grant. However, a failure to consult at all in a case where there was a duty to consult is normally viewed as a serious matter, and the court will normally quash the resultant decision. The court will be more likely to exercise its discretion not to quash a decision where the unlawfulness rose out of inadequacies in a consultation exercise that was carried out, particularly if the inadequacy is one that is capable of being remedied within the consultation process itself. However, even in this context, the courts are likely to be reluctant to allow the resulting decision to stand.

The relevance of these considerations within the planning arena was aired *In the Matter of an Application by HM, a Minor, by PM, her Father and Next Friend for Judicial review* (2007 NICA 2), where it was submitted that the Department of the Environment as the planning authority had not complied with its statutory duty to consult the local council before reaching its decision. However, the Court of Appeal rejected that claim on the evidence before it:

> there is contemporaneous material in the form of correspondence passing between the council and the department about the various concerns that had been raised. Moreover, it is positively asserted by Mr Arnold that these concerns were conscientiously and scrupulously addressed. Unless we were persuaded that the department had cynically resolved to ignore the representations made by the council while creating the false impression that they had done so, it is impossible to accept the appellant's arguments on this issue. There is no basis on which we could possibly reach such a conclusion.

The issue of consultation shortfall as a ground for judicial review is particularly well illustrated by *National Trust's Application* (2013 NIQB 60) in which the National Trust applied for the quashing of planning permission for a golf resort development that included a hotel and lodges some 550 metres removed from the Giants Causeway, a designated World Heritage Site. This is a major tourism attraction in Northern Ireland that is consistently promoted as a must-see experience to international markets. In this case the applicant argued, inter alia, that the respondent had unreasonably failed to consult with the World Heritage Committee in respect of the application for planning permission. Reliance was placed on Northern Ireland planning policy as giving effect to UNESCO (United Nations Educational, Scientific and Cultural Organization) guidelines issued in relation to the overarching 1972 UNESCO World Heritage Convention, as explained by Weatherup J in his judgment:

> The Convention is an International Treaty between the United Nations and the States who subscribe to the Convention. The Convention does not create individual rights for the citizens of the subscribing States. However a State may adopt national measures that apply the standards of the Convention and create individual rights. In the present case the applicant claims that this creation of individual rights has occurred by the Department adopting a planning policy that imposes paragraph 172 of the Guidelines on an application for planning permission affecting the World Heritage Site.

This argued-for provision was key to the applicant's case as Paragraph 172 of the UNESCO Guidelines invites state parties to the Convention to inform the World Heritage Committee of their intention to authorise developments that may affect World Heritage Sites and to do so before making any decisions that would be difficult to reverse so that the Committee may assist in seeking appropriate solutions to ensure the full preservation of the property in question. But having rehearsed the content of the Department of the Environment planning policy relating to the protection of World Heritage Sites in Northern Ireland, Weatherup J concluded:

> None of this imposes any requirement on the Department to consult with the World Heritage Committee (WHC) on a planning application affecting a World Heritage Site. It refers to the criteria for the selection of a site. It states that inclusion on the List highlights the outstanding international importance of the site as a key material consideration in planning decisions. The policy is stated to be a presumption in favour of the preservation of the World Heritage Site. Only in exceptional circumstances will development be permitted that adversely affects the site or its setting. Nor does this domestic policy purport to apply the Guidelines to applications for planning permission affecting the World Heritage Site. It merely states that the Guidelines set the criteria for inclusion in the List. It does not state that the Guidelines are a material consideration for planning purposes. Thus in considering whether or not there was an obligation on the Department to consult with the WHC before determining this planning application it is not to be found in the domestic planning policy … I am satisfied that the advice that operates within the UK, be it right or wrong, is that notice of the planning decision will be given to the WHC once the decision has been made and before it is announced … The domestic requirement is to treat the status of the World Heritage Site as a material consideration for planning purposes. That requirement was acknowledged by the Department but it does not entail a legal obligation to consult with the WHC either under the Convention or the Guidelines or Policy BH5 or by reliance on the statements of the Department officials or the ministerial statements referred to by the applicant.

Nonetheless, what is of added interest in this case is the opinion of the court in regard to its power to intervene. Judicial comment signposts the remit of the relationship between the state and UNESCO in contrast to that of the court:

> I find the outcome on the role of UNESCO to be surprising in a number of respects. First, that the Department does not have to engage with the WHC before making a decision. While there is a Treaty in place and the matter is to be dealt with initially on the international plane, protection might better be afforded by some requirement for engagement with the WHC before a decision is made that may affect the site. However no such domestic obligation has been adopted. Secondly, it is surprising that the Department did not canvass the views of the WHC on the impact of the particular proposal before making its final decision and I am satisfied that they did not do so. The Department reported on the fact of the application having been made but prior to making the final decision the Department did not seek to obtain the views of the WHC. Thirdly, it is surprising that the UK considers that notification of a decision after

> it is made accords with paragraph 172 of the Guidelines. A reading of the Guidelines suggests that the object of the exercise is to engage with the WHC so that they will present a view on the impact of development on the World Heritage Site before the decision is made. I do not know the basis on which this advice has been furnished, nor what the view of the WHC is on this approach but that is a matter between the United Nations and the State. It is not a matter for the Court.

Substantive changes in planning policy without consultation have, finally, been deemed unlawful, as illustrated by *An Application for Judicial Review by Central Craigavon Limited* (2010 NIQB 102). In this case, Minister of the Environment Sammy Wilson made two statements in 2009 before the Northern Ireland Assembly regarding the economic benefits of development proposals and the relationship between major economic development proposals and the development plan scheme in terms of prematurity. While the Minister as respondent claimed that the first did not represent new planning policy and that the second was merely offering clarification, the applicant's case was that new policy was being made and being made unlawfully. In his quashing of the two statements, Treacy J noted:

> The substantive content of planning policy, provided it is lawful, is a matter for the appropriate authorities and not the Court. So that there is no inadvertent misunderstanding as to the Court's role I emphasise that the Court is not concerned with the *merits* of this or that planning policy. The respondent was entitled in principle to introduce policy requiring particular weight to be given, for example, to economic considerations in the determination of planning applications – provided it was done lawfully.
>
> If, in this case, the statements represented a substantive change in planning policy the respondent misdirected himself and, in consequence, unlawfully failed to consult. The requirement to consult is explicitly recognised in para.9 of PPS1 set out above and the respondent did not challenge the requirement for consultation if the statements represented the formulation of planning policy.
>
> But the respondent maintained that he was not changing policy. Of course, whether the impugned statements did effect or represent a change of policy must be assessed on the substance of the statements themselves and not on how they are described by the respondent.
>
> I do not accept that the statements were mere guidance. It went beyond mere advice or information aimed at resolving a difficulty. They were plainly intended, particularly in the case of the economic statement, to bring about a material change in the way planning applications were determined and to influence the outcome. Otherwise there would have been no point in making the statements. It was plainly envisaged that applications that would have been determined one way before the statements were made would (or might), on the same facts, now be determined differently in light of the new policy introduced by the statements. This is an inescapable inference when one considers, in the case of the economic statement, that it required decision makers to give *greater* or *substantial weight* to economic considerations than was previously the case under existing policy.
>
> Some Assembly members appear to have considered the economic statement to be new policy … The Planning Service, contrary to the respondent's case (in relation to the economic statement), also seems to accept that they are policy since it has published

> them in the policy and plans section of its website under the heading of other *policy* publications along with other ministerial statements, with the commentary Ministerial Statements are an expression of Government *policy*. Notwithstanding the respondent's express assertion that the economic statement was not a change in policy and that the prematurity statement was clarification it is manifest that the statements bring about and are designed to bring about a substantive change in the manner in which planning control decisions are taken. As the applicant put it they each usher in a new policy direction.
>
> The respondent recognised in relation to the economic statement that *if* it was a change of policy, it would have *required* widespread consultation etc. The matters addressed in the statements are plainly controversial ... Together the statements have the potential to and may be intended to exercise a significant effect on the planning system and the determination of major planning applications. Provided this objective is lawfully achieved the courts have no role. Had the respondent consulted in respect of these statements it is reasonable to infer that there would have been a variety of views expressed. In accordance with his public law obligations of adequate consultation he would have had to have taken into account these views in formulating planning policy.
>
> The applicant had (in common with others) a legitimate expectation of consultation on policy of this nature arising, inter alia, from the clear and unambiguous representation by the respondent's department to the Northern Ireland public at large that it would consult on new planning policy as per para.9 of PPS1. No such consultation took place and the statements must therefore be quashed.

Error of approach

Error of approach can commence when a planning authority asks itself the wrong question when determining a planning application and then goes on to base its decision on the flawed conclusion to that question. The follow-on implication for judicial review is set out by Girvan J in *Bow Street Mall and Others* (2006 NIQB 28): 'The question for the Court is whether the decision maker asked himself the right question and took reasonable steps to acquaint himself with the relevant information to enable him to answer it correctly'. This clearly overlaps with the need for proper enquiry, as discussed above, but goes further to highlight the importance of establishing the correct position based on the availability of reliable evidence to support a decision. A useful illustration is provided by reference to the retailing sector, where there is often debate around the need for more complete information derived from the completion of a retail impact assessment. The approach of the Department of the Environment to grant planning permission for a new superstore in Portadown was challenged by Tesco on the basis, inter alia, that there was no such assessment that would have informed judgements about the degree to which a new store would draw trade away from the town centre and that in the absence of this assessment the conclusions reached by the Department were irrational (see *Tesco Stores Ltd v Department of the Environment Planning Service* (2011 NICA 23)). However, Morgan LCJ, in delivering the Court of Appeal judgment, concluded in this instance that there was no error of approach or flawed decision making:

> The judgments which the Department made were informed by the pre-existing detailed retail information. There is no express requirement to conduct a retail impact

> assessment in relation to cumulative impact and the criticism in this case is of the planning judgments made by the Department in relation to the effect of the grant of the planning permission. There was no evidence to suggest that the conclusions reached by the Department on the nature of the likely impacts on other centres were flawed. The criticism was that information that would better evaluate the likely extent of those impacts was required. Having determined the nature of the likely impacts it was a matter of judgment for the Department to decide how much more detailed information it required. We see no basis for characterising the judgments made as irrational or even unreasonable and in agreement with the learned trial judge we reject these grounds of appeal.

In contrast, in *Mooreland and Owenvarragh Residents' Associations' Application* (2014 NIQB 130) regarding a challenge to planning permission for the redevelopment of Casement Park, Belfast to a new stadium, a core consideration was whether the Department of the Environment had been correct in basing its determination on the difference between a baseline of 32,600 spectators attending the old Casement Park and 38,000 spectators at the new Casement Park that comprised an increase in capacity of 5,400 spectators. In his judgment, Horner J concluded:

> Impermissibly the Planning Service sought to assess the effects of a capacity crowd attending the new ground based on the difference between 32,600 spectators attending the old Casement Park and 38,000 spectators at the new Casement Park – that is a difference of 5,400 additional spectators. This was neither a fair nor lawful approach because the evidence made it clear that a crowd of 32,600 was never going to attend the present Casement Park now or in the foreseeable future … The Department did not ask, as it should have done, whether there was any prospect of 32,600 ever attending the old Casement Park either at the present time or in the future. This error was compounded, whether innocently or otherwise, by the failure of the Planning Service to tell the Minister that the police who had been consulted and who had responded in some detail, had forecast traffic chaos and risk to life if there was a capacity 38,000 crowd attending the new Casement Park … There was a fundamental flaw in the approach of the Department when it assessed what the effects a capacity crowd attending Casement Park would have. It was an unlawful approach, whether under domestic law or under European law, simply to look at an uplift of 5,400 spectators. This did not reflect the reality.

More recently, the obligation on planning officials to ask the correct question when analysing statutory provisions was addressed *In the Matter of an Application by ABO Wind NI Limited & Energia Renewables Company 1 Limited for Judicial Review and in the Matter of the Chief Planner's Update* 7 (2021 NIQB 96). This case related to advice published by the Department for Infrastructure that Battery Energy Storage Systems (BESS) fall within the meaning of an electricity generating station. The implication of this clarification is that it brings into consideration the Planning (Development Management) Regulations (Northern Ireland) 2015, which defines a hierarchy of developments as comprising major and local categories. In the Schedule to these regulations 'Energy Infrastructure' comprising electricity generating stations is listed in class 2 as major development when the generating capacity is or exceeds 30 megawatts. The court determined that the wrong question had been asked in arriving at this advice. The correct question was stated by Humphreys J as: 'Does a BESS development fall wholly within class 2 of the Schedule to the 2015 Regulations?'

The court concluded that had this correct question been asked, then the answer must be 'no' and that, consequently, a BESS development falls within class 9 of the Schedule and would be subject to a different set of thresholds as to whether a development is local or major. The application for judicial review was successful.

Actual bias/predetermination or apparent bias

The varied dimensions of bias are an important subset of procedural impropriety and essentially decision makers should not take decisions if they are, or seem to be, predisposed towards a particular outcome. As commented by Lord Woolf in *R v Gough* (1993 UKHL 1), actual bias is rare and is contained in the maxim that nobody may be judge in his own cause (*nemo judex in sua causa*), whereas in instances of apparent bias the key principle is that justice must manifestly be seen to be done. In that regard the possibility that a decision maker might unfairly regard (or have unfairly regarded) with favour, or disfavour, the case of a party to the issue under consideration is key. The complexity of bias, which may operate in an invidious manner, is well rehearsed in the observations by Devlin LJ in *R. v. Barnsley Licensing Justices (*1960 2 Q.B. 167): 'Bias is or may be an unconscious thing and a man may honestly say that he was not actually biased and did not allow his interest to affect his mind, although, nevertheless, he may have allowed it unconsciously to do so'. Within the Northern Ireland arena, McCloskey J in *Belfast International Airport Ltd's Application (Leave Stage) and Belfast City Airport Watch Ltd's Application (Leave Stage)* (2011 NIQB 34) has usefully explored the breadth of authorities on this matter and identifies the following sentence from *R v Jones* (2010 NICC 39): 'In every context, the test for apparent bias requires consideration of a possibility, applying the information known to and attributes of the hypothetical observer'. The Court of Appeal in Northern Ireland has similarly drawn on leading authorities from Great Britain as *In the Matter of an Application by William Young for Judicial Review* (2007 NICA 32) wherein the question is posed: 'whether a fair-minded and informed observer, having considered the facts, would conclude that there was a real possibility that the tribunal was biased'. The perception of the informed observer emerges as significant in challenges of apparent bias.

The application of these principles within Northern Ireland has found, for example, in favour of an applicant on the issue of apparent bias by reason of telephone conversation contact between a member of the Planning Appeals Commission and the Department of the Environment (*Re Moore's Application for Judicial Review* (2004 NIQB 49)). It is important to note that the court in this case did not question the actual integrity or impartiality of the Commissioner, but relied on the maxim that justice must be seen to be done. A challenge of bias was also central to argument in *Alternative A5 Alliance's Application for Judicial Review* (2013 NIQB 30), where it was advanced that the inspectors convening a public inquiry into the A5 Western Transport Corridor dual carriageway scheme had demonstrated apparent bias by carrying out site inspections in a car driven by employees of the Department for Regional Development. Stephens J rejected that ground stating:

> Some of the locations being visited were remote and difficult to find though maps and directions could have been used, however in the event the alternative method was to

obtain assistance from an officer of the Department who was not a witness and who knew the area … None of the objectors visited in this way protested at the time and I consider that the fair-minded observer would conclude that the appearance of bias was not something that immediately and forcibly struck the objectors to the extent that they raised the matter with the inspectors immediately or even after some time for reflection … Taking the facts and those circumstances into consideration I do not consider that the fair-minded observer would conclude that there was a real possibility that the inspectors were biased. I decline to set aside the inspectors' report.

Apparent bias and predetermination, finally, are well illustrated *In the Matter of an Application by House of Fraser Ltd and Ors for Leave to Apply for Judicial Review* (2010 NIQB 105). Inappropriate ministerial media comment, in regard to the long-running proposals for shopping development at Sprucefield, on the edge of the Belfast urban area, and particularly in relation to the role of the court in the face of an alleged commercially motivated recourse to judicial review by objectors to the scheme using environmental issues, has drawn stinging judicial comment:

> The *ex tempore* comments made by the Minister on the programme do not on the face of them appear to reflect the distinct roles of the court and the Department. See for example his reference to the courts dictating on decisions. This does not reflect the legal position. Courts decide applications in accordance with the law. They do not decide planning policy and in relation to the review of planning decisions the courts up to the level of the House of Lords and the Supreme Court have made it abundantly clear the courts do not act as appellate courts in relation to planning decisions. They are concerned with questions of the legality, rationality and procedural fairness of planning decisions. Nor do they, in the words of the Minister, interfere with the decision-making process. The courts exercise a supervisory jurisdiction. What the Ministerial statement does not do is to recognise that the Minister's comments should not have been made at all while the litigation was pending. The Minister in his statement has failed to recognise that what happened should not have happened … The Minister's comments in the programme self-evidently lay the basis for the applicants' argument that there is an arguable case of apparent bias and pre-determination. They lay the basis for the applicants' argument that there is an arguable case that the Minister has disqualified himself from making any further decisions in relation to this planning application. They lay the basis for there being an arguable case that the decisions which the Minister has made up to this point in this planning application are tainted by apparent bias and predetermination. Accordingly the court must grant leave for the application in this instance. The consequence of all this is that the inquiry will not now proceed until this matter is resolved. The legitimate rights of the planning applicant to have its application properly processed in a timely and fair manner have been frustrated. Those who have opposed the planning application (as they were legitimately entitled to do) have been held up to vilification by the Minister as have their legal representatives. The whole situation can only be described as lamentable … The gravity of what transpired on Friday morning should not be underestimated. This is a case which should be considered by the Attorney General having regard to his public law duty of protecting the rights of parties to litigate in a fair and dispassionate atmosphere of objectivity. The BBC invited the Minister to comment on pending

> litigation. The BBC should not have done so and the Minister should not have taken the opportunity to do so. The role of each of those parties should be considered by the Attorney General.

These comments usefully draw together the purpose of legal proceedings in this sphere, but they also signal an emergent tension at a wider scale between high-level political preference for less-fettered development decisions and the role of the courts to ensure that decisions are made in accordance with the law. The final section of this chapter examines briefly that matter.

Whither judicial review in the Northern Ireland planning system?

The evidence presented in this chapter points to judicial review as playing a prominent role in the Northern Ireland planning system by providing for scrutiny of public decision making. In this regard, a judiciary independent of government is crucial albeit that it operates as a branch of the state. The Rt. Hon Sir Declan Morgan (2011: 29) has underscored that remit in the following manner:

> The primary function of the judiciary is the fair and impartial resolution of disputes between individuals and the State in accordance with law. No other organ of the State carries out this function. It is evident, therefore, that in order to carry out the function the independence of each judge is of critical importance. It can properly be described as the precondition for the ability to carry out this responsibility. The public must be able to have absolute confidence that the judge hearing a case will do so objectively, according to law, and will not be subject to influence by the parties, the government, a group within civic society or even other judges. The phrase according to law is a very important one. The rule of law is not the rule of judges. Judges like everybody else are subject to the laws as enacted by the legislature. That is the only foundation on which public confidence in the justice system, and in the rule of law can be established.

However, as argued by Elliott (2014), attempts by the UK government through legislative change to reduce the scrutiny to which it is subjected by the courts might suggest that the courts' powers of judicial review are not constitutionally secure. The point here is that legislative change approved by Parliament can alter or attenuate the scope for judicial intervention by reshaping the rule of law. Elliot expresses this more prosaically as: 'an executive government with its hands on the levers of sovereign legislative authority has to suffer only as much scrutiny via judicial review as it is prepared to tolerate' (2014: 2).

Ultimately, in such a scenario much would rest on judicial interpretation of any revised statute. Such matters prompted intensive debate in Parliament during the examination of the provisions in the Criminal Justice and Courts Bill in 2014; at that time the Lord Chancellor and Justice Secretary, Chris Grayling MP, raised the following points in a Parliamentary debate:

> A Minister is confronted by the practical threat of the arrival of a judicial review case virtually every week. It is happening all the time. There are pre-action protocols all the time, and cases are brought regularly. Looking across the majority of a Department's

> activities, Ministers face judicial review very regularly indeed. It happens weeks apart rather than months apart.
>
> *(www.publications.parliament.uk/pa/cm201415/cmhansrd/cm141201/debtext/141201-0002.htm#14120129000001, accessed 15 March 2016)*

In this jurisdiction the drafting of amendments to the Northern Ireland Planning Bill 2013 had sought to limit the right to judicially review certain planning decisions taken by government departments or a future council with the exception of those matters advanced on EU and European Court of Human Rights grounds. The net effect would have been to narrow the scope for judicial review to a very limited number of cases. The matter was debated in the Northern Ireland Assembly and attracted the ire of Ms Anna Lo, the then chairperson of the Environment Committee, when, in debating the Bill at its amendments consideration stage, she stated:

> This amendment does nothing other than take away the mechanism by which people can challenge government decisions on planning applications. It takes away people's right to say and do something about major planning issues. There are a number of judicial reviews that could not and would not have happened if this amendment had been in place when they were taken. To my mind, this amendment is nothing short of an attack on democracy.
>
> *(Northern Ireland Assembly 2013a)*

Following consultation and advice, Minister of the Environment Mark H. Durkan informed the Northern Ireland Assembly in October 2013 that he did not deem it appropriate or sensible to proceed with that particular reform element (Northern Ireland Assembly 2013b).

The chronology of political interest by Parliament and the Northern Ireland Assembly reached a new milestone in March 2021 when the Independent Review of Administrative Law, chaired by Lord Faulks QC, published its report on reforming judicial review. At the same time, the United Kingdom Government published a consultation on judicial review reforms, as a response to the Faulks report (Ministry of Justice 2021). This has been directed primarily at England and Wales, although the other legal jurisdictions across the United Kingdom have been invited to participate in the reform debate. In the context of this chapter, the consultation paper draws attention to the long-running issue of judicial encroachment on matters deemed beyond the purview of the courts and notes:

> a growing tendency for the courts in Judicial Review cases to edge away from a strictly supervisory jurisdiction, becoming more willing to review the merits of the decisions themselves, instead of the way in which those decisions were made. The reasoning of decision-makers has been replaced, in essence, with that of the court. We should strive to create and uphold a system which avoids drawing the courts into deciding on merit or moral values issues which lie more appropriately with the executive or Parliament.
>
> *(ibid. 8)*

It would seem that the tension, identified above, between government initiative and oversight by the courts, endures. But in linking the Faulks review and the consultation paper to future developments in practice, one proposal to emerge from this investigation is the possibility that, in appropriate cases, the remedy of a suspended quashing order might be made on an application for judicial review. A quashing order would then automatically take effect after a certain period of time if certain specified conditions are not met. The 2021 consultation elaborates on this way forward:

> A requirement to make a suspended quashing order when the defect can be remedied provides greater certainty to what rectification the Government will need to undertake if their regulation is successfully challenged. The Government further considers that legal certainty, and hence the Rule of Law, may be best respected by a suspensive quashing of such provisions. Suspended quashing by default would focus remedial action on resolving issues related to the faulty provision.
>
> *(ibid. 32)*

The implications for planning practice could spill over, for example, into the execution of procedures for environmental impact assessment in regard to infrastructure projects with a suspended quashing order allowing for identified defects to be remedied rather than an entire quashing of the project and a subsequent lengthy period that leads to the retaking of the decision. Judicial review is largely a devolved matter and thus, more than probably, we will see the Northern Ireland Assembly engage once again in deliberations around the way forward. But in the final analysis, it can be said that judicial review continues to exercise a significant, independent and legitimate role in ensuring the proper administration of planning governance.

Legislation

European Convention on Human Rights
Judicature (Northern Ireland) Act 1978
The Costs Protection (Aarhus Convention) (Amendment) Regulations (Northern Ireland) 2017
The Planning (Development Management) Regulations (Northern Ireland) 2015
The Rules of the Court of Judicature (NI) 1980
The Rules of the Court of Judicature (Northern Ireland) (Amendment) 2017
The Town and Country Planning (Development Management Procedure) (England) Order 2010
The Urban Waste Water Treatment Regulations (Northern Ireland) 1995

Cases cited

57 Developments Limited and The Department of the Environment for Northern Ireland (2004 NICh 3)
Alternative A5 Alliance's Application for Judicial Review (2013 NIQB 30)
An Application for Judicial Review by Central Craigavon Limited (2010 NIQB 102)
Application by Friends of the Earth for Judicial Review (2006 NIQB 48)
Belfast International Airport Ltd's Application (Leave Stage) and Belfast City Airport Watch Ltd's Application (Leave Stage) (2011 NIQB 34)
Blackburn's (Louisa) Application (Leave Stage) (2010 NIQB 71)
Boddington v British Transport Police (1998 UKHL 13)
Central Craigavon's Application (2010 NIQB 102)

Corbo Properties' Application (2012 NIQB 107)
Council of Civil Service Unions and Others and Minister for the Civil Service (1985 AC 374)
Crusaders Football Club's Application (2013 NIQB 68)
Department of the Environment's Application (2014 NIQB 4)
Gerber v Wiltshire Council and others (2016 EWCA Civ 84)
In Boswell's Application (2009 NIQB 95)
In Doyle's (Ellen) Application (2014 NIQB 82)
In Gibson's (Christine) Application (2016 NIQB 82)
In the Matter of an Application by Christine Gibson for Leave to Apply for Judicial Review and In the Matter of a Decision of Newry, Mourne and Down District Council Made on 29 June 2015 and 25 February 2016 and In the Matter of a Decision by the Department of Agriculture, Environment and Rural Affairs (Marine and Fisheries Division made on 3 June 2016 (2017 NICA 77)
In the Matter of 32 Applications by Rural Integrity (Lisburn 01) Limited and Related Companies for Leave to Apply for Judicial review (2020 NIQB 25)
In the Matter of an Application by Brian Quinn and Michael Quinn (2013 NIQB 24)
In the Matter of an Application by Ciara Patricia Thompson for Judicial Review (2010 NIQB 38)
In the Matter of an Application by Ciara Patricia Thompson for Judicial Review (2010 NIQB 120)
In the Matter of an Application by Coulters Hill Residents Ltd for Leave to Apply for Judicial Review v Planning Appeals Commission (2020 NIQB 1)
In the Matter of an Application by Dean Blackwood, Representing River Faughan Anglers Limited for Judicial Review v Derry City and Strabane District Council (2018 NIQB 87)
In the Matter of an Application by D for Judicial Review (2003 NICA 14)
In the Matter of an Application by John Hill for Leave to Apply for Judicial Review (2007 NICA 1)
In The Matter of an Application by Francis McHenry for Leave to Apply for Judicial Review (2007 NIQB 22)
In the Matter of an Application by HM, a Minor, by PM, her Father and Next Friend for Judicial Review (2007 NICA 2)
In the Matter of an Application by House of Fraser Ltd and Ors for Leave to Apply for Judicial Review (2010 NIQB 105)
In the Matter of an Application by James Stewart Moore for Judicial Review (2007 NIQB 23)
In the Matter of an Application by Kinnegar Residents' Action Group, Park Road and District Residents' Association, Old Stranmillis Residents' Association, Belfast Holyland Regeneration Association and Cultra Residents' Association for Judicial Review (2007 NIQB 90)
In the Matter of an Application by Mr Chris Murphy to Apply for Judicial Review; In the Matter of a Decision by the Department for Infrastructure on 17 August 2016 to Proceed with the Castledawson to Toome Dualling Scheme (2017 NIQB 35)
In the Matter of an Application by Omagh District Council for Judicial Review (2007 NIQB 61)
In the Matter of an Application by Patrick Lynch for Judicial Review (2020 NICA 32)
In the Matter of an Application by Paul McNamara for Leave to Apply for Judicial Review v Lisburn and Castlereagh City Council (2018 NIQB 22)
In the Matter of an Application by Queen's University of Belfast for Judicial Review (1997 Queen's Bench – Crown Side)
In the Matter of an Application by Rural Integrity (Lisburn 01) Limited for Judicial Review (2020 NICA 12)
In the Matter of an Application by Rural Integrity (Lisburn 03) Limited for Leave to Apply for Judicial Review and in the Matter of a Decision of Lisburn and Castlereagh City Council (2021 NIQB 32)

In the Matter of an Application by Sheridan Millennium Limited for Leave to Apply for Judicial Review (2007 NIQB 27)

In the Matter of an Application by Sheridan Millennium Limited for Judicial Review and In the Matter of a Decision by the Department for Social Development and in the Matter of a Decision by Laganside Corporation (2007 NIQB 92)

In the Matter of an Application for Judicial Review by (1) Bow Street Mall Limited (2) Lisburn Chamber of Commerce (3) Multi-Development UK Limited (4) Central Craigavon Limited(5) Belfast City Council (6) Belfast Chamber of Trade and Commerce and In the Matter of a Planning Decision made by the Department of the Environment (2006 NIQB 28)

In the Matter of William Young for Judicial Review (2012 NIQB 15)

In the Matter of an Application by William Young for Judicial Review (2007 NICA 32)

In the Matter of Applications by (1) AM Developments UK Limited (2) Belfast City Council and Belfast Chamber of Trade and Commerce (3) Bow Street Mall Limited and Lisburn Chamber of Commerce (4) Central Craigavon Limited for Judicial Review and In the Matter of a Decision of the RT Hon Lord Rooker, Minister with Responsibility for the Department of the Environment (2006 NIQB 9)

Mooreland and Owenvarragh Residents' Associations' Application (2014 NIQB 130)

Musgrave Retail Partners (NI) Ltd's Application (Leave Stage) (2012 NIQB 109)

National Trust's Application (2013 NIQB 60)

Newry Chamber of Commerce and Trade's Application (2015 NIQB 65)

Quinn's (Brian) Application (2013 NIQB 24)

*R v. Barnsley Licensing Justices (*1960 2 Q.B. 167)

R v Devon County Council, Ex parte Baker (1995 1 All ER 73)

R v Gough (1993 UKHL 1)

R v Jones (2010 NICC 39)

R v Inland Revenue Commissioners, ex parte National Federation of Self-Employed and Small Businesses Ltd. (1982 AC 617)

R v Lord Chancellor, ex parte Child Poverty Action Group (1999 1 WLR 347)

R v Secretary of State ex parte Genova Ltd and another (1992 Queen's Bench Division (Crown Side)

R v Secretary of State for Northern Ireland ex parte Blair (1985 Queen's Bench Division (Crown Side))

Re Moore's Application for Judicial Review (2004 NIQB 49)

River Faughan Anglers Limited's Application (2014 NIQB 34)

Roberts (Gregory) Application (2015 NIQB 20)

Tesco Stores Limited's Application (2010 NIQB 15)

Tesco Stores Ltd v Department of the Environment Planning Service (2011 NICA 23)

References

Anthony, G. 2014. *Judicial Review in Northern Ireland*. Oxford: Hart Publishing.

Auburn, J., J. Moffett and A. Sharland. 2013. *Judicial Review: Principles and Procedure*. Oxford: Oxford University Press.

Deb, A., R. Honey, C. Fegan and M. Anyadike-Danes. 2019. *Judicial Review in Planning and Environmental Cases in Northern Ireland: A Guide for Litigants in Person*. Belfast: Friends of the Earth Northern Ireland.

Department of Justice. 2015a. *Proposal to Change the Time Limits for Bringing a Judicial Review. Consultation*. Belfast: Department of Justice.

Department of Justice. 2015b. *Proposal to Change the Time Limits for Bringing a Judicial Review. Consultation (Summary of Responses)*. Belfast: Department of Justice.

Elliott, M. 2014. 'What a (For Now Failed) Attempt to Curb Judicial Review Tells Us About the UKs Constitution'. Available from http://publiclawforeveryone.com/2014/10/29/what-afor-now-failed-attempt-to-curb-judicial-review-tells-us-about-the-uks-constitutional-architecture/ (accessed 24 February 2016).

Hadfield, B. 1991. 'Judicial Review in Northern Ireland: A Primer', *Northern Ireland Legal Quarterly* 42(4): 332–57.

McKeever, G., L. Royal-Dawson, E. Kirk and J. McCord. 2018. *Litigants in Person in Northern Ireland: Barriers to Legal Participation*. Belfast: Ulster University.

Ministry of Justice. 2021. *Judicial Review Reform: The Government Response to the Independent Review of Administrative Law*. CP408. London: Ministry of Justice.

Northern Ireland Assembly. 2013a. *Northern Ireland Assembly Official Report (Hansard)*. 25 June. Belfast: Northern Ireland Assembly.

Northern Ireland Assembly. 2013b. *Northern Ireland Assembly Official Report (Hansard)*. 22 October. Belfast: Northern Ireland Assembly.

Northern Ireland Courts and Tribunal Service. 2013. *Judicial Review Practice Note 1/2008* (Revised 10 October 2013). Belfast: Northern Ireland Courts and Tribunal Service.

Orbinson, W., G. McGhee and J. Warke. 2009. *Planning and Judicial Review*. Belfast: SLS Legal Publications (NI).

Patterson, F., S. Karim and J. Leslie (eds). 2015. *Judicial Review: Law and Practice*. Bristol: Jordan Publishing Limited.

Rt. Hon Sir Declan Morgan. 2011. 'Judicial Independence in the Twenty-first Century', *Judicial Voices*, SLS Occasional Papers No. 2. Belfast: SLS Legal Publications (NI).

The Independent Review of Administrative Law. 2021. CP407. London: Ministry of Justice.

8

ENVIRONMENTAL LAW IN NORTHERN IRELAND

Brian Jack

Introduction

Northern Ireland presents something of an enigma. Its landscapes and scenery are internationally famous. Yet nationally, as this chapter explores, its environmental governance has been widely criticised. Under the Northern Ireland Act 1998, responsibility for ensuring environmental protection was transferred to the Northern Ireland Assembly and Northern Ireland Executive. Prior to the reshaping of government departments and their functions, which came into effect on 9 May 2016, principal responsibility for environmental matters lay with the Department of the Environment. This Department had broad-ranging responsibilities that extended beyond environmental protection to include road safety and vehicle regulation. It divided its environmental responsibilities across three separate divisions: the Environmental Policy Division, the Northern Ireland Environment Agency and the Marine Division. These divisions were then transferred to a new Department of Agriculture, Environment and Rural Affairs (DAERA), with the exception of the built heritage functions, that were transferred to the Department for Communities.

The Environmental Policy Division is tasked with developing environmental policy and legislation for Northern Ireland. In turn, the Northern Ireland Environment Agency (NIEA) operates as Northern Ireland's principal environmental regulator, charged with implementing and enforcing the legislative framework. These obligations span pollution control, biodiversity protection, waste disposal and the designation and protection of landscapes. NIEA replaced the Environment and Heritage Service, which, until 2008, had carried out these functions. Finally, a Marine Division has been established to develop law and policy in relation to Northern Ireland's marine environment. The protection of the marine environment is considered in Chapter 13 of this book. The Environmental Policy Division, the NIEA and the Marine Division are now clustered as an Environment, Marine and Fisheries Group within DAERA. This chapter concentrates on the functions of the Environmental Policy Division and of the NIEA in developing, implementing and enforcing environmental law within Northern Ireland.

Northern Ireland's legislative framework reflects the province's complex political history. Under the Government of Ireland Act 1920, the United Kingdom Parliament made provision

DOI: 10.4324/9781003218319-8

for the establishment of a devolved parliament and government in Northern Ireland. This enabled the Northern Ireland Parliament to legislate for Northern Ireland on environmental matters, as in the case of the Water Act (Northern Ireland) 1972. Later in 1972, serious civil unrest led to the suspension of the Northern Ireland Parliament and the introduction of direct rule from the United Kingdom Parliament. From that time until the restoration of devolution under the Northern Ireland Act 1998, the United Kingdom Parliament maintained a separate statute book for Northern Ireland by adopting legislation for the province in the form of Orders in Council, such as the Waste and Contaminated Land (Northern Ireland) Order 1997. The restoration of devolved government in 1998 enabled the Northern Ireland Assembly to legislate, for example, adopting the Planning Act (Northern Ireland) 2011. However, in periods after 1998 when direct rule was resumed following the suspension of the Northern Ireland Assembly, the United Kingdom Parliament has again made recourse to the use of Orders in Council. This, for example, is the situation in which the Water (Northern Ireland) Order 1999 and the Environment (Northern Ireland) Order 2002 were adopted. Finally, adding further to the legislative complexity, European Union directives are usually implemented into law in Northern Ireland through secondary legislation in the form of regulations, such as the Planning (Environmental Impact Assessment) Regulations (Northern Ireland) 2015. Environmental impact assessment is explored in Chapter 10 of this book.

Aside from the legislative powers of the Northern Ireland Assembly and, formerly, of the United Kingdom Parliament, the huge influence that the European Union had upon environmental law in Northern Ireland must also be acknowledged. The chapter, therefore, begins by examining this issue. It points to the major role that European Union directives have had in shaping environmental law within Northern Ireland. Additionally, it examines the power that the Court of Justice of the European Union (CJEU) had to impose financial penalties and the influence this had upon policymakers and regulators. The chapter then turns its attention to the impact of the United Kingdom's decision to leave the European Union. The chapter examines the legal frameworks being developed to provide future oversight of environmental governance in Northern Ireland and their practical impact for environmental protection. The chapter then concludes by examining environmental governance within Northern Ireland. In addition to exploring the operation of the environmental regulator, the Northern Ireland Environment Agency, the chapter examines the long-running campaign to establish an environmental regulator that is independent from, rather than integrated within, government in Northern Ireland. The chapter concludes by discussing the need for improved environmental integration in decision making across all of the government departments within the Northern Ireland Executive and for better oversight of their environmental performance.

The European Union and environmental law

Membership of the European Union had a major impact upon environmental law in Northern Ireland. The original 1957 Treaty establishing the European Economic Community (as the European Union was then known) made no direct reference to the environment. However, a liberal interpretation of that Treaty by the European Union institutions and support from the CJEU, in cases such as Case 240/83 *Pro curer de la République v Association de Défense de Brûleurs Usagées*, enabled the European Union to legislate on environmental issues. Any remaining doubts about the European Union's competence to do so were removed when, in 1986, the Single European Act amended the original

Treaty to introduce a specific title on the environment. This can be found today at Articles 191 to 193 of the Treaty on the Functioning of the European Union (TFEU). Ultimately, European Union legislation influenced virtually every avenue of environmental law and policy in Northern Ireland. Within planning law, for example, the Environmental Impact Assessment Directive (Directive 2011/92) and the Directive on Strategic Environmental Assessment (Directive 2001/42) have had an important influence on governance and decision-making processes.

In practice, a directive was the preferred legislative vehicle when the European Union introduced environmental legislation. The TFEU provides that directives are binding on member states as to the result to be achieved, but it leaves to national authorities the choice of form and methods (TFEU: Article 288). Each directive, therefore, created an obligation for member states to introduce national legislation to achieve its objectives. It also set out a time frame within which this was to be achieved, commonly two years, and required member states to provide the European Commission (the Commission) with copies of their national implementing legislation. In Northern Ireland, under the Northern Ireland Act 1998, ultimate responsibility for implementing directives into law in Northern Ireland rested with the Northern Ireland Assembly. However, across the United Kingdom as a whole, the European Communities Act 1972 authorised government departments to use secondary legislation to incorporate European Union obligations into national law (European Communities Act 1972: section 2(2)). This provided a more expedient means to implement European Union directives into national law, in comparison to adopting primary acts of parliament. In Northern Ireland, European Union directives concerning environmental law were usually implemented into national law by means of regulations drafted by the Department of the Environment, which then took effect subject to the approval of the Northern Ireland Assembly (Dickson 2013: 65). One example of this would be the Water Environment (Water Framework Directive) (Northern Ireland) Regulations 2003, which implemented the Water Framework Directive 2000 (Directive 2000/60) into law in Northern Ireland.

It would be wrong, however, to assume that environmental law in Northern Ireland was composed entirely of European Union law. In practice, competence to adopt environmental protection legislation was shared between the member states and the European Union (TFEU: Article 4(2)). In this situation, the Northern Ireland Assembly could adopt its own legislative measures on aspects of environmental protection not already regulated by European Union legislation (TFEU: Article 2(2)). In addition, the CJEU has confirmed that some European environmental directives introduced minimum standards, enabling member states to adopt more stringent national laws (Case C-6/03 *Deponiezweckverbond Eiterköpfe).* It was also possible for national and European Union environmental law to become interlinked. National legislation, originally adopted without intending to implement European Union environmental law, could come to provide the foundation from which to seek to achieve compliance with European environmental legislation. In this way, an existing national law might contribute to the implementation of a subsequent European Union directive. In water law, for example, the Water (Northern Ireland) Order 1999 replaced the earlier Water Act (Northern Ireland) 1972 in regulating water pollution in water bodies across Northern Ireland. Although it is a provision of national law, the Water Order was also an important element within the regulatory framework that Northern Ireland has established with the aim of meeting water quality standards set by the European Union's Water Framework Directive (Directive 2000/60).

The TFEU complements the legislative capacity of the European Union with an enforcement mechanism that seeks to ensure that member states fulfil their obligations. This mechanism enables the Commission to bring member states before the CJEU where it believes they have failed to fulfil their obligations under European Union law (TFEU: Article 258). In 1993 the Treaty on European Union augmented this enforcement process by giving the CJEU power to impose financial penalties. This power is today set out in the TFEU (TFEU: Article 260(2)). It applies where the Commission believes a member state had not fully complied with a judgment of the CJEU. The Commission can then bring that member state back to the CJEU and ask the Court to impose a financial penalty. The 2009 TFEU also gave the CJEU power to impose financial penalties in initial enforcement actions where member states had failed to notify their national legislation, transposing a directive into national law, to the Commission (TFEU: Article 260(3)).

However, the Commission's enforcement actions are concerned with much more than the transposition of directives into national law. They also encompass situations in which the national law implements a directive inaccurately and, more commonly, cases of bad application in which member states do not apply or enforce the directive's requirements effectively. In December 2014, the Commission had 334 enforcement actions open against member states concerning European Union environmental law, of which 86 concerned the failure to notify transposing national legislation to the Commission, 50 alleged inadequacies in national implementing laws and 189 argued that member states had not applied European Union environmental law effectively (European Commission undated). A further nine cases were not classified into any of these three groups. The potential significance of the bad application cases has been further increased by the development of the concept of 'a general and persistent breach'. In *Commission v Ireland*, the CJEU found that Ireland had failed to fulfil its obligations under the 1975 Waste Framework Directive by allowing the operation of twelve illegal waste disposal sites (Case C-494/01 *Commission v Ireland*). Additionally, however, it also accepted the Commission's argument that the existence of so many illegal sites showed that Ireland had been guilty of a general and persistent failure in enforcing European Union environmental law. In other words, for Ireland to avoid the imposition of financial penalties it would not only have to tackle the operation of the twelve illegal sites; it would also have to address the underlying systemic failings in its waste governance that their existence had highlighted.

The risk of receiving a financial penalty from the CJEU provided a powerful incentive for regulators in Northern Ireland to ensure compliance with European Union environmental law (Jack 2011). The CJEU has yet to exercise its new first instance power to apply a financial penalty in a case concerning the failure to notify national transposing legislation. However, a body of case law has developed in which it has done so when member states have failed to comply with previous judgments. Although the TFEU provides for the CJEU to impose 'a lump sum or penalty payment' (TFEU: Article 260(2)), the Court has interpreted the Treaty as enabling it to apply either or both of these punishments in any one case (Case C-304/02 *Commission v France*, para. *82*). It has also applied each in a strategic manner. On the one hand, it has used penalty payments to tackle ongoing non-compliance. In many cases, this means a fine is imposed for each day of continued non-compliance, in others the fine reflects different time periods linked to specific obligations in European Union law. For example, in the first case in which it imposed a financial penalty, the Court ordered Greece to pay EU€20,000 per day until it had complied with an earlier judgment concerning the operation of an illegal landfill site (Case C-387/97 *Commission v Greece*). In its next case, it ordered that Spain should pay €624,500 per annum for each 1 percent of its bathing waters that continued not to meet the standards set by the 1976 Bathing Water Directive (Case 278/01 *Commission v Spain*). This was linked to the Directive's requirement that Spain should provide the Commission with an annual report on its implementation. However, where penalty payments are coercive in effect, lump-sum penalties operate as punitive measures. They punish member state delay in complying with judgments. The attitude of the CJEU is that 'the process of compliance must be initiated at once and completed as soon as possible' (Case C-387/97 *Commission v Greece*, para. 82). Where a breach remains ongoing, this may result in the imposition of both a penalty payment and a lump-sum penalty. For example, France was required to pay a penalty payment of €57,761,250 for each six-month period of continued failure to enforce European Union fishery conservation measures as well as a lump-sum penalty of €20 million (Case C-304/02 *Commission v France*). Each is the largest penalty payment and lump-sum penalty imposed to date. Subsequently, the Court has ordered Ireland to pay financial penalties in two cases arising from European Union environmental law. In the first case, Ireland was ordered to make a penalty payment of €12,000 per day and a lump-sum penalty of €2 million in relation to an ongoing failure to comply with a previous judgment (Case C-188/08 *Commission v Ireland*) establishing its failure to transpose the requirements of the Waste Framework Directive 1975 into national law to regulate the use of septic tanks in rural areas (Case C-374/11 *Commission v Ireland*). In the second case, Ireland was ordered to make a lump-sum payment of €1.5 million as a result of its delay in complying with a judgment (Case C-66/06 *Commission v Ireland*) establishing the inadequate implementation of the Environmental Impact Assessment Directive (Case C-279/11 *Commission v Ireland*).

Ultimately, Northern Ireland's position as a part of the United Kingdom meant that any financial penalty imposed by the CJEU as a result of misapplication of European Union environmental law in Northern Ireland would be levied upon the United Kingdom as a whole. This, however, provided little comfort for environmental regulators within Northern Ireland. As part of the devolution settlement, the devolved governments in Northern Ireland, Scotland and Wales each agreed that they would bear responsibility for any financial penalties levied on the United Kingdom that arise from the implementation or enforcement of European Union law within their territory (United Kingdom Government et al. 2013: para. 20).

Environmental law in Northern Ireland following departure from the European Union

The United Kingdom left the European Union at 11 p.m. on 31 December 2020 (European Union (Withdrawal Agreement) Act 2020: section 39(1).) As a devolved region within the United Kingdom, Northern Ireland also left the European Union at this point. However, the position concerning Northern Ireland's ongoing relationship with the European Union was complicated by the terms of the Protocol on Ireland and Northern Ireland (the Protocol) to the European Union-United Kingdom Withdrawal Agreement 2020 (the Agreement). The Agreement sets out the terms agreed both for the United Kingdom's withdrawal from the European Union and for its future trading relationship with the Union. There was significant concern that Northern Ireland's position outside the European Union and the Republic of Ireland's position within the European Union would result in the need to reintroduce border posts on the island of Ireland to regulate trade between the two territories. The Protocol, however, provides for Northern Ireland to remain within the European Union's single market for the purposes of trade in goods, thus ensuring that the same trade laws continue to apply in both jurisdictions and avoiding any need for border posts. In turn, this necessitated the introduction of inspection measures in Northern Ireland to ensure that goods coming from Great Britain, and therefore entering the single market, comply with European Union trade laws.

At first sight, leaving the European Union may appear to have few implications for Northern Ireland's environmental legislation. As was previously noted, the European Union played a major role in shaping Northern Ireland's statutory framework. Most European Union environmental legislation was adopted in the form of directives. These directives were then incorporated into law in Northern Ireland through domestic legislation. Section 2 of the European Union (Withdrawal) Act 2018 provides for all European Union–derived domestic legislation to remain in force after the United Kingdom's departure from the European Union.

Less frequently, the European Union also adopted environmentally related measures in the form of regulations, such as Regulation 2016/2031 on Plant Health. Unlike directives, regulations were directly applicable and took effect within Northern Ireland without being implemented through domestic legislation. Section 3 of the European Union (Withdrawal Act) 2018 provides for direct European Union legislation, such as regulations, to remain part of domestic law after the United Kingdom has left the European Union.

In practice, leaving the European Union means that domestic environmental legislation implementing European Union directives is now at risk of either being amended to reduce the levels of environmental protection provided or of being repealed altogether. Equally, domestic legislation could also be introduced to limit the protection provided under existing EU regulations. The Protocol provides some limited protection against either eventuality. Article 5 and Annex 2 to the Protocol identify a range of trade-related European Union regulations and directives that the United Kingdom has agreed will continue to apply in Northern Ireland after the United Kingdom's departure. These include regulations and directives concerned with pesticides, waste, environmental protection, genetically modified organisms and plant health. However, these are restricted to trade-related measures and exclude a wide range of European Union directives designed to protect the natural environment. For example, whilst the Protocol requires Northern Ireland to continue applying Directive 94/62 on packaging and packaging waste, it makes no mention of Directive 2008/98, which sets out a legislative framework for waste management in each member state. In contrast, the European Economic Area Agreement requires the European Free Trade Association members (Iceland, Liechtenstein, Switzerland and Norway) to agree to comply with a much more extensive range of European Union legislation concerning the protection of the natural environment in return for providing them access to the European Union single market (Agreement on the European Economic Area 1994: Article 74 and Annex XX).

Aside from legislation, the European Union also had a major influence over environmental policy in Northern Ireland. Then Secretary of State for the Environment, Food and Rural Affairs Michael Gove hailed leaving the European Union as being an opportunity for Britain to shape new environmental policies to better protect the natural world (Defra 2018: i). In practice, policymakers in Northern Ireland proved slow in seeking to grasp this opportunity (Brennan et al. 2019: 90). Ultimately, whilst Scotland and Wales chose to develop their own legislative frameworks for future environmental policies, Northern Ireland has joined England in seeking to do so through the Environment Bill, which in

2021 was under scrutiny before the United Kingdom Parliament. However, some contents of the Bill raise doubts about its ability to achieve the aim of better protecting our natural environment.

The Environment Bill fulfils the obligation, set out in section 16 of the European Union (Withdrawal) Act 2018, to identify the core environmental principles that will guide future environmental law and policy in Northern Ireland (Section 48 and Schedule 2, paragraph 6). In doing so, it lists five of the principles that have shaped European Union environmental law and policy: that environmental protection should be integrated into other policies, that measures should be preventative in nature, precautionary in approach, that environmental damage should be rectified at source and that the polluter should pay (Schedule 2, paragraph 6(5)). These mirror those principles proposed for England, in section 17(5) of the Bill. However, they overlook four other principles that the Withdrawal Act itself stipulated should be included: sustainable development, public access to environmental information, public participation in environmental decision making and access to justice in environmental matters. Equally conspicuous by its absence is any commitment to ensuring that a future environmental policy should aim to achieve a high level of protection. In contrast, this is a core aim of European Union environmental policy (Treaty on the Functioning of the European Union: Article 191(2). It is also surprising that there is no commitment to the principle of non-regression, which would guard against any weakening of the existing legal framework. Similarly, given Northern Ireland's geographical location, a commitment to avoid transboundary harm might also have been expected.

Ultimately, the Environment Bill provides for the Department of Agriculture, Environment and Rural Affairs to prepare a policy statement explaining how the environmental principles should be interpreted and applied by Ministers and government departments in Northern Ireland. Both would then be required to have 'due regard' regard to the policy statement. This falls short of a legal obligation to apply the principles themselves but will at least require that civil servants and Ministers are able to show that the policy statement was fully considered during the decision-making process.

Environmental governance in Northern Ireland

In 1972, the Stockholm declaration highlighted to governments across the globe the urgent need for more effective action to achieve environmental protection (Declaration of the United Nations Conference on the Human Environment 1972). However, as a result of mounting civil unrest, this call went largely unnoticed in Northern Ireland. Political attention here focused on the suspension of devolution and the introduction of direct rule from the United Kingdom. Northern Ireland continued to be governed in this way for most of the subsequent twenty-six years, until devolution was restored by the Northern Ireland Act 1998. The security situation in the province, along with the promotion of economic development, dominated the policy agenda during this period. In contrast, environmental issues were often overlooked. This was highlighted in 1990, in a highly critical report published by the House of Commons Select Committee on the Environment. In particular, the Committee identified the need to transpose a backlog of European Union directives, to modernise environmental laws in areas not covered by European Union environmental law and to bring the Northern Ireland legal framework into line with the rest of the United Kingdom (House of Commons Select Committee on the Environment 1990: paras 32–3).

Over the years, the CJEU has found the United Kingdom to be in breach of its obligations under European Union law on a number of occasions due to the failure to implement environmental directives in Northern Ireland. The cases highlighted the failure to implement the 1976 Bathing Waters Directive (Case C-337/89 *Commission v United Kingdom*), the 1975 Waste Framework Directive (Case C-35/00 *Commission v United Kingdom*), the 1996 Integrated Pollution Prevention and Control Directive (Case C-39/01 *Commission v United Kingdom*) and the 1998 Drinking Water Quality Directive (Case C-63/02 *Commission v United Kingdom*). The continuation of this approach could only lead to the imposition of financial penalties. To avoid this eventuality, the Northern Ireland Executive focused financial resources on a programme of environmental law reform within the Environmental Policy Division that resulted in over forty-five separate pieces of legislation being adopted between 2001 and 2004 (Turner 2006a: 65). In addition, attention also focused on the modernisation of Northern Ireland's own legislative framework. For example, the introduction of the Water (Northern Ireland) Order 1999 upgraded the legal framework in place to tackle water pollution, whilst the Environment (Northern Ireland) Order 2002 did likewise in relation to air pollution, pollution prevention and control and the protection of biodiversity. However, these reforms in themselves have not been sufficient to secure effective environmental governance in Northern Ireland. At one level, gaps still remain in which Northern Ireland's legal framework fails to achieve similar levels of environmental protection to that in place elsewhere in the United Kingdom. For example, in contrast to Great Britain, no legal measures are yet in force to require the identification and remediation of contaminated land within Northern Ireland. The Waste and Contaminated Land (Northern Ireland) Order 1997 sets out the legal framework to do so, but this has never been brought into operation. On another level, the introduction of legislation is not an end in itself. That legislation is of little value if it is not applied and enforced effectively.

Unfortunately, in recent decades, environmental enforcement in Northern Ireland has often been criticised as being weak. This can be seen in publications from scrutiny bodies such as the House of Commons Select Committee on the Environment (1990), the Northern Ireland Audit Office (1997, 2003, 2005), the Northern Ireland Assembly Public Accounts Committee (2001) and the Criminal Justice Inspectorate Northern Ireland (2007, 2015), from professional associations (United Kingdom Environmental Law Association 2004) and from academic commentators (Morrow and Turner 1998; Turner 2006a, 2006b). In 2007 an expert panel appointed to conduct a wide-ranging independent review of environmental governance in Northern Ireland reported that 'the current arrangement of the assets for environmental governance in Northern Ireland is not capable either of resolving the environmental legacies of the past, or of responding to present and emerging environmental pressures' (Burke et al. 2007: 36). One panel member subsequently observed that the panel had become aware of a 'pervasive sense of disregard for the rule of environmental law in Northern Ireland', in which the then regulator, the Environment and Heritage Service, was 'not taken seriously as a credible, consistent or equitable enforcer' (Turner 2007: 439). In 2007, the Criminal Justice Inspectorate also published a highly critical assessment of enforcement by the Environment and Heritage Service. It noted that the Environment and Heritage Service took prosecutions only in exceptional circumstances (Criminal Justice Inspectorate Northern Ireland 2007: 8). This was attributed, in part, to overreliance on a compliance-based approach, which sought to tackle breaches through advice and encouraging behavioural change (Criminal Justice Inspectorate Northern Ireland 2007: 8). Additionally, however, the Criminal Justice

Inspectorate also noted that a lack of resources resulted in many potentially strong cases not being sent to the Public Prosecution Service for potential prosecution (2007: 19). Their report also found that most waste management sites in Northern Ireland were operating without either planning permission or a waste management licence (2007: 9). This was in direct conflict with the requirements of the Waste Framework Directive (now Directive 2008/98) and seems reminiscent of the situation discussed earlier in *Commission v Ireland* (Case 494/01), in which the Court of Justice of the European Union found Ireland to have been guilty of a general and persistent failure to fulfil its legal obligations.

Today, issues remain concerning NIEA's role as the successor to the Environment and Heritage Service. In recent years the discovery of over 500,000 tonnes of waste illegally dumped at Mobuoy, near Derry, has focused public attention upon waste crime. This had also previously been highlighted by cases involving large volumes of waste from the Republic of Ireland that were illegally transported into and dumped within Northern Ireland (Northern Ireland Audit Office 2005). In 2004, the Environment and Heritage Service established a Waste Crime Unit to tackle waste crime. In 2008 then Environment Minister Arlene Foster announced that this would become the Environmental Crime Unit, with additional resources and an intention to broaden its remit to include all areas of environmental crime (Northern Ireland Assembly 2008). Turner and Brennan (2012: 525) point out that the Environmental Crime Unit led the United Kingdom in using the Proceeds of Crime Act 2002 to strip profits from defendants engaged in serious waste crime. The establishment of the Environmental Crime Unit also appeared to meet a core recommendation of the Criminal Justice Inspectorate (Criminal Justice Inspectorate Northern Ireland 2007: 5) that there should be a single environmental enforcement office within the environmental regulator. However, Turner and Brennan (2012: 525) reveal that, in practice, the Environmental Crime Unit continued to focus on serious waste crime, associated with organised crime, whilst paying limited attention to other waste crimes committed by legitimate businesses. The focus on serious waste crime also left the enforcement of other aspects of environmental crime to individual divisions within NIEA, such as the Natural Environment Division. In turn, these divisions did not benefit from the additional resources provided to the Environmental Crime Unit. The Criminal Justice Inspectorate (2007: 5) previously identified a 'silo approach' to enforcement within the Environment and Heritage Service, where it found 'little integration of methods, systems or effort'. It also highlighted the fact that the Environment and Heritage Service lacked sufficient resources to undertake both the monitoring of nature conservation sites and enforcement action. A later review of the Environmental Crime Unit found that, whilst NIEA's own strategy document spoke of a vigorous enforcement policy (NIEA 2012: 3), some staff continued to view enforcement as being a secondary function (Criminal Justice Inspectorate Northern Ireland 2015: 15). This review also noted that inconsistency in approaches to enforcement continued to exist across different divisions within NIEA. NIEA subsequently created a Resources Efficiency Division, linking the Environmental Crime Unit with other units dealing with waste management, water management and with the Drinking Water Inspectorate. This was in response to a recommendation from a review of environmental regulation at the Mobouy site, which proposed that a single directorate be created to incorporate both the regulation and the enforcement of waste crime (Mills 2013: 33). The creation of the Resources Efficiency Division continues to bypass areas such as nature conservation. The introduction of this more unified approach to regulation and enforcement within the waste sector, therefore, came at the expense of a more integrated approach towards the enforcement of environmental crime as a whole.

The review of the environmental regulation at the Mobouy site revealed a number of other weaknesses in the regulatory regime for waste in Northern Ireland (Mills 2013: 78). In addition, it highlighted another area of conflict in relation to environmental regulation. At the time of the review, responsibility for development planning lay with the Planning Service, then a separate division within the Department of the Environment. However, a lack of coordination between planning and environmental law had enabled illegal waste to be disposed of in extensive pits that had been dug without planning permission and then subsequently granted retrospective permission (Mills 2013: 78). Like the Criminal Justice Inspectorate, the review pointed to the need for a more integrated regulatory approach (Mills 2013: 77). Now that responsibility for planning functions has largely been transferred to district councils, under the Planning Act (Northern Ireland) 2011, it is important that NIEA and the councils collaborate to ensure that waste disposal is regulated more effectively.

The review of environmental regulation at the Mobouy site also pointed out that even when cases are successfully prosecuted, the sentences imposed often do not reflect the seriousness of the crime and the costs imposed on both the environment and society (Mills 2013: 79). This repeats comments that have previously been made by the Review of Environmental Governance and by academic commentators within Northern Ireland (Burke et al. 2007: 14; Turner 2006b: 266; Turner 2007: 449). Collectively, these views suggest an institutional attitude that places less importance on environmental crime in comparison with other forms of crime. As a consequence, far from the polluter paying, all too often pollution may be seen to pay for the polluter. The Review of Environmental Governance called for the judiciary to undertake professional development to promote greater awareness of the social, economic and environmental impacts of environmental crime (Burke et al. 2007: 14). It also proposed the establishment of a specialised Environmental Tribunal. The Review also pointed out that the Public Prosecution Service, which prosecutes environmental cases in court, lacks expertise in environmental law. In turn, the Criminal Justice Inspectorate highlighted a similar lack of legal expertise within the then Environment and Heritage Service (2007: 12). It suggested that senior managers within the Environment and Heritage Service favoured the creation of an internal specialised legal unit within NIEA, which would have responsibility for both prosecuting cases involving environmental crime and advising and training enforcement staff (2007: 12). This would be similar to the approach already adopted by the Environment Agency in England and Wales (Criminal Justice Inspectorate Northern Ireland 2007: 12). To date, no specialised Environmental Tribunal has been established. Nor has a specialised legal unit been established within NIEA, whilst the need to promote greater knowledge and expertise in environmental law within the Public Prosecution Service also remains an ongoing issue.

An independent environmental regulator

Like the Environment and Heritage Service before it, NIEA is unique within the United Kingdom in not being independent from central government. Instead, it has operated as an agency, initially within the Department of the Environment and now within DAERA. This situation reflects the centralised structure of governance in general that existed throughout the direct rule period. (Jack 2007: 1) However, its continuation today appears out of step with the requirements of effective environmental governance. Persistent calls have been made for the introduction of an independent environmental regulator in Northern Ireland. The most strident of these came from the Review of Environmental Governance, in which the establishment of an independent regulator was a core recommendation (Burke et al. 2007: 8). The Review (Burke et al. 2007: 50) noted that the position of the regulator within government had resulted in a lack of transparency and accountability in decision-making processes and had contributed to a general loss of public confidence in that regulator. In contrast, environmental regulators elsewhere in the United Kingdom were seeking to promote greater openness and transparency in order to promote public confidence (Macrory 2004: 41). Additionally, the Review pointed out that the lack of independence exposed staff to both real and perceived risks of conflict of interest when carrying out their regulatory functions (Burke et al. 2007: 52). As Turner has noted (2007: 434), the departmental interest is unfortunately not always aligned with the interests of the environment or with the regulator's statutory duties. In this situation, staff may prioritise their obligations to the Department and to their Minister over the interests of the environment. Equally, the lack of independence also prevents the regulator from acting as an effective environmental advocate, for example, publicly criticising policies and procedures adopted within its own department. In recent years environmental non-governmental organisations have stepped into this vacuum. In particular, Friends of the Earth has lodged a number of complaints with the European Commission alleging that Northern Ireland was failing to meet its obligations under a number of European Union environmental directives. However, this should not be considered an adequate substitute for an independent regulator able to take action within Northern Ireland to ensure better environmental protection.

In practice, each of the weaknesses outlined above was amply demonstrated in the so called 'hotspots incident' that occurred in 2002 (Turner 2006a: 75; Turner 2006b: 255). At that time, the European Commission had commenced enforcement action against the United Kingdom, concerning Northern Ireland's failure to comply with the sewage treatment standards required under the Urban Waste Water Treatment Directive. However, economic development within Northern Ireland was also creating demand for new housing. The Northern Ireland Executive initiated a programme to update Northern Ireland's sewage treatment works and meet the standards required under the Directive (Directive 91/271). However, the completion of this work would take some time. In the interim, the Department of the Environment identified fifty-six particularly problematic sewage 'hotspot' areas. Initially, a moratorium was announced on the grant of new planning permissions in these areas whilst a policy review took place. Ultimately, however, this policy review led then Minister for the Environment Dermot Nesbitt to direct that the environmental regulator, then the Environment and Heritage Service, should not object to planning applications within the 'hotspot' areas. As Turner has observed (2006b: 255), the regulator was in effect silenced, whilst economic and social goals were prioritised over environmental protection.

The handling of the 'hotspots incident' acted as a catalyst, sparking a campaign by environmental non-governmental organisations in Northern Ireland to highlight the need for an independent environmental regulator. Participants in the Review of Environmental Governance showed strong support for this objective, with only the Ulster Farmers Union opposing the creation of an independent agency (Turner 2007: 434). Despite this, in May 2008, then Minister for the Environment Arlene Foster rejected the Review's recommendation and announced instead that the NIEA would replace the Environment and Heritage Service (Northern Ireland Assembly 2008). This was essentially a rebranding, with NIEA continuing at that time to be an agency within the Department of the Environment. Turner points to this widely criticised decision as being motivated by the desire to protect the perceived interests of one stakeholder, the Ulster Farmers Union (Turner 2009: 157).

The case for an independent environmental regulator also gained support from the High Court. In *Seaport Investments*, the court examined the requirement imposed by Article 6(3) of the Strategic Environmental Assessment Directive (Directive 2001/42), that member states should designate authorities who would be consulted on the environmental effects of plans and programmes required to undergo environmental assessment under that Directive. The High Court held that it was implicit from the Directive that this consultation body should be external to the body that had created the plan or programme. The court also ruled that this had not been complied with where, as had occurred in that case, the Department of the Environment's Planning Service had produced an area development plan and its Environment and Heritage Service had acted as the consultation body. The Department of the Environment appealed the judgment to the Court of Appeal, which, in turn, asked the CJEU to rule on the obligations imposed by Article 6(3). In Case C-474/10 *Department of the Environment v Seaport (NI) Ltd.*, however, the CJEU, ruled that there was no requirement for the consultation body to be an external body as long as there was a functional separation between it and the body responsible for creating the plan or programme in question. Viewed in that light, the Planning Service and the Environment and Heritage Service did have separate administrative roles as well as their own staff and office buildings.

More recently, the *New Decade, New Approach* agreement, concluded by the British and Irish Governments in 2020 and endorsed by Northern Ireland's political parties as a basis for their return to devolved government in Northern Ireland, contained a specific commitment to establish an independent environmental protection agency in Northern Ireland (United Kingdom Government and Government of Ireland 2020: 44). Public responses to the DAERA consultation document concerning the development of an environmental strategy for Northern Ireland (Department of Agriculture, Environment and Rural Affairs 2019) also revealed continued strong public support for an independent environmental agency (Department of Agriculture, Environment and Rural Affairs 2020: 22).

Ultimately Northern Ireland remains out of step with other parts of the United Kingdom in not having an independent environmental regulator. Irrespective of the judgment of the CJEU, the underlying weaknesses associated with this lack of independence, as identified by the Review of Environmental Governance, continue to exist. The fact that one stakeholder could unilaterally halt progress towards an independent regulator also raises broader concerns for environmental governance generally within Northern Ireland. Equally, the failure to act upon the commitments given in the *New Decade, New Approach* reveals scant regard for democracy in Northern Ireland, particularly given the evident continued public support for reform.

Environmental integration

Successful environmental protection within Northern Ireland depends upon more than an effective regulator and supportive judicial system. Ultimately it also requires broader support across government within Northern Ireland. This was illustrated in the Renewable Heat Incentive scandal that led to the collapse of the Northern Ireland Executive in 2016. In this instance, poor design and management of a green energy subsidy introduced by the then

Department of Enterprise, Trade and Investment created substantial costs for the Executive without delivering environmental returns. As Brennan et al. note, the scheme was ostensibly designed to assist businesses to adopt renewable energy sources, but in practice appeared to focus on providing economic support to industry (2017: 132). In Northern Ireland, until May 2016, the Department of the Environment had principal responsibility for environmental governance. However, Turner (2007: 428) highlighted that aspects of environmental governance also spanned nine of the twelve departments that made up the then Northern Ireland Executive. She identified these as being the Department of the Environment; the Department for Regional Development; the Department of Agriculture and Rural Development; the Department of Culture, Arts and Leisure; the Department of Health, Social Services and Public Safety; the Department of Enterprise, Trade and Investment; the Department of Finance and Personnel; the Office of the First and Deputy First Minister and the Department for Social Development. In practice, effective environmental governance often required cooperation between two or more of these departments. However, the nature of the power-sharing government in Northern Ireland makes such policy integration difficult to achieve. As Turner has argued (2013: 208), the departments were not structured with notions of policy cohesion in mind, but with the aim of promoting conflict resolution by creating equally sized departments that divide power across Northern Ireland's political parties. Consequently, party political rivalry can impede agreement on environmental protection.

The reduction in the number of government departments in Northern Ireland from twelve to nine in 2016 (outlined in Chapter 1 of this book) certainly has helped to reduce the scale of this problem (Northern Ireland Assembly 2015). This is particularly true of the amalgamation of the Department of the Environment and Department of Agriculture and Rural Development (to become the Department of Agriculture, Environment and Rural Affairs), given agriculture's environmental footprint upon the Northern Ireland countryside – both negative and positive. The key to effective environmental protection remains integration across government as a whole. In that regard, Northern Ireland's government departments have a statutory duty to promote the achievement of sustainable development (Northern Ireland Miscellaneous Provisions Act 2006: section 25). However, the core governance structure, in which ministerial portfolios are divided across the political parties, will continue to make it difficult to ensure that environmental protection concerns become integrated within broader economic development goals.

Oversight of environmental governance in Northern Ireland

Environmental governance in Northern Ireland has largely been undermined by a lack of strong oversight mechanisms within the Northern Ireland constitutional framework. The Northern Ireland Assembly's Agriculture, Environment and Rural Affairs Committee operates with an oversight function in relation to DAERA. Elsewhere, the Assembly's Public Accounts Committee, the Northern Ireland Audit Office and the Criminal Justice Inspectorate have all published influential reports challenging aspects of the performance of the Environment and Heritage Service and, more recently, the NIEA. However, the remit of each of these latter bodies exceeds environmental protection. They oversee all aspects of government and regulation. This leaves them only limited ability to devote resources to environmental issues. In reality, there is currently no body specifically charged with assessing the environmental performance of each of Northern Ireland's government departments or of ensuring that they meet their statutory duty to promote sustainable development. This should be contrasted with the position within the United Kingdom Parliament, where the House of Commons established the Environmental Audit Select Committee in 1997. The Committee's role is to consider and report on the extent to which the policies and programmes implemented by all government departments contribute to environmental protection and sustainable development and to audit their performance against sustainable development and environmental protection targets (Environmental Audit Committee undated). Elsewhere, in Canada, an Environmental Audit Select Committee is supported by an independent environment audit commissioner for the environment and sustainable development, who is attached to Canada's Office of the Auditor General (Ross 2000: 192). In Northern Ireland, the Review of Environmental Governance called for the establishment of an Environmental Audit Committee within the Northern Ireland Assembly (Burke et al. 2007: 12). However, this recommendation has not been acted upon. As a result, a vacuum has existed as regards the accountability of the Northern Ireland Executive for its environmental performance. The European Union's enforcement mechanisms, examined above, played an essential role in requiring the Northern Ireland Executive to pay attention to implementation and application of European Union environmental laws. This was by no means perfect, particularly since it had no remit in relation to purely national environmental law and policy. Yet even this safeguard has now been lost. However, the Environment Bill proposes to introduce a new oversight body, the Office for Environmental Protection (OEP), to oversee environmental governance in both Northern Ireland and England.

Then Secretary of State for the Environment Food and Rural Affairs Michael Gove promised that the OEP would be a world-leading, independent environmental watchdog that would take over the role that the European Commission and the Court of Justice of the European Union had played in holding government to account on its environmental performance (Department for Environment Food and Rural Affairs 2018: i). The introduction of a dedicated environmental oversight body in Northern Ireland has been strongly welcomed (DAERA 2021: 9). Unfortunately, however, Mr Gove seems to have overstated its potential effectiveness.

The Environment Bill tasks the OEP with contributing to environmental protection and to the improvement of the natural environment (section 23). The Bill requires DAERA to prepare an environmental improvement plan setting out the steps that it intends to take to improve Northern Ireland's natural environment (Section 49 and schedule 3). The Department would also be required to provide the Northern Ireland Assembly with an annual report on its progress in implementing the plan and to publish this report. In turn, the OEP would be required to monitor the Department's progress and to provide the Northern Ireland Assembly and the public with an annual monitoring report. In addition, the OEP would be tasked with monitoring the implementation of environmental law in Northern Ireland, providing advice on proposed changes to Northern Ireland environmental law and taking action against public authorities in Northern Ireland who had failed to comply with their obligations under environmental law.

The introduction of these measures would help to fill a gaping hole in Northern Ireland's environmental governance structures. However, they would need further improvement before they can truly be described as world-leading. The OEP's role would enable it to exercise oversight functions on the Northern Ireland Environment Agency and the application of planning and environmental laws by district councils. The OEP's role in monitoring the delivery of environmental improvement plans would also result in a more sustained focus on outcomes. However, in other respects the OEP's role should be further enhanced. The Bill would limit the OEP to advising on Northern Ireland environmental law, yet the integration principle requires government departments to integrate environmental protection into all decision making. Equally, the OEP could be truly innovative if this included a power to represent the natural environment in such circumstances, including as a statutory consultee in relation to planning applications. However, it is the OEP's proposed role in seeking to ensure public authorities' compliance with environmental law that perhaps appears most underwhelming. As with the European Union environmental law, individuals would be able to lodge complaints to the OEP where they believed a public authority had failed to comply with environmental law. Where the OEP believes there are reasonable grounds for suspecting a serious failure to comply with environmental law it can serve an information notice requiring a public authority to provide information. Ultimately, subject to the same caveats, it can issue a decision notice describing the failure by the public authority in question, why it considers this serious and setting out the action required to remedy, or mitigate, it or prevent its reoccurrence. However, the Bill provides that the OEP's ultimate power to enforce compliance with a decision notice, or indeed environmental law generally, is to bring an application for judicial review to court. This falls far short of the European Court of Justice's powers to impose financial penalties. In practice, judicial review is poorly suited to act as an arbiter of environmental governance. It is usually concerned with the process through which particular decisions were taken, rather than the environmental merits of the decisions themselves (Pedersen 2019). Ultimately, a decision notice must be binding upon a public authority, which should have the power to appeal that notice. In the absence of an appeal or full compliance, the OEP should have power to bring the matter before the courts, which, in turn, must then be empowered to impose effective financial penalties.

Conclusion

Despite the international reputation of its landscape and scenery, environmental protection has for too long been a secondary consideration in Northern Ireland. The current drive to promote economic development overlooks one of the region's prime assets, its environment. The absence of an Environmental Audit Committee, which can vet the environmental credentials of individual departments within the Northern Ireland Executive and of their agencies remains a distinct weakness. It should be playing a central role in ensuring that the Northern Ireland Executive's Sustainable Development Strategy becomes a practical reality (Northern Ireland Executive 2010). In recent decades a concerted effort has been made to transpose a raft of European Union environmental directives into law in Northern Ireland and to update an antiquated legislative framework. This may have brought Northern Ireland up to date in terms of implementing European Union environmental law, but the failure to introduce a contaminated land regime illustrates that our environmental law is not yet on a par with that in Great Britain.

Equally, introducing legislation is arguably the easy part, but implementing and enforcing that legislation is more challenging. Reports by scrutiny bodies and commentators suggest that this remains a central weakness within environmental governance in Northern Ireland. They tell a tale of concentration on serious waste crime to the detriment of enforcement in other areas of environmental crime; of a need to promote environmental law expertise, preferably by creating a specialised legal unit within the regulator or, at the very least, within the Public Prosecution Service; and of ineffectual penalties when cases are prosecuted successfully. Northern Ireland also continues to lack an independent regulator, which, freed from departmental control, would be entrusted with both enforcement of environmental law and advocacy on behalf of the environment. In the recent past, the threat of financial penalties prompted the Northern Ireland Executive to move promptly to remove a backlog of unimplemented European Union environmental directives. Following the United Kingdom's departure from the European Union, this threat has now largely been removed. The measures proposed in the Environment Bill, before the United Kingdom Parliament in 2021, represent a partial closure of this vacuum. The introduction of the Office for Environmental Protection is a welcome innovation, as is its proposed role in monitoring DAERA's progress in implementing environmental improvement plans. However, it will become an effective environmental oversight body only if it is entrusted with real teeth to pursue errant public authorities.

Legislation

Agreement on the European Economic Area 1994

Council Directive 91/271/EEC of 21 May 1991 concerning urban waste water treatment (1991) Official Journal L135/40

Directive 94/62/EC of the European Parliament and of the Council of 20 December 1994 on packaging and packaging waste (1994) Official Journal L365/10

Directive 2000/60/EC of the European Parliament and of the Council of 23 October 2000 establishing a framework for Community action in water policy (2000) Official Journal L327/1

Directive 2001/42/EC of the European Parliament and of the Council of 27 June 2001 on the assessment of the effects of certain plans and programmes on the environment (2001) Official Journal L197/30

Directive 2008/98/EC of the European Parliament and of the Council of 19 November 2008 on waste (2008) Official Journal L312/3

Directive 2011/92/EU of the European Parliament and of the Council of 13 December 2011 on the assessment of the effects of certain public and private projects on the environment (2011) Official Journal L26/1

Environment Bill 2021European Communities Act 1972

European Union-United Kingdom Withdrawal Agreement 2020

European Union (Withdrawal) Act 2018

European Union (Withdrawal Agreement) Act 2020

Government of Ireland Act 1920

Marine Act (Northern Ireland) 2013

Northern Ireland Act 1998

Planning Act (Northern Ireland) 2011

Regulation 2016/2031 of the European Parliament and of the Council of 26 October 2016 on protective measures against pests of plants (2016) Official Journal L317/4.

Single European Act 1986

Treaty Establishing the European Economic Community 1957

Treaty on the Functioning of the European Union (the TFEU) 2009

Waste and Contaminated Land Order (Northern Ireland) 1997, Statutory Instrument 1997 No. 2778 (NI.19)
Water Act (Northern Ireland) 1972
Water Environment (Water Framework Directive) Regulations (Northern Ireland) 2003 Statutory Rule 2003/544
Water (Northern Ireland) Order 1999, Statutory Instrument 1999 No. 662 (NI 6)

Cases cited

Case law of the Court of Justice of the European Union

Case 240/83 *Procurer de la République v Association de Défense de Brûleurs Usagées* [1985] ECR 531
Case C-337/89 *Commission v United Kingdom* [1992] ECR I-6103
Case C-387/97 *Commission v Greece* [2000] ECR I-5047
Case C-35/00 *Commission v United Kingdom* [2002] ECR I-53
Case C-39/01 *Commission v United Kingdom* [2002] ECR I-2513
Case C-278/01 *Commission v Spain* [2003] ECR I-14141
Case C-494/01 *Commission v Ireland* [2005] ECR I-3331
Case C-63/02 *Commission v United Kingdom* [2003] ECR I-821
Case C-304/02 *Commission v France*, [2005] ECR I-6263
Case C-6/03 *Deponiezweckverband Eiterköpfe v Land Rheinland-Pfalz* [2005] ECR I-2753
Case C-66/06 *Commission v Ireland*, [2008] ECR I-158
Case C-188/08 *Commission v Ireland*, [2009] ECR I-172
Case C-474/10 *Department of the Environment v Seaport (NI) Ltd., and others* [2011] ECR I-10227
Case C-279/11 *Commission v Ireland*, not yet reported
Case C-374/11 *Commission v Ireland*, not yet reported

Northern Ireland case law

Re Application for Judicial Review by Seaport Investments Limited and Others [2007] NIQB 62

References

Brennan, C., M. Dobbs and V. Gravey. 2019. 'Out of the Frying Pan, into the Fire? Environmental Governance Vulnerabilities in Post-Brexit Northern Ireland', *Environmental Law Review* 21: 84–110.
Brennan, C., R. Purdy and P. Hjerp. 2017. 'Political, Economic and Environmental Crisis in Northern Ireland: The True Cost of Environmental Governance Failures and Opportunities for Reform', *Northern Ireland Legal Quarterly* 68: 123–57.
Burke, T., S. Turner and G. Bell. 2007. *Foundations for the Future: The Review of Environmental Governance. Final Report.* Belfast: Department of the Environment.
Criminal Justice Inspectorate Northern Ireland. 2007. *Enforcement in the Department of the Environment.* Belfast: Criminal Justice Inspectorate Northern Ireland.
Criminal Justice Inspectorate Northern Ireland. 2015. *A Review of the Northern Ireland Environment Agency's Environmental Crime Unit.* Belfast: Criminal Justice Inspectorate Northern Ireland.

Department of Agriculture, Environment and Rural Affairs. 2019. *Environment Strategy for Northern Ireland: Public Discussion Document*. Belfast: Department of Agriculture, Environment and Rural Affairs.

Department of Agriculture, Environment and Rural Affairs. 2020. *An Environment Strategy for Northern Ireland: Responses to the 2019/20 Public Discussion Document- Summary Report*. Belfast: Department of Agriculture, Environment and Rural Affairs.

Department of Agriculture, Environment and Rural Affairs. 2021. *Environmental Plans, Principles and Governance for Northern Ireland: Public Discussion Document: Synopsis of Responses*. Belfast: Department of Agriculture, Environment and Rural Affairs.

Department for Environment Food and Rural Affairs. 2018. *Environmental Principles and Governance after the United Kingdom Leaves the European Union*. London: Department for Environment Food and Rural Affairs.

Dickson, B. 2013. *Law in Northern Ireland*. Second Edition. Oxford: Hart Publishing.

Environmental Audit Committee. Undated. *Role – Environmental Audit Committee*. Available at https://committees:parliament.uk/committee/62/environmental-audit-committee/role/ (accessed 1 September 2021).

European Commission. Undated. *Legal Enforcement: Statistics on Environmental Infringements*. Available at https://ec.europa.eu/environment/legal/law/statistics.htm (accessed 1 September 2021).

House of Commons Select Committee on the Environment. 1990. *Environmental Issues in Northern Ireland* (HC 39 Session 1990–91). London: HMSO.

Jack, B. 2007. 'Environmental Governance in Northern Ireland: Returning to the Drawing Board', *Environmental Law Review* 9: 1–5.

Jack, B. 2011.'Enforcing Member State Compliance with EU Environmental Law:A Critical Evaluation of the Use of Financial Penalties', *Journal of Environmental Law* 23: 73–95.

Macrory, R. 2004. *Transparency and Trust: Reshaping Environmental Governance in Northern Ireland*. London: University College London.

Mills, C. 2013. *A Review of Waste Disposal at the Mobouy Site and the Lessons Learnt for the Future Regulation of the Waste Industry in Northern Ireland*. Belfast: Department of the Environment.

Morrow, K. and S. Turner. 1998. 'The More Things Change, the More They Stay the Same? Environmental Law, Policy and Funding in Northern Ireland', *Journal of Environmental Law*10: 41–59.

NIEA (Northern Ireland Environment Agency). 2012. *Our Passion, Our Place: NIEA Strategic Priorities 2012–2022*. Belfast: NIEA (Northern Ireland Environment Agency).

Northern Ireland Assembly. 2008. *Official Report of the Northern Ireland Assembly, 27 May 2008*. Belfast: Northern Ireland Assembly. Available at http://archive.niassembly.gov.uk/record/reports2007/080527.htm#4 (accessed 1 September 2021).

Northern Ireland Assembly. 2015. *Civil Service Departments: Reduction/Names/Functions. Statement to the Northern Ireland Assembly by the First Minister*, 2 March, Hansard, Session 2014–15, 102(5): 1. Available at http://data.niassembly.gov.uk/HansardXml/plenary-02-03-2015.pdf (accessed 1 September 2021).

Northern Ireland Assembly Public Accounts Committee. 2001. *Control of River Pollution in Northern Ireland, (NIA 6/00)*. London: HMSO.

Northern Ireland Audit Office. 1997. *The Control of River Pollution in Northern Ireland* (HC 693, Session 1997–98). London: HMSO.

Northern Ireland Audit Office. 2003. *Areas of Special Scientific Interest* (HC499, Session 2003–04). London, HMSO.

Northern Ireland Audit Office. 2005. *Northern Ireland's Waste Management Strategy* (HC 88, Session 2005–06). London: HMSO.

Northern Ireland Executive. 2010. *Everyone's Involved: Sustainable Development Strategy*. Belfast: Northern Ireland Executive.

Pedersen, O. 2019. 'A Study of Administrative Environmental Decision-Making Before the Courts', *Journal of Environmental Law* 31: 59–82.

Ross, A. 2000. 'Greening Government: Tales from the New Sustainability Watchdog', *Journal of Environmental Law* 12: 175–96.

Turner, S. 2006a. 'Transforming Environmental Governance in Northern Ireland. Part One: The Process of Policy Renewal', *Journal of Environmental Law* 18: 55–87.

Turner, S. 2006b. 'Transforming Environmental Governance in Northern Ireland: Part Two: The Case of Environmental Regulation', *Journal of Environmental Law* 18: 245–75.

Turner, S. 2007. 'Laying the Foundations for a Sustainable Northern Ireland: The Review of Environmental Governance', *Northern Ireland Legal Quarterly* 58: 422–58.

Turner, S. 2009. 'Devolution as a Barrier to Environmental Reform: Assessing the Response to the Review of Environmental Governance in Northern Ireland', *Environmental Law Review* 11: 150–60.

Turner, S. 2013. 'Committing to Effective Climate Governance in Northern Ireland: A Defining Test of Devolution', *Journal of Environmental Law* 25: 203–34.

Turner, S. and C. Brennan. 2012. 'Modernising Environmental Regulation in Northern Ireland: A Case Study in Devolved Decision Making', *Northern Ireland Legal Quarterly* 63: 509–32.

United Kingdom Environmental Law Association. 2004. *Environmental Regulation in Northern Ireland: An Agenda for Reform*. London: United Kingdom Environmental Law Association.

United Kingdom Government, Scottish Ministers, the Welsh Ministers and the Northern Ireland Executive Committee. 2013. *Memorandum of Understanding and Supplementary Agreements Between the United Kingdom Government, Scottish Ministers, the Welsh Ministers and the Northern Ireland Executive Committee.* Available at https://assets.publishing.service.gov.uk/government/uploads/system/uploads/attachment_data/file/316157/MoU_between_the_UK_and_the_Devolved_Administrations.pdf (accessed 1 September 2021).

United Kingdom Government and Government of Ireland. 2020. *New Decade, New Approach*. Available at https://assets.publishing.service.gov.uk/government/uploads/system/uploads/attachment_data/file/856998/2020-01-08_a_new_decade_a_new_approach.pdf (accessed 1 September 2021).

United Nations. 1972. *Declaration of the United Nations Conference on the Human Environment*. Stockholm: United Nations.

9

ENVIRONMENTAL HEALTH AND PROTECTION

Sean MacIntyre

Introduction

The roles of the spatial planning, environmental protection and public health regimes are interconnected and to an extent complementary. All are involved with a range of similar issues and share the aims of achieving more sustainable communities by protecting the environment and health, preventing and mitigating nuisance or loss of amenity and facilitating healthier lifestyles. The links between the built environment and health are long established, with advances in public health in western societies having been extensively driven by improvements undertaken since the 19th century in sanitation, housing conditions, air and water quality. The current planning, environmental health and building control professions all sprang from the Public Health Acts of the Victorian era indicating the fundamental relationship between these functions and the impacts their roles have on the quality of the built and natural environment. An appreciation of the importance of the wider determinants of health has also gained increasing attention in recent years with, for example, the development of the 'Health Map' (Barton and Grant 2006), which enables spatial planners to recognise that 'the urban development process and the design and planning of settlements, resides in one sphere – the built environment, thus planners can see their place in determining health. In direct terms, they can affect the quality of that environment' (ibid: 3). These links with the wider determinants of health parallel those of the World Health Organisation's definition of the role of environmental health which is to

> address all the physical, chemical, and biological factors external to a person and all the related factors impacting behaviours. It encompasses the assessment and control of those environmental factors that can potentially affect health. It is targeted towards preventing disease and creating health-supportive environments.
>
> *(World Health Organisation 2016)*

In addition, the publication of the United Nations seventeen sustainable development goals (https://sdgs.un.org/goals), which draw together themes such as Sustainable Cities and

DOI: 10.4324/9781003218319-9

Communities, Good Health and Wellbeing, Climate Action, Reduced Inequalities and Life on Land and Below Water, illustrate the interconnectedness of health and the environment.

In Northern Ireland responsibility for discharging the environmental health function lies with local government comprising eleven district councils. Environmental Health Department duties are broadly based around the functions of food safety, health and safety and consumer safety, but it is around the core functions of environmental protection and public health/nuisance that environmental health interacts most closely with spatial planning. This chapter explores more fully that relationship.

Amenity and statutory nuisance

The terms amenity and nuisance are widely used within planning policy with one of the broad aims of the planning system being to mitigate against adverse impacts on amenity and to determine when incompatibility between differing types of development is likely to cause nuisance. While the terms 'loss of amenity' and 'nuisance' are often used interchangeably, within the planning and environmental health spheres there are marked differences in meaning. Loss of amenity and nuisance are both material planning considerations. However, the relevant environmental health legislation only addresses statutory nuisance.

The concept of amenity is very broad and is not clearly defined within planning legislation and guidance. It is accepted as referring to the pleasantness or convivial nature of a particular location. While the term is not defined, guidance does indicate that amenity 'is a matter of fact and degree and common sense, each case will be different and what would not be considered amenity in one part of a planning authority's area might well be considered so in another' (ODPM 2005: 8). The *Strategic Planning Policy Statement for Northern Ireland* (SPPS) states 'there are a wide range of environment and amenity considerations, including noise and air quality, which should be taken into account by planning authorities when proposing policies or managing development' (DoE NI 2015b: 17). The SPPS then expands upon this by providing examples of how planning can impact amenity/environmental protection issues:

> the planning system has a role to play in minimising potential adverse impacts, such as noise or light pollution on sensitive receptors by means of its influence on the location, layout and design of new development. The planning system can also positively contribute to improving air quality and minimising its harmful impacts.
>
> *(DoE NI 2015: 17)*

The environmental impacts highlighted above are also regulated by environmental health legislation but, while amenity or the loss of it is a material planning consideration, the same loss of amenity may not amount to a statutory nuisance within the remit of environmental health legislation.

Definition of a statutory nuisance

The concept of the statutory public health nuisance emerged from common law nuisance and was incorporated into the Public Health Acts of the 19th century, the local statute being the Public Health (Ireland) Act 1878, which remained in force until 2011. Nuisance provisions are now contained within the Clean Neighbourhoods and Environment Act (Northern

Ireland) 2011. Section 63 of the Act sets out those factors that can be determined as being a statutory nuisance, namely:

- any premises in such a state as to be prejudicial to health or a nuisance;
- smoke emitted from premises so as to be prejudicial to health or a nuisance;
- fumes or gases emitted from premises so as to be prejudicial to health or a nuisance;
- any dust, steam, smell or other effluvia arising on industrial, trade or business premises and being prejudicial to health or a nuisance;
- any accumulation or deposit which is prejudicial to health or a nuisance;
- any animal kept in such a place or manner as to be prejudicial to health or a nuisance;
- any insects emanating from relevant industrial, trade or business premises and being prejudicial to health or a nuisance;
- artificial light emitted from premises so as to be prejudicial to health or a nuisance;
- noise emitted from premises so as to be prejudicial to health or a nuisance;
- noise that is prejudicial to health or a nuisance and is emitted from or caused by a vehicle, machinery or equipment in a street;
- any lake, watercourse, privy, urinal, cesspool, drain or ash pit which is so foul or in such a state as to be prejudicial to health or a nuisance;
- any part of a watercourse which is so choked or silted up as to obstruct or impede the proper flow of water and thereby to cause a nuisance, or give rise to conditions prejudicial to health;
- any private dwelling so overcrowded as to be prejudicial to the health of those living there or a nuisance.

As can be seen, many of the above classes of statutory nuisance are also material planning considerations. It is therefore necessary to identify what factors are considered in determining the existence of a statutory nuisance and how this links to safeguarding amenity.

What constitutes a statutory nuisance?

Statutory nuisance is not defined in the 2011 Act but, in essence, it is regarded as an unreasonable interference with the personal comfort or enjoyment of property. While there is no objective definition as to what constitutes a nuisance,

> it has been said that there is a scale between mildly irritating and intolerable and in each case the determination of whether a nuisance exists is a matter of judgement. In addition, the determination is based upon the test of what ordinary, decent people would find unacceptable and unreasonable.
>
> *(DoE NI 2012: para 5.4)*

In determining the existence of a nuisance, a local authority must be of the opinion that, either substantial personal discomfort or a health effect must exist and the following key issues must be evaluated in determining the existence of a statutory nuisance:

a) Impact – this is a measure of the impact of the alleged nuisance on the receptor.
b) Locality – the potential for amenity interference is largely related to the character of the neighbourhood.

c) Time – many nuisances have a significant impact because of the time at which the nuisance occurs and the degree of impact changes depending upon the time of occurrence.
d) Frequency – nuisances that occur frequently or continuously are more likely to be determined to be a nuisance (depending to some degree on the impact).
e) Duration – in general, short-term events would be regarded differently to longer period or continuous impact.
f) Convention – convention is important when determining what a reasonable person would find objectionable.
g) Importance – the importance of an activity in respect of the community is often a key consideration.
h) Avoidability – even though an activity may have social importance there should be a balance as to whether reasonable steps have been taken to minimise the impact. (DoE NI 2012: para 5.6)

In establishing if a statutory nuisance is present, environmental health officers will assess the issues germane to the complaint against the above criteria. If conditions are such that a district council is satisfied that a statutory nuisance exists, or is likely to occur or recur, the council shall serve a notice under section 65 of the Clean Neighbourhoods and Environment Act (Northern Ireland) 2011. The notice is served upon the person responsible for the nuisance, requiring the abatement of the nuisance or prohibiting or restricting its occurrence or recurrence and necessitating the execution of such works as may be necessary for any of those purposes. The Act also provides a 'best practical means' defence to the offence in certain instances, where the person responsible for the nuisance is able to prove that the best practicable means have been used to prevent or to counteract the effects of the nuisance. This defence is principally applied in relation to commercial premises. In considering planning proposals, environmental health will assess proposals against their potential to cause or be subject to nuisance.

The existence of planning permission for facilities or activities on a site is also of relevance to the determination of nuisance complaints. In the case of *Coventry and others (Respondents) v Lawrence and another (Appellants)* (UKSC 2014), the Supreme Court confirmed that the grant of planning permission for a particular use is potentially relevant to a nuisance claim in two ways:

> First, the grant … of a planning permission may permit the noise which is alleged by the claimant to constitute a nuisance. In such a case, the question is the extent to which the planning permission can be relied on as a defence to the nuisance claim. Lord Carnwath indicated that in exceptional cases a planning permission may be the result of a considered policy decision leading to a fundamental change in the pattern of uses, which cannot sensibly be ignored in assessing the character of the area against which the acceptability of the defendant's activity is to be judged … Secondly, the grant … of a planning permission may permit the defendant's property or another property in the locality to be used for a certain purpose, so that the question is how far that planning permission can be relied on by the defendant as changing the character of the locality. The Supreme Court indicated that it is wrong in principle that, through the grant of planning permission, a planning authority should be able to deprive a property owner of a right to object to what would otherwise be a nuisance, without providing compensation … Furthermore, the Supreme Court said, the mere fact that the activity which

> is said to give rise to the nuisance has the benefit of a planning permission is normally of no assistance to the defendant in a claim brought by a neighbour who contends that the activity causes a nuisance to her land in the form of noise or other loss of amenity.
>
> Nevertheless, there will be occasions when the terms of a planning permission could be of some relevance in a nuisance case: the fact that the planning authority takes the view that noisy activity is acceptable after a certain time in a particular locality may be a real value as a starting point in a case where the claimant contends that the activity gives rise to a nuisance if it starts at a different time. In summary, the existence and terms of a planning permission are not irrelevant as a matter of law.
>
> *(Freshfields Bruckhaus Deringer LLP 2014: 2–3)*

The planning and environmental protection regimes

It is clear that both spatial planning and environmental health deal with many of the same issues that impact on amenity but the questions may be asked as to what distinguishes the approach and scope of the powers of both regimes, and why the discharging of their roles is essential for environmental protection, the safeguarding of amenity and the prevention of nuisance?

Northern Ireland *Planning Policy Statement 11: Planning and Waste Management* sets out the relationship between the planning and environmental protection regimes as follows:

> The planning and pollution control systems are separate but complementary systems of control and regulation designed to protect the environment from harm as a result of development and related operations. Planning control focuses primarily on whether the development itself is an acceptable use of the land, rather than on the control of the processes or substances involved and on regulating the location of the development in order to avoid or minimise adverse effects on people, the use of land and the environment.
>
> *(DoE NI 2002: 15)*

Guidance in Britain adds to this:

> Pollution control is concerned with preventing pollution through the use of measures to prohibit or limit the release of substances to the environment to the lowest practicable level. It also ensures that ambient air and water quality meet standards that guard against impacts to the environment and human health.
>
> *(ODPM 2004: 4)*

It was the view of the Department of the Environment of Northern Ireland (DoE NI) that:

> planning controls should not duplicate other statutory controls or be used to achieve objectives relating to other legislation. Planning must make its planning decisions on the basis that the pollution control regimes will be properly applied and enforced. The relevant expertise and statutory responsibility for pollution control rests with the relevant pollution control authorities.
>
> *(DoE NI 2002: 15)*

The *National Planning Policy Framework* for England has concurred with this opinion and has contextualised it by embracing the aim of seeking to deliver sustainable development:

> the planning system has a key role to play in protecting and improving the environment. Local Plans can direct potentially polluting developments to appropriate locations and facilitate the redevelopment of polluted sites. Development management decisions can prevent or mitigate potentially polluting developments, protecting the environment and human health and wellbeing. Land use decisions should complement pollution control decisions to help deliver sustainable development.
>
> (*Environment Agency 2012: 2*)

However, the separation between each regime can be indistinct. While planning should not duplicate pollution control provisions, it is also the case that matters relevant to environmental protection may also be material considerations in the planning process. It should be borne in mind that the extent of coverage of issues that the pollution and planning regimes have may differ in relation to specific planning proposals. In the case of a waste incinerator, for example, while its impact on air quality is regulated under the Pollution Prevention and Control (Industrial Emissions) Regulations (Northern Ireland) 2013, pollution from the vehicles delivering waste to the plant would not be covered by this legislation. However, the vehicle movements together with their air quality and noise impacts would be, together with emissions from the incinerator, material planning considerations as set out under *Planning Policy Statement 11: Planning and Waste Management*.

Planning guidance in some instances is very specific in addressing the types of amenity/nuisance issues that can be controlled via planning conditions. *Development Control Advice Note 4 (DCAN 4) Restaurants, Cafés and Fast Food Outlets*, for example, highlights planning conditions that can be attached to approvals in order to minimise noise disturbance. These include restrictions on hours of opening and Sunday opening, sound-proofing premises and the prevention or restriction of live or amplified music (The Planning Service 2002: 15). DCAN 4 also indicates that planning conditions can be used to prevent odour nuisance by requiring 'approval of a scheme for the extraction of cooking odours'.

The *Creating Places, Achieving Quality in Residential Developments* planning guidance also addresses nuisance issues such as noise and dust and states that 'consideration will need to be given to the amelioration of noise and other nuisances, either by means such as mounding and buffer planting or to their prevention at source, through discussion and agreement with Environmental Health Officers' (DoE NI and DRD NI 2000: 24). The *Creating Places* guidance emphasises the importance of design considerations in mitigating potential noise by suggesting that bedrooms and living rooms should be located away from facades that overlook noise sources such as roads (DoE NI and DRD NI 2000: 66).

It is the case, therefore, that both the planning and environmental protection regimes share responsibilities for regulating many of the same issues. However, an important factor in the operation of the planning and pollution control regimes is that planning controls are anticipatory in effect and can be used to prevent pollution and disturbance arising, whereas statutory nuisance controls have, until recently, been retrospective and based upon experience of persistent specific problems having arisen. Anticipatory 'likely to occur' provisions were introduced under Section 65 of the Clean Neighbourhood and Environment Act (Northern Ireland) 2011 but the implications of this change remain to be fully explored by regulators in relation to planning proposals. It may be the case that, given time, a 'doctrine of anticipatory nuisance' similar to one that exists in the United States, whereby a defendant is restrained from undertaking an activity, where it is highly probable that it will give rise to a nuisance, will emerge in Northern Ireland with the potential associated implications for the planning process. In summary, in examining the two regimes, the role of spatial planning is to determine the acceptability of specific land uses in particular locations, while the environmental protection regime regulates the facilities and operations to ensure compliance with nuisance provisions and/or environmental standards and objectives.

The role of environmental health in the planning system

The environmental health departments of district councils are statutory consultees in the development management process and may also have a role in both local plan-making and planning enforcement. Environmental health departments are consulted on the majority of planning applications lodged within their local authority area. Advice issued by the Chief Environmental Health Officers Group (CEHOG 2007) lists those matters that should be explicitly addressed by environmental health during the planning consultation process. This is to ensure that key aspects of a development proposal are consistently considered and that a record of the decision-making process underlying the recommendation to planning is retained. The advice requires that the following matters be considered during the development management consultation process:

a) Noise including vibration – does the proposal have the potential to adversely impact on amenity due to noise or is the proposal likely to be adversely affected by noise?
b) Air Pollution – does the proposal have the potential to adversely impact on amenity due to air pollution, odour or dust or is the proposal likely to be adversely affected by air pollution?
c) Ambient Air Quality – does the proposal have the potential to have an adverse impact on ambient air quality or is the proposal likely to be adversely affected by ambient air quality?
d) Land Quality – is the site potentially contaminated either as a result of
    - the previous use of the site
    - the use, or previous use, of adjacent land where there may be the possibility of migration of contaminants on to the application site or
    - naturally occurring levels of contamination?
e) General Amenity – does the proposal have the potential to adversely impact on amenity due to vermin, insect or litter problems or is the proposal likely to be adversely affected by vermin, insects or litter, etc.?
f) Other Considerations – does the proposal have the potential to adversely impact on local amenity due to other environmental health considerations? (CEHOG 2007)

Environmental health is also required on occasion to examine environmental statements submitted in conjunction with planning applications, to consider third-party objections associated with proposals and to advise if other agencies, such as the Northern Ireland Environment Agency (NIEA), or the Health and Safety Executive, should be consulted on proposals.

Having considered the proposal within the context of the above issues, environmental health has to make a decision on a planning application on the following basis. In broad terms, the response to the consultation may be no objection to the proposal or that further information may be required from the applicant in order to arrive at a recommendation. There might also be no objection to the proposal with an informative attached or no objection subject to planning conditions. There is also the option of objecting to proposals on environmental health grounds. In some cases, depending upon the remit of the Environmental Health Department, comments in the form of an informative may also touch upon issues such as the presence of a right of way, open space provision within residential areas, the existence of Tree Preservation Orders and wider sustainability issues, as well as core food safety and health and safety issues. In arriving at one of the above determinations, each of the issues under consideration will be assessed in the light of relevant guidance and legislative requirements, the underlying purpose of which is environmental protection in relation to site-specific issues, air quality, noise, etc. together with ensuring public health and safety, the identification of hazards and the minimising and/or management of risks associated with the proposal.

Environmental issues

Contaminated land

Contaminated land is defined in the Waste and Contaminated Land Order (Northern Ireland) 1997 Article 49 as:

> land which appears to a district council … to be in such a condition, by reason of substances in, on or under the land, that (a) significant harm is being caused or there is a significant possibility of such harm being caused; or (b) pollution of waterways or underground strata is being, or is likely to be, caused.

Part of this definition of contaminated land was changed by the Waste and Contaminated Land (Amendment) Act (Northern Ireland) 2011 to read: '(b) significant pollution of waterways or underground strata is being caused or there is a significant possibility of such pollution being caused'. The issue of contaminated land is of particular relevance to planning as it is commonly encountered on brownfield sites. Brownfield sites are those that have been previously developed, 'the term may encompass vacant or derelict lands, infill sites, land occupied by redundant or underused buildings, a piece of industrial or commercial property that is abandoned or underused and often environmentally contaminated' (DRD NI 2010: 106). The *Northern Ireland Regional Development Strategy 2035* has a regional target of 60 percent of new housing to be located in appropriate 'brownfield' sites within the urban footprint of settlements greater than 5,000 population (DRD 2010: 40). As a result of this policy of urban containment, planning proposals for the redevelopment of areas potentially affected by contamination are common.

Contaminants are defined as substance or substances that are in, on or under the land and which have the potential to cause significant harm to a receptor (people, flora, fauna, ecosystems) or to cause significant pollution of controlled waters. The purpose of land quality assessment and remediation is to identify and break the 'pathways' (routes via which the contaminant may come into contact) with possible receptors, the pathways including ingestion, inhalation and dermal contact. Contaminants can be solid, liquid or gaseous in form, they can be widely dispersed across sites or concentrated in hot spots, visible on surface layers or buried deep within made ground.

In Great Britain local authorities are required under Part 2A of the Environmental Protection Act 1990 Section 78B to 'cause its area to be inspected from time to time for the purpose of identifying contaminated land'. While a similar provision exists in Northern Ireland under Article 50 of the Waste and Contaminated Land Order 1997, Part 3 of the Order that contains these provisions, has yet to be enabled. There is also extensive planning guidance in the *National Planning Policy Framework* on dealing with contaminated sites in England, but the *Strategic Planning Policy Statement for Northern Ireland* is largely silent on the issue, although it does state that some previously developed sites may be subject to land contamination, and that such areas should be identified in consultation with relevant bodies and where investment in site assembly, remediation, infrastructure and environmental improvement will enable successful economic development in the longer term (DoE NI 2015b: 59). The obligation to carry out remediation schemes are also incorporated in a few of the site-specific requirements of some local development plans. In the Belfast Metropolitan Area Plan, for example, in relation to the Sirroco Works site, it is stated that: 'a remediation scheme for the decontamination of the site shall be submitted to and approved in writing by the Department in consultation with the Health and Environmental Services Department' (DoE NI 2015a: 41). The current legislative and policy situation in Northern Ireland means that sites affected by contamination are being dealt with only as they are brought forward through planning applications. The onus, therefore, rests with both applicants and regulators to consider the issue of land quality and its implications in the early stages of the planning process. It is inappropriate to leave land quality issues to be considered post-planning approval via the use of planning conditions, as information on land quality is required during the determination process to establish whether or not the site can be feasibly remediated to a standard suitable for the proposed end use.

In assessing land quality issues on planning proposals environmental health will consider the current and previous use of the site and land adjacent to the application site. This is to ascertain if the site's land use history may have given rise to potential contamination. The site history can be determined from a variety of sources, such as historical maps, aerial photographs, local knowledge, GIS databases and NIEA's land quality database. Land uses that might indicate potential land quality issues are many and diverse, but common problematic uses include former petrol stations, factories, gas works, engineering and chemical works and garages. The DOE GB has produced industry profile guidance booklets that provide information on processes, materials and wastes associated with specific industries for contaminated land risk assessment and these are available via the Department of Agriculture, Environment and Rural Affairs (DAERA) NI website. In addition, the Department for Environment, Food & Rural Affairs (DEFRA) and the Environment Agency have produced guidance titled *Model Procedures for the Management of Contaminated Land 2014* that sets out the requirements for dealing with contaminated sites.

In Northern Ireland during the development management consultation process, if a site is considered to be potentially contaminated, environmental health will request, via the planning department, that the applicant or their consultant produce a preliminary risk assessment for the site. The purpose of this is 'to develop an initial conceptual model of the site and establish whether, or not, there are potentially unacceptable risks' (Environment Agency 2014: 16). This is usually achieved by conducting a desk-top study of the site together with a site visit. If the preliminary assessment indicates that contamination is likely, the guidance sets out the procedures to be followed to bring the site to a condition suitable for the proposed end use. This includes producing a conceptual model of the site, carrying out intrusive investigations, determining the type and extent of contamination, developing a remediation strategy, its implementation, monitoring and verification. The implementation, monitoring and verification of a remediation strategy are routinely specified as planning conditions in planning approvals. The investigation and remediation process can take months or longer and may involve considerable cost, but it should be borne in mind that if land quality issues are not identified and addressed at this early stage, post-development site investigations and remediation are extremely expensive and time consuming. In one instance a former brownfield site in Northern Ireland, which had post-development site investigations and remediation carried out, accrued costs that were reported to have been in the region of £3,000,000 for the site investigation alone.

Waste management

The disposal of waste to landfill is a potential source of pollutants which can impact on land, air and water quality and produce greenhouse gases. Currently, the United Kingdom is required by the EU Landfill Directive, as transposed into UK law by The Waste (Circular Economy) (Amendment) Regulations 2020, 'to prevent, or reduce as far as possible, negative effects on the environment, in particular on surface water, ground water, soil, air, and on human health from the landfilling of waste by introducing stringent technical requirements for waste and landfills' (European Commission 2016). The focus of waste management policy has been the removal of material from the waste stream and the minimisation of the amount of waste going to landfill. The legal obligation to conform to the waste hierarchy was introduced into Northern Ireland by The Waste Regulations (Northern Ireland) 2011, which, under Regulation 17, place a duty on any person producing, carrying, collecting etc. waste to 'apply the waste hierarchy priority'. The waste hierarchy prioritises five possible ways of dealing with waste to minimise its environmental effects and optimise resource efficiency. These are, first, waste prevention; then in order of preference: preparing for reuse, recycling, other recovery (for example, energy recovery) and finally, disposal, which is the least favourable option in the hierarchy.

In Northern Ireland, substantial progress has been made with regard to reducing the amount of waste being landfilled. In 1995 some 1,225,000 tonnes of municipal waste were landfilled across the region, while in 2019–20 Northern Ireland's local authorities collected 998,985 tonnes of waste with only 240,220 tonnes (24 percent) being sent to landfill (DoE NI 2015c; NISRA 2020). Despite this improvement, significant quantities of waste are sent to landfill each year and facilities to accommodate this disposal are a necessity. Councils are obliged to assess the likely extent of waste management facilities to be required within their districts and to identify sites within the provisions of local development plans, including their key site requirements. Areas regarded as suitable candidates for waste management facilities include industrial or port areas, quarries, derelict or contaminated land. In the case of civic amenity facilities, these should be conveniently situated in terms of neighbourhood proximity while avoiding unacceptable adverse impact on the character, environmental quality and amenities of the local area.

Landfill sites are subject to regulation under both the pollution control and planning regimes. The planning system interfaces with waste management issues on a number of levels.

First, at a strategic level, the *Northern Ireland Regional Development Strategy 2035* in its long-term policy direction highlights the importance of managing waste sustainably. The strategy sets out that this is to be achieved by applying a waste hierarchy. This:

> aims to encourage the management of waste materials in order to reduce the amount of waste materials produced, and to recover maximum value from the wastes that are produced. It encourages the prevention of waste, followed by the reuse and refurbishment of goods, then value recovery through recycling and composting.
>
> *(DRD NI 2010: 47)*

In considering specific facilities, landfill sites and waste transfer and management facilities require planning approval and these facilities must be designed, from the planning perspective, so as to 'not cause demonstrable harm to human health or result in an unacceptable adverse impact on the environment and to be compatible with the character of the surrounding area and adjacent land uses' (DoE NI 2002: 19). Finally, at a household or commercial/industrial premises level, the layout of these buildings must include areas with adequate storage for waste and waste receptacles, which are suitably designed so as not to cause loss of amenity or nuisance to neighbouring residents.

The *Strategic Planning Policy Statement for Northern Ireland* (DoE NI 2015b: 111) outlines the regional strategic objectives for waste management as being to:

> promote development of waste management and recycling facilities in appropriate locations; to ensure that detrimental effects on people, the environment, and local amenity associated with waste management facilities (for example, pollution) are avoided or minimised and to secure the appropriate restoration of proposed waste management sites for agreed after-uses.

Planning Policy Statement 11: Planning and Waste Management, Policy WM 1, requires that a proposed waste management facility will not cause demonstrable harm to human health or result in an unacceptable adverse impact on the environment; that it be designed to be compatible with the character of the surrounding area and adjacent land uses; be acceptable in the landscape and that the development will not have an unacceptable visual impact on any area designated for its landscape quality. In addition, the access to the site and the nature and frequency of associated traffic movements must not prejudice the safety and convenience of road users or constitute a loss of amenity to neighbouring residents by virtue of noise, dirt and dust. The public road network must also be able to satisfactorily accommodate, or be capable of upgrading, to accommodate the traffic generated. Adequate arrangements must be provided within the site for the parking, servicing and circulation of vehicles. The development must also not have an unacceptable adverse impact on nature conservation or archaeological/built heritage interests. The types of waste to be deposited or treated and the proposed method of disposal or treatment must not pose a serious environmental risk to air, water or soil resources that cannot be prevented or appropriately controlled by mitigating measures. The proposal should avoid (as far as is practicable) the permanent loss of the best and most versatile agricultural land and, in the case of landfilling, the proposal must include suitable, detailed and practical restoration and after-care proposals for the site (DoE NI 2002: 19).

Landfill sites are also required to be licenced under the provisions of Article 4 of the Waste and Contaminated Land (Northern Ireland) Order 1997. In addition to this, the Landfill Regulations (Northern Ireland) 2003 set out a pollution control regime for landfills for the purpose of implementing the Landfill Directive 99/31/EC in Northern Ireland. The Landfill Regulations require that the location of a landfill must take into consideration requirements including the distances from the boundary of the site to residential and recreational areas; the existence of groundwater, coastal water or nature protection zones in the area; the geological or hydro-geological conditions in the area; the risk of flooding, subsidence, landslides or avalanches on the site; and the protection of the natural or cultural heritage in the area. The Regulations require that appropriate design arrangements be made with regard to the characteristics of the landfill and prevailing meteorological conditions in order to control rainwater entering the landfill body, prevent surface water or groundwater from entering into landfilled waste and to collect and treat contaminated water and leachate to the appropriate standard. Soil, groundwater and surface water are also to be protected by the use of a geological barrier combined with a bottom liner during the operational phase of the landfill and a top liner following closure and during the after-care phase. Measures must also be taken in order to control the accumulation and migration of landfill gas and to minimise the nuisances arising from the landfill in relation to emissions of odours and dust; wind-blown materials; noise and traffic; birds, vermin and insects; and fires. Additional controls are also contained within the Landfill (Amendment) Regulations (Northern Ireland) 2004, as amended by the Landfill (Amendment No2) Regulations (Northern Ireland) 2007, which require that waste can be accepted only at landfill sites where its acceptance would not 'result in unacceptable emissions to groundwater, surface water or the surrounding environment; jeopardise environment protection systems (such as liners, leachate and gas collection and treatment systems) at the landfill; put at risk waste stabilisation processes within the landfill or endanger human health'.

The NIEA is the regulating body for waste licencing and much of the enforcement activity surrounding waste management. Environmental protection authorities are:

> consulted on planning applications for waste management facilities in order to take account of the scope and requirements of the relevant pollution controls. In assessing such proposals, the weight attached to pollution issues associated with waste management will be reduced to the extent that they are capable of being addressed by the pollution control authority in carrying out its statutory responsibilities and will depend on the circumstances of each case.
>
> *(DoE NI 2002: 17)*

It is recommended that developers, as part of their pre-application discussions for planning permission, raise issues such as integrated pollution prevention and control permits, licences, consents to discharge and authorisations for proposed waste management facilities with all the relevant authorities prior to submission of proposals. In relation to the illegal disposal of waste, NIEA is empowered to take enforcement action in relation to the operation of unlicensed waste disposal sites under Article 4 of the Waste and Contaminated Land (Northern Ireland) Order 1997 and both NIEA and local authorities may take action under Article 28 in relation to the illegal depositing of waste.

Regarding waste storage facilities at residential and commercial premises, the increasing amounts of waste being recycled and the shift to fortnightly collection schedules have led to an increase in the number of waste receptacles being used by households. These may now include grey, brown and blue 240 litre capacity bins and/or red and black Kerbie recycling boxes. The issue of bin storage is addressed in the guidance *Avoiding Rubbish Design, Providing for Bin Storage on New Housing Developments* (NHBC 2015), which recommends that 'bins are kept in purpose-built stores' designed to minimise issues such as visual impact, odour, noise, fires and vermin. The design of the bin stores should also allow adequate space for the number and size of bins and other containers, be low maintenance, convenient to use and minimise the impact on the use of pavements and streets for bin storage. The issue of bin storage should be addressed in the assessment of planning proposals as a matter of routine (ibid.: 7).

Air quality

Poor air quality remains an issue in many parts of the UK including Northern Ireland. As recently as 2015, DEFRA stated:

> Medical evidence shows that many thousands of people still die prematurely every year because of the effects of air pollution. Air pollution from man-made particles is currently estimated to reduce average UK life expectancy by six months, air pollution contributes to thousands of hospital admissions per year … and exposure to nitrogen dioxide is increasing mortality by the equivalent of 23,500 deaths per year.
>
> *(DEFRA 2015: 4)*

Air pollution giving rise to ill health and premature death is, however, not a recent issue, as previous incidents such as the London smog in 1952 caused thousands of deaths and prompted the enactment of the Clean Air Acts 1956 and 1968. These introduced powers for local authorities to declare smoke control areas, control emissions of grit and dust from furnaces, prohibit the emission of dark smoke from chimneys and required the approval of chimney heights. This legislation was replicated in Northern Ireland via the Clean Air (Northern Ireland) Order 1981. While these pieces of legislation are located within the environmental protection regime, planning also has a role in relation to air quality. DoE NI has stated:

> The planning system can positively contribute to the improvement of air quality and in minimising its harmful impacts on health and well-being. In exercising their planning functions, planning authorities should consider the location of development which may give rise to air pollution and ensure that other developments are, as far as practicable, not adversely affected by major existing, or potential future, sources of air pollution.
>
> *(DoE NI 2015b: 18)*

The system of air quality management and assessment is based on the EU Air Quality Directives and on the UK Air Quality Strategy which set air quality standards and objectives. Air quality standards are 'the concentrations of pollutants in the atmosphere which can broadly be taken to achieve a certain level of environmental quality' (DEFRA 2007: 13). These are based on the effects of pollutants on health including the effects on sensitive subgroups and ecosystems. The air quality objectives 'are policy targets often expressed as a maximum ambient concentration not to be exceeded, either without exception or with a permitted number of exceedences, within a specified timescale' (DEFRA 2007: 13). The Air

Quality Standards Regulations (Northern Ireland) 2010 and the Environment (Northern Ireland) Order 2002 set out the responsibilities in relation to air quality management for local authorities within Northern Ireland.

The Air Quality Standards Regulations (Northern Ireland) 2010 require under Regulation 11 that 'district councils shall cause a review to be conducted of the quality for the time being, and the likely future quality of air within the council's district'. If this review indicates that the air quality standards or objectives are not going to be achieved then, under Regulation 12, the council shall 'designate as an Air Quality Management Area (AQMA) all or any part of its district in which it appears that those standards or objectives are not being achieved, or are not likely to be achieved within the relevant period'. The air quality review process involves assessing the ambient levels of the following pollutants against the air quality standards and objectives: sulphur dioxide (SO_2), nitrogen dioxide (NO_2), particulates (PM_{10} and $PM_{2.5}$), lead, benzene, carbon monoxide, benzo(a)pyrene, ozone polycyclic aromatic hydro carbons and some metals. The review and assessment process involves monitoring and modelling the level of these pollutants produced by local emission sources. In locations where the air quality objectives are not met, local authorities are required to declare an AQMA and to then produce an air quality action plan setting out how the standards will be met within a specified period. In 2021 there are 19 active AQMAs across nine district councils in Northern Ireland (see https://www.airqualityni.co.uk).

Spatial planning has an important role to play in addressing local air quality management issues:

> At the strategic level, planning can provide for more sustainable transport links between the home, workplace, educational, retail and leisure facilities, and identify appropriate locations for potentially polluting industrial development. The pattern of land use determines the need for travel, which is in turn a major influence on transport related emissions.
>
> *(IAQM 2015: 6)*

This is of particular importance in Northern Ireland because, while significant progress has been made in tackling SO_2 from domestic and industrial sources, the same cannot be said for NO_2 from road transport sources; hence, the number of AQMAs still in force regarding transport-related sources. When making development management decisions, planning authorities need to consider if the proposal falls within an AQMA. New development proposals can impact on AQMAs either by introducing new residents to an area, where the air quality already fails to meet the air quality standards, or by allowing additional emission sources into an area, thereby increasing the level of pollutants in locations with already poor air quality. Development proposals should also be compatible with the provisions of air quality action plans in order that improvements to air quality can be achieved. Design considerations can also help improve air quality, for instance, specifying the type of fuel to be used for heating in residential applications (gas heating has considerably fewer emissions than solid fuel) or ensuring that 'new developments should not create a 'street canyon' or a building configuration that inhibits effective pollution dispersion; they should also be designed to minimise public exposure to pollution sources, for example, by locating habitable rooms away from busy roads' (IAQM 2015: 15).

In addition, both planning design criteria and building regulations that promote higher energy efficiency in buildings can reduce energy consumption and the burning of fossil fuels and, therefore, lead to lower emissions of pollutants into the atmosphere. Energy efficiency measures also have the added benefit of cutting greenhouse gas emissions, as well as impacting positively on levels of fuel poverty.

Water quality

Society and the economy are dependent on the provision of potable, reliable water supplies and adequate wastewater treatment. While waterborne diseases, such as cholera and dysentery, the control of which prompted the introduction of the Public Health Acts in the 19th century, no longer present the same hazard in the developed world, issues such as eutrophication from nitrate and phosphate pollution remain a threat to water quality. In addition, access to drinking water supplies and adequate wastewater treatment facilities are material considerations in the development management process impacting on the acceptability of planning proposals and in the assessment of housing fitness. The *Regional Development Strategy for Northern Ireland 2035* in policy RG12 seeks to promote a more sustainable approach to water and sewerage service provisions. Policy RG12, requires that 'Land-use planning should be informed by current water and sewerage infrastructure and future investment programmes. This will involve close cooperation between planning authorities and the water industry in the preparation of local development plans and long-term water strategies'. (DRD NI 2010: 51)

The regulation of surface water quality is set out in The Water Environment (Water Framework Directive) Regulations (Northern Ireland) 2017. The provisions of these regulations prevent the deterioration of surface or groundwater water bodies, support the achievement of environmental objectives (such as good ecological and chemical status in waterbodies) and reduce pollution from a specified list of persistent, bio-accumulative and toxic priority hazardous substances, such as benzene, cadmium and lead. Regulation 3 also requires the designation of River Basin Districts, based on geographical water catchment areas, as the basis on which water quality is managed, via the production of River Basin Management Plans. Resulting from this, Northern Ireland has three designated River Basin Districts: the Neagh Bann, North Western, and North Eastern.

Wastewater treatment provision is regulated under The Urban Waste Water Treatment Regulations (Northern Ireland) 2007. Regulation 7 requires that urban settlements with population equivalents of between 2,000 and 10,000 that discharge into freshwaters and estuaries, together with all discharges from agglomerations with a population equivalent of more than 10,000, must have secondary treatment of sewage. In 2006 the environmental pressure group, Friends of the Earth, brought a successful legal case to the European Court of Justice against the Department of the Environment for Northern Ireland regarding inadequate treatment provision at nine wastewater plants. This resulted in the Department, and subsequently NI Water, not accepting any new connections to its wastewater network at specified locations, where sewage treatment was at capacity. While this situation arose in 2006, it remained an issue for seventy sewage treatment plants in 2018, with estimates that it could take a decade to bring water treatment infrastructure up to the required standard. In 2021 NI Water lists at least sixty wastewater treatment plants with 'network capacity issues'. This has directly impacted upon the development management process in relation to residential proposals as commented on by the *Northern Ireland Regional Development Strategy 2035*: 'it will also be important to phase and manage housing growth taking account of capacity (including water supply and waste-water treatment infrastructure) or if it can be provided ahead of development without environmental harm' (DRD NI 2010: 74). The availability of adequate wastewater treatment infrastructure, therefore, has been and remains a constraint on the approval of residential planning proposals in parts of the region and is a material consideration in emerging Local Development Plans. NI Water in relation to wastewater treatment capacity issues advises that developers use the Pre-Development Enquiry process to ascertain the current availability of treatment and that in those instances where treatment is at capacity, new planning applications may be declined.

While wastewater treatment at the settlement level remains an issue, it is also the case that on-site wastewater treatment at individual properties can be problematic. A substantial number of dwellings in the rural areas of Northern Ireland are not connected to main sewage treatment works with estimates that approximately 18 percent of the total housing stock is reliant on non-mains sewage treatment. In circumstances where dwellings cannot be connected to the sewerage network and main treatment works, they are reliant on the use of on-site treatment in the form of either small package sewage works or, more commonly, septic tanks. NIEA in 2015 indicated that a total of 113,254 discharge consents for septic tanks had been issued, with an estimate of a further 16,000 unconsented systems also being in use (Northern Ireland Assembly Research and Information Service 2015: 1). Furthermore between 2015 (when the planning management function was in the main returned to local government) and 2019, out of 14,994 new build residential applications, 10,291, or 68.8 percent, were outside development limits in greenfield locations, indicating that the likely number of septic tanks is approximately 140,000 (Haran et al. 2019: 18).

Septic tanks consist of a two-chambered tank buried in the ground adjacent to dwellings with a drainage network or soakaway connected to it. Household wastewater flows into the tank where it undergoes some biological treatment. The effluent then drains into a soakaway to be absorbed into the ground while the 'solids' fall to the bottom of the tank and have to be emptied, usually annually, and the sewage sludge disposed of. The effluent treatment that takes place in a septic tank is not to the same standard as takes place at main wastewater treatment works. There is a legal requirement under Article 9 of The Water (Northern Ireland) Order 1999 that septic tanks must obtain a 'consent to discharge' from NIEA. While a properly situated, constructed and maintained septic tank poses a low risk of causing water pollution, the high number of septic tanks in Northern Ireland, combined with poor maintenance, increases the likelihood of pollution incidents. The main pollutants of concern from septic tanks discharges are phosphates (derived from detergents), nitrates from sewage together with bacterial and viral pathogens and parasites. Part N of The Building Regulations (Northern Ireland) 2012 specifies the construction details for septic tanks and BS6297:2007 code of practice sets out the parameters for tank location and drainage field (soakaway) design within proposal sites. These parameters include matters such as tanks being sited at least 7m from dwellings, 50m from a well or spring and 2m from site boundaries. The strategic context for the impact of development relying on non-mains sewage treatment is addressed in *Planning Policy Statement 21 (PPS 21: Sustainable Development in the Countryside)* which acknowledges that:

> it will not be feasible for many buildings in the countryside to connect to a public water borne sewerage system and they will rely instead on some means of 'on-site sewage treatment', such as a septic tank or package treatment plant.
>
> *(DoENI 2010: 40)*

Policy CTY 16 of PPS 21 sets out the constraints on proposals relying on non-mains sewerage by stating:

> planning permission will only be granted for development relying on non-mains sewerage, where the applicant can demonstrate that this will not create or add to a pollution problem. Applicants will be required to submit sufficient information on the means of sewerage to allow a proper assessment of such proposals to be made. In those areas identified as having a pollution risk, development relying on non-mains sewerage will only be permitted in exceptional circumstances.

PPS21 goes on to state that:

> Planning permission will be refused for development relying on non-mains sewerage where the physical arrangements proposed for on-site sewage treatment are unsatisfactory or in cases where 'Consent to Discharge' under the Water Order is unlikely to be forthcoming due to pollution risks. The Northern Ireland Environment Agency may identify certain areas where a pollution risk exists sufficient to warrant no further development relying on non-mains sewerage arrangements. In appropriate circumstances these will be highlighted in the relevant development plan for the area.
>
> *(DoENI 2010: 41)*

The relatively large proportion of properties served by septic tanks in Northern Ireland, resulting from our dispersed settlement pattern, can impact water quality and must call into question the sustainability of the high rate of planning approvals in rural areas.

Noise

Noise, or unwanted sound, is a reality of life, especially within urban areas. While exposure to noise is inescapable and usually has little impact on quality of life, in certain instances it can be so intrusive as to cause significant adverse effects on health. DEFRA, in discussing the impact of noise on health, stated:

> it is recognised that noise exposure can cause annoyance and sleep disturbance both of which impact on quality of life. It is also agreed by many experts that annoyance and sleep disturbance can give rise to adverse health effects and there is emerging evidence that long term exposure to some types of transport noise can additionally cause an increased risk of direct health effects.
>
> *(DEFRA 2010: 7)*

Noise can be categorised as 'environmental noise – noise from transportation and industrial sources; neighbour noise – noise from inside and outside people's homes; and neighbourhood noise – noise arising from within the community such as from entertainment premises, trade and business premises, construction noise and noise in the street' (DoE NI 2014a: 3). Noise is also a material planning consideration and 'the planning system has a role to play in minimising the potential for adverse impact upon health and well-being through noise, by means of its influence on the location, layout and design of new development and consideration of the amenity impacts' (DoE NI 2015b: 117).

The planning system can reduce the potential for detrimental noise impact by considering within local development plans, the zoning of land for uses likely to generate significant levels of noise together with land zonings for noise-sensitive uses, such as housing. Where 'the potential for adverse noise impact exists, the Local Development Plan should seek to mitigate this through the application of appropriate key site requirements to new zonings' (DoE NI 2015b: 117). Additionally, the development management process has a role in mitigating potential noise impacts:

> Noise is a material consideration in the determination of planning applications for proposals likely to give rise to noise that would impact adversely upon residential / public amenity and/or the character of the locality. In cases where noise is identified as a significant material consideration there should be consultation with the relevant Local Authority Environmental Health Department.
>
> *(DoE NI 2014a: 7)*

The development management consultation process is likely to generate a request from environmental health that a noise impact assessment is carried out for the proposal. This will assess the current noise background level for the neighbourhood and then determine the noise impact of the proposal on the area post-development and commissioning. Noise impact assessments should be carried out taking cognisance of the following:

- for industrial noise, British Standard 4142 *Methods for Rating and Assessing Industrial and Commercial Sound*;
- for transport impacts, the World Health Organization threshold of noise-induced awakenings of 35 dB (A);
- for noise from wind turbines, ETSU 97 *The Assessment and Rating of Wind Turbine Noise*.

If the assessments conclude that noise exposure is likely to give rise to a nuisance, a range of physical measures can be implemented or installed to attenuate the noise. These include increasing the physical distance between the noise source and receptors, as doubling the distance will half the intensity of the noise. In addition, measures such as the use of acoustic double glazing and ventilation; noise barriers, such as walls; the redesign of dwellings so that bedrooms face away from noise sources; and anti-vibration trenches can all reduce the impact of noise.

Noise from premises or land that gives rise to complaints is investigated by environmental health departments, under the Clean Neighbourhoods and Environment Act (Northern Ireland) 2011. Section 63 (i) identifies 'noise emitted from premises so as to be prejudicial to health or a nuisance'. But 'there is no fixed level of noise which constitutes a statutory nuisance, each case is judged on its merits' (DoE NI 2014: 3). In determining if a noise amounts to a statutory nuisance, Environmental Health will consider the reaction of the average person to the nuisance, taking account of its volume, when and how often the noise occurs and the duration of the occurrence. 'Where a nuisance exists, the district council may serve an Abatement Notice requiring the noise to be stopped altogether or limited to certain times of the day. A notice can be served on the person responsible for the noise or the owner or occupier of the premises' (DoE NI 2014b: 3).

While nuisance legislation deals with noise from specific sources, the Environmental Noise Directive 2002/49/EC is the main legislative instrument to identify noise pollution levels and trigger the necessary action across the EU. The Environmental Noise Directive was transposed into Northern Ireland legislation by the Environmental Noise Regulations (Northern Ireland) 2006, which focus on three action areas: the assessment of exposure to environmental noise; the provision of information on environmental noise and its effects on the public; and preventing and reducing environmental noise and preserving environmental noise quality where it is good. This is being achieved by a phased exercise in noise mapping that now requires strategic noise maps to be produced for areas with roads with more than three million vehicle passages annually, railways with more than 30,000 train passages annually, airports with more than 50,000 movements annually and urban areas with more than 100,000 inhabitants. Based on the noise maps, competent authorities must prepare action plans, which set out how the requirements of the Regulations will be delivered. To date, a number of candidate noise management areas (CNMAs) have been identified in Northern Ireland. In relation, for example, to road noise, the Department for Infrastructure and planning authorities will determine if developments are within CNMAs and will assess mitigation measures as part of the design and planning process. There is a recognition that, in order to deliver the aims of action plans, 'noise issues need to be carefully considered through the planning process for specific development proposals and, at a more strategic scale, through the production of Local Development Plans' (DRD NI 2013: 6).

Light pollution

Legislative provisions for the control of light pollution were introduced under the provisions of the Clean Neighbourhoods and Environment Act (Northern Ireland) 2011. This states, under section 63 (h), that 'artificial light emitted from premises so as to be prejudicial to health or a nuisance' constitutes a statutory nuisance. While light pollution can be in the form of 'sky glow' created by light directed upwards from large urban conurbations, it is mainly the issues of light trespass, glare and waste light that will give rise to statutory light nuisance. In these instances, artificial light in the form of light trespass illuminates an unwanted area, such as a security light illuminating a bedroom, or causes glare by the excessive or uncontrolled brightness of lighting, or generates waste light by the illumination being directed upwards causing sky glow, or illuminating areas where lighting is not required. Many types of development, including domestic and commercial security lights along with the exterior lighting of buildings and sports facilities have the potential to generate light pollution. It is therefore 'preferable to address these potential statutory nuisances at the planning stage. The courts have ruled that lighting itself is not "development". However, planning permission is required for lighting if it alters the material appearance of a building and local authorities can decide to regulate lighting under planning permission and set planning obligations for lighting to prevent light pollution' (DEFRA 2006: 10). Examples of lighting issues that may be mitigated by the use of planning conditions include 'limiting the time of use of the lighting, specifying lamps, luminaires and columns, the design, height and position/angle of the lighting, the retention of screening vegetation and the use of planting and bunding to contain lighting effects' (Colchester Borough Council 2012: 20). The *Strategic Planning Policy Statement for Northern Ireland*, in paragraph 4.11 Safeguarding Residential and Work Environs, acknowledges the role of planning in minimising loss of amenity from light pollution by stating: 'the planning system has a role to play in minimising potential adverse impacts, such as noise or light pollution on sensitive receptors by means of its influence on the location, layout and design of new development' (DoE NI 2015: 17). The issue of the potential for planning proposals to generate light nuisance should therefore be taken into consideration during the development management process as it constitutes a material planning consideration and should be addressed along with an assessment of other potential sources of nuisance and loss of amenity.

Pollution prevention and control

The Pollution Prevention and Control (Industrial Emissions) Regulations (Northern Ireland) 2013 (PPC) apply an integrated approach to the regulation of a wide range of industrial and other activities, including operations as diverse as petrol stations, agricultural plants, cement works and dry cleaning. The regulations require that emissions to air, water and land must be considered in an integrated manner and installations that fall under the PPC regime must have a permit to operate. Permits include conditions that are set out with the aim of achieving compliance with the regulations and a high level of protection for the environment. Best Available Techniques (BAT), specified in industry-specific process guidance notes, are implemented to prevent and minimise emissions with the aim of balancing costs to the operator against the benefits to the environment. Part Two of The Pollution Prevention and Control (Industrial Emissions) Regulations (Northern Ireland) 2013 requires the operators of specified industrial and other installations to obtain a permit to operate. An application must be made to the appropriate regulatory body (either NIEA or local authorities dependent upon the scale and type of installation), which then decides, on the basis of compliance with the provisions of the applicable Process Guidance Notes, whether to issue or refuse a permit. If a permit is issued, it will include conditions aimed at minimising and preventing pollution. The aim of the regime is to protect the environment and human health, to deliver permitting and compliance effectively and efficiently, to minimise the administrative burden on both regulators and operators and to encourage the promotion of best practice in the operation of installations. Again, if there is potential for planning proposals to require a PPC permit, discussions should take place with the appropriate regulatory body.

Contextualising and prioritising decision making on environmental planning issues

A public authority in Northern Ireland must, under section 25 of the Northern Ireland (Miscellaneous Provisions) Act 2006, 'in exercising its functions, act in the way it considers best calculated to contribute to the achievement of sustainable development in Northern Ireland, except to the extent that it considers that any such action is not reasonably practicable in all the circumstances of the case'. This broad duty has been translated into a series of principles, objectives and priorities via the Northern Ireland Executive (2011) *Sustainable Development Strategy and Sustainable Development Implementation Plan*. In relation to environmental protection issues, public authorities are required, under Priority Area Four, to:

> strike an appropriate balance between the responsible use and protection of natural resources in support of a better quality of life and a better quality environment and

> ensure an appropriate policy and legislative framework is in place supported by a regulatory regime which will deliver statutory environmental standards in respect of air, water and other environmental pollution.

While this objective is common to regulators within the spatial planning and environmental protection regimes, there is a difference in approach and in the interpretation of this matter by both groups, despite the fact that, in many cases, they are dealing with the same or linked issues. Planning is essentially the political control of what were previously private property rights, by subjecting them to regulation in the public interest. The very nature of the decision-making process is legalistic and takes into consideration a broad range of social, economic and environmental issues, balancing one against the other. In contrast, the environmental protection regime is largely quantitative and technical in nature, while at the same time requiring judgements to be made on the basis of acceptability or reasonableness. It is concerned with achieving standards that mitigate or minimise environmental and health impacts, for example, are specific air quality standards being achieved, is land remediated to a level that complies with soil assessment criteria or is noise attenuated to a specific decibel level?

While spatial planning examines a broad range of material considerations in the public interest, from the environmental protection perspective not all issues pose the same level of risk or hazard and some issues have the potential to cause substantially greater health and environmental impacts than others. The challenge for regulators is to recognise the potential impacts of development and ensure that they are given sufficient weight within the decision-making process. Figure 9.1 outlines an approach whereby the increasing severity of nuisance or environmental impacts are set against the likelihood of their occurrence and their worst-case environmental or health consequences in relation to public health and safety. This approach weighs issues such as aesthetic design considerations or visual intrusion against issues of increasingly severe nuisance and environmental impact in order to highlight their potential consequences during the decision-making process. It is an attempt to emphasise that not all material planning considerations are equal in terms of their potential impacts and, in turn, it links issues of a technical and quantitative type to those of an essentially qualitative nature within the spectrum of the impacts or hazards under consideration. It is envisaged that a precautionary approach would be taken with regard to those high-probability, high-impact issues as a means of discharging broader sustainability and environmental protection responsibilities.

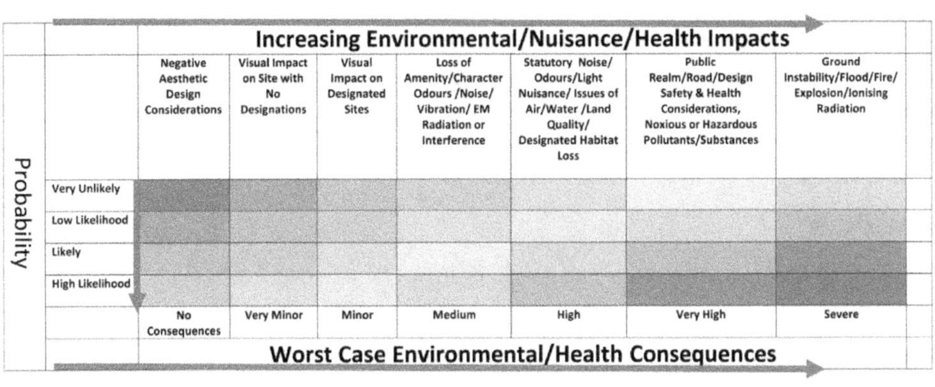

FIGURE 9.1 An approach to environmental assessment

Conclusion

The spatial planning and environmental protection regimes share many common themes. Environmental protection operates to mitigate and minimise a range of emissions and impacts from a diverse selection of commercial, industrial and residential sources. However, no process or activity can be 100 percent efficient and, as a consequence, they generate effects on health and on the built and natural environment. It is the role of spatial planning to ensure that development takes place in appropriate locations and that the potential impacts of development are anticipated and balanced against the public interest of the proposal. When applying for planning permission, it is essential that applicants consider both the environmental impacts of their proposals and the effects of neighbouring premises on the proposed development. It is, therefore, important that these issues be raised during pre-application discussions and, where appropriate, that a dialogue takes place between applicants and the relevant environmental protection agency, be it local authority environmental health departments or the Northern Ireland Environment Agency, at an early stage of the application process. It should be borne in mind that the early identification of issues not only facilitates better environmental, safety and health outcomes but also, in many cases, proves to be the most economical route for applicants.

Legislation

Clean Air (Northern Ireland) Order 1981
Clean Neighbourhoods and Environment Act (Northern Ireland) 2011
Climate Change Act 2008
Environmental Protection Act 1990
Northern Ireland (Miscellaneous Provisions) Act 2006
The Air Quality Standards Regulations (Northern Ireland) 2010
The Environment (Northern Ireland) Order 2002
The Environmental Noise Regulations (Northern Ireland) 2006
The Landfill (Amendment) Regulations (Northern Ireland) 2004
The Landfill (Amendment No.2) Regulations (Northern Ireland) 2007
The Landfill Regulations (Northern Ireland) 2003
The Pollution Prevention and Control (Industrial Emissions) Regulations (Northern Ireland) 2013
The Urban Waste Water Treatment Regulations (Northern Ireland) 2007
The Waste and Contaminated Land (Northern Ireland) Order 1997
The Waste Regulations (Northern Ireland) 2011
The Water Environment (Water Framework Directive) Regulations (Northern Ireland) 2017
The Water (Northern Ireland) Order 1999
Waste and Contaminated Land (Amendment) Act (Northern Ireland) 2011

Case cited

Coventry and others v Lawrence and another (2014 UKSC 46)

References

Barton, H. and M. Grant. 2006. 'A Health Map for the Local Human Habitat', *The Journal of the Royal Society for the Promotion of Health* 126: 252.

CEHOG (Chief Environmental Health Officers Group). 2007. *Planning Application Consultation Check List*. Northern Ireland: Unpublished Guidance.

Colchester Borough Council. 2012. *External Artificial Lighting Planning Guidance Note*. Colchester: Colchester Borough Council.

DEFRA (Department for the Environment, Food and Rural Affairs). 2006. *Statutory Nuisance from Insects and Artificial Light*. London: DEFRA.

DEFRA (Department for the Environment, Food and Rural Affairs). 2007. *The Air Quality Strategy for England, Scotland, Wales and Northern Ireland (Volume 1)*. London: DEFRA.

DEFRA (Department for the Environment, Food and Rural Affairs). 2010. *Noise Policy Statement for England*. London: DEFRA.

DEFRA (Department for the Environment, Food and Rural Affairs). 2015. *Air Pollution in the UK Report 2015*. London: DEFRA.

DoE NI (Department of the Environment of Northern Ireland). 2002. *Planning Policy Statement 11 (PPS 11) Planning and Waste Management*. Belfast: DoE NI.

DoE NI (Department of the Environment of Northern Ireland). 2010. *Planning Policy Statement 21 (PPS 21) Sustainable Development in the Countryside*. Belfast: DoE NI.

DoE NI. 2012. *Statutory Nuisances Guidance to District Councils on Part 7 (Statutory Nuisances) of the Clean Neighbourhoods and Environment Act (Northern Ireland) 2011*. The Clean Neighbourhoods and Environment Team. Belfast: DoE NI.

DoE NI. 2014a. *Noise Policy Statement for Northern Ireland*. Belfast: DoE NI.

DoE NI. 2014b. *Air and Environmental Quality Team, Noise Legislation Applicable in Northern Ireland*. Belfast: DoE NI.

DoE NI. 2015a. *Belfast Metropolitan Area Plan (Volume 2)*. Belfast: DoE NI.

DoE NI. 2015b. *Strategic Planning Policy Statement*. Belfast: DoE NI.

DoE NI. 2015c. *Northern Ireland Local Authority Collected Municipal Waste Management Statistics*. Belfast: DoE NI.

DoE NI and DRD NI (Department for Regional Development Northern Ireland). 2000. *Creating Places Achieving Quality in Residential Developments Incorporating Guidance on Layout and Access*. Belfast: DoE NI and DRD NI.

DRD NI (Department for Regional Development Northern Ireland). 2010. *Regional Development Strategy 2035*. Belfast: DRD NI.

DRD NI (Department for Regional Development Northern Ireland). 2013. *Department for Regional Development Northern Ireland – Roads – Environmental Noise Directive Round Two – Noise Action Plan 2013–2018*. Belfast: DRD NI.

Environment Agency. 2012. *National Planning Policy Framework – Preventing Unacceptable Risks from Pollution*. London: Environment Agency.

Environment Agency. 2014. *Model Procedures for the Management of Land Contamination Contaminated Land Report 11*. Bristol: Environment Agency.

Haran, M., J. Berry, P. Davis, D. Lo, S. MacIntyre and M. McCord. 2019. *Creating a More Effective and Sustainable Housing Development Model for Northern Ireland*. Belfast: Forum for a Better Housing Market NI.

IAQM (Institute of Air Quality Management). 2015. *Land-Use Planning and Development Control: Planning for Air Quality*. London: IAQM.

NHBC (National House Building Council). 2015. *Avoiding Rubbish Design, Providing for Bin Storage on New Housing Developments*. Milton Keynes: IHS BRE Press, NHBC Foundation.

Northern Ireland Assembly Research and Information Service. 2015. *The Use of On-Site Wastewater Treatment Systems in Northern Ireland*. Briefing Paper 110/15. Belfast: Northern Ireland Assembly Research and Information Service.

Northern Ireland Executive. 2011. *Focus on the Future Sustainable Development Implementation Plan 2011–14*. Belfast: Northern Ireland Executive.

NISRA (Northern Ireland Statistical Research Agency). 2020. *Northern Ireland Local Authority Collected Municipal Waste Management Statistics, Annual Report 2019/20*. Belfast: NISRA.

ODPM (Office of the Deputy Prime Minister). 2004. *PPG 23 Planning and Pollution Control.* London: ODPM.

ODPM (Office of the Deputy Prime Minister). 2005. *Town and Country Planning Act 1990 Section 215, Best Practice Guidance.* London: ODPM.

The Planning Service. 2002. *Restaurants, Cafes and Fast Food Outlets. Development Control Advice Note 4.* Belfast: The Planning Service.

UKSC (United Kingdom Supreme Court). 2014. *Coventry and Others (Respondents) v Lawrence and Another (Appellants).* Judgment. London: UKSC.

Web links

European Commission. 2016. *Environment/Waste.* http://ec.europa.eu/environment/waste/landfill_index.htm.

Freshfields Bruckhaus Deringer LLP. 2014. *Supreme Court Gives its View on Noise Nuisance.* www.freshfields.com/uploadedFiles/SiteWide/Knowledge/Noise%20nuisance.pdf.

NI Water. https://www.niwater.com/wastewater-capacity/.

United Nations Sustainable Development Goals. 2015. https://sdgs.un.org/goals.

World Health Organization. 2016. *Health Topic/Environmental Health.* www.who.int/topics/environmental_health/en.

10

ENVIRONMENTAL IMPACT ASSESSMENT AND DEVELOPMENT MANAGEMENT

Neil Galway and Stephen McKay

Introduction

The 1960s was a decade that saw increasing concern about the impact of development upon both the natural and the built environment. In this context, it was recognised that a need existed to develop appropriate mechanisms to calibrate the associated effects. Whilst assessment toolkits existed, those in vogue were not appropriate to deal with the expansive and varied raft of consequences. Techniques such as cost-benefit analysis, whilst relevant in specific situations, did not have the capacity to cover the range of impacts that needed to be assessed. Furthermore, reducing all issues to monetary value, for example, when applied to matters such as aesthetics, biodiversity or ecosystem health is problematic and leaves such methodologies open to the accusation of false accounting. For the most part, whilst many approaches worked on paper, they did not work in reality. What was required was a systematic approach that would take into account both the quantitative and the qualitative impacts of development. Quantitative approaches are normally relatively straightforward, identifying specific thresholds below which development is acceptable or, alternatively, if a threshold is exceeded, citing remediation measures to be put in place before the development is permitted. The assessment of qualitative impacts is, however, frequently much more complex, demanding the evaluation of criteria such as rarity, sensitivity and fragility. Here, Environmental Impact Assessment (EIA) has much to contribute as it provides a framework that allows expertise and balanced judgement to make an analysis of the consequences of development on the environment. EIA is, therefore, fundamentally a tool for development management. The purpose of this chapter is to scrutinise the process and demonstrate the value that it can make to planning decision making. In the first instance, the process will be contextualised with the spotlight falling on the legislative, policy and guidance framework. Attention will then turn to scrutiny of the key stages that underpin EIA. Finally, a number of case illustrations will seek to demonstrate the implications of EIA procedures for planning practice in Northern Ireland.

DOI: 10.4324/9781003218319-10

Legislative development

The impetus for the development of an EIA toolkit gathered considerable momentum from the National Environmental Policy Act in the United States in 1970, which made it mandatory for agencies to submit an Environmental Statement (ES) when seeking consent for federal projects. The role of the ES was to consider the ramifications of environmental as well as economic and technical issues generated by the proposed development and, consequently, EIA emerged as the preferred methodology for identifying and evaluating social, economic, environmental and technical implications during the decision-making process. EIA is now a key instrument in development management globally, although the degree of rigour and application varies significantly from continent to continent. There is, however, some consensus that, whilst the techniques employed differ from jurisdiction to jurisdiction, the principles are universally accepted. In this context, there are eight guiding principles, which are as follows:

- Participation: an appropriate and timely access to the process for all interested parties;
- Transparency: all assessment decisions and their basis should be open and accessible;
- Certainty: the process and timing of the assessment should be agreed in advance and followed by all participants;
- Accountability: the decision-makers are responsible to all parties for their action and decisions under the assessment process;
- Credibility: assessment is undertaken with professionalism and objectivity;
- Cost-effectiveness: the assessment process and its outcomes will ensure environmental protection at the least cost to society;
- Flexibility: the assessment process should be able to adapt efficiently with any proposal and decision-making situation;
- Practicality: the information and outputs provided by the assessment process are readily usable in decision making and planning (IAIA 1999).

In Europe, the need for such a process, with the ability to scrutinise and evaluate environmental factors, arose in the early 1970s, and, in 1977, a draft directive requiring member states of the European Economic Community (now the European Union) to incorporate EIA into specific project appraisal frameworks was drawn up. The statutory requirement for EIA eventually came into force in 1985 with European Council Directive 85/337/EEC 'on the assessment of the effects of certain public and private projects on the environment'. In 1997, the directive was amended by European Council Directive 97/11/EC, which stated that:

> the assessment procedure is a fundamental instrument of environmental policy as defined in Article 130r of the Treaty and of the Fifth Community Programme of policy and action in relation to the environment and sustainable development … Community policy is based on the precautionary principle and the principle that preventative action should be taken, that environmental damage should as a priority be rectified at source and that the polluter should pay.
>
> *(European Council Directive 97/11/EC, 1997)*

This was subsequently amended under Directive 2011/92/EU, which specifically concerned the assessment of the environmental consequences of those public and private projects likely to have significant effects on the environment. The Directive brought in the requirement for member states to adopt all measures necessary to ensure that, before consent is given, projects likely to have significant impacts on the environment are required to undergo an assessment with regard to their effects. It specified that EIA must identify, describe and assess the following factors: (a) human beings, fauna and flora; (b) soil, water, air, climate and the landscape; (c) material assets and the cultural heritage; (d) the interaction between these factors. Council Directive 2011/92/EU of the European Parliament and of the Council on the assessment of the effects of certain public and private projects on the environment was amended by Council Directive 2014/52/EU of the European Parliament and of the Council of 16 April 2014 to simplify the rules for assessing the potential effects of projects on the environment. Driven by the aspiration to introduce smarter regulation, it aimed to reduce administrative burdens and improve the level of protection afforded to the environment with the goal of making decisions on public and private investments more sound, predictable and sustainable. The key amendments are as follows:

- Member states now have a mandate to simplify their different environmental assessment procedures;
- Timeframes are introduced for the different stages of environmental assessments; screening decisions should be taken within 90 days (although extensions are possible) and public consultations should last at least 30 days. Members states also need to ensure that final decisions are taken within a 'reasonable period of time';
- The screening procedure, determining whether an EIA is required, is simplified. Decisions must be duly motivated in the light of the updated screening criteria;
- EIA reports are to be made more understandable for the public, especially regarding assessments of the current state of the environment and alternatives to the proposal in question;
- The quality and the content of the reports will be improved. Competent authorities will also need to prove their objectivity to avoid conflicts of interest;
- The grounds for development consent decisions must be clear and more transparent for the public. Member states may also set timeframes for the validity of any reasoned conclusions or opinions issued as part of the EIA procedure;
- If projects do entail significant adverse effects on the environment, developers will be obliged to do what is necessary to avoid, prevent or reduce such effects. These projects will need to be monitored using procedures determined by the member states. Existing monitoring arrangements may be used to avoid duplication of monitoring and unnecessary costs (https://ec.europa.eu/environment/eia/review.htm).

These requirements have been transposed into Northern Ireland law via The Planning (Environmental Impact Assessment) Regulations (Northern Ireland) 2017, which came into force on 16 May 2017 and which replace The Planning (Environmental Impact Assessment) Regulations (Northern Ireland) 2015 for cases being dealt with after the effective date of these new regulations. The key changes introduced by the 2017 regulations include:

- A revision of the information developers need to provide when requesting a screening direction or scoping opinion. The information planners need to take into account when making a screening determination and giving a scoping opinion has also been revised.

- A new requirement to use competent experts. The developer must ensure that their environmental statement is prepared by competent experts, while the consenting authority must ensure that it has, or has access as necessary to, sufficient expertise to examine the environmental statement.
- The use of coordinated procedures when a development also requires assessment under the Habitats/Wild Birds Directive.
- The introduction of monitoring measures for developments which could potentially have significant negative effects on the environment. (Department for Infrastructure 2017)

To ensure that these regulations would continue to be operable following the exit from the European Union of Northern Ireland as part of the United Kingdom, enabling legislation was progressed in the form of The Planning (Environmental Assessments and Miscellaneous Amendments) (EU Exit) (Northern Ireland) Regulations 2018 and The Planning (Environmental Assessments and Technical Miscellaneous Amendments) (EU Exit) Regulations (Northern Ireland) 2020. These made no substantive changes to the 2017 EIA instrument.

The environmental impact assessment process in development management

EIA is defined as 'a systematic process that examines the environmental consequences of development actions in advance' (Glasson and Therivel 2019: 4) with the emphasis on prevention of detrimental impacts. It endeavours to ensure that the likely effects of a proposed development are identified, comprehensively understood and considered prior to the issuing of planning consent. Arguably, it is a more proactive approach to planning than the normal development management system as the onus is placed upon the developer to demonstrate from the outset that the project is acceptable. In short, EIA, in theory, not only provides a more complete knowledge base for decision making but also adds to the transparency of the process, engages the public and identifies potential problems at an early stage. It also allows for a fuller consideration of impacts, enables remedies to be crafted, improves project design and reduces detrimental environmental impact. The EIA process in development management follows a number of distinct stages: 1. Project screening; 2. Scoping; 3. Assessment and preparation of the Environmental Statement; 4. Submission of the Environmental Statement and consultation; 5. Examination of Environmental Information and further information; 6. The decision and the EIA decision notice (Department for Infrastructure 2021a). An additional post-decision monitoring stage has value and can be regarded as part of a more complete EIA procedure.

1. ***Project screening***

This initial stage determines whether a proposed development constitutes EIA development. Projects that have few or no impacts are 'screened out' and proceed through the normal planning process, whilst those that may have a significant adverse impact may be subject to an EIA. Screening will take into account the 2017 EIA Regulations. Projects of a type listed in Schedule 1 of the Regulations will always be subject to EIA, due to their scale and potential environmental impact. These include oil refineries, power stations, steelworks, chemical installations, major airport developments, waste disposal installations, major oil/gas pipelines, major quarries and mining installations. For many other projects, the screening decision may be less clear. Guidance is give in Schedule 2 of the Regulations, which sets out the descriptions of development, applicable thresholds and criteria that would warrant an EIA. These tend to be smaller-scale projects within the agriculture, extractive industry, energy, infrastructure and tourism/leisure sectors and can include the metal, mineral, chemical, food and textile industries. Prior to submitting a planning application, an applicant can make a screening request to the planning authority and subsequently an opinion will be issued on whether the proposal should be subject to an EIA. If the applicant does not concur with the opinion, an appeal can be submitted to the Planning Appeals Commission. Similarly, if a planning application is submitted without an EIA, the planning authority can determine that it should be subject to the screening process. Again, if the applicant disagrees there is an option to appeal to the Planning Appeals Commission. The Regulations specify that, when making a request for a screening determination, an applicant shall provide the following information:

(a) a plan sufficient to identify the land;
(b) a description of the development, including, in particular, a description of the physical characteristics of the whole development and, where relevant, of demolition works; a description of the location of the development, with particular regard to the environmental sensitivity of geographical areas likely to be affected; a description of the aspects of the environment likely to be significantly affected by the development; to the extent the information is available, a description of any likely significant effects of the development on the environment resulting from expected residues and emissions and the production of waste, where relevant and the use of natural resources, in particular soil, land, water and biodiversity.

An applicant, when making a request for a screening determination, may also provide a description of any features of the proposed development or any measures envisaged to avoid or prevent what might otherwise have been significant adverse effects on the environment. At the same time that a screening determination is being requested from the planning authority, an applicant may also request an opinion as to the scope and level of detail of the information to be provided in a subsequent Environmental Statement to be submitted with an EIA application (a 'scoping opinion'). Where the planning authority, or following a 'hear and report' engagement by the Planning Appeals Commission, confirms that as a result of project screening an Environmental Statement is required, it should be submitted within a period of six months from the date of the determination or such extended period as may be agreed in writing between the parties. If the applicant does not submit an Environmental Statement in accordance with this provision, the application for planning permission (or any subsequent application) shall be deemed to be refused and a deemed refusal by a council shall not give rise to an appeal to the Planning Appeals Commission.

2. ***Scoping***

If, following a screening process, it is concluded that a planning application proposal is EIA development, scoping determines the environmental issues that must be addressed within the Environmental Statement that will accompany the planning application. Possible impacts may be scoped in order to identify those considered the most significant, those whose significance is not clear and those that are insignificant. Those that are not significant are eliminated from the process while those whose significance is unclear should be explored as part of the process. Rigour in the scoping process is crucial because, if an issue of concern is omitted or its impact underestimated, not only is the efficacy of the planning system called into question but the result could have serious consequences for the environment. While there is no formal basis within the EIA regulations for a planning authority to engage with statutory consultees to inform a screening decision, this is not the case with scoping, where they are required to consult with statutory consultees to help determine the content of an Environmental Statement (Department for Infrastructure 2021b). Accordingly, a planning authority can consult with other bodies holding environmental responsibilities, including the Northern Ireland Environment Agency Development Management Team within the Department of Agriculture, Environment and Rural Affairs and which can advise if records are available about the site, proposed methods of assessment and what information and level of detail will need to be provided by the developer. The overall process can be facilitated by the use of checklist criteria and / or matrices, such as those stylised in Figure 10.1, in order to identify key issues and endeavour to ensure that none of these have been omitted.

	Environmental impact						
	Characteristic		**Scale**	**Degree and duration of impact**			**Project Phase**
Environmental component	beneficial,detrimenal, neutral	direct, indirect, cumulative	local, regional, national, trans-national	long-term, short-term	irreversible, reversible	major, minor	pre-development, construction, operational, decommissioning
Human population							
Transportation							
Natural heritage							
Built heritage							
Cultural heritage							
Flora							
Fauna							
Water							
Air, atmosphere and climate							
Soils and geology							
Agriculture							
Noise							

FIGURE 10.1 Matrix of issues to inform a scoping opinion

When, following the receipt of a scoping opinion, an applicant wishes to proceed with the submission of an Environmental Statement, the applicant shall by notice in writing inform the council or, as the case may be, the Department to such effect within four weeks of the date of the scoping opinion.

3. ***Assessment and preparation of the Environmental Statement***

The preparation of an Environmental Statement is the responsibility of the applicant for planning permission, who must ensure that the significant effects of a proposed development are subject to an assessment undertaken by competent experts. It should be based on the scoping opinion and include:

(a) a description of the proposed development comprising information on the site, design, size and other relevant features of the development;
(b) a description of the likely significant effects of the proposed development on the environment;
(c) a description of any features of the proposed development, or measures envisaged in order to avoid, prevent or reduce and, if possible, offset likely significant adverse effects on the environment;
(d) a description of the reasonable alternatives studied by the applicant, which are relevant to the proposed development and its specific characteristics, and an indication of the main reasons for the option chosen, taking into account the significant effects of the development on the environment;
(e) a non-technical summary of the information.

More specifically, an Environmental Statement should discuss the physical characteristics of the whole development and the land use requirements during the construction and operational phases. It should also include a description of the main characteristics of the production processes and an estimate, by type and quantity, of expected residues and emissions. In addition, an outline of the main alternatives studied and an indication of the main reason for the preferred choice, taking into account the environmental effects, should be provided. The Environmental Statement should include a description of the aspects of the environment likely to be affected by the development, including, in particular, population, fauna, flora, soil, water, air, climatic factors, material assets, including the architectural and archaeological heritage, landscape and the interrelationship between the above factors. It should also describe the significant effects of the development on the environment, not only the direct impacts but also any indirect, secondary, cumulative, short, medium and long term, permanent and temporary, positive and negative effects of development resulting from the existence of the development; the use of natural resources; the emission of pollutants, the creation of nuisances and the elimination of waste; and the description of the forecasting methods used to assess the effects on the environment. In addition, it should include a description of the measures envisaged to prevent, reduce and, where possible, offset any significant adverse effects on the environment (Friends of the Earth 2014: 12). In short, the Environmental Statement should aim to respond to the minimum requirements set out in Regulation 11 and Schedule 4 of the Planning (Environmental Impact Assessment) Regulations (Northern Ireland) 2017.

4. ***Submission of the Environmental Statement and consultation***

Once prepared, a developer will submit the Environmental Statement along with the planning application to the District Council or Department for Infrastructure as appropriate. Both the planning application and the Environmental Statement are then advertised by way of notice in line with the 2017 Regulations, which specify:

(a) that an application for planning permission or a subsequent application has been made to the council or, as the case may be, the Department and is accompanied by an environmental statement;
(b) the address or location and nature of the proposed development;
(c) in the case of an application for planning permission, that the environmental statement is available for inspection free of charge and the times and places at which, and the means by which, it is available for inspection;
(d) in the case of a subsequent application, that a copy of the planning permission and supporting documents for the development in respect of which the application has been made, may be inspected by members of the public at all reasonable hours at the relevant office of the council or, as the case may be, the Department;
(e) details of a website maintained by or on behalf of the council or, as the case may be, the Department on which the environmental statement and other documents have been made available;
(f) a postal address (within the locality in which the land proposed to be developed is situated) at which copies of the environmental statement may be obtained from the applicant, so long as stocks last, and if a charge is to be made for a copy, the amount of the charge; and
(g) that any person wishing to make representations about the application should make them in writing to the council or, as the case may be, the Department, and the latest date by which representations should be made (being a date not less than 30 days from the date on which the notice is first published).

Additionally, the planning authority will initiate a consultation process with other government bodies holding environmental responsibilities. Moreover, where a project is likely to have significant transboundary environmental effects (most commonly with the Republic of Ireland), a council shall immediately send to the Department for Infrastructure a copy of the planning application and Environmental Statement and any supporting documents. The Department will then be responsible for taking consultative action in regard to council referrals and applications submitted to it in the first instance by:

(a) publishing a notice in the *Belfast Gazette* giving the address of the proposed development, stating that it is accompanied by an environmental statement and that it is likely to have significant effects on the environment of another European Economic Area (EEA) State and giving an address at which additional information may be obtained;
(ii) sending to the EEA State the relevant particulars; and
(iii) giving the EEA State a reasonable time in which to indicate whether it wishes to participate in the EIA procedure.

Should the EEA State wish to participate, the Department will then consult with the EEA State concerned regarding the potential significant effects of the development on the environment of that EEA State and the measures envisaged to reduce or eliminate such effects; and determine, in agreement with the other EEA State, a reasonable period of time for the duration of the consultation period. In due course, the Department will inform the EEA State of the decision. Transboundary consultation also operates in the reverse direction whereby the Department for Infrastructure will receive notification from a planning authority (including An Bord Pleanála) in the Republic of Ireland on planning applications accompanied by Environmental Statements that are likely to have significant effects on the environment in Northern Ireland. These notifications will be advertised by way of notices that will give information on viewing documentation and making submissions.

5. ***Examination of Environmental Information and further information***

The consultation replies and the Environmental Statement collectively make up what is referred to as 'Environmental Information'; it is the examination of this material that will facilitate a planning authority in drawing its conclusions on the likely significant effects of the proposed development on the environment. To add to the information base that, in turn, will allow the planning authority to reach a reasoned conclusion, supplementary information can be requested from the applicant by the planning authority. The regulations also allow for a public local inquiry or hearing to be convened. Essentially this stage is moving the planning authority towards a situation that can integrate a reasoned conclusion into the decision to grant or refuse planning permission.

The evaluation and assessment of significance gauge the relative importance of the predicted impacts so that the principal impacts can be focused on. The impact of a proposed development is the predicted change to the existing state of the environment likely to result if it takes place. This can rely on value judgements, which may involve placing a subjective weighting on the environment likely to be affected by the change. Those impacts identified at the scoping stage will be evaluated, whilst additional environmental effects may emerge in the assessment process (Scottish National Heritage 2013). The duration, timing and extent of effects, the degree of certainty in the prediction of impacts and the likelihood that the changes will take place are all evaluated when assessing the significance of the change. Whilst every effort should be made to collect and evaluate information relevant to the EIA, it is recognised that there is an element of prediction in the process that, in itself, may be based on assumptions rather than scientific facts. As guidance in dealing with such eventualities, the Precautionary Approach articulated by the 1992 Rio Declaration is helpful:

> Where there are threats of serious or irreversible damage, lack of full scientific certainty shall not be used as a reason for postponing cost-effective measures to prevent environmental degradation.
>
> *(Scottish National Heritage 2013: 134)*

Therefore, even if it is deemed unlikely that the effect will take place, if that effect would be very serious or irreversible, then it may be considered significant.

In coming to a conclusion based on the Environmental Information, the planning authority will also have regard to mitigation, which involves consideration of measures that will avoid, cancel, reduce or compensate for any major adverse impacts. Avoidance measures are intended to stop or prevent effects from occurring, or to remove the risk of them occurring, perhaps by relocating a project away from a sensitive area or removing from a project the element that may cause an adverse effect. If avoidance measures are implemented successfully, there will be no adverse effect. In some cases, the 'do nothing' option may be the most effective avoidance measure (Scottish National Heritage 2013: 95). Cancellation measures are designed to cancel out the effects of the project before they can have a negative impact, whilst reduction measures seek to minimise those negative effects that have not been avoided or cancelled. Compensatory measures are implemented after a decision has been made. They are intended to at least try to recompense, or offset, the adverse effects that have been anticipated. They come into play when it has been decided that the project should nevertheless go ahead, and the compensatory measures try to make amends (Scottish National Heritage 2013: 95). There is an established order to the application of mitigating measures and thus planning authorities will seek to:

- avoid or cancel the adverse effects of the project;
- minimise or at least reduce the effects where it has not been possible to avoid or cancel them;
- where it has not been possible to avoid or significantly reduce the adverse impact, compensation should be offered; and
- developers may also be encouraged to include environmental enhancement measures in the EIA in an attempt to offset any negative effects of the development.

(Scottish National Heritage 2013: 98)

6. ***The decision and the EIA decision notice***

The reasoned conclusions on the significant environmental effects arising from the EIA development along with any measures needed by way of mitigation will be set out in the Planning Officers' report with the aim of demonstrating that all relevant considerations have been taken into account and that any irrelevant considerations have not been given weight. Where planning permission is granted, it is commonplace for conditions to be attached, say in regard to mitigation and future monitoring. There are, finally, attendant publicity requirements with the 2017 Regulations specifying that where an EIA application is determined by a council or, as the case may be, the Department for Infrastructure, it shall promptly inform relevant environmental bodies of the decision in writing and inform the public of the decision by local advertisement or by such other means as are reasonable in the circumstances. Additionally, where an EIA development is determined by the Department, it shall also notify the council in whose district the proposed development is to be situated of its decision and provide the council with a statement containing:

(i) the main reasons and considerations on which the decision was based, including information about the arrangements taken to ensure the public had the opportunity to participate in the decision-making procedures;
(ii) a summary of the results on the consultations undertaken and information gathered in respect of the application and how those results, in particular comments received from an EEA State pursuant to any such consultation, have been addressed.

7. ***Post-decision monitoring***

The purpose of monitoring is to assess the actual environmental impacts of a project that has been granted planning permission and has been developed. Where mitigating measures have been identified, it is vital that the effectiveness of mitigation is monitored to ensure that it meets the standards set out in the decision. This is not a mandatory requirement of the legislation but is, nevertheless, highly desirable since it has the potential for verifying predictions of impact that can improve forecasting techniques for the future. It will also allow the authorities to assess whether the residual impacts need additional mitigation, and whether the original mitigating measures are effective or need adjustment in order to ensure that there is the least possible impact on the environment. It may also be necessary, where no mitigation was proposed or required, because the proposal was not expected to cause significant environmental change (EPOA 2007: 86). Monitoring and review should be clearly dealt with in the EIA process. We would argue that an Environmental Statement should clearly explain how mitigating measures, monitoring and review should be carried out and should give a clear commitment and readiness to accept conditions and legal agreements to ensure they are implemented at the right time and in appropriate ways. Monitoring may be delegated to developers or their consultants.

Case law insights

EIA procedures and outputs have informed a broad sweep of regulatory decision making in Northern Ireland that includes not only planning application decisions but also, for example, confirmations of road schemes, marine construction licensing, flood alleviation management, water resource projects and afforestation. Within the planning sphere, the courts have exercised a supervisory jurisdiction that has been reported in published judgments and which gives guidance for best practice.

As noted above the EIA process commences with the determination of a screening decision and the issue that has arisen is the need to revisit an earlier negative screening decision in subsequent planning applications for an amended development project on the same site. *In the Matter of an Application by Colum Sands for Judicial Review v Newry, Mourne and Down District Council* (2018 NIQB 80), the proposal comprised a nursing home and residential apartments in Rostrevor and it was common case across the parties that this was a Schedule 2 development within the compass of the EIA Regulations. The formal screening decision to the effect that an Environmental Statement is not required was issued by the Department of the Environment in October 2009 with the reasons stated being:

> The application has included several reports namely Transport Assessment Report, Flora and Fauna Survey and Bat Survey. It is considered that much of what would be included in an Environmental Statement has been considered in these reports and any further information could be requested and considered through the processing of the application. Consideration has also to be given to the fact that this site is currently developed.

However, this first planning application remained undetermined, prompting the developer to lodge a second planning application in May 2012, and, subsequently, with the first and second applications remaining undetermined, a third planning application was lodged in December

2016. The applicant for judicial review contended that the second and third planning applications should have been treated as new applications rather than mere amendments of the first application, thus triggering the requirement to make a fresh screening decision. The court ruled against this proposition with McCloskey J stating:

> The question of whether a subsequent planning application simply amends an earlier application or is to be treated as constituting a new one requires an exercise of evaluative judgement on the part of the decision maker. The formulation of this proposition points to the *Wednesbury* principle as being the appropriate standard of review … The juridical truism is that a negative screening decision is not necessarily a once and for all act of indefinite legal validity. Rather, it is implicit in the EIA Directive and its domestic law counterpart that decisions of this kind may, in certain circumstances, require reconsideration and remaking … What is the trigger for the legal duty to revisit an earlier negative screening decision? The decided cases, unsurprisingly, have shrunk from any attempt to provide a prescriptive answer. The common theme is the open textured test of whether a new formal screening process would generate a realistic prospect of a different outcome … there is no evidence of any material environmental change either within the site or in the environs of its boundaries in any of the areas foreseeably affected by the development; the total information accumulated by November 2017 was substantially greater than that available some eight years previously; the proposal ultimately determined by the Council had been twice reduced in scale during the intervening period in a manner which entailed diminished physical construction with associated reduced human and vehicular usage and occupancy; both objectively and sensibly the latter pointed in the direction of shrinking, rather than increasing, environmental impact; the issue of possible land contamination was addressed extensively by specialised consultants; there was nothing in the responses from statutory consultees, who included agencies with environmental expertise, indicating the desirability of revisiting the 2009 screening decision; and there was nothing to suggest that the 2009 decision was suffering from some flaw or deficiency and/or had not been made in accordance with the EIA Regulations … Balancing all of these considerations I conclude that if the Council had applied its mind to the question any ensuing decision not to make a fresh screening decision would have withstood an irrationality challenge.

A frequent matter that has prompted litigation relates to the quality and depth of Environmental Information that has been taken into account during the EIA process. To recap, the EIA process is underpinned by the submission of the Environmental Statement, the key purpose of which is to identify the significant impacts of the proposed development and, where appropriate, suggest remedies that can be put in place to allow the project to proceed. The Environmental Statement can then be supplemented with information on the project submitted by other government bodies with relevant expertise and by the public. The availability of Environmental Information should enhance transparency, thereby providing an opportunity for scrutiny and the development of a deeper understanding of all material planning considerations by everyone affected. Weatherup J in *An Application for Judicial Review by the National Trust for Places of Historic Interest or Natural Beauty* (2013 NIQB 60) has provided a valuable summary of the wide sweep of this information imperative:

> The Environmental Statement is but a part of the environmental information that the decision maker must take into account … In relation to 'environmental information' a number of matters may be noted. First of all, the 'environmental information' gathered by the Department from the developer, the consultees and others will be wider than the 'Environmental Statement' provided by the developer. The Environmental Statement must contain the information required by the schedule to the Regulations and will fail if it cannot be regarded as in substance an Environmental Statement. As well as the Environmental Statement and further information from the developer the Department will also be able to rely on the environmental information obtained from consultees and others. For this purpose the representatives of the Northern Ireland Environment Agency are consultees. Secondly, it is for the Department to decide if there are likely significant effects, that is, if there is an effect, if it is significant and if it is likely. Thirdly, the Department can take into account mitigation measures, thus concluding that what would otherwise be a likely significant effect will, by reason of mitigation, not be so. In order to so conclude the mitigation measures should be clearly established and easily achievable. Fourthly, the Department cannot postpone the decision on likely significant effects or on whether mitigation measures will mean that there is no likely significant effect. If a conclusion on likely significant effects requires a survey then the survey must be done. Nor can the Department impose conditions instead of making the assessment. Fifthly, there must be sufficient information for the Department to decide on the likely significant effects and mitigation. It is for the Department to decide if there is sufficient information and the Department may require further information from a planning applicant, with the required publicity for such further information, and may obtain additional information from consultees or from members of the general public.

In this regard, Lord Carnwath has helpfully opined in *R (Jones) v Mansfield District Council* (2003 EWCA Civ 1408):

> It needs to be borne in mind that the EIA process is intended to be an aid to efficient and inclusive decision making in special cases, not an obstacle race. Furthermore, it does not detract from the authority's ordinary duty, in the case of any planning application, to inform itself of all relevant matters, and take them properly into account in deciding the case.

Another issue relates to the reliance by the planning authority on the developer to produce an Environmental Statement. The Regulations specify the need to draw on competent experts and the preface of Environmental Statements will usually contain those professional and experience details. Again case law has endorsed this practice. *In the Matter of an Application by William Donnelly for Judicial review and in the Matter of a Decision by the Department of the Environment for Northern Ireland made on 27 July 2015* (2017 NIQB 84) McBride J ruled:

> Firstly … I find it is not unlawful for the Department to rely on the environmental statement provided by a developer. This is not an abdication of its discretion under the Regulations. This is standard practice and one which the Regulations clearly envisages by requiring the developer to provide the Environmental Statement. Secondly, as

> Weatherup J stated in National Trust Application (2013 NIQB 60) … 'the adequacy of the contents of an Environmental Statement is a matter for the Department, subject to the *Wednesbury* rule that the decision must be rational, take into account relevant considerations and leave out of account irrelevant considerations'.

Where a proposed development in Northern Ireland is subject to an EIA determination, and is likely to have significant effects on another European Economic Area, interstate consultation is required to inform the decision-making process. This matter arose in *Christine Gibson's Application* (2017 NICA 77), which contested a proposal to construct ferry terminal facilities on each side of Carlingford Lough (Greenore in the Republic of Ireland and Greencastle in Northern Ireland) related to the operation of a vehicular ferry service. The plaintiff claimed that no transboundary notification had been submitted. In this instance, the Court of Appeal reported high-level engagement on the basis of affidavit evidence that:

> transboundary consultation did take place in relation to the original application for planning permission. Louth County Council notified the relevant authorities within the Republic of Ireland and issued a consultation response that the proposed development would not cause any likely significant effects in its area. The Minister for the Environment (ROI) concurred.

Projects that evolve over time, perhaps involving a number of iterations, raise the issue regarding the need to update the information used to inform decision making. In this vein, the identification of mitigation measures can be regarded as a dynamic process in the lead-up to implementation and comprise project review and adjustments, if necessary. *In the Matter of an Application by Mr Chris Murphy to Apply for Judicial Review, In the Matter of a Decision by the Department for Infrastructure on 17 August to Proceed with the Castledawson to Toome Dualling Scheme* (2017 NIQB 35) the court considered documentation that updated the test of significance and appropriate assessment relating to a road scheme. Over a period from 2007 to 2016, involving intermittent review, it had been concluded that this road scheme would not have an adverse effect on the integrity of the Lough Neagh and Lough Beg Special Protection Area and Ramsar site. In commenting on the last iteration that helped inform decision making by the Minister, Keegan J notes that 'the mitigation plans did change over the years in tandem with the ongoing process' and that 'it makes sense that updates to the appropriate assessment would occur'. The judgment elaborates as follows:

> Such a process must be with a view to assessing if the original assessment should remain in place. But it must also mean that there must be an openness to see if the original assessment should be revised, in other words this process must contemplate not proceeding if the information is unfavourable … In this case what the Minister has done is consider whether in the light of updating information, the appropriate assessment was correct. He has relied upon current methods and methodologies in the particular circumstances. The reviews represent a further iteration of the original assessment and lead to a confirmation that it was correct. In my view it was appropriate to review pursuant to national law obligations. Fundamentally, this illustrates the Department's commitment to the appropriate assessment and its adherence to promises previously made during the process. Indeed it seems to me that there would be a valid criticism

> if the Department had reneged upon these commitments. The updating process was not completed in a vacuum or behind closed doors. There were swan surveys and other updating information and consultation with relevant bodies about the mitigation measures. This included the RSPB and the Irish Whooper Swan Group. It must be remembered that the requirement in Article 6(3) to conduct an assessment in line with the best scientific knowledge can never be absolutely certain. However in this instance the Department has gathered together the best information available.

This case concerned an Appropriate Assessment under the Conservation (Natural Habitats, etc.) Regulations (Northern Ireland) 1995 and related to a decision by the Minister to proceed with a road scheme. The findings of the court in this instance regarding 'best information available' have relevance for EIA development planning applications.

Finally, planning permission for an EIA development can include by way of condition the submission of a monitoring plan to be approved by the planning authority prior to the commencement of the development. *In the Matter of an Application by William Donnelly for Judicial review and in the Matter of a Decision by the Department of the Environment for Northern Ireland made on 27 July 2015* (2017 NIQB 84) the applicant alleged that accepting a monitoring plan for acidic rock testing post-approval for a valuable mineral mining development in the Sperrins showed a gap in the Environmental Assessment and, furthermore, that the requirement for such testing showed that the planning authority is no longer convinced that the rock waste is inert. McBride J ruled, however:

> that the imposition of the condition for a monitoring plan does not indicate any deficiency in the assessment. Rather it was a prudent approach taken to ensure protection of the aquatic environment in the future to ensure that effects did not arise above those assessed, or if unforeseen effects did arise, that there was mitigation in place to address them.

Conclusion

EIA is an important tool in developing a deeper understanding of the issues at stake in planning decision-making processes. Whilst simple in theory and underpinned by a series of coherent procedures, effective implementation is a much more complex equation. Whilst regulatory frameworks and good practice guides are invaluable, those charged with applying the legislation to specific proposals must remain cognisant of the pitfalls that can occur in failing to transpose the regulations appropriately. It is hoped that this chapter has shed some light on the context, procedures and challenges underpinning the process.

Legislation

European Council Directive 85/337/EEC, 1985
European Council Directive 97/11/EC, 1997
European Council Directive 2011/92/EU
European Council Directive 2014/52/EU
National Environmental Policy Act in the United States in 1970
Environmental Impact Assessment (EIA) Directive 2014/52/EU

The Conservation (Natural Habitats, etc) Regulations (Northern Ireland) 1995
The Planning (Environmental Impact Assessment) Regulations (Northern Ireland) 2015
The Planning (Environmental Impact Assessment) Regulations (Northern Ireland) 2017
The Planning (Environmental Assessments and Miscellaneous Amendments) (EU Exit) (Northern Ireland) Regulations 2018
The Planning (Environmental Assessments and Technical Miscellaneous Amendments) (EU Exit) Regulations (Northern Ireland) 2020

Cases cited

An Application for Judicial Review by the National Trust for Places of Historic Interest or Natural Beauty (2013 NIQB 60)
Christine Gibson's Application (2017 NICA 77)
In the Matter of an Application by Colum Sands for Judicial Review v Newry, Mourne and Down District Council (2018 NIQB 80)
In the Matter of an Application by Mr Chris Murphy to Apply for Judicial Review, In the Matter of a Decision by the Department for Infrastructure on 17 August to Proceed with the Castledawson to Toome Dualling Scheme (2017 NIQB 35)
In the Matter of an Application by William Donnelly for Judicial review and in the Matter of a Decision by the Department of the Environment for Northern Ireland made on 27 July 2015 (2017 NIQB 84)
R (Jones) v Mansfield District Council (2003 EWCA Civ 1408)

References

Department for Infrastructure. 2017. 'Planning Legislation' in *Chief Planners Update*, 20 September. Belfast: Department for Infrastructure.

Department for Infrastructure. 2021a. *Unauthorised Environmental Impact Assessment (EIA) Development, Development Management Practice Note 9a*. Belfast: Department for Infrastructure.

Department for Infrastructure. 2021b. *Planning Forum Internal Advice Note – Consultations in the Planning Application Process: Operating Principles for Planning Authorities and Consultees*. Belfast: Department for Infrastructure.

EPOA (Essex Planning Officers' Association). 2007. *The Essex Guide to Environmental Impact Assessment*. Brentwood: Essex County Council and EPOA.

FOE (Friends of the Earth). 2014. *Environmental Impact Assessment: A Briefing*. London: Friends of the Earth.

Glasson, J. and R. Therivel. 2019. *Introduction to Environmental Impact Assessment*. London: Routledge.

IAIA (International Association for Impact Assessment). 1999. *Principles of Environmental Impact Assessment Best Practice*. Fargo, ND: IAIA.

Scottish National Heritage. 2013. *A Handbook on Environmental Impact Assessment: Guidance for Competent Authorities, Consultees and Others Involved in the Environmental Impact Assessment Process*. Inverness: SNH.

11

PLANNING AND RENEWABLE ENERGY

Stephen McKay and Michael Murray

Introduction

On 26 October 2020 Infrastructure Minister Nichola Mallon announced that planning permission had been refused for the Doraville Wind Farm project comprising the erection of thirty-three wind turbines to the west of Draperstown. In addressing the many considerations attached to the scheme, it was concluded that its economic and environmental benefits would not outweigh the considerable harm that would be caused to the landscape interest of a large part of the Sperrins Area of Outstanding Natural Beauty. The decision was reached following a public inquiry and a recommendation from the Planning Appeals Commission to refuse planning permission that included comment in its report that the scale of the proposal and the sheer magnitude of the associated change has the potential to detract from local people's sense of place and connection to the land. There were 2,160 representations received in relation to this regionally significant planning application, of which 2,153 were objections.

Subsequently, on 16 March 2021 Infrastructure Minister Nichola Mallon announced the granting of planning permission for Corlacky Wind Farm west of Swatragh, within the Sperrins Area of Outstanding Natural Beauty, and comprising eleven wind turbines. A hearing into the Department's Notice of Opinion to refuse planning permission had been previously held in October 2019 with the Planning Appeals Commission recommending that planning permission be granted. It had concluded that 'the proposed wind farm would bring economic benefits during the construction and operational phases, help Northern Ireland move from an overdependence on imported fossil fuels, and contribute to meeting longer term renewable energy and greenhouse gas reduction targets. In the absence of any sustained objection to the development, these benefits strongly reinforce the case for granting approval' (Planning Appeals Commission 2020: 14). In this instance, the Minister was satisfied that the environmental and economic benefits are in favour of the proposal and outweigh any potential impact on archaeological sites in the vicinity.

These two illustrations of decision making in regard to wind farm projects demonstrate the interaction between planning procedures discussed in Chapter 2 of this book and the balancing of localised material considerations within a wider strategic context of a public policy

DOI: 10.4324/9781003218319-11

commitment to renewable energy. Essentially, they help to frame the discussion below, which deals with the following matters: the Northern Ireland context for energy and renewable electricity; the regulatory planning arena for renewables; wind energy development management; solar farm development management; renewable electricity and the transmission system; and emergent investment in battery electricity storage systems.

The Northern Ireland context for energy and renewable electricity

In 2004 only around 3 percent of electricity consumed in Northern Ireland was generated from indigenous renewable sources. At that time the Department of Enterprise, Trade and Investment had published its first Strategic Energy Framework document, which set a target of 12 percent for the proportion of electricity that would be generated from renewable sources by 2012 in an effort to reduce the region's high dependency on fossil fuels. Fast forward to the twelve-month period April 2020 to March 2021 and data from the Department for the Economy indicate that just over 46 percent of total electricity consumption in Northern Ireland was sourced from renewable sources located in Northern Ireland of which almost 84 percent was generated from wind (Department for the Economy 2021a:1). Other types of renewable electricity generation included biogas (5.8 percent), biomass (4.6 percent), solar photovoltaic (3.6 percent) and landfill gas (1.5 percent); non-wind renewable electricity generation volumes in Northern Ireland have been increasing from a low base.

This drive to expand renewable electricity is set against a backdrop of ambition to transform the ways by which energy is produced and consumed. So, for example, small-scale Distributed Generation (below 5MW) is being hailed as an innovative technology that can help reduce transmission and distribution losses, avoid transmission upgrade costs, facilitate grid infill, allow users to generate and deploy their own electricity, enhance security of supply and support the rural economy, not least the farming sector within which there is increasing emphasis on decarbonisation (KPMG 2021). As part of a Distributed Generation model, renewable electricity can connect directly into the distribution system or be located adjacent to demand without having to use the grid network. Decarbonising the power sector, as a contribution to moderating climate change, is now centre stage in the evolving collaborative relationships among government, renewables developers and grid operators. In this vein, the 2030 energy strategy for Northern Ireland (Northern Ireland Executive 2021) has proposed meeting at least 70 percent of electricity consumption from a diverse mix of renewable sources by that time horizon and doubling the size of the region's low carbon and renewable energy economy by 2030. Interestingly, the prospect of a more diversified technology mix is canvassed. In particular, offshore wind technologies are identified as having considerable potential in helping to reach these targets, which, in turn, are aligned with an overall goal of net zero carbon energy in Northern Ireland by 2050. Accordingly, the energy strategy places emphasis on facilitating pre-commercial and demonstration sites for marine developments during the 2020s. Nonetheless, a significant proportion of new renewable generation is likely to be from onshore wind energy infrastructure, while interest in solar production and energy storage is also gaining developer attention. These latter modes will continue to engage the terrestrial regulatory planning system.

The regulatory planning arena for renewable energy

Extant planning policy for this sector in Northern Ireland at the time of writing is set out in *Planning Policy Statement 18 Renewable Energy*, initially published by the Department of the Environment (2009a). This details the Department for Infrastructure planning policy for development generating energy from renewable resources and that requires the submission of a planning application. Simply put, the aim of this Statement, as retained planning policy prior to later replacement, is to facilitate the siting of renewable energy generating facilities in appropriate locations within the built and natural environment in order to achieve Northern Ireland's renewable energy targets and to realise the benefits of renewable energy. Key considerations for decision making are set out in *Policy RE1 – Renewable Energy Development* and indicate that proposals to generate energy from renewable resources will be permitted provided the proposal, and any associated buildings and infrastructure, will not result in an unacceptable adverse impact on:

(a) public safety, human health, or residential amenity;
(b) visual amenity and landscape character;
(c) biodiversity, nature conservation or built heritage interests;
(d) local natural resources, such as air quality or water quality; and
(e) public access to the countryside.

Additional criteria and their justification are set out in this document for wind energy development along with projects for integrated renewable energy and passive solar design. Policy RE1 in PPS 18 states that 'significant' weight should be accorded to the wider environmental, economic and social benefits attached to proposed developments. Further information to be read in conjunction with this Planning Policy Statement is contained in *Best Practice Guidance to PPS 18 Renewable Energy* (Department of the Environment 2009b), which outlines a wide range of considerations that need to be addressed across nine renewable energy technologies. Supplementary Planning Guidance is also detailed in the publication *Wind Energy Development in Northern Ireland's Landscapes* (Northern Ireland Environment Agency 2010). These policy considerations have subsequently been transposed in the main into the *Strategic Planning Policy Statement for Northern Ireland* (Department of the Environment 2015) with mention that the wider environmental, economic and social benefits of all proposals for renewable energy projects are material considerations that will be given 'appropriate' weight in determining whether planning permission should be granted.

Meanwhile, in exercising its supervisory jurisdiction over planning decisions, the High Court has neatly summed up the policy spaces being mediated by the planning system within the renewable energy sector. *In the Matter of an Application by RES UK and Ireland Limited for Judicial Review and in the Matter of Decisions of the Planning Appeals Commission dated 26 June 2017* (2018 NIQB 16), Keegan J, in commenting on that case more generally, opined that it:

> highlights the tension between two policy drivers namely, environmental protection and renewable energy. Neither trumps the other but they are not always easy companions. In the absence of local planning policy there are different policy strands and policy documents with subtly different wordings. It is not a simple task to settle upon a fixed and consistent meaning.

A key ground for this judicial review was the misinterpretation of planning policy by the presiding Commissioner relating to the protection of priority blanket bog habitats and, in particular, when and how mitigation and compensatory measures could be taken into account when applying the policy to a wind farm proposal. The decision of the Planning Appeals Commission to refuse planning permission was quashed and reconsideration was invited in the light of the judgment.

Accordingly, it is clear that the development management policy base for renewable energy is not only multistranded but also of considerable vintage and thus, unsurprisingly, Infrastructure Minister Chris Hazzard announced in September 2016 a review of planning policies to take account of rapidly changing technologies and the increasing scale of renewable energy development with special mention of wind turbines and solar farms. External consultants were appointed by the Department for Infrastructure to provide specialist research and policy advice for onshore renewable energy development, including a critical assessment of extant policy and matters to be addressed through new Local Development Plans or through supplementary planning guidance. This initiative was followed in April 2021 by an announcement from Infrastructure Minister Nichola Mallon that the strategic planning policy framework on renewable and low carbon energy would be reviewed in parallel with the Northern Ireland Executive's then emerging 2030 energy strategy. Given the often-contested nature of renewable energy projects, the Minister emphasised that 'there will be extensive engagement with stakeholders and the public to ensure that the new policy is up to date, fit for purpose and has the support of local communities' (Department for Infrastructure 2021a). In December 2021 the Department for Infrastructure (2021b) published a consultation *Issues Paper* as part of this review of strategic planning policy on renewable and low carbon energy and listed specific planning matters germane to that review:

- energy targets and strategic planning policy;
- locational considerations;
- siting new wind farms in perpetuity;
- wind turbines and amenity considerations;
- decommissioning and site restoration for new development;
- solar farms and agricultural land;
- co-locating renewable, low carbon and supporting infrastructure;
- re-powering existing wind farms; and
- emerging technologies.

Accordingly, the relationship between planning and renewable energy has reached a new and significant milestone. The Department for the Economy in its *Energy Strategy for Northern Ireland Consultation on Policy Options* (2021b 97) had expressed this more pithily:

> Planning was a key issue raised in the Call for Evidence as a barrier that could impact on Northern Ireland meeting its renewables targets and therefore our contribution to net zero emissions.

When set against the observation in a detailed background report prepared for the Department for the Economy on the future of renewables in Northern Ireland (Cornwall Insight and Ionic Consulting 2020) that planning at a regional scale is but part of a wider inter-departmental policymaking arena that interacts with investors, local government, consumers and communities, it would seem that mediating between environmental protection and the expansion of renewable energy outputs is urgently awaiting high-level policy succession in Northern

Ireland, while holding on to a fundamental principle of planning practice that the need for renewable energy does not ***automatically*** override environmental protections (Department for Communities and Local Government 2013). Following a review of generating electricity from renewables, the Northern Ireland Audit Office (2020: 32) has specifically recommended that the Department for the Economy should take a lead role in strengthening and formalising partnership arrangements across all relevant public bodies, to ensure that any future renewable electricity or energy schemes are supported by a more proactive and joined-up approach to accreditation, monitoring and enforcement.

Wind energy development management

Much of the investment interest in renewable electricity can be linked to the financial support available through the Northern Ireland Renewables Obligation scheme, which, over the period from 2005, allowed operators of accredited renewable generating stations to receive Renewables Obligation Certificates (ROCs) for each megawatt-hour of eligible output. Apart from grace period exceptions, the scheme closed to new applicants in 2017. These certificates could then be traded and bought by electricity suppliers to demonstrate compliance with the government's renewable obligation level, with the income from selling ROCs giving producers an income stream for a period of twenty years or until 31 March 2037, whichever is earlier (Northern Ireland Audit Office 2020). Onshore wind turbines have generated most renewable electricity, comprising 5 percent of all renewable generating stations in 2019, but 85.4 percent of renewable electricity generated in 2019–20 (ibid.: 16). These stations can vary from small-scale, single microturbines to multi-MW wind farms with multiple turbines and, as shown in Table 11.1, have given rise to a high volume of planning applications with single wind turbines peaking in their development profile during the period prior to closure of the Northern Ireland Renewables Obligation scheme. Over the period 2002–20 some 66 percent of all single wind turbine planning applications were approved, rising to 83 percent when 'withdrawn' and 'no decision recorded' applications are removed from the data set. The corresponding approval rates for the much lower number of wind farm planning applications were 61 percent and 77 percent.

Planning appeals provide a useful source for uncovering the issues embedded in the determination of wind energy development proposals and for purposes of this chapter the content of wind farm–related Planning Appeals Commission reports was examined. Some of the significant principles for planning practice that emerge may be summarised as follows:

- development plans that are silent in relation to renewable energy do not provide assistance or, at best, limited assistance, in considering wind farm appeal proposals (2015/A0102; 2014/A0180);
- moratoria on determining applications for renewable energy development, whilst Local Development Plans are being prepared or updated, are not appropriate (2013/A0220);
- the Environmental Impact Regulations provide for additional environmental information to be provided by appellants and others as appeals are being processed and the Planning Appeals Commission is under a duty to take into account the totality of environmental information presented to it (2015/A0102);
- alternative locations for a wind farm suggested by objectors are deemed to be outside the remit of an appeal, with the Planning Appeals Commission being required to consider the proposal before it (2015/A0102);

TABLE 11.1 Wind energy planning applications in Northern Ireland, 2002–2020

Year	*Single Wind Turbines*					*Wind Farms*				
	Total lodged	*Approved*	*Refused*	*Withdrawn*	*No decision recorded*	*Total lodged*	*Approved*	*Refused*	*Withdrawn*	*No decision recorded*
2002	16	15	0	1		4	1	0	3	
2003	34	33	0	1		5	3	0	2	
2004	48	40	5	3		12	9	1	2	
2005	72	68	1	3		17	13	1	3	
2006	119	107	9	3		24	19	3	2	
2007	166	145	9	12		15	10	2	3	
2008	254	221	19	14		17	9	6	2	
2009	134	97	18	19		15	14	0	1	
2010	489	298	80	111		14	10	2	2	
2011	689	445	95	149		18	11	2	5	
2012	628	424	92	112		8	5	3	0	
2013	582	392	76	114		19	6	10	3	
2014	412	252	89	70	1	6	2	3	1	
2015	324	146	73	103	2	14	7	3	3	1
2016	52	35	1	16		4	0	0	3	1
2017	22	15	3	3	1	5	2	1	1	1
2018	51	42	4	1	4	5	5	0	0	
2019	71	36	1	10	24	4	0	0	0	4
2020	88	7	0	2	79	1	0	0	0	1
Total	**4251**	**2818**	**575**	**747**	**111**	**207**	**126**	**37**	**36**	**8**

Source: Department for Infrastructure. 2021c. *Renewable Energy Planning Applications List, April 2002 to December 2020*. Belfast: Department for Infrastructure.

- should wind farm development with a visual impact on residents at a property be considered unacceptably overbearing to justify refusal of the appeal development? (2015/A0102);
- environmental assessments conducted by statutory consultees outweigh objections by third parties that are general in nature (2013/A0220, 2015/A0102);
- the planning considerations that have relevance to a wind farm go well beyond renewable energy policy and engage with, for example, natural heritage and tourism (2017 /A0023);
- the stated environmental, economic and social benefits of a wind farm proposal can be concluded as substantial, and significant weight can be attached to these benefits in the determination of an appeal (2017 /A0023);
- it is commonplace for appellants to commit to a community benefit fund, but the practice convention adopted is that such payments are not related to the development proposed on the appeal site, do not serve any planning purpose and that weight cannot be attached to this (2014/A0180; 2015/A0102; 2017/A0023). This is in line with the unanimous decision subsequently handed down by the UK Supreme Court in *R (on the application of Wright) (Respondent) v Resilient Energy Severndale Ltd and Forest of Dean District Council (Appellants)* (2019 UKSC 53), whereby it was determined that the grant of planning permission for a wind turbine linked to a community fund donation condition has rightly been quashed as the Council relied on matters that do not qualify as material considerations.
- objections to a wind farm that are withdrawn during the appeal process do not inform the appeal decision (2017 /A0023);
- does the appeal proposal, when added to the existing approved and built wind farms or single turbines in the area, give rise to any significant increase in the impacts of wind farm development in the area or give rise to an overwhelming or unacceptable cumulative impact? Does a wind farm proposal detract any further than existing turbines from the wild character of the surrounding area? (2017 /A0023)
- Noise-related informatives are unenforceable, thus leaving statutory nuisance legislation as the only way to address noise issues emanating from turbines (2013/A0169);
- general criticisms of wind power are inappropriate for consideration in the context of a specific appeal, for example, whether wind turbines are more or less efficient or cost-effective relative to other power sources; general concerns about the green credentials of wind farms; economic viability is a matter for the developer (2013/A0169).

Solar farm development management

As noted by the Northern Ireland Audit Office (2020), there were 23, 690 renewable generating stations accredited to the Northern Ireland Renewables Obligation scheme in March 2019, of which some 94 percent were accounted for by solar stations. However, these were mostly micro-generators contributing just over 3 percent of renewable electricity generated in the year April 2019 to March 2020. The installation of solar farms has the potential to dramatically change the role of solar energy. So, for example, in August 2018, Lightsource BP opened a 6.42MW solar installation in Mallusk, providing power to a manufacturing company just over a mile away. The installation will produce enough renewable electricity, to provide 14 percent of the company's annual demand, in turn saving an estimated 2,000 tonnes of carbon emissions each year. Another example comprises the completion of a solar farm to supply the energy needs of the Dunore Point Water Treatment Works on the shores of Lough Neagh. The project involved the installation of 24,000 solar panels across a 33-acre site and cost approximately £7million to complete. This installation will provide a peak output of 4.99MW, with spare capacity going to the grid. It is expected to save approximately 2,000 tonnes of carbon every year, in addition to saving the company over half a million pounds annually in energy costs (Action Renewables 2019).

Solar farms constitute development and, as shown in Table 11.2, most projects have received planning permission. Clearly, the financial inducement to investors from the Northern Ireland Renewables Obligation scheme, prior to its closure to this technology in March 2017, has influenced the playing out of this development profile with lodged planning applications peaking in 2015, but falling off thereafter. Clearly, the sector has been mindful of the often-lengthy period of time required for submitting and determining planning applications and which, given the large range of material considerations to be addressed, has at times been perceived as a delay in decision making.

Only a few planning decisions for solar farms have resulted in a planning appeal to the Planning Appeals Commission, but one which was the subject of a hearing in October 2017 and a decision in May 2018 (2016/A0240) offers interesting insights into both procedure and reasoning. The proposed development comprised the construction of a solar farm on the fringe of Antrim, originally with a total generating capacity of 25MW. The Council's reasons for refusal centred on visual impact, the partial location of the development on a flood plain and the effects of run-off water, and the failure of the proposal to demonstrate its impact on the surrounding natural environment. In the light of these concerns, a revised scheme was submitted to the Council involving a reduction in the site area, a reduction in the generating capacity, the removal of solar panels associated with watercourse floodplains, an addendum to the flood risk assessment and landscaping details. These amendments were submitted prior to the Council's Planning Committee meeting, but they were not presented to the Planning Committee, which issued its decision on the basis of the original scheme. Subsequently, and in order to assist the Planning Appeals Commission, the Council consulted on the amendments and then withdrew its reasons for refusal related to flood risk concerns and natural environment impact. The presiding Commissioner ruled that, because the Council had an opportunity to review the amendments, it would not be prejudiced by consideration of these amendments in the planning appeal. However, the Council also raised concerns regarding

TABLE 11.2 Solar farm planning applications in Northern Ireland, 2002–2020

Year	*Solar farm planning applications*				
	Total lodged	*Approved*	*Refused*	*Withdrawn*	*No decision recorded*
2013	2	2	0	0	
2014	7	7	0	0	
2015	21	15	2	3	1
2016	4	3	1	0	
2017	1	1	0	0	
Total	**35**	**28**	**3**	**3**	**1**

Source: Department for Infrastructure. 2021c. *Renewable Energy Planning Applications List, April 2002 to December 2020*. Belfast: Department for Infrastructure.

potential prejudice to third parties, who would not have had an opportunity to comment on the proposed adjustments. The Commissioner decided otherwise:

> I consider the following matters to be relevant:
>
> - The development on the appeal site has been considerably reduced in size;
> - A number of solar panels have been removed as a consequence of the reduced site;
> - The size, height and scale of the remaining solar panels and associated infrastructure is unchanged from the original proposal;
> - The proposed amendments to the scheme do not bring the impacts of the development any closer to the neighbouring residential development in the area.
>
> Taking these matters into account I am satisfied that the proposed amendments are not so substantial as to represent a different proposal from that which was the subject of the Council's decision dated 15 December 2016. I am satisfied that the third parties who have not had opportunity to comment would not be prejudiced by the acceptance of these amendments ... Furthermore the local community have the right to become involved in the processing of the planning application and this appeal. I am therefore not persuaded that the local community has not been properly consulted or been given an opportunity to raise concerns (p. 3).

In short, what we see here is the exercising of procedural discretion to hear an appeal that is based on an examination of the facts of each case and an overarching concern to act with fairness to all parties – the Council, the developer and third-party community interests. Nonetheless, the Department for Infrastructure (2022: 56–7) has more generally indicated that it will bring forward proposals that would disallow the variation of a development proposal at appeal on the basis that, if an applicant is aware that a council is minded to refuse planning permission, it may encourage early alteration of the development proposal before a decision is made in the knowledge that it cannot be varied thereafter at appeal stage.

Turning now to the reasoning behind a decision to grant this appeal, three main considerations, with potentially wider relevance, emerge from the Commissioner's report:

- *Matters not relevant or lacking persuasive evidence* – the report notes that criticism of solar energy versus other forms of renewable energy is not a matter to be adjudicated and that issues raised in regard to the appellant's company are not a relevant planning matter. What also comes through as a principle in decision making is that 'general assertions' on concerns such as the loss of agricultural land, effects on property value, the loss of recreational amenity to those who spend leisure time within the area, the abandonment of the solar panels to rust and giving rise to leakage, the safety of children, the personal health of a local resident, driver distraction, wind noise pollution, the impact on the local commercial economy and tourism sector, and wildlife issues cannot be sustained without persuasive evidence.
- *Environmental, economic and social benefits* – the *Strategic Planning Policy Statement for Northern Ireland* states that the wider benefits of all renewable energy projects are material considerations to be given appropriate weight and thus it is now routine that an appellant will give details on employment creation, carbon emission savings, financial returns to

the economy as well as more generic contributions to renewable energy targets and consistency with government strategies. This case adopts that rubric and has allowed the Commissioner to accept that the benefits are substantial and attract significant weight. Community gain is also offered as an additional benefit. However, as noted in the case of wind farms above, it is held that this does not serve any planning purpose and is a matter for the appellant. Accordingly, it would seem that the commonplace stance is to simply note this financial contribution and to conclude that the environmental, economic and social benefits of the project stand regardless of whether or not a community fund is in place.

- *The testing of empirical evidence* – the detailed evidence presented at this appeal included an assessment of the visual impact of the solar farm on the surrounding landscape and the setting of Antrim, What is significant, however, is that the Commissioner undertook her own assessment of visibility using the appellant's viewpoints, as well as a road-based transient assessment and an examination of the cumulative visual impact of wind and solar infrastructure in the surrounding area. The Commissioner also made a visual assessment of the appeal proposal in regard to its impact on the residential amenity of ten properties. The important point here is that this testing of empirical evidence provides for a robust balancing of material considerations in reaching a decision as well as informing the use of planning conditions to satisfactorily mitigate some of the identified impacts from the solar farm.

Renewable electricity and the transmission system

The planning and operation of the electricity grid in Northern Ireland is the responsibility of the transmission system operator, SONI Ltd, which brings electricity from where it is generated to the Northern Ireland Electricity Networks' distribution infrastructure that interfaces directly with consumers. It is a wholly owned subsidiary of EirGrid plc, registered in Dublin, comes under the remit of the Northern Ireland Utility Regulator and is a key player in the all-island single electricity market. In preparing to help meet the 70 percent at least renewable electricity target by 2030 mentioned above, SONI Ltd has stated its position as follows:

> The grid will need to carry more power and most of this power will come from renewable generation that varies depending on the weather. Where we can, we will use the existing grid to meet this goal. However, given the scale of change, we will also need to plan for a great deal of new grid infrastructure – such as underground cables, pylons and substations.
>
> *(p. 2)*

Clearly, project investment will interface with development management but in giving consideration to a more strategic approach that can shape the future of the electricity grid, SONI Ltd published a consultation paper in 2021 that set out four options: generation-led by putting clean electricity generation close to where most power is used; developer-led by letting developers decide where to locate clean electricity generation; technology-led by trying new ways to move clean electricity across Northern Ireland; and demand-led by putting electricity users close to sources of clean electricity generation (SONI Ltd 2021a). A similar set of approaches has been articulated for the Republic of Ireland with the Irish Planning Institute

(2021) arguing that one approach alone is inadequate and that, instead, a blended approach across all these options is to be preferred. From the perspective of planning practice what comes through is the spatiality of grid development with implications for development plans regarding the location of large-scale renewable energy sources, settlement planning and economic growth. SONI Ltd takes the view that regardless of a decision being taken on the way forward, there remains a fundamental challenge of delivering new projects to strengthen the grid and it assumes that a new North South Interconnector will go ahead 'as planned and on schedule' (p.9). The Transmission Development Plan Northern Ireland 2020-2029, approved by the Utility Regulator in March 2021, envisages the Interconnector being operational in 2025 (SONI Ltd 2021b). This strategically significant infrastructure has attracted considerable public attention with its merits and decision making being hotly contested within the regulatory planning and legal arenas. There were, for example, in excess of 6,000 letters of objection to the original proposal. Accordingly, it requires brief consideration in this chapter.

The North South Interconnector project comprises the addition of a new 400kV overhead line that will connect the electricity grids of Northern Ireland and the Republic of Ireland. When commissioned, the Northern Ireland part of the line will run from a substation near Moy in County Tyrone and through County Armagh – a distance of some 34 kms with just over 100 towers – and then into Counties Monaghan, Cavan and Meath. That part of the project involving development along approximately 100 kms in the Republic of Ireland was the subject of separate planning procedures in that jurisdiction, which included the granting of planning permission by An Bord Pleanála in December 2016 and subsequent legal challenges. This consent was upheld by the Supreme Court in February 2019. Within Northern Ireland the planning process formally commenced with the submission of planning applications by Northern Ireland Electricity in December 2009 (O/2009/0792/F) and April 2013 (O/2013/0214/F). In 2011 the Department of the Environment requested that the Planning Appeals Commission hold a public local inquiry into the proposed development associated with the initial planning application. This was convened in 2012 but was adjourned to allow for updating and proper advertising of the Environmental Statement. The Department then renewed its request in 2014 for a public inquiry into the 2009 application that would also take on board the later 2013 planning application, but, given the passage of time and changes to the proposal, the Planning Appeals Commission determined that it was necessary to recommence the process with fresh evidence. The interconnector project thus became the subject of a two-stage public inquiry dealing with legal and procedural issues in June 2016 and the planning merits in February 2017. The substantive report with conclusions and a recommendation that full planning permission (with conditions) be granted was tendered to the Department for Infrastructure in November 2017. SONI Ltd, which had assumed responsibility in April 2014 for progressing the planning applications, received planning approval in January 2018. However, a legal challenge that planning permission could not be granted in the absence of the Infrastructure Minister was conceded and the decisions were quashed by the High Court in February 2019 and remitted back to the Department for redetermination. Following the restoration of the Northern Ireland Executive in January 2020, the Minister of the Department for Infrastructure approved planning permission in September 2020 for that part of the North South Interconnector project within Northern Ireland. But that was not the end of the matter.

In March 2021 the community-based coalition of Safe Electricity for Armagh and Tyrone (SEAT), which had been an objector at the public inquiry, was granted leave to apply for judicial review by Scoffield J. with detailed arguments to be heard thereafter. A key ground was the assertion that the planning decision was a cross-cutting matter, involving more than one Minister, and thus required the participation of the Northern Ireland Executive. It was contended that the Minister did not refer the significant and / or controversial decision to the Executive for discussion and to agree whether the planning applications should be granted or refused in compliance with her obligations under the Ministerial Code. A point at issue was the relationship between the then extant 2006 Ministerial Code and an amendment introduced by the Executive Committee (Functions) Act (Northern Ireland) 2020, which removes a requirement to refer a matter to the Executive for discussion and agreement and gives discretion to the relevant Minister to determine the matter. The Ministerial Code at the time of this hearing had not yet been aligned with the amended statutory position secured by the 2020 Act albeit there was evidence before the court that some updating work had been conducted by officials. This fell short, however, of full compliance with the formal statutory procedure for amendment to the Ministerial Code, which includes cross-community Assembly support before such an amendment takes effect. Accordingly, the court ruled that the Minister's obligations under the existing Code had not been properly discharged requiring the referral of a significant and controversial matter to the Executive Committee for its consideration, but that this was not a 'wilful or reckless disregard of the Code in circumstances where Executive referral was likely to make a difference to the outcome in substance' (2021 NIQB 93). An order of *certiorari* to quash the planning permissions was deemed by Scoffield J as inappropriate and a declaration was granted with a view that the Executive Committee proceed 'as expeditiously as possible with amendment to the Ministerial Code' to reflect the new and modified requirements of the 2020 legislation. The court concluded:

> Had the necessary amendments been made at the appropriate time, it is unlikely that these proceedings would have been brought.

While the applicants in this High Court litigation did not pursue any grounds of challenge relating to the planning merits of the development proposals, the 2017 report of the Planning Appeals Commission provides a useful source document to review the issues examined in reaching its conclusion to recommend approval of planning permission. This is worthy of brief mention. The guiding principle for the Planning Appeals Commission role in this case is cogently stated as being 'not to make planning policy but to interpret and apply it only in relation to the proposed development' (p. 26). Accordingly, what watermarks the approach taken here is the recurrent return to development plans and planning policy statements against which to benchmark the evidence of all the parties contributing to the public inquiry. So, for example, considerable attention at the outset is given to an analysis of energy policy, which is acknowledged as being within the remit of the Department of Enterprise, Trade and Investment / Department for the Economy. But the important point here is that this is specifically interpreted within the context of Policy PSU 2 of the Planning Strategy for Rural Northern Ireland, which defines the appropriate planning test as being 'whether there is an overriding national or regional need for the proposed development' (p. 151). The Planning Appeals Commission concludes:

> On the basis of the evidence … the test is satisfied on grounds of competition, security of supply, the facilitation of the use of renewable energy and the technical limitations of

> the existing transmission system. There is also evidence that the proposed development enjoys cross-border governmental support and would yield long-term benefits on an all-island basis. Its designation as a Project of Common Interest identifies it as being of strategic trans-European importance ... Therefore, these considerations amount to imperative reasons of over-riding public interest. The policy-driven case of need for this project is compelling and potential negative environmental impacts must be balanced against this material consideration as permitted by statute and required by Policy PSU 2.

Moreover, the benefits of the proposal related to contributing to the regional economy that include boosting employment in the renewable sector and giving confidence to investors within the sector are acknowledged as material considerations attracting considerable weight. When conjoined with this overriding public interest, the Commission concludes that these material considerations command 'determining weight'. On the other hand, determining weight is not attached to the significant localised adverse impact upon residential visual amenity and landscape character. It is argued that the 'degree of interference is necessary in the public interest and is proportionate' (p.153). In short, while the evidence that the proposed development will produce some unavoidable environmental impacts that cannot be adequately mitigated is embraced within the balancing exercise, the Planning Appeals Commission has determined that 'there are no persuasive grounds for recommending refusal of the planning applications on the basis of environmental impact' and that it is 'satisfied that it represents the best achievable balance between environmental impacts, technical requirements and economic limitations' (p. 155). To conclude, over and above its intrinsic relevance to renewable energy infrastructure, this report deserves close reading in order to appreciate the multiple issues that are evaluated against planning policy in the balancing of material considerations and the weight that is given to each in reaching a recommendation that can inform decision making.

Battery Electricity Storage Systems

One category of development related to renewable energy that has raised planning issues, largely underplayed within the extant policy framework, comprises Battery Energy Storage Systems (BESS). This infrastructure, more generally, is part of a growing investment in electricity storage facilities that seek to offset, in particular, the intermittent and unpredictable nature of wind generation. Essentially it allows the technology to support the grid at times when renewable generators are producing high levels of electricity but when demand is low. Key to its future role will be the availability of long-duration storage. When situated adjacent to an electricity substation, a battery storage facility can connect easily to the supply system. Queries have arisen as to whether this form of development should be considered electricity generating and while the Planning Appeals Commission in granting planning permission for a battery energy storage facility at Kells (see 2018/A0248) held that it was not, the Department for Infrastructure (2020) subsequently confirmed by way of *Chief Planner's Update* 7 that for purposes of planning in Northern Ireland, electricity storage development does fall within the meaning of an electricity generating station. The implication of this clarification is that it brings into consideration the Planning (Development Management) Regulations (Northern Ireland) 2015, which defines a hierarchy of developments as comprising major and local categories. In the Schedule to these regulations 'Energy Infrastructure' comprising electricity generating stations is listed in Class 2 as major development when the generating capacity is or exceeds 30 megawatts. This, in turn, introduces mandatory consultation with the Department for Infrastructure to determine if the proposed development is one of regional significance, a requirement for pre-application community consultation, and the possibility of a planning application made initially to a district council being called-in by the Department for Infrastructure. One consequence of this *Update* provided for purposes of clarification is that it has created 'a recipe for administrative chaos' with planning officers of district councils faced with a choice of following the Planning Appeals Commission decision or the Chief Planner's advice. This matter was judicially reviewed (2021 NIQB 96), with Humphreys J concluding in favour of the applicant that:

> it was simply not open to him (ie. the Chief Planner), as a matter of constitutional propriety and *Wednesbury* rationality, to depart from the PAC decision. I recognise that the respondent is not bound by the PAC decision as a matter of strict precedent but I concur with the view expressed by Mr Orbinson that such decisions must either be accepted and respected or challenged through the courts. I would add that it is always open to the respondent to seek to amend legislation through the Assembly or to introduce new planning policies. Both these routes would ensure an appropriate level of industry and community wide consultation which can only ensure better decision-making.

Previously in that judgment the court also declared on the basis of statutory interpretation that while BESS does act as an electricity generation device, it does not do so wholly within any single class of development as set out in Classes 1 to 8 of the 2015 Development Management Regulations. It should be regarded, therefore, as 'all other development' within Class 9 of the Schedule and is subject to the thresholds contained therein, which are: development that comprises 5,000 square metres or more gross floor space; or the area of the site is or exceeds 1 hectare. If neither of these criteria is met, then the proposed development is local. Accordingly, the advice given in *Chief Planner's Update* 7 was deemed wrong in law. However, it was also noted that the Planning Act (Northern Ireland) 2011, Section 25 (3) specifies that the Department for Infrastructure may direct that a specific local development is to be dealt with as if it were a major development. More than probably, in regard to the harnessing of renewable energy, we will see planning policy refinement applied to BESS technology in Northern Ireland.

Conclusion

The publication in 2021 of the United Nations Intergovernmental Panel on Climate Change report has renewed societal awareness of the need to deliver a low carbon future. Embracing renewable energy in an aspired-to global commitment to transition away from fossil fuels is now centre stage in the attack on ecological breakdown. The public policy scope of renewable energy is wide-ranging with regulatory planning tasked with mediating project decisions within a diverse set of environmental, economic and social considerations. This chapter throws light on some of these matters in what is a fast-evolving sector and where new technologies have the potential to outrun the capacity of the planning system to engage effectively on behalf of the public interest. There is an urgency attached to planning policy succession in Northern Ireland that can respond to the challenges of moving away from a dependency on fossil fuels, helping to deliver a low carbon future and promoting greater community acceptance in this quest for a more complete renewable energy reality.

Legislation

Planning Act (Northern Ireland) 2011
The Planning (Development Management) Regulations (Northern Ireland) 2015

Cases cited

In the Matter of an Application by ABO Wind NI Limited & Energia Renewables Company 1 Limited for Judicial Review and in the Matter of the Chief Planner's Update 7 (2021 NIQB 96)

In the Matter of an Application by RES UK and Ireland Limited for Judicial Review and in the Matter of Decisions of the Planning Appeals Commission dated 26 June 2017 (2018 NIQB 16)

In the Matter of an Application by SAFE Electricity A & T Limited and Patrick Woods for Judicial Review and In the Matter of Decisions of the Minister for the Department for Infrastructure and the Executive Office (2021 NIQB 93)

Planning Appeals Commission. 2014. *Appeal by Drumadarragh Wind Farm Limited Against the Refusal of Full Planning Permission, Drumadarragh Hill, Ballyclare.* Commission Reference: 2013/A0169.

Planning Appeals Commission. 2015. *Appeal Against the Non-determination of an Application for Full Planning Permission for Wind Turbines at Bessy Bell Wind Farm, County Tyrone.* Commission Reference: 2013/A0220.

Planning Appeals Commission. 2017. *Regionally Significant Planning Application by SONI Ltd.* Report on the Second Stage of the Public Inquiry. Commission References: 2011/C001 and 2014/C004

Planning Appeals Commission. 2017. *Appeal Against the Non-determination of an Application for Full Planning Permission for the Erection of Three Wind Turbines at Seegronan Wind Farm, Killeter, Strabane.* Commission Reference: 2014/A0180.

Planning Appeals Commission. 2018. *Construction and Operation of a Solar Farm East of Dunsilly Roundabout, Antrim.* Commission Reference: 2016/A0240.

Planning Appeals Commission. 2018. *Appeal Against the Refusal of Full Planning Permission for Proposed Extension to Dunmore Windfarm at Dunmore, Limavady.* Commission Reference: 2017/A0023.

Planning Appeals Commission. 2019. *Appeal Against the Refusal of Full Planning Permission, Barr Cregg Wind Farm, County Londonderry.* Commission Reference: 2015/A0102.

Planning Appeals Commission. 2019. *Battery Energy Storage Facility at Doagh Road, Kells.* Commission Reference: 2018/A0248.

Planning Appeals Commission. 2020. *Regionally Significant Planning Application by Renewable Energy Systems Limited – Proposed Wind Farm Comprising 11 Three- blade Wind Turbines on Land Approximately 3 Kilometres to the West of Swatragh, Accessed off Corlacky Road, in the Townlands of Corlacky, Knockoneill, Half Gayne and Tirkane, County Derry/Londonderry (LA09/2016/0232/F).* Commission Reference: 2018/C009.

R (on the application of Wright) (Respondent) v Resilient Energy Severndale Ltd and Forest of Dean District Council (Appellants) (2019 UKSC 53)

References

Action Renewables. 2019. *What's the Latest with Solar Farms in Northern Ireland?* Available at actionrenewables.co.uk/news-events/post.php?s=whats-the-latest-with-solar-farms-in-northern-ireland (accessed 16 August 2021).

Cornwall Insight and Ionic Consulting. 2020. *Future of Renewables in Northern Ireland*. Belfast: Department for the Economy.

Department for Communities and Local Government. 2013. *Planning Practice Guidance for Renewable and Low Carbon Energy*. London: Department for Communities and Local Government.

Department for Infrastructure. 2020. *Battery Energy Storage Systems (BESS), Chief Planner's Update 7*. Belfast: Department for Infrastructure.

Department for Infrastructure. 2021a. *Mallon Gives 'Green Light' to Renewable Energy Planning Review*. 21 April. Belfast: Department for Infrastructure.

Department for Infrastructure. 2021b. *Issues Paper: Review of Strategic Planning Policy on Renewable and Low Carbon Energy*. Belfast: Department for Infrastructure.

Department for Infrastructure. 2021c. *Renewable Energy Planning Applications List, April 2002 to December 2020*. Belfast: Department for Infrastructure.

Department for Infrastructure. 2022. *Review of the Implementation of the Planning Act (NI) 2011, Report*. Belfast: Department for Infrastructure.

Department for the Economy. 2021a. *Electricity Consumption and Renewable Generation in Northern Ireland: Year Ending March 2021*. Belfast: Department for the Economy.

Department for the Economy. 2021b. *Energy Strategy for Northern Ireland: Consultation on Policy Options*. Belfast: Department for the Economy.

Department of the Environment. 2009a. *Planning Policy Statement 18 Renewable Energy*. Belfast: Department of the Environment.

Department of the Environment. 2009b. *Best Practice Guidance to PPS 18 Renewable Energy*. Belfast: Department of the Environment.

Department of the Environment. 2015. *Strategic Planning Policy Statement for Northern Ireland*. Belfast: Department of the Environment.

Irish Planning Institute. 2021. *Comment on Eirgrid's Shaping Our Electricity Future Roadmap*. 14 June. Dublin: Irish Planning Institute.

KPMG. 2021. *An Analysis on the Future Role of Distributed Generation in Northern Ireland. A Report for RenewableNI*. Available at https://renewableni.com/wp- content/uploads/2021/07/KPMG-Analysis-of-Future-for-Distributed-Generation.pdf.

Northern Ireland Audit Office. 2020. *Generating Electricity from Renewable Energy*. Belfast: Northern Ireland Audit Office.

Northern Ireland Environment Agency. 2010. *Wind Energy Development in Northern Ireland's Landscapes: Supplementary Planning Guidance to Accompany Planning Policy Statement 18 Renewable Energy*. NIEA Research and Development Series 10/01. Belfast: Department of the Environment.

Northern Ireland Executive. 2021. *The Path to Zero Net Energy: Energy Strategy for Northern Ireland*. Belfast: Northern Ireland Executive.

SONI Ltd. 2021a. *Shaping Our Electricity Future: Preparing for At Least 70% Clean Electricity by 2030*. Belfast: SONI Ltd.

SONI Ltd. 2021b. *Transmission Development Plan Northern Ireland 2020–2029*. Belfast: SONI Ltd.

12

PLANNING AND PLACEMAKING

Mura Quigley

Introduction

Placemaking as a concept and practice was integrated into spatial planning in the United Kingdom in the 1990s and has since become widespread in its application across towns and cities. The global outbreak of COVID-19 in 2020 has propelled placemaking to become the focus of attention for making town and city centres more vibrant, authentic and diverse. Originating from the embedding of the social sciences into the public space discourse, its roots can be traced back to studies of public space and life in the 1960s by Jane Jacobs and William 'Holly' Whyte. Since that time, placemaking across the world has accelerated at pace as part of a new cultural regeneration approach, as a way to counter the negative consequences of car domination, and as a return to human-centred development, walkable neighbourhoods and vibrant public spaces.

Northern Ireland has not been unaffected by this momentum towards placemaking. Once considered a peripheral community action, placemaking has been adopted by the rubric of formal planning in order to articulate a new direction for the profession. Greater awareness of having greener, cleaner, safer and healthier public spaces has gained legitimacy. Alternative narratives for city and town high streets, struggling to adapt beyond retail and to deal with the costs of congestion and air quality, are calling for the strengthening of social cohesion and social capital in order to build local resilience. Nonetheless, progress on the ground has been slow as the priorities for government in the wake of protracted civil conflict continue to be focused on 'recovering' rather than 'transforming' or 'adapting'. This has resulted in predominantly top-down, formalised approaches to placemaking and limited engagement with authentic placemaking in cities, towns and villages where the private car always wins in the battle for public space. However, with district councils having latterly gained some place-related functions, the green shoots of optimism are emerging that can facilitate more bottom-up community action, a new active citizenship and the creation of places that work for people. Climate and health agendas are helping to shape this trajectory in line with the traction being gained for more availability of local walking and cycling opportunities.

DOI: 10.4324/9781003218319-12

It is against this background that this chapter explores the evolution of the placemaking concept more widely and then links it to the current practice of placemaking in Northern Ireland. The institutional context for placemaking is reviewed along with varied approaches to delivery, including the emergent contribution of local government. Reflections are provided on the challenges and opportunities for placemaking, not least in the context of COVID-19 and its effects on placemaking. The key message in this chapter is that Northern Ireland needs to overcome an institutional inertia and adapt its governance towards more networked, open and collaborative partnership working across sectors and with local communities in order to be successful in its placemaking endeavours.

Perspectives on the evolution of placemaking

The concept of placemaking became popular during the 1960s as a reaction to large-scale infrastructure and urban development that was taking place in the United States. At that time a perception of increased 'placelessness' was being articulated in the writing of Jane Jacobs (1961), who felt that planning had become too top-down, lacking in context and unable to manage the complexity of the urban environment. Furthermore, the increased speed of change to accommodate an urbanising world and the rise of the car was resulting in neighbourhoods where local character, identity and culture were being erased. What they were being replaced with was feeling increasingly clinical, devoid of life and 'place-less'. Jacobs captured the moment and accorded relevance to the civic voice within planning, but, most importantly, she began to articulate the distinctive qualities of places that work and advocated a number of ideas that have informed placemaking today, such as mixed use, vibrancy and natural surveillance.

Around the same time, William 'Holly' Whyte led a *Street Life Project* in New York that studied the behaviour of people in newly developed public spaces (Whyte 1980). This focus on streets and behaviour in public spaces continued to the 1990s with calls for a more human-centred approach to planning and design that employed serial vision, ethnographic studies and observational research. Danish architect and urban designer Jan Gehl popularised the ethnographic approach of Whyte, as did the *Project for Public Spaces*, which was set up following involvement by its founder, Fred Kent, in the *Street Life Project*. William 'Holly' Whyte and Jan Gehl are celebrated pioneers of people-centred approaches through their focus on street life, how people behave in public spaces and planning at the human scale. Today, Gehl Architects and the *Project for Public Spaces* are in the global leadership vanguard of placemaking approaches, which embrace the human, neighbourhood scale and which integrate sociology, culture and anthropology into placemaking and urban design.

In the United Kingdom, placemaking gained popularity during the 1990s and early 2000s and the process began to be formalised within government policy and programmes. This coincided with a European spatial planning agenda (Faludi and Waterhout 2002; Royal Town Planning Institute 2007) and the communicative turn in planning (Healey and Hillier 1996). The post-industrial decline of urban centres was much in evidence and, thus, regeneration moved into the planning mainstream, along with a call for better-designed places. This was thrown into sharp focus by the publication of the Urban Task Force report *Towards an Urban Renaissance*, which developed key principles and ideas around sustainable urban design such as mixed-use, density and movement interactions (Rogers 1999). The Urban Task Force report constituted a watershed moment for urban design and placemaking as it focused attention on sustainable regeneration, the reuse of brownfield sites and the concept of walkable communities. The report, however, has been criticised for its neoliberal tendencies and its support for place competitiveness in urban policy with questions at the time being raised about the contribution of placemaking to gentrification (Punter 2007).

Also in 1999 the Commission for Architecture and the Built Environment (CABE) was established as a non-departmental public body funded by two government departments (the Department for Culture, Media and Sport and the Department for Communities and Local Government) and which, until its demise and integration into the Design Council during the Conservative Government austerity cuts of 2011, was influential in advocating for more design-led approaches to planning in England. Elsewhere in Great Britain this initiative was complemented by the setting up of Architecture and Design Scotland and the Design Commission for Wales. During its existence CABE established the practice of design reviews, campaigned for green infrastructure, coordinated the delivery of *Building for Life* standards, improved quality of design in public procurement, and it produced important research highlighting the economic, environmental and social value of design and placemaking (CABE 2001). CABE also contributed to design policy within the planning system and, shortly after the publication of the Urban Task Force report, it promulgated urban design policy guidance, *ByDesign* (CABE 2000), that had the aim of improving the planning system through better design. While the word 'placemaking' does not appear in this document, *ByDesign* sets out important urban design objectives, such as quality of the public realm, adaptability and diversity, that tie in with ideas around place quality and how to achieve it.

The period post-2000 across the United Kingdom heralded an urban design renaissance with an accompanying sharper focus in planning. The influence of the *European Spatial Development Perspective* is important in that regard because it opened minds to the potential of networked governance, collaborative spaces, fuzzy boundaries and a wider 'place-shaping' agenda that could break down operational silos and promote working collectively for better place outcomes. These ideas have been carried forward by practices that seek to respond to sustainability, resilience and climate change challenges. Within the United Kingdom many of the concepts of placemaking have been adopted into formal planning policy, but at varying speeds and in accordance with political context and governance systems that put differing emphases on aspects of health, economy and community engagement. In Scotland there is now policy and guidance to help support more people-centred planning, exemplified by the *Place Standard* tool, developed jointly by NHS Scotland and Architecture and Design Scotland, but used widely now as best practice in collaborative placemaking (Architecture and Design Scotland et al. 2017). An emphasis on better long-term planning and coordination across health and planning is also apparent in Wales. The Well-being of Future Generations (Wales) Act 2015 requires public bodies to think about wellbeing in public spaces and how, instead of being reactive to public health issues, they might be addressed through better integration of health into placemaking and planning. One outworking of this shift towards healthy places has been the establishment of a *Placemaking Wales Partnership*, a multidisciplinary group representing professions and organisations working within the built and natural environment (Design Commission for Wales 2020a). The Partnership developed a *Placemaking Charter* inviting professionals to sign up to a series of six key placemaking principles of people and community, movement, public realm, location, mix of uses and identity (Design Commission for Wales 2020b). While it is too early to determine if this Charter is improving the quality of placemaking, the very act of coming together (albeit at a national level) under a shared vision and set of principles underpins the ethos of what good placemaking should champion. As well as the Charter, a placemaking guide was published in 2020 (Design Commission for Wales 2020c) that aims to encourage a greater 'whole place' approach to the design and planning of places.

Placemaking and the Northern Ireland planning system

The integration of 'place' into the Northern Ireland planning system followed closely the cultural changes happening in the rest of the United Kingdom and can be viewed as part of the reaction to increasing globalisation and homogenisation of towns and city centres into 'samey streets' of chain stores. This corporatisation of towns was slower to develop in Northern Ireland, but it is present and is denoted by a standard design and branding for shops and stores that do not integrate into their surrounding townscape context. Instead, they impose a feeling of being 'anywhere', whether along the high street or in out-of-town retail centres. In Northern Ireland, at the same time as the 'CABE experiment' was kicking off in England, the former Department of the Environment (through the Planning Service) and the Department for Regional Development (through Roads Service) collaborated to create the first integrated design guidance for the region in 2000, named *Creating Places*. This reinforced the previous aspirations of the Quality Initiative in Planning introduced by the Department of the Environment in 1996, which emphasised the need to secure high-quality design, layout and landscaping for new residential development that would also incorporate open space, walking and cycleways, and the promotion of public transport provision. *Creating Places* represented a significant move away from heavy engineering solutions for residential roads and called for a clear design concept, process and focus on the 'place function' of streets in residential neighbourhoods. It continues to be used as a material consideration for the determination of planning applications, notwithstanding the emergence of additional placemaking and design guidance in the interim.

At around the same time as this new design guidance was published for Northern Ireland, a series of regeneration projects, spearheaded by the former Department for Social Development through its urban regeneration remit, saw the preparation of town and city centre masterplans, many of which subsequently led on to public realm schemes as key priorities in their delivery. However, the main criticism of these is that they were largely top-down and, while there was consultation with local interests, there was little community ownership. The slow progress on placemaking in Northern Ireland is, in part, due to this centralised power and function relationship that continues to exist. While local authorities in England, Scotland, Wales and the Republic of Ireland have remits and funding for the delivery of placemaking, Northern Ireland district councils do not have responsibility for roads nor do they have full regeneration powers, thus making the task of placemaking more complex and difficult to achieve. The contextual approaches required to achieve places that are distinctive and that draw on local community involvement have been largely absent. The result is an often placeless character from central government department-led placemaking projects. Moreover, this historical lack of 'localness' in Northern Ireland has led to a dearth of local expertise, knowledge and capacity for civil society involvement. This has been evidenced by the poor response from the public during the master planning process, as well as a number of subsequent public realm projects that at the time of development have experienced strong community opposition. A further challenge is the need for better stewardship in the delivery of well-designed streets. Utility providers, district councils and others have highlighted that these placemaking projects can have problems of legacy as, in many cases, there are subsequent digging works followed by surface replacement with standard tarmac.

The reform of planning in Northern Ireland over the period from 2011 has allowed for a deeper embedding of place, placemaking and stewardship by declaring a plan-led system and the adoption of a policy focus that has shifted towards sustainable development. In 2014 the Department of the Environment published *Living Places: An Urban Stewardship and Design Guide for Northern Ireland*. This guidance breaks new ground within the United Kingdom by bringing together the process of stewardship and concepts of urban design within planning policy. It identifies and illustrates ten qualities that represent the ingredients of successful places, advocates a unity of purpose between urban design and placemaking and defines placemaking as the creation of somewhere with a distinct identity. This document was closely followed by the issuing of the *Strategic Planning Policy Statement for Northern Ireland* (Department of the Environment 2015), which sharpened the focus on a plan-led system and sustainable development by setting out five core planning principles to assist plan-making and decision taking. One of these in the context of this chapter is especially important: 'Supporting Good Design and Positive Placemaking'. Three paragraphs under that heading capture the spirit and methodology of placemaking:

> Placemaking is a people centred approach to the planning, design and stewardship of new developments and public spaces that seeks to enhance the unique qualities of a place, how these developed over time and what they will be like in the future. Key to successful placemaking is identifying the assets of a particular place as well as developing a vision for its future potential. This involves a collaborative approach that recognises the value of cooperating and communicating with the people that use a place. Placemaking is not simply the remit of architects but requires collaborative working between architects, town planners, transport planners, urban designers, landscape architects, transportation and roads engineers, developers, public space providers, communities, individuals and everyone else affected by the development process.
>
> Successful placemaking also acknowledges the need for quality, place-specific, contextual design, whilst at the same time making responsible use of limited resources. Successful placemaking promotes accessibility and inclusivity for all, acknowledges the importance of creating hospitable and safe places, and emphasises the contribution that vibrancy, adaptability, and diversity of use can make to the viability of place and how it will endure into future.
>
> Key to successful placemaking is the relationship between different buildings; the relationship between buildings and streets, squares, parks, waterways and other spaces; the nature and quality of the public domain itself; the relationship of one part of a village, town or city with other parts and the patterns of movement and activity that are thereby established. The compatibility of a development with its immediate and wider context, and the settlement pattern of a particular area, is an important consideration.
>
> *(Department of the Environment 2015: 21)*

Other initiatives dealing with governance have sought to enhance placemaking in Northern Ireland. In 2007 a Ministerial Advisory Group for Architecture and the Built Environment was established by the former Department of Culture, Arts and Leisure (DCAL), a forerunner of the current Department for Communities. This led on from the publication of the *Architecture and the Built Environment for Northern Ireland* policy (DCAL 2006). In declaring a commitment to good design, the implementation strategy of that document called for the setting up of appropriate institutional arrangements and noted that, for Northern Ireland, the Government had agreed to establish an influential Ministerial Advisory Group to advise the Minister, who would, in turn, become the Ministerial Design Champion. A main role of the Ministerial Advisory Group is to appoint Design Review Panels composed of experts to examine the design aspects of significant projects, and, at present, these reviews occur at the pre-planning application stage through post-planning application referrals and by referrals from funders. Reviews can be called for by developers and promoters, project and design teams, district council planning authorities and the Department for Infrastructure.

While the statutory planning system in Northern Ireland is currently overseen by the Department for Infrastructure, the Department for Communities is responsible for architecture and built environment policy, as well as urban regeneration, and so an overlap occurs in the operation of these two Departments in the sphere of placemaking. Collaborative working is necessary and is evidenced by the establishment of a Strategic Design Group as an informal partnership with a membership of people 'passionate about making places better' (Department for Infrastructure 2021). The Strategic Design Group was jointly established by the Department for Infrastructure and the Ministerial Design Group under the auspices of the Department for Communities to encourage more knowledge-sharing across government, professional bodies and the voluntary sector. Its purpose is to work together 'to promote successful, inclusive, well-designed places which inspire civic stewardship and have an enduring positive impact on people's lives' (Department for Infrastructure 2021).

To demonstrate the continued relevance of the 2014 *Living Places* guide, the Strategic Design Group initiated a pilot 100-day intervention project in 2018 along a city centre street to test placemaking principles and to share learning outcomes. The rationale for what was coined the *Better Bedford Street* project can be summarised as follows:

- the street is a key gateway to the city centre and forms part of a main walking corridor from the south of the city;
- it is dominated by the car and demonstrates competing priorities between pedestrians, parked and moving vehicles;
- it is a wide street that can be challenging for pedestrians to cross and navigate, especially for those with disabilities, because of street clutter;
- people do not seem to spend time there;
- it has transferable learning potential.

The plan involved the removal of car parking bays and footpath widening, the introduction of seating, greenery and artwork, street music, heritage walking tours and outdoor dining. These elements were trialled with varying degrees of success. The temporary footway extension helped to rebalance the street in favour of pedestrians and the introduction of trees created a more pleasant environment. Learning included recognition of the challenges for departmental collaboration in the face of regulatory requirements, the capacity to challenge structures and norms and the dependency of better placemaking initiatives on the involvement of all relevant parties and users (Department for Infrastructure 2020).

Another recent placemaking initiative that was locally driven is the *Ormeau Parklet* along Ormeau Road in south Belfast. The project was developed in 2020 as a collaborative venture involving Queen's University Belfast, the community and voluntary sectors, local businesses, the Department for Infrastructure and Belfast City Council. The purpose of the pilot was to carry out a modest change to the streetscape by removing a number of car parking spaces and introducing planting and seating. Co-design and a collaborative approach involving the project partners, local businesses and residents have generated a beneficial outcome (Sustainable NI 2020). In short, small-scale projects, such as *Better Bedford Street* and *Ormeau parklet*, can contribute greatly to placemaking policy formulation and implementation.

The context for placemaking has also evolved following local authority reform whereby merged district councils were allocated planning powers in regard to development plans, development management and enforcement. However, the heritage management and urban design resourcing within local authority planning departments is low and they have very little specialist expertise. The Local Government Act (Northern Ireland) 2014 charged councils with community planning responsibilities that are centred on improving citizen wellbeing. Through their community planning duties, some councils are progressing placemaking plans for their towns that are facilitating new ways of working with local stakeholders in order to improve the quality of the built environment. In Armagh City, for example, a new plan has identified a range of strategies to promote positive placemaking in the city. This included a UK City of Culture 2025 bid, commitments on the location of services from statutory bodies that they will contribute to placemaking by taking a town centre–first approach, and the establishment of a green and blue infrastructure network across the city that will create new people-focused places aimed at walkers and cyclists. Additionally, district councils are taking direction from the *Strategic Planning Policy Statement for Northern Ireland* by embedding good design and positive placemaking policies into their Local Development Plans. Some are recruiting staff to work primarily to produce a place-shaping and regeneration brief by managing the delivery of their Community Plan, Local Development Plan and Corporate Plan, and assisting with the discharging of statutory planning responsibilities.

Healthy placemaking in a post-COVID-19 world

The COVID-19 pandemic has become a watershed moment for spatial planning across the world (Song et al. 2021) and has accelerated the placemaking agenda on many fronts. It has brought public space sharply into focus (Sepe 2021). Furthermore, the pandemic has exposed urban vulnerabilities, such as liveability concerns in city and town centres, food security, lack of diversity in the high-street, car dominance in urban environments, unsustainable commuter patterns, the effect of poor air quality on public health, perceptions of public space safety and the social resilience of communities in being able to cope with disruption. The wellbeing agenda has accelerated globally (Scott 2020), and a belief is held that the 'new normal' will require a reorganisation of urban systems.

Public space has become a significant asset for building the resilience of people during the COVID-19 lockdowns, and the value of public space has been more deeply appreciated by a wider tranche of society. Placemaking has become 'mainstream' on the international stage with high-profile ambitions being set for cities that include Paris and Barcelona, which have sought to reclaim urban space from the private car. Public spaces have been reshaped to allow people to meet, shop, eat and get exercise outdoors. Places have very quickly been reconceived through changes in how people use them. The shift to working from home has strengthened local neighbourhood centres and brought the previous heavy reliance on retail and office use in city centres into question. The pandemic has accelerated trends, such as online shopping and home deliveries. Streets have become experimental testbeds, with pilot projects and 'pop-ups' providing new and improved amenity for pedestrians and cyclists. Community efforts have transformed car parking spaces into parklets. While the longer-term impact of COVID-19 remains uncertain, the evidence points to an acceleration of a previously much called-for integration of public health and planning (see Crawford 2010), the building of climate resilience, investing in the stewardship of places and the advancement of digital skills and tools that can interact with experiences of public space.

The walkable neighbourhood, or fifteen-minute city, is one example of a planning initiative that embraces these changed circumstances. Popularised in Paris by Mayor Anne Hidalgo, the fifteen-minute city is a walkable neighbourhood where inhabitants have access within their vicinity to all the services they need to live, learn and thrive (Moreno et al. 2021). In terms of placemaking, the fifteen-minute city seeks to flip the priority towards pedestrians and cyclists and away from cars through traffic calming, a connected street network as well as hierarchies of green and open spaces. The fifteen-minute city concept can be traced back to much earlier spatial planning concepts, such as garden cities (Howard 1902), the neighbourhood unit (Perry 1929) and the urban village (Aldous 1992). It is more recently encapsulated by ideas of the walkable neighbourhood (Weng et al. 2019) and the twenty-minute city (Da Silva et al. 2019) that act as viable models for modern cities to address their underlying place-related challenges and to build a resilience needed both in the current and in a post-COVID-19 world.

Within Northern Ireland several programmes of investment have emerged that respond to calls for positive placemaking during the pandemic. Investment by central government is taking place in green and blue infrastructure, COVID-19 revitalisation and active travel. The regeneration of the high street, by addressing vacancy and dereliction, is now being tackled through new powers regarding landlords and investment in adapting former retail spaces to create the mixed uses and vibrancy needed for successful places. Also interesting is a project, developed during 2020, known as *Relink*, a transformative health and wellbeing project for the Westlink dual carriageway throughpass in Belfast. This draws on stakeholder group-working across the Public Health Agency, Belfast Health and Social Care Trust, Belfast City Council, Police Service of Northern Ireland, the Department for Infrastructure and the Department for Communities and aims to transform inner-city road bridges in order to improve the physical and mental health of the communities that live close-by. Green bridges are being developed as initial pilots to assist with understanding place perception and to improve air quality. An ambitious transformation project is being informed by extensive engagement and co-design with residents, the key stakeholders and academia, and will seek to reduce the dominance of cars in the surrounding neighbourhoods (see https://relinkbelfast.com). The determination to maintain momentum and to realise a longer-term vision will test the extent to which the impact of COVID-19 has shaped a new climate, health and resilience agenda for placemaking in the city. In this regard, the *Belfast Resilience Strategy* places children and young people at the heart of a framework for change (Belfast City Council 2020). Meanwhile, Derry and Strabane District Council are making substantial commitments to increasing blue and green infrastructure across the city region (Derry and Strabane District Council 2019). New climate action plans across local government districts in Northern Ireland will also focus on securing better places. Nonetheless, while there has been a range of responses in the context of the pandemic that involve adapting public spaces, the overall pace of change has been slow. This may well reflect the complex placemaking governance arrangements in Northern Ireland and which, as noted above, is largely divided between central government departments and local authorities.

Conclusion

Formal contemporary placemaking in Northern Ireland comprises a combination of design guidance and policy, programmes of investment in place and regulatory regimes that shape public space and how it is used, maintained and managed (for example, through pavement licensing). Informal placemaking operates in organised community settings for the improvement of spaces and can range across markets, festivals and events, to guerrilla gardening or artistic interventions on streets, on vacant sites and within derelict buildings. Creative placemaking has gathered momentum in recent years as the arts and cultural sectors find ways of expressing themselves, their places and identities.

It is highly probable that placemaking will continue to evolve its concepts and practices as the impact of the COVID-19 pandemic is assessed and longer-term responses are introduced into public policy. The health, wellbeing and climate agendas will ensure continuity of place-based regeneration initiatives that emphasise the importance of localities, their identities and derived experiences of active and sociable living. However, the ongoing incorporation of bottom-up placemaking into the mainstream is dependent on the collection and dissemination of best practice evidence regarding the social, economic and environmental benefits of well-designed and vibrant public spaces. The urgent task ahead for planners is to facilitate a wide range of bottom-up, as well as more strategic placemaking endeavours, that can introduce greater diversity into urban environments and support more walkable communities. Northern Ireland may well have some distance to travel in its acceptance of a European café culture lifestyle, but the green shoots of change identified in this chapter offer encouragement towards a new vigour that can shape better places for everyone.

Legislation

Local Government Act (Northern Ireland) 2014
Well-being of Future Generations (Wales) Act 2015

References

Aldous, T. 1992. *Urban Villages: A Concept for Creating Mixed-Use Urban Developments on a Sustainable Scale.* London: Urban Villages Group.

Architecture and Design Scotland, NHS Scotland and the Scottish Government. 2017. *Place Standard Tool.* Available at https://placestandard.scot/place-standard.pdf (accessed 14 October 2021).

Belfast City Council. 2020. *The Belfast Resilience Strategy.* Belfast: Belfast City Council.

Commission for Architecture and the Built Environment (CABE). 2000. *By Design: Urban Design in the Planning System: Towards Better Practice.* Tonbridge: Thomas Telford.

Commission for Architecture and the Built Environment (CABE). 2001. *The Value of Urban Design.* Tonbridge: Thomas Telford.

Crawford, J. 2010. 'Health at the Heart of Spatial Planning', *Planning Theory and Practice* 11(1): 91–113.

Da Silva, D.C., D.A. King and S. Lemar. 2019. 'Accessibility in Practice: 20-Minute City as a Sustainability Planning Goal', *Sustainability* 12(1): 129.

Department for Communities. 2021. *The Strategic Design Group*. Available at https://www.infrastructure-ni.gov.uk/articles/strategic-design-group-sdg (accessed 14 November 2021).

Department for Culture, Arts and Leisure. 2006. *Architecture and Built Environment Policy for Northern Ireland*. Available at https://www.communities-ni.gov.uk/sites/default/files/publications/dcal/dcal-architecture-and-built-environment-policy-document.pdf (accessed 26 September 2021).

Department for Infrastructure. 2020. *Learning from a Collaborative Placemaking Project*. Belfast: Department for Infrastructure.

Department for Infrastructure. 2021. *The Strategic Design Group (SDG)*. Belfast: Royal Town Planning Institute.

Department of the Environment. 2014. *Living Places*. Belfast: Department of the Environment.

Department of the Environment. 2015. *Strategic Planning Policy Statement for Northern Ireland*. Belfast: Department of the Environment.

Department of the Environment and Department for Regional Development. 2000. *Creating Places: Achieving Quality in Residential Developments*. Belfast: Department of the Environment and Department for Regional Development.

Derry and Strabane District Council. 2019. *Green Infrastructure Plan*. Available at http://www.derrystrabane.com/getmedia/3e64f0d9-7b23-4d03-9f4b-344e0c1ef793/www_Green-Infrastructure-Plan-2019-2032.pdf.

Design Commission for Wales. 2020a. *Building Better Places: The Planning System Delivering Resilient and Brighter Futures*. Available at https://gov.wales/sites/default/files/publications/2020-07/building-better-places-the-planning-system-delivering-resilient-and-brighter-futures.pdf (accessed 12 November 2021).

Design Commission for Wales. 2020b. *Placemaking Charter for Wales*. Available at https://dcfw.org/placemaking/placemaking-charter/ (accessed 21 October 2021).

Design Commission for Wales. 2020c. *Placemaking Guide for Wales*. Available at https://dcfw.org/wp-content/themes/dcfw-child/assets/PlacemakingGuideDigitalENG.pdf (accessed 10 September 2021).

Faludi, A. and B. Waterhout. 2002. *The Making of the European Spatial Development Perspective: No Masterplan*. Abingdon: Routledge.

Healey, P. and J. Hillier. 1996. 'Communicative Micropolitics: A Story of Claims and Discourses', *International Planning Studies* 1(2): 165–84.

Howard, E. 1902. *Garden Cities of Tomorrow*. Cambridge, MA: MIT Press.

Jacobs, J. 1961. *The Death and Life of Great American Cities*. New York: Vintage.

Moreno, C., Z. Allam, D. Chabaud, C. Gall and F. Pratlong. 2021. 'Introducing the '15-Minute City': Sustainability, Resilience and Place Identity in Future Post-Pandemic Cities', *Smart Cities* 4(1): 93–111.

Perry, C.A. 1929. *Neighbourhood and Community Planning*. New York: Russell Sage Foundation.

Punter, J. 2007. 'Design-Led Regeneration? Evaluating the Design Outcomes of Cardiff Bay and Their Implications for Future Regeneration and Design', *Journal of Urban Design* 12(3): 375–405.

Rogers, R. 1999. *Towards an Urban Renaissance: Final Report of the Urban Task Force Chaired by Lord Rogers of Riverside*. London: Department of the Environment, Transport and the Regions.

Royal Town Planning Institute. 2007. *Shaping and Delivering Tomorrow's Places: Effective Spatial Planning in Practice*. London: Royal Town Planning Institute.

Scott, M. 2020. 'Covid-19, Placemaking and Health', *Planning Theory and Practice* 21(3): 343–48.

Sepe, M. 2021. 'Covid-19 Pandemic and Public Spaces: Improving Quality and Flexibility for Healthier Places', *Urban Design International* 26: 159–73.

Song, X., M. Cao, K. Zhai, X. Gao, M. Wu and T. Yang. 2021. 'The Effects of Spatial Planning, Well-Being and Behavioural Changes During and After the COVID-19 Pandemic', *Frontiers in Sustainable Cities* 3: 686706. https://doi.org/10.3389/frsc.2021.686706.

Sustainable NI. 2020. 'The Ormeau Parklet: A Test Model to Expand Public Space', *Newsletter*.

Weng, M., N. Ding, J. Li, X. Jin, H. Xiao, Z. He and S. Su. 2019. 'The 15-Minute Walkable Neighbourhoods: Measurement, Social Inequalities and Implications for Building Healthy Communities in Urban China', *Journal of Transport and Health* 13: 259–73.

Whyte, W.H. 1980. *The Social Life of Small Urban Spaces*. Washington, DC: Conservation Foundation.

13

MARINE SPATIAL PLANNING

Heather Ritchie, Joseph Onwona Ansong and Wesley Flannery

Introduction

Due to technological advances, globalisation and lax governance frameworks, the industrialisation of the marine environment has rapidly increased (Berkes et al. 2006) with projections indicating that this is likely to accelerate in the coming decades (Douvere 2008). Global aquaculture production, for example, is projected to increase from 26 million tonnes in 2018 to 109 million tonnes by 2030 (FAO 2020). The global offshore wind market grew by 30 percent per year between 2010 and 2018 (IEA 2019) and in Europe is expected to increase from 25,014 MW installed at the end of 2020 to 300 GW by 2050 based on the target set by the Offshore Renewables Energy Strategy (Wind Europe 2021).

Increasing coastal and marine activity is having a negative impact on marine biodiversity. Assessments of the state of global and regional marine environments reveal the transformation of food webs and growing marine pollution due to the increasing industrialisation of our coasts and seas (Worm et al. 2006; Halpern et al. 2008; Nellemann et al. 2008). Additionally, climate change is radically altering the marine environment, leading to marine biodiversity loss, sea-level rise, acidification, habitat alteration and changes in ocean circulation and stratification (Worm and Lotze 2021). Growth in maritime sectors is increasing the demand for marine space (Jay 2010) and is generating the potential for conflict amongst marine users (van Tatenhove 2013). Some marine industries are unsuited to being located in close proximity with one another and can result in competition for marine space. Spatial conflicts between, for example, fisheries and the submarine cable sector have resulted in revenue loss for both sectors due to damage and loss of fishing gear and the substantial costs associated with cable repair (Coffen-Smout and Herbert 2000). The array of possible conflicts has grown in recent decades due to new marine activities, such as wind farms and liquefied natural gas terminals being located offshore (Crowder et al. 2006). In Northern Ireland, for example, discussion is ongoing about a proposed tidal turbine development along the north Antrim coast and the potential conflicts that may arise between it, fishing and recreation interests. There is also a proposed 500 million cubic metres of gas underground natural gas storage facility at Islandmagee below Larne Lough to be commissioned by 2024 (Infrastrata 2021). The project

DOI: 10.4324/9781003218319-13

is expected to contribute to net-zero transition initiatives by transforming natural gas into hydrogen. And latterly, there has been a political push for the construction of a fixed bridge (Irish Sea Link) between Northern Ireland and Scotland, although it was formally shelved in November 2021.

For coastal nations, the task of accommodating new marine uses while ensuring the continuation of longstanding marine sectors, without degrading the marine environment, causing conflict or jeopardising their ability to meet legally binding conservation obligations, will necessitate the implementation of new integrated management strategies. Integrated management refers to 'integrating the management of previously separate activities, for example oil exploration, fishing and mining, in a way that promotes conservation and the sustainable use of resources' (Curtin and Prellezo 2010: 826). Integrated marine management approaches can resolve the issues highlighted above by connecting the governance of various marine sectors and illuminating their cumulative impacts on marine ecosystems.

Over the past fifteen years, marine spatial planning[1] (MSP) has been advanced as a marine management process that can address integration issues. MSP is promoted as 'a means of managing human uses of the sea in a sustainable manner and described as a planning process that enables integrated, forward looking, and consistent decision making on the human uses of the sea' (Ehler and Douvere 2007: 8). It is argued that by focusing on marine space, MSP enables sectoral integration, incorporates hierarchical policies from the supranational to the local and seeks to anticipate and address future resource demands in a sustainable manner (Claydon 2006). MSP can replace the current piecemeal, sectoral approach and provide a mechanism for a strategic and integrated plan-based approach to managing 'current and potential conflicting uses, the cumulative effects of human activities and marine protection' (Douvere 2008: 766). The centrality of the principle of integration in MSP has been a major factor in its rapid uptake (Kidd 2013). The adoption of MSP appeals to those concerned with the degradation of marine ecosystems as it allows for the integrated management of ecosystem stresses, while it also appeals to those who wish to address management inefficiencies arising from existing labyrinthine marine governance frameworks (Kidd 2013).

Both the United Kingdom and the Republic of Ireland have introduced marine legislation and policy to underpin an integrated system of marine governance, which includes MSP as a core delivery mechanism. However, questions have been raised about the degree of integration that will be achieved through the implementation of these new interventions within England, in the devolved administrations of Northern Ireland, Scotland and Wales and in the Republic of Ireland (Boyes and Elliott 2015; Ansong et al. 2021). The impact of Brexit and the outworking of the 2019 Protocol on Ireland / Northern Ireland may well lead to further marine governance fragmentation on the island of Ireland and between Great Britain and Northern Ireland.

It is against this multifaceted background that this chapter outlines the key changes to marine governance that have been introduced through these processes, explores how these are given practical effect in a Northern Ireland context and evaluates the impact of these changes in facilitating integrated marine planning. More specifically, integrated management is outlined in the first instance, attention is drawn to the fragmented nature of marine governance in Northern Ireland and the evolving new marine governance regime for the region is discussed. Developments in the Republic of Ireland are noted. The chapter concludes with a call for a more coherent approach to marine issues in Northern Ireland.

Integrated management

Integrated management is typically considered to have at least two key dimensions: a vertical dimension and a horizontal dimension. The vertical dimension concerns integration across different tiers of government, from national to local levels. The horizontal dimension focuses on integrated management across policy domains and across territories and includes the integration of related government departmental competencies (Holden 2012). A number of normative goals have also been identified for integrated management, including, for example, cross-sectoral integration, inter-agency integration, strategic integration and operational integration. Usefully, Kidd and Shaw (2007) have categorised these various goals into three broad strands: sectoral integration, territorial integration and organisational integration.

Sectoral integration

Sectoral integration is concerned with the integration of specific sectoral policies with other policy domains and their actors within a defined space (Kidd and Shaw 2007). The goal of sectoral integration is to address the issues arising from the adoption of the traditional, sectoral approach to natural resource management. The sectoral approach has been criticised as being ineffective as it fails to recognise complex interactions between different environmental planning policy areas. The adoption of sectoral integration is necessary so as to facilitate greater synergy among related governance and policy areas in order to resolve natural resource management issues (Kidd and Shaw 2007). In terms of marine management, the adoption of a sectoral approach has been unsuccessful in ensuring sustainable development of marine resources because it focuses solely on specific sectors and does not take account of the cumulative impacts that these activities have on the ecosystem as a whole (Guerry 2005). Marine sectors, such as shipping, fisheries, tourism and coastal development, are largely managed on a separate basis (Curtin and Prellezo 2010). All of these activities exert pressure on the ecosystem through, for example, eutrophication, pollution and habitat loss (Curtin and Prellezo 2010). The sectoral management approach results in these impacts being viewed in isolation from one another and understood only in the context of the sector from which they emanate (Curtin and Prellezo 2010). Separate, unconnected sectoral management regimes can resolve problems within individual sectors but are unlikely to be able to resolve problems across sectors. This is because cross-sectoral decision making is often ad hoc, with no entity given authority to resolve cross-sectoral conflicts. Without sectoral integration, these conflicts may not be realised until policies have been implemented (Jay 2012).

Territorial integration

The category of territorial integration is based upon the fact that natural systems and management regimes are seldom coterminous (Young et al. 2007). Often resource management regimes do not comprehensibly cover the biogeophysical scale of the resource they seek to manage. This can result in a resource or system being managed by a variety of institutions across governance scales and across a number of territories (Crowder et al. 2006). For example, straddling fish stocks[2] are often subject to a number of territorially distinct management regimes, with each regime having various hierarchical dimensions. Effective and sustainable resource management, therefore, depends upon integration across governance levels *and* across territorial divides (Kidd and Shaw 2007). Territorial integration is, therefore, concerned with both vertical integration (coherence across governance scales) and horizontal integration (coherence across territories that form part of the same natural system) (Kidd and Shaw 2007). Under the traditional marine governance model, separate parts of the marine environment were managed by different authorities, leading to poor resource management. For example, in California, the full implementation of marine protected areas around the Channel Islands was delayed due to complexities arising from the fragmented governance of the marine environment (Crowder et al. 2006). The adoption of territorial integration approaches encourages management regimes to ignore socially constructed boundaries and to manage the marine ecosystem as a whole.

Organisational integration

Organisational integration is a core component of both sectoral and territorial integration, as both these categories are premised on the need for greater integration between parties that may not have cooperated in the past. It is recognised that the willingness amongst parties to cooperate will have a direct impact on the level of integration achieved (Kidd and Shaw 2007). There are various forms of organisational integration, including strategic integration, which focuses on the integration of strategies, programmes and plans, and operational integration, which focuses on the integration of delivery mechanisms (Kidd and Shaw 2007). Whichever form of organisational integration is pursued, it is likely to be enhanced where there is some form of central overview mechanism that can facilitate integration (Stead and Meijers 2009). Due to the predominance of the sectoral approach, many marine resource management agencies have overlapping and sometimes conflicting governance frameworks. This can be addressed through the organisational integration of marine management governance entities.

Fragmented marine governance in Northern Ireland

The spatial fragmentation of Northern Ireland's marine areas is, in part, derived from the 1982 United Nations Convention on the Law of the Sea (UNCLOS), which established the current international legal framework for maritime governance. UNCLOS delimits the spatial limits of maritime zones (Figure 13.1), measured from baselines[3], including internal waters, territorial seas, an exclusive economic zone (EEZ) and the continental shelf, and establishes coastal states' rights and obligations in each zone. Internal waters are all waters and waterways on the landward side of the baseline. Coastal states are free to set laws, regulate use and harness any resource within this area. Vessels have no right of passage within these waters. Territorial seas cover the marine area from the baseline seaward to the 12 nautical mile (M) limit. Again coastal states are free to set laws, regulate use and utilise any resource in this area, while vessels are given the right of innocent passage within this area. The contiguous zone is an area that extends for another 12M beyond the 12M limit. Within the contiguous zone, coastal states can continue to enforce laws in four specific areas: customs, taxation, immigration and pollution. The EEZ extends 200M seawards from the baseline. Within EEZs, states have sovereign rights for the purpose of exploring and exploiting, conserving, and managing the living and non-living natural resources of the seabed and subsoil and the suprajacent water, including energy from water, currents and winds. The continental shelf is defined as the natural extension of the land territory to the continental margin's outer edge, or 200M from baseline, whichever is

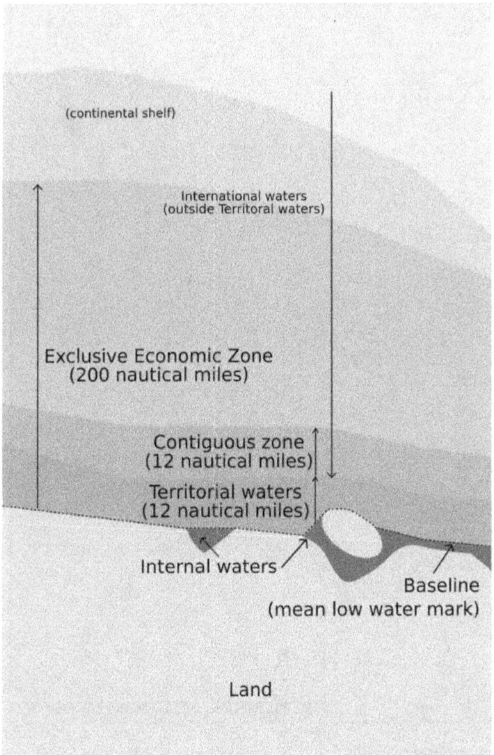

FIGURE 13.1 Maritime zones under UNCLOS. *Source*: Historicair (2006).

greater. Coastal states have exclusive control over living resources 'attached' to the continental shelf, but not to creatures living in the water column beyond the limit of the EEZ.

The fragmented nature of rights bestowed by UNCLOS is reflected in the governance of Northern Ireland's marine area (Figure 13.2). The Northern Ireland marine area is comprised of an 'inshore' and an 'offshore' region as defined at s322 of the Marine and Coastal Access Act 2009 (c.23). The inshore region is the area of sea within the seaward limit of the territorial sea of the United Kingdom adjacent to Northern Ireland (12M). More specifically, Northern Ireland's internal waters begin at the high-water mark and end at the 12M territorial limit. Northern Ireland's offshore region is the area from 12M to 40M (where applicable), the boundary of the 'Northern Ireland Zone'[4], meaning the sea within British Fishery Limits (BFL) which is adjacent to Northern Ireland. Under the BFL, which extends to 200M from baseline[5], the Northern Ireland Zone is defined as being the sea within the BFL that is adjacent to Northern Ireland, and lying between the territorial sea around Northern Ireland and that of Scotland, the Isle of Man and the Republic of Ireland. Under s.98(8) of the Northern Ireland Act 1998:

> her Majesty may by Order in Council[6] determine, or make provision for determining, for such purposes of this Act as may be specified, any boundary between a) the waters or parts of the sea which are to be treated as adjacent to Northern Ireland: and (b) those which are not, and may make different determinations or provisions for different purposes.

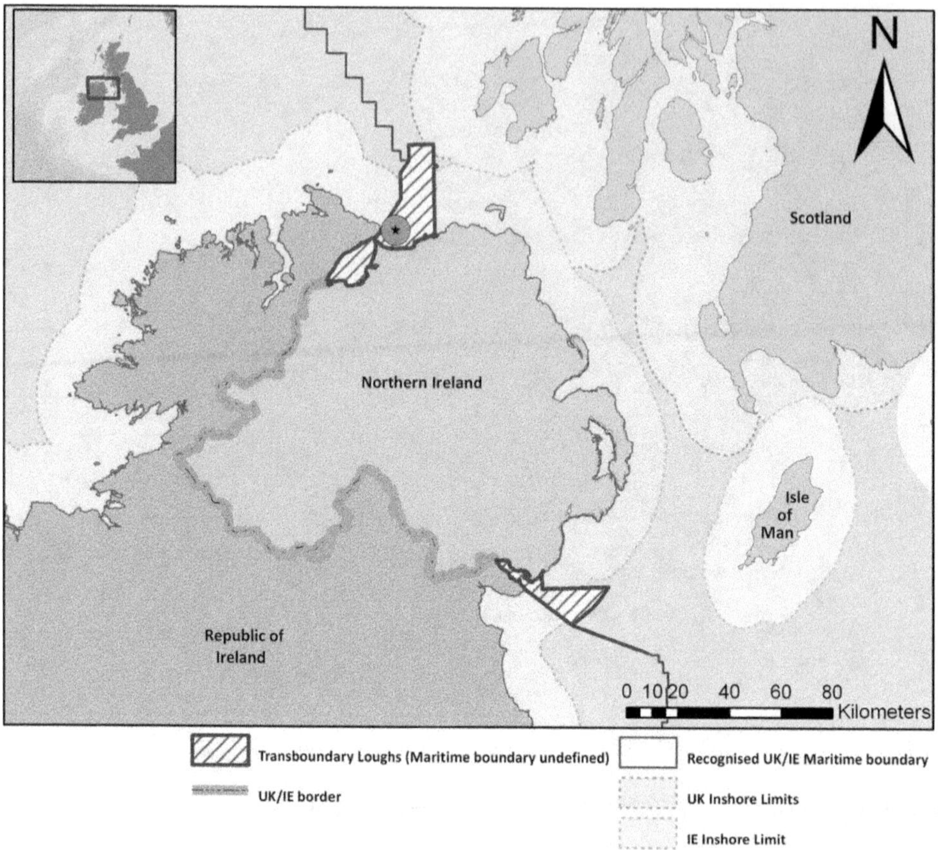

FIGURE 13.2 Marine boundary areas and approximate location of the proposed Tunes Plateau Offshore Wind Farm.

The boundary between the Northern Ireland marine area and other United Kingdom marine areas is clearly demarcated. However, in terms of delineating the territorial sea between the United Kingdom (i.e., Northern Ireland waters) and the Republic of Ireland, no boundary has been formally agreed upon, other than through ad hoc Memoranda of Understanding for potential renewable energy ventures. Joint management of the two sea loughs (Carlingford and Foyle) was established under the Belfast/Good Friday Agreement 1998. Joint management is facilitated through the Loughs Agency, whose specific functions include promoting the development of Lough Foyle and Carlingford Lough for commercial and recreational purposes in respect of marine, fishery and aquaculture matters; and the management, conservation, protection, improvement and development of the inland fisheries of the Foyle and Carlingford areas. This unique cross-border remit for the Loughs Agency relating to economic interests, such as fisheries and aquaculture, was expected to reduce the challenges of not having defined boundaries in the cross-border loughs. However, the inability to resolve border issues within these two loughs still has practical legislative and policy implications for marine planning and management. For example, in Lough Foyle, the Loughs Agency is prevented from exercising statutory powers over marine aquaculture licensing because of the undefined border and contested ownership of Lough Foyle. The Foyle Fisheries Act (Northern Ireland) 1952 focussed on the licensing and regulation of the salmon and trout fisheries on the Lough and made no specific reference to shellfish and aquaculture. The Foyle and Carlingford Fisheries (Northern Ireland) Order 2007 and the Foyle and Carlingford Fisheries Act 2007 (enacted by the Oireachtas) provided a new aquaculture regulatory and proposed licensing system. These two pieces of legislation also covered the transfer of existing aquaculture licensing provisions from the responsible departments to the Foyle, Carlingford and Irish Lights Commission (FCILC) and, by default, to the Loughs Agency. However, the failure to implement the full provisions of these legal provisions due to ownership and jurisdictional issues has stalled efforts towards having a management agreement between the United Kingdom Government, the Irish Government and the Crown Estate for the Loughs Agency in respect of a licensing system for shellfish.

The above is just one example of how marine governance in Northern Ireland (as part of the United Kingdom) is sectorally and organisationally fragmented. The fragmentation of Northern Ireland's marine governance is due to the nature of the devolution settlement for Northern Ireland and the complex mix of devolved and non-devolved functions. Up until 2016, thirteen United Kingdom Government and Northern Ireland departments had marine functions within Northern Ireland's inshore and offshore regions. At that time, the number of central government departments in Northern Ireland was reduced. The Department of Agriculture, Environment and Rural Affairs (DAERA) was formed from an amalgamation of the former Department of Environment and the former Department of Agriculture and Rural Development. Marine governance remits such as marine planning and licensing, nature and conservation, sustainable fisheries management, integrated maritime policies and legislation now fall under the remit of DAERA. However, other marine governance remits fall under three different departments, for example, protection of historic and archaeological sites (Department for Communities), licensing for oil and gas exploration, coastal and marine tourism (Department for the Economy) and ports and sea flood defences (Department for Infrastructure). A number of United Kingdom government departments with their sponsored agencies lead on various marine remits affecting Northern Ireland, including the Department for Transport (for example, maritime safety, marine accident investigations, wreck and salvage), the Home Office (for example, immigration, crime and security), the Ministry of Defence (for example, the regulating of sea areas for military defence, the collection of hydrographic data for the Royal Navy and merchant shipping), and HM Revenue and Customs (for example, customs clearance and excise duty on oil for maritime voyages). While the Northern Ireland Executive works closely with the United Kingdom government departments and has agreed on common position statements (for example, unexploded ordnance clearance in the marine environment in 2021 involving DAERA), some activities taking place offshore (outside of the 12M) may be reserved, non-devolved, matters that operate under the Marine and Coastal Access Act 2009, where the Marine Management Organisation (MMO) is responsible.

The Crown Estate also has a distinct role in the management of the Northern Ireland Marine Area. Although not a government body, the Crown Estate has statutory functions under the Crown Estate Act 1961. The Crown Estate is charged by Parliament with managing the property interests belonging to the Sovereign as part of the hereditary possessions of the Crown. Crown land is considered a 'moveable freehold' (movable because of the changing tides) (McElduff and Ritchie 2021). The Marine Estate is one of the four constituent estates that make up the property interests managed by the Crown Estate on behalf of Her Majesty the Queen. The Marine Estate includes the entire United Kingdom seabed out to 12M as well as the rights vested in the Crown to explore and utilise the natural resources of the United Kingdom continental shelf, which extends to 200M from the coast. The statutory duty of the Crown Estate is to maintain and enhance the land and property rights under its remit while having regard to the requirements of good management. The Crown Estate manages most of the foreshore[7] in Northern Ireland, the remainder is owned privately or is owned and protected by the National Trust or district councils. Coastal land above the high-water mark is usually in private ownership. The Crown Estate acts as leasing agent for the seabed and, under Schedule 3 of the Northern Ireland Act 1998, the foreshore and seabed are both reserved matters. Therefore, projects that span the marine and terrestrial environments (for example, an offshore wind farm with its grid connection) will need relevant consents from regulatory authorities under both marine and terrestrial planning regimes.

The recent and emerging legislation and policy in regard to the marine environment has implications for the way that planning decisions are made and implemented in the coastal zone. In the rest of the United Kingdom policy guidance exists for development in coastal areas, namely through tools such as Coastal Change Management Areas (England), Technical Advice Notes (Wales) and Planning Circulars (Scotland). However, there is no comparable coastal-specific policy guidance for Northern Ireland, despite numerous calls. The terrestrial planning system and the marine planning system are legally and functionally separate but overlap in the intertidal area. Policies and plans should be complementary in the coastal zone. There is an identified need across the United Kingdom for better coherence between terrestrial and marine planning, planning within the coastal zone and between the various jurisdictions (McElduff and Ritchie 2021).

Given appreciation of the need to move away from piecemeal management of the coastal zone and adopt a more holistic and strategic approach to sustainably manage coastal and marine areas more flexibly, Integrated Coastal Zone Management (ICZM) has long been mooted as the solution. As a strategic management process for dealing with coastal areas and issues of shared interest between regional marine plans and terrestrial development plans, ICZM has been introduced in various guises across the United Kingdom. The adoption by Council and Parliament of the EU ICZM recommendation 2002/413/EC, which explained how the European Commission will be promoting ICZM through the deployment of Community instruments and programmes, was an initial catalyst that was then followed by separate ICZM strategies, for example, *An integrated coastal zone management strategy for Northern Ireland 2006–2026* was published by the former Department of the Environment in 2006. However, these were non-binding and were not implemented. Whilst championed at local level on a relatively ad hoc basis by coastal partnerships, there have been no formal frameworks for coordination of these strategies across the United Kingdom (Hewett and Fletcher 2010). The lack of ICZM progress in the United Kingdom has been widely criticised and the extent to which the objectives of the regional strategies have been achieved to date is limited (McElduff and Ritchie 2021). Fletcher et al. (2014) suggest that the momentum around ICZM within the United Kingdom has been hindered by the legislative and policy developments associated with the operation of the Marine and Coastal Access Act 2009. However, progress being made with marine plans might revive the ICZM process and, if that were to be the case, this would be a step forward towards territorial integration (Ritchie et al. 2019).

Issues arising from fragmented governance

The implementation of legislative instruments in the marine and coastal environment is somewhat hampered by the large number of governance actors with marine-related competencies. In the recent past, for example, the bifurcation in Northern Ireland of conservation (within the former Department of the Environment) and fisheries management (within the former Department of Agriculture and Rural Development) resulted in the poor application of conservation measures in Strangford Lough, which is one of the most protected marine areas in Europe. It is part of the wider Strangford and Lecale Area of Outstanding Natural Beauty designated in the 1960s and then redesignated in 2010. Strangford Lough was designated a Marine Nature Reserve in 1995 and, under the Marine Act (Northern Ireland) 2013, it became Strangford Lough Marine Conservation Zone. As an internationally important wetland environment for wintering and breeding birds, Strangford Lough is a Ramsar site. It is part of the Europe-wide network of Natura 2000 and European Maritime Sites, is covered by Special Area of Conservation (SAC) and Special Protection Area (SPA) designations and has several coastal Areas of Special Scientific Interest. The oversight structure for Strangford Lough is provided by the Strangford Lough and Lecale Partnership comprising a management group of statutory authorities that is assisted by an advisory committee of sectoral interests. However, the combination of these conservation provisions failed to protect the Lough's horse mussels, which were found in 2001 to have suffered serious decline. Whilst the former Department of the Environment was responsible for maintaining the SAC features of Strangford Lough in a favourable condition, it did not have the authority to manage the fisheries that caused damage to the horse mussel beds, thus illustrating the challenges associated with multiple departments involved in marine management (Yates et al. 2012).

The lack of defined maritime boundaries between the Republic of Ireland and Northern Ireland has resulted in misunderstanding over jurisdictional competencies. There has also been perceived confusion affecting marine planning within the cross-border lough setting. The abandonment of a considerable offshore wind development proposal off the north coast of Northern Ireland is illustrative of the former issue (Figure 13.2). In this instance, a project known as the Tunes Plateau Offshore Wind Farm was being developed by a consortium consisting of B9 Energy, Powergen and Renewable Energy Systems (RES). The project involved the construction of up to eighty-five turbines, with a capacity of 150 to 250 MW. In June 2002, the Crown Estate, presuming that it was the owner of the proposed development site on the seabed, granted a licence to the consortium. This was opposed by the Irish Government, which claimed that it owned part of the proposed development site and that the project would, therefore, need its approval. In October 2002, DETI (the former Department of Enterprise, Trade and Investment) acknowledged the possibility that part of the proposed development could be within the Republic of Ireland's jurisdiction. In September 2004, Powergen withdrew from the consortium. Although some assessment of potential environmental impact has been undertaken in the interim, Powergen's withdrawal has effectively ended the proposed development.

Perceived confusion has arisen over transboundary consultation requirements as exemplified by the construction of a CO_2 gas import and distribution terminal at Warrenpoint port. Marine works have also been required to facilitate the transfer of CO_2 gas from ship to storage tanks. Community and local environmental groups in Northern Ireland have been concerned about this expansion at Warrenpoint port due to its visual impact on the beauty of a gateway to the Mournes. Residents in the Republic of Ireland, along Omeath's lower shore, have asserted that they were not consulted during the landscape and noise assessment process because they live in a different jurisdiction, although the liquid CO_2 storage tanks are less than a few hundred metres from their homes. Officials in Louth County Council have indicated that they did not receive any transboundary notification and consultation for the CO_2 facility (O'Connell 2020). The project is located within fifteen kilometres of a Natura 2000 designation, straddling a site that is in Northern Ireland and the Republic of Ireland and thus concerns have been raised by community interests, local politicians and local environmental groups that a Habitats Regulations Assessment was carried out only in 2019 to inform a Marine Licence Application, but after the planning application for the terminal facility had been approved in October 2015 (Visit Warrenpoint 2020). The planning application (LA07/2015/0056/F) had, however, been screened against Environmental Impact Assessment criteria that included noise, pollution, impact on the Area of Outstanding Natural Beauty, views from the Republic of Ireland and transboundary transport of gases, and it was concluded that there would be no likely significant environmental effects. Had a different conclusion been reached, there would have followed the preparation of an Environmental Statement and mandated notification to authorities in the Republic of Ireland regarding development likely to have significant effects on the environment in another EEA state under Part 7 of the Planning (Environmental Impact Assessment) Regulations (Northern Ireland) 2015. No written representations or petitions were submitted in regard to the planning application. The Loughs Agency as a cross-border statutory consultee raised no objections. Moreover, Newry, Mourne and Down District Council were alive to the issue of the protected sites within Carlingford Lough and had noted that a future Habitats Regulations Assessment should consider the Natura 2000 designation in the Republic of Ireland. This subsequently determined that 'there are no likely significant adverse effects in relation to designated areas of Carlingford Lough Special Area of Conservation and Special Protection Area as a result of the proposal' (Eolas Ecology 2019: 19). A marine licence under the Marine and Coastal Access Act 2009 was then issued on 31 December 2019 by the Department of Agriculture, Environment and Rural Affairs (Marine and Fisheries Division) for the construction of the ship-to-shore transfer facility for CO_2. Trees and shrubs have been planted to mitigate the visual impact of the storage tanks as required by the planning permission conditions and in response to an engagement by the port authority with the local community. Arguably, the perceived confusion and voiced concerns in this case point to a wider issue regarding public understanding of planning practice in a very complex marine governance setting.

Marine legislation and policy for integrated management

A number of legislative and policy provisions with relevance to Northern Ireland have been put in place for integrated marine management, including the Marine and Coastal Access Act 2009 (MCAA); the UK Marine Policy Statement (2011) and the Marine Act (Northern Ireland) 2013.

Marine and Coastal Access Act 2009

The MCAA was the first piece of comprehensive legislation in the United Kingdom focused on the governance of the marine and coastal environment. It provides for the exercise of controls over territorial waters and offshore waters out to the EEZ. The aim of the MCAA is to ensure 'clean healthy, safe, productive and biologically diverse oceans and seas, by putting in place better systems for delivering sustainable development of the marine and coastal environment'. In this manner, the MCAA was responding to the fact that marine and coastal governance frameworks needed to be updated to better fit the complex 21st-century demands of a rapidly expanding marine sector (Boyes and Elliott 2015). The MCAA introduced a new integrated marine governance regime (Figure 13.3) but has had a complicated geographical impact, whereby its legislative provisions provided for MSP in inshore and offshore waters of England and Wales, and the offshore waters of Northern Ireland and Scotland. The MCAA assigned the Secretary of State as the planning authority for the inshore and offshore areas of England, the Welsh Ministers as the planning authority for the inshore and offshore areas of Wales, the Scottish Ministers for the offshore area of Scotland and the then Department of the Environment for the offshore area of Northern Ireland. Subsequently, Northern Ireland and Scotland each introduced their own legislation to provide for MSP in their inshore waters. In Northern Ireland, this was done through the Marine Act (Northern Ireland) 2013, which appoints DAERA as the marine authority for the inshore area. This means that two marine plans will have to be prepared, one for the inshore area and another for the offshore area of Northern Ireland.

The UK Marine Policy Statement 2011

The United Kingdom Administrations published a set of high-level marine objectives (HLMO) in 2009 to ensure consistency of approach towards a shared vision of having clean, healthy, safe, productive and biologically diverse oceans and seas. The HLMO are concerned with promoting sustainable economic development, enabling the United Kingdom's move towards a low-carbon economy in order to mitigate the causes of climate change and ocean acidification and adapt to their effects, ensuring a sustainable marine environment which

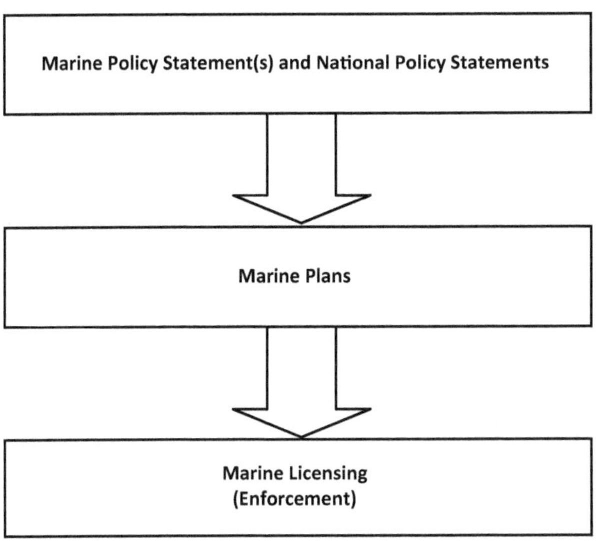

FIGURE 13.3 New marine governance regime

promotes healthy, functioning marine ecosystems and protects marine habitats, species, and our heritage assets and contributing to the societal benefits of the marine area including the sustainable use of marine resources to address local social and economic issues. This led, in turn, to the production of the United Kingdom Marine Policy Statement (H M Government 2011) as the high-level policy context within which national and sub-national marine plans are to be developed, implemented, monitored and amended. It is designed to ensure appropriate consistency in marine planning across the United Kingdom marine area and sets the direction for marine licensing. Guidance was subsequently published in 2020 on behalf of all the United Kingdom Administrations by the Department of Environment, Food and Rural Affairs explaining how references to EU law in the UK Marine Policy Statement should be interpreted as 'retained law' from 1 January 2021 following the United Kingdom's withdrawal from the European Union.

Marine Act (Northern Ireland) 2013

This Act provides for the preparation of marine plans in regard to the Northern Ireland inshore territory, allows for the designation of marine conservation zones within this area and makes additional provision relating to marine licensing for certain electricity works. Section 4 of the Marine Act (Northern Ireland) 2013 states that a 'marine plan is a document which states the policies of the relevant Northern Ireland departments (however expressed) for and in connection with the sustainable development of the area'. In relation to the United Kingdom Marine Policy Statement, the Marine Act (Northern Ireland) 2013 states that the marine plan should: 'Unless relevant considerations indicate otherwise … be in conformity with any marine policy statement which governs marine planning for the marine plan area'. Similarly, the MCAA states that a 'marine plan must be in conformity with any MPS which governs marine planning for the marine plan area unless relevant considerations indicate otherwise'. This deviates with the approach traditionally adopted in the terrestrial regime, wherein lower-tier plans are to be 'in *general* conformity' with higher-tier plans (Lyness 2010). This may be an area that produces case law, as it has under the terrestrial planning system (Lyness 2010).

Marine planning in Northern Ireland

The first marine plan for Northern Ireland was published as a draft for public consultation in 2018 (DAERA 2018) to advance efforts towards an integrated spatial planning approach in Northern Ireland's inshore and offshore areas regarding the resources, activities and interactions present in those areas. As noted above, the governance arrangements in place have required the preparation of two plans, one for the inshore region under the 2013 Marine Act and one for the offshore region under the 2009 MCAA. However, as the offshore region is small in relation to the rest of the United Kingdom waters, it was decided for practical and administrative purposes to have the two plans (for inshore and offshore waters) combined within a single document to be collectively known, when adopted, as the Marine Plan for Northern Ireland. The marine plan will inform and guide the regulation, management, use and protection of the Northern Ireland marine area.

The draft plan was brought forward following liaison with other United Kingdom marine plan authorities, departments and agencies, transboundary cooperation with the Republic of

Ireland and the Isle of Man and wide-ranging stakeholder engagement with Northern Ireland interests. It was accompanied by a Sustainability Appraisal, a Habitats Regulations Assessment, an Equality Impact and Human Rights Screening, a Rural Needs Impact Assessment and a Partial Regulatory Impact Assessment. The content of the draft plan (Table 13.1) comprises a marine plan vision, a set of eight objectives that reflect the United Kingdom's high-level marine objectives, core policies (covering the general issues to be considered for all proposals) and key activity policies (to support or safeguard a particular activity without undue impact on the marine area, its ecosystem services and the users that rely on them). It contains provisions relating to retained functions (i.e., matters that are not fully devolved to Northern Ireland).

Following an eight-week public consultation on the draft Marine Plan for Northern Ireland in 2018, DAERA published a summary of responses report in April 2021 (Department of Agriculture, Environment and Rural Affairs 2021). Over seventy organisations and individuals had made submissions. Stakeholders raised several integration challenges that the plan could address, including the lack of a defined border in the Foyle and Carlingford loughs and the lack of a clear statement regarding the application of the plan in the shared waters. It was also noted that more emphasis should be placed on the ecosystem-based approach, ecosystem services and synergies with key activities. The need to ensure integration between the marine plan and terrestrial planning documents, including statutory Local Development Plans, shoreline management plans and the ICZM plan, was mentioned. At the time of writing this chapter the progression period leading to adoption and publication remains unclear. The most likely next stage will be the preparation of a statement detailing how DAERA intends to proceed with the Marine Plan for Northern Ireland. The plan will then be adopted only when the Northern Ireland Executive (with agreement of the Secretary of State regarding retained functions) decides to publish the plan.

Marine licensing in Northern Ireland

DAERA is the authority responsible for marine licensing in Northern Ireland's inshore area. Under the 2009 MCAA, the Marine Management Organisation (MMO) carries out licensing and enforcement on behalf of the Secretary of State for Northern Ireland's offshore regions and is responsible for most defence matters in Northern Ireland's inshore waters. Activities that could require a marine licence include construction and removal; dredging; sea disposal; and plans, projects or policies with potential to affect protected areas and historic assets. Separately, marine wildlife licensing is used to authorise what would otherwise be an offence under nature conservation legislation. Marine wildlife licensing exists to safeguard the conservation and welfare of marine species, such as whales, dolphins, porpoises, seals and basking sharks. Whilst these species are protected under EU retained law and national legislation, a marine wildlife licence may be obtained for certain activities that may have a negative impact on protected species and normally would be prohibited under Article 18 of the Wildlife (Northern Ireland) Order 1985 and Regulation 40 of the Conservation (Natural Habitats etc.) Regulations (Northern Ireland) 1995. For example, pile driving or rock blasting would create noise disturbance and would normally be prohibited where it could have an impact on certain marine species but may be permitted through a marine wildlife licence where there is no satisfactory alternative and the activity will not be detrimental to the maintenance of the population of the species concerned. DAERA is the authority responsible for

TABLE 13.1 The content of the draft Marine Plan for Northern Ireland, 2018

Vision	*Objectives*	*Core policies*	*Key activity policies*
A healthy marine area that is managed sustainably for the economic, environmental and social prosperity of present and future generations.	1. To promote the sustainable development of productive activities, which support employment at all skill levels while fully considering the requirements of other marine interests. 2. To help realise the potential of energy resources and energy storage within the marine area, while fully considering the requirements of other marine interests. 3. To promote the development of vibrant, accessible and sustainable coastal communities. 4. To promote the marine resource, its recreational value and its wider economic, environmental and social benefits to all. 5. To promote the preservation and enjoyment of marine-related heritage assets. 6. To promote a healthy, resilient and adaptable marine ecosystem and an ecologically coherent network of Marine Protected Areas. 7. To contribute towards climate change mitigation and adaptation measures. 8. To continue to develop a sound marine evidence base in a coordinated manner, to increase understanding and to support the development, monitoring and review of marine plans.	1. Stakeholder engagement. 2. Air quality. 3. Climate change. 4. Coastal processes. 5. Coexistence. 6. Cumulative impacts. 7. Heritage assets. 8. Invasive alien species. 9. Land and sea interaction. 10. Marine litter. 11. Marine noise. 12. Natural heritage. 13. Seascape. 14. Use of evidence. 15. Water quality.	1. Aquaculture. 2. Carbon capture and storage 3. Commercial fishing. 4. Defence and national security. 5. Dredging. 6. Energy. 7. Marine aggregates. 8. Ports, harbours and shipping. 9. Telecommunications cabling. 10. Tourism and recreation.

Source: DAERA 2018

marine wildlife licensing in the inshore area and the MMO is the authority for the offshore area. The Northern Ireland Environment Agency (NIEA) is responsible for wildlife licences in the inshore area for the protected species of terrestrial animals, birds and plants.

The United Kingdom Marine Planning Statement and the Marine Plan for Northern Ireland (once adopted) will be material considerations for planning authorities for any development management and enforcement decisions relating to the marine area. Responsibility is shared between central government and local government planning authorities around the coast at the intertidal area, meaning that certain activities or development will require both a marine licence and planning permission. In addition, certain activities will also need an environmental impact assessment (EIA) under the Marine Works (Environmental Impact Assessment) Regulations 2007, the Planning (Environmental Impact Assessment) Regulations (Northern Ireland) 2017, the Planning (Environmental Assessments and Miscellaneous Amendments) (EU Exit) (Northern Ireland) Regulations 2018 that came into force in 2019 and / or a Habitats Regulations Assessment (HRA). In 2015 the former Department of the Environment published *Planning in the Coastal Area*, setting out illustrations of developer requirements for, inter alia, a marine licence, a marine wildlife licence, planning permission, EIA and HRA.

The Crown Estate provides a lease (of the seabed) for projects, but such projects also need the requisite consents from the relevant Northern Ireland departments. For example, the Crown Estate may provide a lease for a renewable energy project, but this project will be required to obtain relevant marine consents and a generating consent from central government departments. The Marine Act (Northern Ireland) 2013 provides for these consent applications relating to generating stations to be considered together. Both the application for consent under the Electricity (Northern Ireland) Order 1992 and the application for a marine licence will go through the same administrative procedure to ensure that they are dealt with in parallel by each of the departments. Furthermore, if the proposed project is in, or partly in, the offshore area it will also be required to obtain consent from the MMO. Although there were initial ambitions for marine planning to provide a one-stop-shop approach to gaining consents and licences, this rubric of organisational integration has not yet filtered down to practice.

Marine planning and marine licensing in the Republic of Ireland

The Republic of Ireland's first marine plan, the National Marine Planning Framework (NMPF), was published on 1 July 2021. The plan sets outs a vision, objectives, planning policies and key references for the planning of various maritime sectors. The marine planning policies are outlined under overarching themes: 'Environmental – Ocean Health' (biodiversity and protected marine sites, non-indigenous species, water quality, sea floor and water column integrity, marine litter, underwater noise, air quality and climate change), 'Economic – Thriving Maritime Economy' (coexistence and infrastructure) and 'Social – Engagement with the Sea' (access, employment, heritage assets, rural coastal and island communities, seascape and landscape, social benefits and transboundary). These matters are followed by discussion of key sectoral/activity policies. The NMPF was developed under the Planning and Development (Amendment) Act 2018. However, the enabling legislation for MSP is now under the Maritime Area Planning Act, which was enacted on 23 December 2021. This legislation allows in Part 2 for the development of more detailed regional and sectoral plans known as Designated Maritime Area Plans, which will require inclusive public participation, consultations and environmental assessment in the preparation process.

Part 4 of the Maritime Area Planning Act 2021 streamlines the marine consent process to a two-stage process comprising Maritime Area Consent (MAC) and development permission by aligning the foreshore consent system with the planning system and providing for a single Environmental Impact Assessment for projects. Under the new planning and licensing system, a MAC must first be issued by the Maritime Area Regulatory Authority (MARA) (to be established in 2022 to replace the role of the Minister). The MAC process will check for maritime space conflicts and examine public and state interests along with the financial and technical capabilities of developer(s). A MAC is a prerequisite to obtaining development permission. An Bord Pleanála (the Planning Appeals Board) will be the decision-making body for obtaining permission for developments classed as 'strategic infrastructure development' and/or subject to Environmental Impact Assessment/Appropriate Assessment within this nearshore area as well as developments beyond the nearshore, i.e., within the foreshore, EEZ and continental shelf. The coastal planning authorities will be the decision-maker for development permissions that are located in the near shore and do not comprise a strategic infrastructure project. Accordingly, what we see in the Republic of Ireland is a busy MSP policy arena: the NMPF was published in 2021, the Maritime Area Planning Act was enacted in 2021 and a new regulatory authority (MARA) will be established in 2022. This has led to confusion on the part of stakeholders and agencies regarding the role of the marine plan within the wider governance system.

Conclusion: are we moving towards integrated marine planning?

Recent legislation and policy initiatives in the United Kingdom are indicative of a major, if underappreciated, milestone in environmental regulation. The MCAA, for example, has been hailed as one of the most important and long-awaited pieces of conservation legislation. The MCAA, the United Kingdom Marine Policy Statement and the Marine Act (Northern Ireland) 2013 have established a new regulatory and planning framework for Northern Ireland's marine environment. The new marine governance regime, described in this chapter, has taken a significant step towards achieving integrated marine planning. The marine plan, once published in its adopted format, is likely to achieve a greater degree of sectoral integration across the relevant Northern Ireland government departments. This should enable DAERA to identify sectoral conflicts and synergies while facilitating deeper coordination between conservation and fisheries governance. This new regime also achieves territorial integration insofar as there will now be a set of coordinated plans for Northern Ireland's inshore and offshore areas. Moreover, the outworking of Brexit may well offer new opportunities for stakeholders to input directly to marine policy processes that can contribute to the emergence of progressive policy solutions for all-island maritime governance. The new regime also facilitates greater integration between the terrestrial and marine planning and licensing regimes. Nevertheless, the extant complex consenting arrangements illustrate just how fragmented marine planning remains; they fall well short of facilitating a unified, simplified approach to marine licensing. Northern Ireland should explore the development of a single marine body with responsibility for all marine issues, similar to the approach being adopted by the Republic of Ireland with its proposal for MARA.

Legislation

European Council 2002/413/EC
Foreshore Act 1933
Foyle and Carlingford Fisheries Act 2007

Foyle Fisheries Act (Northern Ireland) 1952
Marine Act (Northern Ireland) 2013
Marine and Coastal Access Act 2009
Maritime Area Planning Act 2021
Northern Ireland Act 1998
Planning and Development (Amendment) Act 2018
The Adjacent Waters Boundaries (Northern Ireland) Order 2002
The Conservation (Natural Habitats, etc.) Regulations (Northern Ireland) 1995
The Electricity (Northern Ireland) Order 1992
The Foyle and Carlingford Fisheries (Northern Ireland) Order 2007
The Limitation (Northern Ireland) Order 1989
The Marine Works (Environmental Impact Assessment) Regulations 2007
The Planning (Environmental Assessments and Miscellaneous Amendments) (EU Exit) (Northern Ireland) Regulations 2018
The Planning (Environmental Impact Assessment) Regulations (Northern Ireland) 2017
The Wildlife (Northern Ireland) Order 1985
United Nations Convention on the Law of the Sea (UNCLOS), 1982

Notes

1 Known as marine planning in the United Kingdom and maritime spatial planning in the EU context.
2 The United Nations defines straddling stocks as stocks of fish, such as pollock, which migrate between, or occur in both, the economic exclusion zone of one or more states and the high seas.
3 A baseline predominately follows the low water line, but when the coastline is highly indented, has adjacent islands or is highly unstable, straight baselines may be used.
4 As defined by section 98 (1) and 98(8) of the Northern Ireland Act 1998 (c.47).
5 The baseline is measured from the low water mark around the coast.
6 The Adjacent Waters Boundaries (Northern Ireland) Order 2002 defines the boundary between UK waters that are adjacent to Northern Ireland and those waters that are not adjacent to Northern Ireland, namely, the waters around Scotland, Wales and England.
7 The foreshore is defined within the Limitation (Northern Ireland) Order 1989 (section 2(2)) as 'the bed and shore, below the line of high-water of ordinary or medium tides, of the sea and of every tidal river and tidal estuary and of every channel, creek and bay of the sea or any such river or estuary'. In the Republic of Ireland, the foreshore means 'the bed and shore, below the line of high-water of ordinary or medium tides, of the sea and of every tidal river and tidal estuary and of every channel, creek, and bay of the sea or of any such river or estuary (Definition according to the Foreshore Act 1933 and Maritime Area Planning Act 2021).

References

Ansong, J.O., L. McElduff, and H. Ritchie. 2021. 'Institutional Integration in Transboundary Marine Spatial Planning: A Theory-Based Evaluative Framework for Practice', *Ocean and Coastal Management* 202: 105430.

Berkes, F., T.P. Hughes, R.S. Steneck, J.A. Wilson, D.R. Bellwood, B. Crona, C. Folke, L.H. Gunderson, H.M. Leslie, J. Norberg, M. Nystrom, P. Olsson, H. Osterblom, M. Scheffer and B. Worm. 2006. 'Globalization, Roving Bandits and Marine Resources', *Science* 311: 1557–58.

Boyes, S.J. and M. Elliott. 2015. 'The Excessive Complexity of National Marine Governance Systems – Has this Decreased in England since the Introduction of the Marine and Coastal Access Act 2009?', *Marine Policy* 51: 57–65.

Claydon, J. 2006. 'Marine Spatial Planning: A New Opportunity for Planners', *Town Planning Review* 77: 1–6.

Coffen-Smout, S. and G. Herbert. 2000. 'Submarine Cables: A Challenge for Ocean Management', *Marine Policy* 24: 441–48.

Crowder, L.B., G. Osherenko, O.R. Young, S. Airamé, E.A. Norse, N. Baron, J.C. Day, F. Douvere, C.N. Ehler, B.S. Halpern, J. Langdon, K.L. McLeod, J.C. Ogden, R.E. Peach, A.A. Rosenberg and J.A. Wilson. 2006. 'Resolving Mismatches in U.S. Ocean Governance', *Science* 313: 617–18.

Curtin, R. and R. Prellezo. 2010. 'Understanding Marine Ecosystem-Based Management: A Literature Review', *Marine Policy* 34: 821–30.

DAERA (Department of Agriculture, Environment and Rural Affairs). 2018. *Draft Marine Plan for Northern Ireland, Public Consultation*. Belfast: DAERA.

DAERA (Department of Agriculture, Environment and Rural Affairs). 2021. *Public Consultation Report, Summary of Responses*. Belfast: DAERA.

Douvere, F. 2008. 'The Importance of Marine Spatial Planning in Advancing Ecosystem-Based Sea Use Management', *Marine Policy* 32: 762–71.

Ehler, C. and F. Douvere. 2007. *Visions for a Seachange: Report of the First International Workshop on Marine Spatial Planning*. Paris: UNESCO.

Eolas Ecology. 2019. *Habitats Regulation Assessment (HRA) Stage 1 Test of Likely Significance: Nippon Gases UK*. Derrylin: Eolas Ecology.

FAO (Food and Agriculture Organisation). 2020. *The State of the World Fisheries and Aquaculture 2021*. Rome: FAO.

Fletcher, S., R. Jefferson, G. Glegg, L. Rodwell and W. Dodds. 2014. 'England's Evolving Marine and Coastal Governance Framework', *Marine Policy* 45: 261–68.

Guerry, A.D. 2005. 'Icarus and Daedalus: Conceptual and Tactical Lessons for Marine Ecosystem-Based Management', *Frontiers in Ecology and the Environment* 3: 202–11.

Halpern, B.S., S. Walbridge, K.A. Selkoe, C.V. Kappel, F. Micheli, C. D'Agrosa, J.F.. Bruno, K.S. Casey, C. Ebert, H.E. Fox, R. Fujita, D. Heinemann, H.S. Lenihan, E.M.P. Madin, M.T. Perry, E.R. Selig, M. Spalding, R. Steneck and R. Watson. 2008. 'A Global Map of Human Impact on Marine Ecosystems', *Science* 319: 948–52.

Hewett, T., and S. Fletcher. 2010. 'The Emergence of Service Based Integrated Coastal Management in the UK Area', *Marine Policy* 17(2): 118–29.

Historicair. 2006. *Marine Zones*. Available at https:/commons.wikimedia.org/wiki/File:Zonmar-en.svg (accessed 13 December 2021).

HM Government. 2011. *United Kingdom Marine Policy Statement*. London: The Stationery Office.

Holden, M. 2012. 'Is Integrated Planning any More than the Sum of its Parts? Considerations for Planning Sustainable Cities', *Journal of Planning Education and Research* 32: 305–18.

Infrastrata. 2021. *Island Magee Gas Storage Project*. Available at https://www.infrastrataplc.com/projects/islandmagee-energy/.

International Energy Agency. 2019. *Offshore Wind Outlook*. Available at https://www.iea.org/reports/offshore-wind-outlook-2019.

Jay, S. 2010. 'Planners to the Rescue: Spatial Planning Facilitating the Development of Offshore Wind Energy', *Marine Pollution Bulletin* 60: 493–99.

Jay, S. 2012. 'Marine Space: Manoeuvring Towards a Relational Understanding', *Journal of Environmental Policy and Planning* 14: 81–96.

Kidd, S. 2013. 'Rising to the Integration Ambitions of Marine Spatial Planning: Reflections from the Irish Sea', *Marine Policy* 39: 273–82.

Kidd, S. and D. Shaw. 2007. 'Integrated Water Resource Management and Institutional Integration: Realising the Potential of Spatial Planning in England', *Geographical Journal* 173: 312–29.

Lyness, S. 2010. *Marine and Coastal Access Act 2009: Marine Planning*. London: Landmark Chambers.

McElduff, L. and H. Ritchie. 2021. 'United Kingdom', in R. Alterman and C. Pellach (eds) *Regulating Coastal Zones: International Perspectives on Land Management Instruments*. Abingdon: Routledge, pp. 31–65.

Nellemann, C., S. Hain and J.E. Alder. 2008. *In Dead Water – Merging of Climate Change with Pollution, Over-Harvest and Infestations in the World's Fishing Grounds*. Arendal: UNEP/GRID-Arendal.

O'Connell, R. 2020. *Louth County Council was not Consulted Over Plans for CO2 Gas Facility in Warrenpoint*. Available at https://www.lmfm.ie/news/lmfm-news/louth- county-council-was-not-consulted-over-plans-for-co2-gas-facility-in-warrenpoint/.

Ritchie, H., W. Flannery, A.M. O'Hagan, S. Twomey and C. O'Mahony. 2019. 'Marine Spatial Planning, Brexit and the Island of Ireland', *Irish Geography* 52(2): 213–33. https://doi.org/10.2014/igj.v52i2.1402.

Stead, D. and E. Meijers. 2009. 'Spatial Planning and Policy Integration: Concepts, Facilitators and Inhibitors', *Planning Theory and Practice* 10: 317–32.

Van Tatenhove, J.P.M. 2013. 'How to Turn the Tide: Developing Legitimate Marine Governance Arrangements at the Level of the Regional Seas', *Ocean & Coastal Management* 71: 296–304.

Visit Warrenpoint. 2020. *Local Concern of the CO2 Gas Terminal in Warrenpoint*. Available at: https://www.visitwarrenpoint.com/local-concern-of-the-co2-gas-terminal-in-warrenpoint/.

Wind Europe. 2021. *Offshore Wind in Europe, Key Trends and Statistics*. Available at https://windeurope.org/intelligence-platform/statistics/.

Worm, B., E.B. Barbier, N. Beaumont, J.E. Duffy, C. Folke, B.S. Halpern, J.B.C. Jackson, H.K. Lotze, F. Micheli, S.R. Palumbi, E. Sala, K.A. Selkoe, J.J. Stachowicz and R. Watson. 2006. 'Impacts of Biodiversity Loss on Ocean Ecosystem Services', *Science* 314: 787–90.

Worm, B. and H.K. Lotze. 2021. 'Marine Biodiversity and Climate Change', in T. Letcher (ed) *Climate and Global Change: Observed Impacts on Planet Earth*. Amsterdam: Elsevier, pp. 445–64.

Yates, K.L., A. Payo and D.S. Schoeman. 2012. 'International, Regional and National Commitments Meet Local Implementation: A Case Study of Marine Conservation in Northern Ireland', *Marine Policy* 38: 140–50.

Young, O.R., G. Osherenko, J. Ekstrom, L.B. Crowder, J. Ogden, J.A. Wilson, J.C. Day, F. Douvere, C.N. Ehler, K.L. McLeod, B.S. Halpren and R. Peach. 2007. 'Solving the Crisis in Ocean Governance: Place-Based Management of Marine Ecosystems', *Environment* 49: 20–32.

14

CROSS-BORDER SPATIAL PLANNING

Gavan Rafferty

Introduction

Contemporary spatial planning across the two jurisdictions on the island of Ireland has a similar provenance and has evolved on paths that are parallel and, on occasion, overlap. In theory, greater symmetry and convergence should provide the basis for stronger cooperation across border geographies that have gained increased attention within many academic disciplines, notably human geography, economics, political science, anthropology and, indeed, spatial planning (Jacobs 2016). In Europe, much of the interest in and policy focus on national borders has been driven by EU-sponsored planning concepts and EU-funding programmes to promote territorial and social cohesion.

The island of Ireland comprises two jurisdictions: Northern Ireland, which is a part of the United Kingdom (UK) and the Republic of Ireland (hereafter, Ireland) being an independent parliamentary republic. While partition was quintessentially conceived as a solution to political contestation, recent decades have witnessed concerted efforts to navigate away from the cessation of protracted conflict, especially in Northern Ireland, towards building a shared island. In part, as this chapter will illuminate, the achievements towards a post-conflict shared society within the Irish borderlands have also been influenced by European planning initiatives and the injection of significant EU Structural Funds.

The Irish border is 310 miles in length, but it is largely invisible and open with 208 formal – and a likely similar number of informal – crossing points (NIAC 2018). Coakley and O'Dowd (2007) have noted how a dynamic pattern of interaction and collaboration between the two parts of the island had developed by the end of the 20th century. This is evidenced by the transboundary flows on the island with an estimated 23,000–30,000 cross-border workers (Centre for Border Studies 2016) and 37 percent of Northern Ireland's total exports going to south of the border in 2018 (NISRA 2020). Nevertheless, social and economic disadvantage affecting border communities remains an issue of debate across state and civil society actors, including those with an interest in spatial planning, that has been heightened by the United Kingdom's referendum-based decision to exit the European Union.

DOI: 10.4324/9781003218319-14

This chapter provides insight into aspects of the contemporary character of the spatial planning system on each side of the Irish border. In mapping the trajectories of planning in each jurisdiction, the chapter outlines how the degree of convergence that has emerged in recent times demonstrates greater institutional 'thickness'. This provides for better appreciation of the multiple transboundary geographies that are being nurtured by cross-border relationships.

North-South cooperation

Cross-border cooperation takes many forms. In its broadest understanding, such cooperation can include any type of concerted action between public and/or private institutions of the border regions of two or more states (Sousa 2013). However, Sousa (2013) fails to acknowledge the significant contribution offered by non-institutional, informal arenas towards cross-border cooperation. Such cooperation relies on relational dynamics across networks of stakeholders with different power relations. The activity can involve a range of actors: politicians, policy-makers, planners and the public, to name a few. The geography of cooperation demands a recognition of the socio-spatial relationships that exist across jurisdictional borders in order to appreciate the dynamics of transboundary regionalism that manifests within functional territories. Reflecting on Sousa's (2013) conceptualisation, the prospect of advancing cross-border cooperation is contingent on the degree of institutional 'thickness' (Rafferty and Blair 2019) between central and regional/sub-regional authorities, and the extent to which local government convergence exists around policy and planning instruments. As such, cooperation may involve the creation of new institutions and/or experimentation in the synergistic utilisation of transboundary links of public and private agencies at various levels. Therefore, the governance architecture facilitating and guiding cooperation may consist of a blend of formal ('hard') institutions and informal ('soft') networks. However, as Scott and Collins (1997) remark, a core question surrounding cross-border cooperation and, indeed, spatial planning is the extent to which a mix of formal institutions and informal networks can promote integration and achieve socioeconomic cohesion.

Cross-border cooperation on the island of Ireland has been influenced by earlier European integration, supported by non-statutory frameworks to guide cross-border planning and operationalised via cross-border networks and varying degrees of local authority interaction (see, for example, Greer 2001). Cross-border cooperation has been deepening as a result of two primary interventions: the 1998 Good Friday Agreement (also referred to as the Belfast Agreement) and EU policy/funding programmes. These have nurtured a change in public discourse and the creation of new socio-spatial imaginaries that have enabled actors, particularly in the spatial, economic and community development sectors, to consider the island of Ireland as a functional space (Walsh 2015). Recognising the shared history and geography across the island, the 1998 Good Friday Agreement established a three-stand approach as a meta-governance arrangement to better deal with Northern Ireland. *Strand One* concerns the devolved institutions, chiefly the internal arrangements, which consist of the Assembly and a power-sharing Executive. *Strand Two* focuses on North-South relations by establishing the North/South Ministerial Council (NSMC) to bring together those with executive responsibilities in Northern Ireland and the Irish Government. *Strand Three* concerns 'East-West relations' and which, under the auspices of the British-Irish Council (BIC), brings together Ministers from the United Kingdom and Irish Governments, the devolved administrations in

Northern Ireland, Scotland and Wales and the Isle of Man Government, Government of Jersey and Government of Guernsey. Collaborative spatial planning is an important work theme of the North-South and East-West institutions, and, in line with the Good Friday Agreement, unprecedented positive relationships have been nurtured across these territories. These have induced 'de-bordering' effects on the island of Ireland, i.e., facilitating the dismantling of hard border infrastructure and supporting the free movement of people and goods across the Irish frontier that builds on membership of the Single European Market (established in 1993) and which, in turn, has allowed for the standardising of rules and regulations on either side of the Irish border (de Mars et al. 2018). Rafferty and Blair (2019) argue that this has produced a shift to inclusive cooperation and shared responsibility across some sectoral policies, as illustrated by the establishment of cross-border organisations, for example, InterTrade Ireland, Waterways Ireland and Tourism Ireland. These formal entities represent progress in synchronising policy development and delivering mutual benefits on an all-island basis by acknowledging larger functional geographies that operate beyond administrative boundaries.

The laws, institutions and policy landscape in each jurisdiction have differed over the century of partition. The historical evolution of legislation, structures and policy agendas is significant; it is not static or passive but exhibits a dynamic and changing morphology that is capable of nurturing mutual benefits. In effect, the nature of the symbiosis among laws, institutions and policy will determine the extent and quality of cross-border cooperation and planning. Political will is also important. Against this changing context, experiments with regional development have tended to focus on reducing geographical disparities within and, to a lesser extent, across both jurisdictions through economic and infrastructure interventions (for example, based on EU funded strategies for arterial drainage and tourism in the Erne Catchment and for communications in Derry/Donegal in the 1980s). The journey of change has continued and has produced a greater degree of symmetry between the various systems and structures of governance in each jurisdiction. One outcome is the coming together of specific arrangements for spatial planning and other place-shaping instruments that have amplified the potential of cross-border cooperation.

Spatial planning on the island of Ireland

Spatial planning does not have a single definition (Morphet 2011) and can be considered a 'slippery' concept (Allmendinger and Haughton 2009). Nonetheless, spatial planning (as a concept, practice and process) encompasses the regulation and forward management of land and property development in the broader public interest. However, much of the academic literature points towards spatial planning going beyond purely land use management to embrace other sectoral policies in a new form of integrated spatial governance (Morphet 2011, 2019; Baker and Hincks 2009). Spatial planning can be viewed as a mechanism of coordination, integration and connectivity for guiding the evolution of places and how they function and for managing socio-spatial change (Rafferty and Blair 2019). Furthermore, some argue that spatial planning goes further, ensuring the implementation of plans in line with the principles of sustainable development (Baker and Hincks 2009; Morphet 2019). These considerations are relevant for considering spatial planning on the island of Ireland.

The evolution of spatial planning in Ireland and Northern Ireland has been influenced by two features of particular interest when considering cross-border functionality, cooperation and planning. First, there is the influence within both jurisdictions of European spatial designs and a wider body of EU environmental law. The application of strategic spatial planning at the European level, and the appreciation of regional disparities, extends over many decades (Roberts 1997), culminating with the emergence of the European Spatial Development Perspective (ESDP) in 1999. This marked a milestone in the evolution of spatial planning in Europe, introducing new perspectives and spatial concepts aimed at guiding balanced and polycentric development (Murray 2004; Walsh 2005; Jensen and Richardson 2006). These ideas cascaded downwards, stimulating planners to consider how to translate and operationalise them at national, regional and local scales within their member states. The evolution of European strategic spatial planning has played a role in fostering cooperation in local and regional development within the Irish border region and beyond (Blair, et al. 2007; ICLRD 2006, 2010; ICBAN 2013; Murray 2004; Walsh 2015). In addition to new spatial logics, both spatial planning systems on the island have been shaped by European directives that guide environmental law, related, for example, to the management of water, habitats and species resources. However, Walsh and Rafferty (2019: 131) note how the 'history of the EU is characterised by a tension between two concepts of spatiality: that of a transnational European space of relational flows and connections across borders on the one hand and that of the EU as a mosaic of self-contained nation-states on the other hand.' Likewise, Murray (2004: 228) articulates how the two jurisdictions on the island of Ireland have 'shared an uneasy public policy co-existence with a considerable range of shared interests filtered frequently through the lenses of political interpretation and symbolism.' This has presented challenges for cross-border cooperation and it has been exposed more recently by the outworking of Brexit.

Second, the constitutional positions on the island of Ireland have created different political discourses, legislative trajectories and planning traditions. The decades following the establishment of the border in 1921 represented a period of limited cross-institutional interaction, causing the divergence of policy approaches vis-à-vis economic development, health and education (Greer 2003), and, in the Northern Ireland case, a high concentration of economic activity in the eastern half of Northern Ireland, dominated by the industrial base of linen, shipbuilding and engineering in Belfast (Ellis and Neill 2006). This had detrimental effects on socioeconomic progress in the Irish border region. The 1950s and 1960s witnessed the strengthening of cooperation between officials from Ireland and Northern Ireland, particularly in relation to transboundary infrastructure (for example, railways), shared environmental assets (for example, rivers) and economic relationships (for example, trading tariffs). However, the progress in nurturing further cooperation and, indeed, a degree of institutional thickening, was tempered by escalating terrorist violence and political unrest in the late 1960s and early 1970s that became known as the 'Troubles'. The political priorities of subsequent years were heavily focused on security within the Irish borderlands. Progress on cease-fires, the ratification of the Good Friday Agreement and the introduction of the Northern Ireland Act 1998 provided the foundations for new institutions, as noted above, along with opportunities to strive for new policy outcomes. Of particular relevance to spatial planning was the establishment of the British-Irish Council (BIC) with a group on spatial planning created under its auspices in 2009. However, as Morphet (2019: 151) comments, the BIC has not matured, and it has 'taken longer to develop an institutional character' in comparison to the North South Ministerial Council. However, recent research identifies the considerable value of the spatial planning group for exchanging ideas, sharing political and practical issues associated with spatial planning and acting as an advisor across jurisdictions (Clifford and Morphet 2015; Morphet and Clifford 2018).

The strategic planning context in Ireland

Planning legislation in Ireland emerged with the Town and Regional Planning Act 1934, with amendments in 1939, but with limited application and negligible impact. At a time of economic growth in the 1960s, the state introduced the Local Government (Planning and Development) Act, 1963. However, Williams and Nedović-Budić (2020: 6) reflect on how the then system 'placed a disproportionate emphasis on development control and paid little attention to implementation and strategic issues.' In parallel, Boylan (2005) acknowledges how, in Ireland, an industrialisation dominated postwar development, ambition imparted an unwarranted urban bias, while an increasing emphasis on trade liberalisation was strongly aligned with national economic objectives. This combination of factors caused a retrenchment in regional policy and meaningful consideration of border geographies.

It was not until the 1990s, with an appreciation for European spatial planning concepts and the rise of the much vaunted 'Celtic Tiger', that a national consensus emerged towards the need for regional and national spatial frameworks to provide a coordination role that transcended the local authority scale (Williams and Nedović-Budić 2020; Boylan 2005). Boylan (2005) argues how the rediscovery of interest in regional policy was exclusively influenced by European developments (as noted earlier) and the ability to draw down monies from Structural Funds (e.g., INTERREG). In 1994 eight new Regional Authorities (Figure 14.1) were created under the Local Government Act 1991. The Border Regional Authority covered Counties Donegal, Leitrim, Sligo, Cavan, Monaghan and Louth and provided an interface with local authorities in Northern Ireland. While supported financially via the EU's INTERREG programme in the early 1990s, Greer (2003) highlights a key criticism of early operations being the implementation of only a few truly cross-border projects, with the majority being stand-alone projects occurring on either side of the border that demonstrated minimal development impact on the region as a whole. The learning from these early days of INTERREG provided the impetus to strengthen partnership working during subsequent funding periods, with proposals being assessed for their contribution towards cross-border cooperation and their degree of regional and local authority involvement.

Planning in Ireland, during the late 1990s and early 2000s, witnessed the maturing of a more coherent strategic spatial logic for territorial development (Murray 2004; Scott 2009), new legislation and an expanded remit for the regional authorities, all of which, in theory, provided the ideal foundation for enriching cross-border dynamics. The passage of the Planning and Development Act 2000 provided the basis for the contemporary planning system in Ireland (for a further discussion see Grist 2013). Also significant was the publication of the National Development Plan (NDP) in 2000, which articulated balanced regional development as a national objective. In 2002 the National Spatial Strategy (NSS) emerged as part of the NDP to reach out to a planning horizon of 2020 that would secure more balanced regional development and address a pattern of uneven growth denoted by the dominance of the Greater Dublin Area. In the context of this chapter, it differed from the Regional Development Strategy for Northern Ireland by giving greater profile to the spatial interdependencies of each jurisdiction (Murray 2004). The 2000 Act required the Regional Authorities to provide long-term strategic planning frameworks for sustainable development across their geographies, including the preparation of Regional Planning Guidelines (RPGs), that would be aligned with the National Spatial Strategy (NSS) in order to secure a better balance of social, economic and physical development. Significantly, both the NSS and the RPGs articulated

FIGURE 14.1 Regional Authorities in Ireland, 1994.

potential cooperation on key strategic planning issues with Northern Ireland, embedded new spatial concepts of 'gateways' and 'hubs' and provided the parameters to better apply the functional geographies that transcend council boundaries and the Ireland–Northern Ireland border. The eight Regional Authorities were subsequently streamlined into three Regional Assemblies in 2015 comprising the Northern and Western Region, the Eastern and Midlands Region and the Southern Region (Figure 14.2). These regional structures, arguably supra-regions in scale, play a vital role in coordinating strategic spatial planning frameworks in the form of a Regional Spatial and Economic Strategy (RSES) and EU funding for their constituent local authorities.

The Irish Government's *National Planning Framework* (NPF) (DoHPLG 2018) acknowledges the functional realities that straddle the border geographies of Ireland and Northern Ireland. It rolls forward the thinking of the earlier NSS, presents a shift beyond traditional land use regulation towards strategic spatial planning (Lennon et al. 2018) and strengthens the spatial logics contained in its predecessor. The NPF's dedicated section on 'Working with Our Neighbours' acknowledges the potential of Brexit to significantly change the nature of

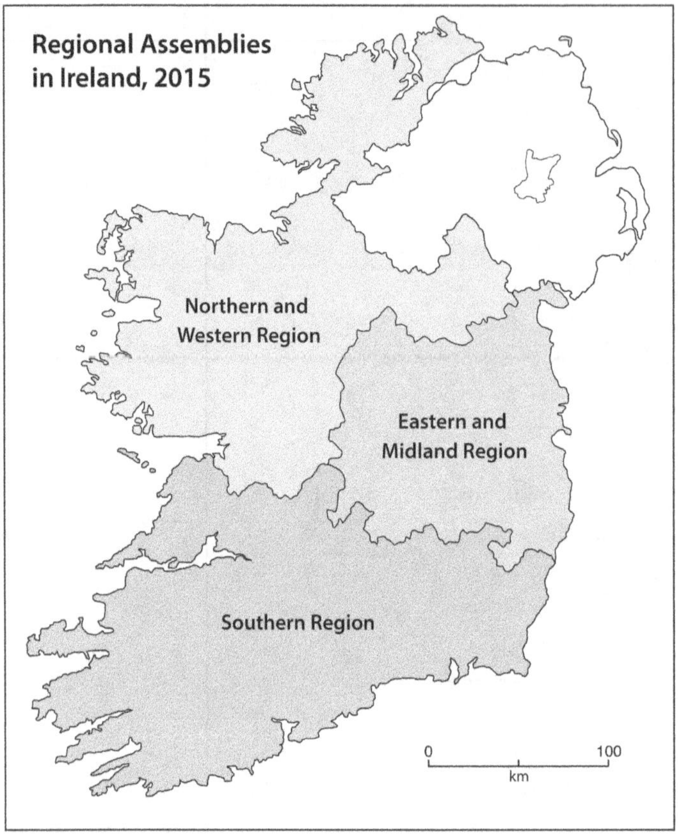

FIGURE 14.2 Regional Assemblies in Ireland, 2015

the relationship with the United Kingdom nations and, in particular, the Northern Ireland jurisdiction. In responding to these circumstances, the NPF contains specific spatial designs and policy objectives to develop the economic potential of the Dublin-Belfast Corridor and to support the northwest of the island by deeper institutional thickening between Donegal County Council (Ireland) and Derry City and Strabane District Council (Northern Ireland) with their co-designed North-West City-Region concept. Both spatial planning initiatives reflect the functional realities that operate across jurisdictional boundaries. More recently, the *2021 National Development Plan* (Government of Ireland 2021) devotes an entire chapter to a 'shared island' that is better connected, prosperous and sustainable and underpinned by capital funding for infrastructure projects that strengthen connectivity. Its priorities include enhanced high-quality transport links, strengthened third-level education access, harnessing the potential in renewable energies and cross-border health care provision. Thus in November 2021 Ireland Minister for Transport Eamon Ryan and Northern Ireland Minister for Infrastructure Nichola Mallon launched a public consultation on the future of rail on the island of Ireland as part of the All-island Strategic Rail Review. This is in line with the commitments by the Irish Government to the delivery of the *New Decade, New Approach* agreement of January 2020.

The strategic planning context in Northern Ireland

The emergence of spatial planning in Northern Ireland in recent years has similarly been influenced by European perspectives. Moreover, since 2015 the planning system has been restructured from a unitary system to a new two-tiered system comprising, at the regional scale, the Department for Infrastructure (DfI), which is responsible for regional planning policy, strategic (regionally significant) planning applications and giving support to local planning authorities. The second tier consists of district councils, which act as local planning authorities with responsibility for Local Development Plans, development management and planning enforcement. District councils have themselves been reduced in number from twenty-six to eleven by a process of merger. (See Chapter 1 in this book for a further contextualisation of spatial planning in Northern Ireland.)

The publication of the Regional Development Strategy for Northern Ireland, *Shaping Our Future* (Department for Regional Development 2001), has been hailed as 'the first part of the United Kingdom to develop a systematic approach to spatial planning' (Morphet 2011: 23). Conceptually and similar to the NSS it also drew upon principles set out in the European Spatial Development Perspective. Nevertheless, when this planning framework is located within the context of the island of Ireland, a degree of uneasiness is revealed in giving recognition to cross-border and all-island development linkages. This stands in marked contrast to a more expansive perspective from the NSS for Ireland. Put simply, *Shaping Our Future* emphasised policy guidance for spatial development *within* Northern Ireland (Murray 2004). The revised Regional Development Strategy, *Building a Better Future* (Department for Regional development 2012), articulates cross-border relationships more fully, for example, by developing a strong North West on the island, by acknowledging the peripherality of this region and by stating the need for councils to collaborate with their counterparts in County Donegal. While this indicates a degree of progress towards cross-border planning, commentary at that time highlights how, with a ministerial change, the 'final version contains less of an all-island emphasis than in the draft' (AgendaNI 2012). Political ideology is always at play in steering the extent to which cross-border spatial relationships might be nurtured. Nevertheless, research by Walsh (2015) reveals how having an all-island dimension explicitly mentioned in the terms of reference of the Regional Development Strategy for Northern Ireland provided 'political cover' for departmental spatial planners to collaborate with their counterparts across the border.

Enabling cross-border planning and development

Decades of de-bordering associated with peace-building initiatives and underpinned by the benefits of EU membership have strengthened cross-border contact, communication and cooperation among public, private and third (voluntary and community) sectors (Taillon 2018). These stakeholders have been able to explore 'soft' cooperation opportunities based on mutual interest, work across 'fuzzy' boundaries, and appreciate the added value of operating beyond the 'hard' structures and boundaries associated with the machinery of government. This is very much in line with a wider academic commentary that recognises the emergence of 'soft spaces' that reflect the complex relational world existing beyond the shackles of formal regulatory apparatus and territory (see Allmendinger and Haughton 2009; Walsh 2015; ESPON 2017). Cross-border planning and development on the island of Ireland has been enabled by four significant modes of cooperation: bilateral spatial planning, network planning, community planning and partnership planning.

Bilateral spatial planning

As an addition to the spatial planning strategies for Ireland and Northern Ireland mentioned above, quiet cross-institutional working has produced a non-statutory expression of ambition to guide cooperation. The *Framework for Cooperation: Spatial Strategies of Northern Ireland and the Republic of Ireland* (Department for Regional Development and Department of Environment, Heritage and Local Government 2013) comprises a bilateral spatial planning framework for joint working on the island and is described in its terms of reference as:

> a non-statutory approach to providing advice and guidance at relevant spatial or geographical scales. It seeks to encourage policy makers in the public sector to take account of the wider impact of their work, to recognise and exploit opportunities for a wider perspective and to avoid 'back to back' planning. (5)

The framework articulates the need for stronger cross-border planning and supports the construction of 'soft spaces' for policymakers and planners to pragmatically manage development and infrastructure across the border. In effect, it presents a 'positive re-imagining of cross-border regionalism through a number of layers of spatial public diplomacy' (Peel and Lloyd 2015: 2223). Viewed as an enabler for mediation, Rafferty and Blair (2019) argue the possibility for the framework to intensify cooperation and integration by overcoming any 'fuzziness' associated with boundaries. It has the potential to more effectively navigate spatial development in functional geographies derived from decades of socioeconomic interaction and exchange. The framework builds on formal cross-border institutions, which, as noted above, have been fruitful in managing special landscapes, rivers and habitats in order to support certain sectors, for example, tourism. Nonetheless, cooperative capacity is still required to tackle the multiple disparities that pertain along the Irish border and which were greatly illuminated by the deliberations and testimonies recorded in the Houses of the Oireachtas (2019) *Report of the Joint Committee on Brexit and the Border: The Impact on Rural Communities*. The strategic planning framework emanating from central government departments in Northern Ireland and Ireland provides a crucial first step towards an enabling environment for addressing barriers in policy responsiveness.

Network planning

Across the Irish border region, a number of networks have been established to support communication and cooperation between local authorities, namely the East Border Region (EBR) in 1976, the North-West Region Cross Border Group (NWRCBG) also in 1976 and the Irish Central Border Area Network (ICBAN) in 1995. In their formative years these networks, particularly EBR and NWRCBG, encountered challenges in operating with an agenda that was perceived as politically contentious (see Greer 2003) and which spilled over in relation to building trust and reconciling different priorities. Furthermore, Greer (2003) notes that the networks adopted a reactive approach to cooperation that resulted in piecemeal development. However, the maturing of these networks has facilitated a degree of institutional thickening between their constituent local authorities and they now represent a 'soft' arena for promoting distinctive integrated spatial frameworks for sub-regional development (Rafferty and Blair 2019). Changes to the public administration geography of Northern Ireland as a result of district council amalgamations in 2015 have had implications for the configuration of these networks. In the case of NWRCBG the incorporation of Magherafelt District Council into Mid Ulster District Council has widened the spatial reach of ICBAN (Figure 14.3), while Limavady Borough Council has been absorbed into the Causeway Coast and Glens District Council. But the loss of these two NWRCBG members has also afforded the opportunity to create a new cross-border alliance between Derry City and Strabane District Council and Donegal County Council, discussed below. Meanwhile, Meath County Council has joined EBR, thus giving shape to a more complete corridor of development cooperation in the local government space between Dublin and Belfast (Figure 14.4). EU funding, not least from INTERREG, has been key to the effectiveness of these networks and thus it is not surprising that the impact of Brexit should loom large in their ongoing commitment to collective working. In 2020, for example, the six local authority members of EBG co-signed a highly symbolic Charter outlining detailed working arrangements for strategic, cross-border cooperation as their

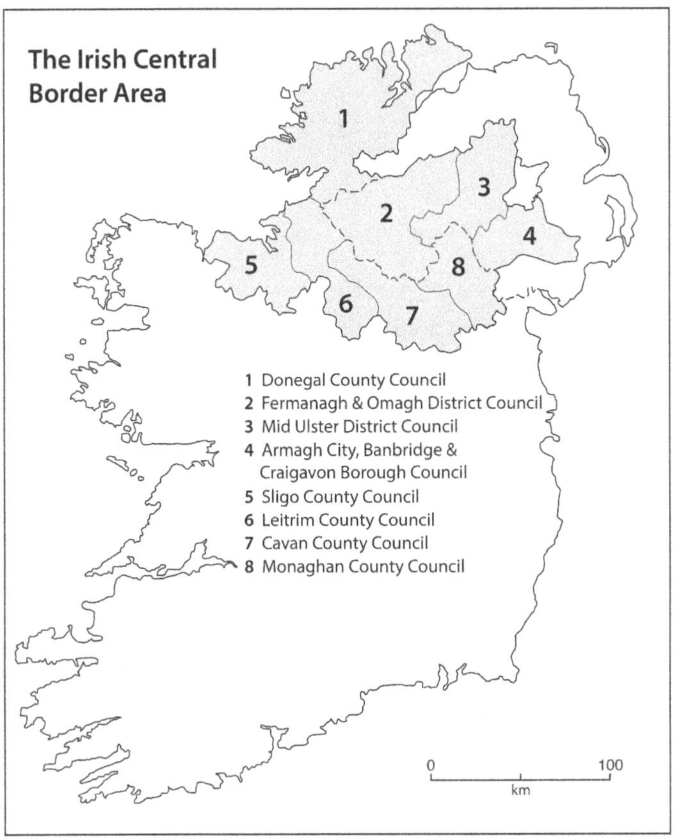

FIGURE 14.3 The Irish Central Border Area

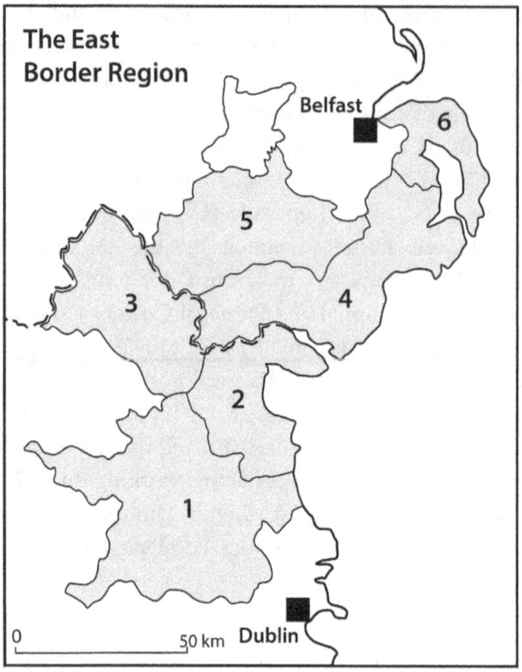

1 Meath County Council
2 Louth County Council
3 Monaghan County Council
4 Newry, Mourne & Down District Council
5 Armagh, Banbridge, Craigavon Borough Council
6 Ards & North Down Borough Council

FIGURE 14.4 The East Border Region

response to Brexit under the banner of securing 'a smart, competitive, sustainable, and inclusive cross-border region'.

Community planning

The objective of collaboration to enrich social and economic wellbeing in places along the border has been reinforced of late by complementary community planning measures in each jurisdiction. The Local Government Act (Northern Ireland) 2014 introduced a new statutory duty on local authorities to lead and coordinate service delivery through a process of community planning. The aim of this new planning instrument is for district councils to work in partnership with statutory bodies (for example, the Northern Ireland Housing Executive, the Police Service of Northern Ireland) and support organisations (for example, local colleges, universities and businesses) to promote social, economic and environmental wellbeing in line with citizen needs. The production of a Community Plan identifies a set of outcomes that should then be strategically managed by a Community Planning Partnership. At the same time the Local Government Reform Act 2014 in Ireland introduced new public administration structures that included Local Community Development Committees (LCDCs) that would bring together elected representatives and officials, state agencies and a range of local development and community development organisations. The preparation of a Local Economic and Community Plan (LECP) would allow for thematic policy objectives to be outlined. These are comparable to Community Plans in Northern Ireland whereby the partners coordinate, manage and oversee the implementation of local economic and community development programmes, thus enhancing the links between policy formation and service delivery. The important point in the context of this chapter is that community planning in each jurisdiction has the potential to reach across the border as evidenced by the content of border-facing plans:

- the Derry City and Strabane Community Plan canvasses the North West City Region proposal as being critical to the future wellbeing of everyone living within Northern Ireland and the North West of the island of Ireland that requires support from investment in connectivity infrastructure, cross-border collaboration in health and social care and North West Cross-Border Partnership arrangements.
- the Fermanagh and Omagh Community Plan recognises its juxtaposition with Counties Donegal, Cavan, Leitrim and Managhan.
- the Armagh City, Banbridge and Craiagavon Community Plan articulates an aspiration for the borough to become an island of Ireland and Northern Ireland internationally renowned tourism destination, while its arts venues currently attract cross-border audiences.
- the Newry, Mourne and Down Community Plan identifies implementation of the district Tourism Strategy 2017–21 as a 'focused priority' linked to 'prosperous communities' that includes a renewed application for a cross-border UNESCO Global Geopark designation, development of the Carlingford Lough Greenway, harnessing the potential of Newry as a major gateway on the Dublin-Belfast Corridor and capitalising on the opportunities afforded by the Greenore–Greencastle ferry.
- the Donegal Local Economic and Community Plan identifies Letterkenny as a linked gateway with Derry that is attracting people and should be promoted as the primary growth centre for the region, cites a cross border mobile library service as a positive cultural resource and commits to the development of opportunities for peace-building and conflict resolution both within the county and on a cross border basis.
- the Leitrim Local Economic and Community Plan has a specific section on 'cross-border dimensions' that references 'ongoing opportunities for cross-border collaboration and development in the areas of agriculture, business, education, energy, health and care, infrastructure, leisure, regeneration, regional planning, telecommunications, tourism, transportation and utilities'. ICBAN and Waterways Ireland are identified as key cross-border enablers.
- the Cavan Local Economic and Community Plan comments that its border location gives rise to positive and negative economic impacts on the county. Actions include participation in cross-border micro-enterprise and small business networks, resourcing innovation through support from InterTrade Ireland, linking the County Museum with its World War 1 Trench Experience and Peace Garden to groups on a cross-border basis and working collaboratively with Fermanagh and Omagh District Council to further develop the Marble Arch Caves UNESCO Global Geopark.
- the Monaghan Local Economic and Community Plan confirms a willingness to embrace cross-border collaboration, including, for example, development of the Ulster Canal project and expanding the Greenway from Clones to Armagh.

- the Louth Local Economic and Community Plan underlines the significance of a Memorandum of Understanding between Louth County Council and Newry, Mourne and Down District Council as a strategic alliance to promote economic development in the cross-border region in which they are situated.

While the cross-border dimension may be more muted within the Northern Ireland plans, this tableau of possibilities and relationships comprising policies, projects and enablers points, nonetheless, to a progressive journey of cooperation across many sectors. Collectively, the plans demonstrate a welcome shift in planning practice by going well beyond regulatory land use control and facilitating the formation of new transboundary geographies that are reshaping the Irish borderlands.

Partnership planning

The North West Strategic Growth Partnership, established in 2016, is a cross-border regional development structure jointly led by Derry City and Strabane District Council and Donegal County Council whose aims include driving economic growth and investment, promoting physical and environmental development and improving community cohesion and wellbeing. Its distinctive character derives from the participation of representatives from government departments in Northern Ireland and Ireland along with financial support from both the Executive Office and the Irish Government. The partnership is mandated to deliver *The North West Strategic Growth Plan – a Model for Regional Development and Growth of the North West of the Island of Ireland* (ICLRD 2016), whose terms of reference comprise the following statement:

> The North West Region of the Island of Ireland is the fourth largest urban agglomeration on the island of Ireland; it is the only functional economic region of such scale on the island which experiences a national jurisdictional border. As such, specific arrangements involving regional and local government leadership and partnership with central government are required in order for the region to release its full and considerable potential as a net contributor to the economy, North and South and in an East-West context... An opportunity has been created, through the alignment of local government functions in a place-based leadership model, for a new arrangement which allows central government to coordinate its resource planning and spending as they relate to the region - leading to a corresponding impact on the region's growth potential and optimal value for money as regards public investment in both jurisdictions. (1)

The key spatial planning concept driving this collaborative approach to cross-border regional development is the North West as a 'City Region' with an urgency for action underlined by the challenges of Brexit for the economies of Ireland and Northern Ireland. Early proposals include the clustering of business support across the main urban centres of Letterkenny, Derry-Londonderry and Strabane involving coordination of capital and programme resources, collaborative delivery of third-level education, enhanced air access and route development, building rural resilience through extended broadband connectivity and delivering better health and wellbeing outcomes.

Conclusion

Cross-border cooperation, particularly in relation to informal ('soft') networks or spaces, has matured over recent decades, but the extent of well-established, meaningful and productive collaboration between the formal ('hard') planning instruments on the island of Ireland is still in its infancy. While policy articulation has become much stronger in acknowledging cross-border dynamics, the structural parameters and regulatory legacy of planning remain challenging. Planning has become more plural and differentiated within the United Kingdom, providing an opportunity in Northern Ireland for its system to evolve along a path that adopts a changing morphology that further aligns, if not fully integrates, with the planning system south of the border. The prospect of greater coherence and an inclusive arena within which to include institutional and civil society interests that span sectoral boundaries and territorial borders is a powerful reminder of the enduring need for spatial planning to guide the social, economic and environmental outcomes in functional geographies that people recognise and value. This will require leadership, engagement and support from key actors within central and local government in Northern Ireland and Ireland to ensure that the forms of cross-border working, discussed in this chapter, can realise their potential.

Legislation

Local Government Act, 1991
Local Government Act (Northern Ireland) 2014
Local Government (Planning and Development) Act, 1963
Local Government Reform Act 2014
Northern Ireland Act 1998
Planning and Development Act, 2000
Town and Regional Planning Act, 1934
Town and Regional Planning (Amendment) Act, 1939

References

AgendaNI. 2012. *Northern Ireland's New Regional Development Strategy*. Available at https://www.agendani.com/northern-irelands-new-regional-development-strategy/ (accessed 30 August 2021).

Allmendinger, P. and G. Haughton. 2009. 'Soft Spaces, Fuzzy Boundaries, and Metagovernance: The New Spatial Planning in the Thames Gateway', *Environment and Planning A* 41: 617–33.

Baker, M. and S. Hincks. 2009. 'Infrastructure Delivery and Spatial Planning: The Case of English Local Development Frameworks', *Town Planning Review* 80(2): 173–99.

Blair, N., A. Adair, B. Bartley, J. Berry, C. Creamer, J. Driscoll, S. McGreal and F. Vigier. 2007. 'Delivering Cross-Border Spatial Planning: Proposals for the Island of Ireland', *Town Planning Review* 78(4): 485–509.

Boylan, T. 2005. 'From Undeveloped Areas to Spatial Strategies: Reflections on Irish Regional Policy', in M. McEldowney, M. Murray, B. Murtagh and K. Sterrett (eds) *Planning in Ireland and Beyond: Multidisciplinary Essays in Honour of John V. Greer*. Belfast: Queens University Belfast, pp. 91–106.

Centre for Border Studies. 2016. *EU Referendum Briefing Papers Briefing Paper 4, The UK Referendum on Membership of the EU: Citizen Mobility*. Available at http://crossborder.ie/site2015/wp-content/uploads/2016/06/CCBS-and-Cooperation-Ireland-EU-Referendum-Briefing-Paper-4.pdf (accessed 23 August 2021).

Clifford, B. and J. Morphet. 2015. 'A Policy on the Move? Spatial Planning and State Actors in the Post-Devolutionary UK and Ireland', *The Geographical Journal* 181(1): 16–25.

Coakley, J. and L. O'Dowd. 2007. 'The 'New' Irish Border: Changing Political, Economic and Social Dimensions', in J. Coakley and L. O'Dowd (eds) *Crossing the Border: New Relationships between Northern Ireland and the Republic of Ireland*. Dublin: Irish Academic Press, pp. 3–28.

de Mars, S., C. Murray, A. O'Donoghue and B. Warwick. 2018. *Bordering Two Unions: Northern Ireland and Brexit*. Bristol: Policy Press.

Department for Regional Development (DRD). 2001. *Shaping Our Future: Regional Development Strategy for Northern Ireland 2025*. Belfast: DRD.

Department for Regional Development (DRD). 2012. *Regional Development Strategy: Building a Better Future*. Belfast: DRD.

Department for Regional Development and Department of Environment, Heritage and Local Government (DRD and DEHLG). 2013. *Framework for Cooperation. Spatial Strategies of Northern Ireland and the Republic of Ireland*. Belfast: Department for Regional Development and Dublin: Department of Environment, Heritage and Local Government.

Department of Housing, Planning and Local Government (DoHPLG). 2018. *Project Ireland 2040 National Planning Framework*. Dublin: DoHPLG.

Ellis, G. and W.J.V. Neill. 2006. 'Spatial Governance in Contested Territory: The Case of Northern/ North of Ireland', in M. Tewdwr-Jones and P. Allmendinger (eds) *Territory, Identity and Spatial Planning: Spatial Governance in a Fragmented Nation*. London: Routledge, pp. 123–38.

ESPON. 2017. *ESPON ACTAREA: European Atlas of Soft Territorial Cooperation*. Available at https:// www.espon.eu/actarea (accessed 07 March 2018).

Government of Ireland. 2021. *National Development Plan 2021–2030*. Dublin: Department of Public Expenditure and Reform.

Greer, J. 2001. *Partnership Governance in Northern Ireland - Improving Performance*. Farnham: Ashgate.

Greer, J. 2003. 'Planning and Development Networks in the Borderlands of Ireland', in J. Greer and M. Murray (eds) *Rural Planning and Development in Northern Ireland*. Dublin: Institute of Public Administration, pp. 241–58.

Grist, B. 2013. *An Introduction to Irish Planning Law*. Dublin: Institute of Public Administration.

Houses of the Oireachtas. 2019. *Report of the Joint Committee on Brexit and the Border: The Impact on Rural Communities*. Dublin: Houses of the Oireachtas.

International Centre for Local and Regional Development (ICLRD). 2006. *Spatial Strategies on the Island of Ireland: Development of a Framework for Collaborative Action*. Newry: InterTradeIreland. Available at http://iclrd.org/documents/SpatialStrategiesontheIslandofIreland.pdf.

International Centre for Local and Regional Development (ICLRD). 2010. *Developing a Strategic Alliance between Newry & Mourne District Council and the Louth Local Authorities: Background Report*. Armagh: ICLRD.

International Centre for Local and Regional Development (ICLRD). 2016. *The North West Strategic Growth Plan - A Model for Regional Development and Growth of the North West of the Island of Ireland*. Available at http://meetings.derrycityandstrabanedistrict.com/documents/s27182/North%20West%20-%20Appendix%201.pdf.

Irish Central Border Area Network (ICBAN). 2013. *Regional Strategic Framework for the Central Border Region: 2013–2027*. Available at http://icban.com/site/wp-content/uploads/2018/03/Regional-Strategic-Framework-Nov-13.pdf.

Jacobs, J. 2016. 'Spatial Planning in Cross-Border Regions: A Systems-Theoretical Perspective', *Planning Theory* 15(1): 68–90.

Jensen, O.B. and T. Richardson. 2006. 'Towards a Transnational Space of Governance? The European Union as a Challenging Arena for UK Planning', in M. Tewdwr-Jones and P. Allmendinger (eds) *Territory, Identity and Spatial Planning: Spatial Governance in a Fragmented Nation*. London: Routledge, pp. 47–63.

Lennon, M., M. Scott and P. Russell. 2018. 'Ireland's New National Planning Framework: (Re)Balancing and (Re)Conceiving Planning for the Twenty-First Century?', *Planning Practice and Research* 33(5): 491–505.

Morphet, J. 2011. *Effective Practice in Spatial Planning*. London: Routledge.

Morphet, J. 2019. *Changing Contexts in Spatial Planning: New Directions in Policies and Practices*. London: Routledge.

Morphet, J. and B. Clifford. 2018. 'Who Else Would We Speak To?' National Policy Networks in Post-Devolution Britain: The Case of Spatial Planning', *Public Policy and Administration* 33(1): 3–21.

Murray, M. 2004. 'Strategic Spatial Planning on the Island of Ireland', *Innovation: The European Journal of Social Science Research* 17(3): 227–42.

Northern Ireland Affairs Committee (NIAC). 2018. *The Land Border between Northern Ireland and Ireland*, HC329. London: HMSO.

Northern Ireland Statistics and Research Agency (NISRA). 2020. *Northern Ireland Broad Economy Sales and Exports Statistics: Goods and Services Results 2018*. Available at www.nisra.gov.uk/sites/nisra.gov.uk/files/publications/BESES-Goods-Services-Publication-2018-%20Headline-results.PDF.

Peel, D. and G. Lloyd. 2015. 'Towards a Framework for Cooperation: Spatial Public Diplomacy on the Island of Ireland', *European Planning Studies* 23(11): 2210–226.

Rafferty, G. and N. Blair. 2019. 'Emergent Cross-Border Functional Geographies on the Island of Ireland: Pre- and Post-Brexit', *Irish Geography* 52(2): 75–191. https://doi.org/10.2014/igj.v52i2.1400.

Roberts, P. 1997. 'Strategies for the Stateless Nation: Sustainable Policies for the Regions in Europe', *Regional Studies* 31: 875–82.

Scott, J.W. and K. Collins. 1997. 'Inducing Transboundary Regionalism in Asymmetric Situations: The Case of the German-Polish Border', *Journal of Borderlands Studies* 12(1–2): 97–121.

Scott, M. 2009. 'Regional Planning and Sustainability', in T. Varley, S. Shortall and J. McDonagh (eds) *A Living Countryside? The Politics of Sustainable Development in Rural Ireland*. London: Routledge, pp. 85–106.

Sousa, L. 2013. 'Understanding European Cross-Border Cooperation: A Framework for Analysis', *Journal of European Integration* 35(6): 669–87.

Taillon, R. 2018. 'Cross-Border Issues in Ireland: Lessons for the Anglo-Scottish Border', *Journal of Borderlands Studies* 33(1): 85–102.

Walsh, C. 2015. 'Creating a Space for Cooperation: Soft Spaces, Spatial Planning and Cross-Border Cooperation on the Island of Ireland', in P. Allmendinger, G. Haughton, J. Knieling and F. Othengrafen (eds) *Soft Spaces in Europe*. London: Routledge, pp. 192–212.

Walsh, C. and G. Rafferty. 2019. 'Creating Spaces for Cooperation: Crossing Borders and Boundaries before and after Brexit', *Irish Geography* 52(2): 127–35.

Walsh, J. 2005. 'Spatial Planning Frameworks for Ireland: Critical Reflections on the Dublin-Belfast Corridor', in M. McEldowney, M. Murray, B. Murtagh and K. Sterrett (eds) *Planning in Ireland and Beyond: Multidisciplinary Essays in Honour of John V. Greer*. Belfast: Queens University Belfast, pp. 125–35.

Williams, B. and Z. Nedović-Budić. 2020. 'Transitions of Spatial Planning in Ireland: Moving from a Localised to a Strategic National and Regional Approach', *Planning Practice and Research*. https://doi.org/10.1080/02697459.2020.1829843.

15

EQUALITY LAW, GOOD RELATIONS AND PLANNING

Diane O'Neill and Brendan Murtagh

Introduction

Planning policy and practice routinely adopts the language of equality as justification for a range of interventions in strategic and development planning and in its connections with urban regeneration, housing management and infrastructure policy (Gale and Thomas 2018). The lack of legal precision about what equality means for practice and how (or whether) planning can deliver it in any coherent way is a contested arena especially in Northern Ireland. The significance of territory, the control and ownership of land and resource competition, especially between interface communities, have placed equality and planning and the relationship between them at the heart of wider conflict transformation processes (Coakley and Todd 2020). The Northern Ireland Act 1998, which enshrined the main elements of the Belfast/Good Friday Peace Agreement (1998), brought in a devolved power-sharing Assembly, institutional arrangements to protect the various rights of Unionists and Nationalists and some of the most formidable equality legislation seen anywhere in Europe at that time (McLaughlin 2007).

Section 75 of the Act set out requirements to tackle discrimination and promote good relations supported by a new Equality Commission with extensive regulatory and investigative powers. Equality Plans described how all public agencies would ensure and strengthen equality of opportunity and Equality Impact Assessments (EQIAs) aimed to evaluate the likely effects of legislation, major decisions and programmes on nine groups, including people of different religions. The equality provisions were the culmination of decades of affirmative action and anti-discrimination legislation, especially in relation to employment and the delivery of public services. However, they also aimed to build community relations and were viewed as a necessary and ultimately successful component of conflict transformation and stabilising the fragile political institutions. Ruane and Todd (2012: 182) concluded that 'egalitarian policies (of which affirmative action programmes were an important part), introduced late and against strong resistance, have … lessened inequality and changed and moderated national conflict; resistance to these policies has lessened and the issue of communal economic inequality is now depoliticised'. For others, the practice of equality has been much less

DOI: 10.4324/9781003218319-15

impressive and, in reality, has made little actual difference to the way in which public agencies formulate, deliver or evaluate their services (Hughes 2004; Murtagh et al. 2019). Indeed, the Equality Commission (ECNI 2014) itself pointed out that Northern Ireland has slipped significantly behind the rest of the United Kingdom in equality protection, especially on disability, age and race relations.

This chapter reviews the implications of equality legislation for spatial planning and, in particular, how it connects the profession to the territorial legacies of segregation, interfacing and poverty. Critics point out that equality has been slow to progress because of a dominant techno-rational ideology within the planning profession and because avoiding, not engaging the conflict has insulated the system from accusations of discrimination and bias (Ellis 2001; McDowell et al. 2017). However, there are more complex reasons why planners fail to embed the Section 75 provisions in mainstream development planning and management decisions. The chapter reflects on surveys with planners and case studies to show that a lack of skills, the sheer complexity of problems and an absence of guidance on the interpretation of what at times appear to be contradictory policy aims are all part of the problem. Murtagh et al. (2019) also argue that neoliberalism and the primacy of markets, rolling back the state and deregulating the private sector have all undermined equality, especially in areas such as planning, where it has been stripped back to a narrow set of regulatory functions. The chapter concludes that by building on strategies for shared housing and interface planning, the profession might help deliver the provisions of the Northern Ireland Act across areas, practices and spatial programmes. In particular, developing skills, sharing best practices and stronger interdisciplinary working among planning, equality groups and the community relations sector could be at the heart of a more progressive approach to learning and professional support.

Political and legal context

It is important to set equality legislation in the context of the Northern Ireland conflict and the importance of land within it. The Northern Ireland Constitution Act 1973, which abolished the suspended Parliament of Northern Ireland, also made it unlawful for a public authority carrying out its functions to discriminate or to aid or incite another person to discriminate, against a person or class of persons on the grounds of religious belief or political opinion. The Act also established the Standing Advisory Committee on Human Rights (SACHR) as an advisory body in relation to policy matters as well as transposing international treaties to Northern Ireland. SACHR had an important impact on equality legislation by illustrating the scope and endemic nature of direct and indirect discrimination, the different economic and social status of Catholics and Protestants and the need for 'parity of esteem' as the basis of any constitutional settlement (Aughey 2005).

Pressure to comply with European and international obligations, the strength of the gender equality debate in Britain and the need for guarantees for investors, especially from the USA, gradually raised the importance of *equality mainstreaming* beyond the employment arena (Donaghy 2004). The emphasis on monitoring effects, ex-ante assessments of policy impacts and evaluating progress towards closing the economic differential between Catholics and Protestants led to the introduction of *Policy Appraisal and Fair Treatment (PAFT)* in 1994. The PAFT guidelines, in which all government departments, agencies and public bodies had to assess and track the effects of their services on the two main religions, had little legal weight but were important in developing equality policy and assessment methods. As a proto *Equality Impact*

Assessment (EQIA) process, it highlighted a particular concern for analysing how resources, laws and investments could achieve more inclusive as well as fairer outcomes (Central Secretariat Circular 5/93 1993; McCrudden et al. 2004). McCrudden (1998) pointed out that there was little guidance about how to conduct a PAFT assessment and that, given the intense political pressure from the Irish-American lobby, its function was 'primarily window-dressing for the government, a necessary symbolic gesture, but not much more' (McCrudden 1998:1713). In particular, there was a lack of understanding about who was responsible for conducting PAFT, how enforceable it was and how the differential impact between the two communities would be measured and, if necessary, adjusted (Osborne 2007). SACHR's own review concluded that:

> The implementation of PAFT has been inadequate to the task of giving effect to the aspirations expressed for it and we question the effectiveness of the body with overall responsibility for its operation (and) implementation of the policy to date has fallen far short of what might reasonably have been expected in an area of such importance.
>
> *(SACHR 1997: section 5.35)*

Part of the problem was that the British and Irish Governments perceived equality to be primarily a confidence-building measure for the Catholic community and, as momentum grew around the cease-fires and the Belfast/Good Friday Peace Agreement, equality, the economic differential, parity of esteem and how to strengthen community relations were conflated with the political settlement itself. For Knox and Carmichael (2005: 63) the Agreement became an exercise in 'creative ambiguity' especially as it was translated into the Northern Ireland Act 1998 itself. In this respect, its vagueness suited the nature of the peace process and progress toward a political settlement:

> Two equality agendas were addressed in the Agreement. One agenda related to national equality between the two different allegiances, one Irish, the other British. Another related to social equality between communities defined by other characteristics: religion, ethnic affiliation, race, disability, gender, and so on. Both issues were recognised as requiring significant protection. The Agreement sought to enshrine both agendas, irrespective of whether there would be a united Ireland, or continued membership of Northern Ireland in the United Kingdom.
>
> *(McCrudden 1998: 1734)*

Hughes and Donnelly (2004) have argued that the Belfast/Good Friday Agreement assembled an impressive architecture of not always consistent or complementary legislative guarantees for both traditions, including the adoption of the European Convention on Human Rights; the establishment of a Human Rights Commission and the Equality Commission; arrangements for promoting, facilitating the use of and protecting the Irish and Ulster/Scots language; and a programme of policing reform. In their retrospective on Northern Ireland, Tonge and Evans (2020) show that this was an essential stage in the wider process of democratisation, which also locked in and legitimised a binary politics in which Unionist and Nationalist identities, claims and rivalries shaped the political mainstream.

> The nature of consociational arrangements which constitute power-sharing, where key decisions are taken on a cross-community basis through parallel consent or a weighted

> majority, means that either the DUP and/or Sinn Féin can exercise a mutual veto. This has led to gridlock in the decision-making process over contentious issues and marginalization of other political parties in the Executive (SDLP and Ulster Unionist Party).
>
> *(Knox and Carmichael 2010: 82)*

Similarly, Murray et al. (2009) argued that the institutional arrangements are fundamentally 'elitist, consociational, top-down models of power-sharing' that leave little scope for other identities, social issues or interest groups to exert influence or attract attention. The sclerotic effects of equality are viewed in the lack of progress on poverty, support for community peace-building and the failure to transpose rights legislation (such as same-sex marriage) to Northern Ireland (Morrow 2016). The politicisation of rights, guarantees that Protestants and Catholics would be treated equally. How this, in particular, contributed to peace-building and the promotion of good relations was not well developed in either the legislation or the subsequent guidance to policymakers.

Northern Ireland Act 1998 and equality planning

Agocs and Osborne (2009) emphasise that the Northern Ireland Act 1998 was especially significant for bringing in formidable institutions, policies and systems to secure equality of opportunity and attracted global attention for its scope and ambition. Its main implication for the planning system was the introduction of a statutory duty on all public bodies to have:

> Due regard to the need to promote equality of opportunity –
>
> (a) between persons of different religious belief, political opinion, racial group, age, marital status or sexual orientation;
> (b) between men and women generally;
> (c) between persons with a disability and persons without;
> (d) between persons with dependents and persons without (Section 75(1)).

In carrying out their functions, public authorities must also 'have regard to the desirability of promoting good relations between persons of different religious belief, political opinion or racial group' (Section 75(2)). Government departments and agencies had to prepare *Equality Schemes* that showed how these objectives would be met through current policies and *Equality Impact Assessments (EQIAs)* were introduced to proof all new programmes against the needs of the nine groups identified in Section 75(1).

The key aim of the Act was to mainstream and embed equality practice in routine policy formulation, implementation and review processes. Coakley (2009) warned that there were concerns that this might result in the fragmentation of equality, especially if it displaced or became an alternative to traditional anti-discrimination practice, which might enable some departments to devalue its role in policymaking. Within the Equality Scheme, organisations are required to identify all policies and distinguish between those having limited equality concerns and those having probable equality implications. The latter were then to be subject to the EQIA process in order to identify any adverse or discriminatory effects and thus lock in policymakers to a proper scrutiny of their programmes and decisions. The EQIA is based on the concept of Environmental Impact Assessment and aims to anticipate the effects of a plan or policy on the nine specified groups, supported by structured consultations with each set of interests:

> The primary function of the EQIA is to determine the extent of differential impact of a policy upon the groups and in turn whether that impact is adverse, that is, whether it has a negative impact on one or more of the nine equality categories. If it is decided that the policy has an adverse impact, the public authority must consider measures which might mitigate the adverse impact, and alternative policies which might better achieve the promotion of equality of opportunity.
>
> *(ECNI 2005: 4)*

The process of preparing an EQIA takes place through seven integrated stages:

1. Defining the aims of the policy;
2. Consideration of available data and research;
3. Assessment of impacts;
4. Consideration of:
   - measures which might mitigate any adverse impact; and
   - alternative policies which might better achieve the promotion of equality of opportunity;
5. Consultation;
6. Decision by public authority and publication of report on results of EQIA; and
7. Monitoring for adverse impact in the future and publication of the results of such monitoring.

The process is data and consultation intensive, building up a strong quantitative and qualitative assessment on how the policy relates to each group in turn. If an adverse impact is identified, the guidance advises the policymaker to assess whether it is *unlawfully* discriminatory as some programmes may involve affirmative action to address historic disadvantages. Even when a policy is not deemed unlawful, the public body still needs to outline a strategy to mitigate any adverse matters identified. However, if it is intended to address the needs of a particular group, then the differential may well be justified and even necessary to promote equality, access to services or redistributive outcomes. Notwithstanding such effects, the organisation is advised to consider alternative measures that would achieve the policy aim, without risking a differential impact. There is no 'test of statistical significance' to assess the *adverse threshold* but the Commission sets out critical questions and case study examples as guides to determine the nature and scope of uneven and unreasonable consequences (ECNI 2005: 22).

The guidance also recommends that the EQIA includes both formal and informal consultations, with the latter taking place at key stages *throughout* the EQIA process. While it is cautious about formally recommending consultees as the process is designed to be flexible, the guidance does set a minimum of two months for the overall consultation process. It suggests that different methods should be applied and tailored to the needs, experiences and expectations of the nine groups, especially racial minorities, children and people with disabilities. It also recognises that it may be necessary to allocate resources to build the capacity of stakeholders in order to avoid the participatory optics that give the illusion of involvement but, in reality, provide little opportunities to challenge or change policy delivery (Rafferty 2012).

The *mitigation* of adverse impacts and consequent remedies should also be a continuous process with justification provided if alternatives suggested or tested have not been accepted. Finally, auditing systems need to be set in place to monitor, at least on an annual basis, the impact of the policy on the relevant groups and how their conditions have changed. The tested policy may then need to be adjusted or even terminated if the consequences are adverse or discriminatory, although the methodological lacunae on what constitutes adverse remain problematic for evaluators (CAJ 2013).

For a number of critics, the whole equality regime has failed to live up to its reformist potential, especially in contributing to local peace-building, addressing residential segregation or improving the lives of the poorest communities. McLaughlin and Faris (2004) argued that there is little integration between equality strategies in areas where there are clear interdependencies, and Dickson and Osborne (2007) pointed out that policymakers often lacked the skills to evaluate and measure effects in reliable and consistent ways. Similarly, McLaughlin (2007) argued that the guidance lacks technical specificity in terms of how to determine key elements of the process, including adverse effects, mitigation or how to adjust complex policy interventions. Although it has considerable enforcement and sanction powers, the Equality Commission has rarely dealt with *policy* breaches or failures to comply with the legislation and the *statistication* of monitoring has precluded a deeper analysis of whether or how the duty has created better policies (Birrell 2009). Some of the most significant criticisms have come from community relations practitioners, who argued that equality has now been reduced to a narrow set of regulatory procedures rather than helping to build a genuinely inclusive society (Morrow 2016). Indeed, McDowell et al. (2017) claimed that the approach has simply reproduced a segregationist mentality by constantly focusing on the rights and entitlements of rival ethno-national groups, when it is clear that these identities are not fixed and that other processes, such as ageing and migration, also impact on equality of opportunity.

Section 75: Equality and good relations

This relationship between equality and community relations, reflected in the distinction between Section 75(1) and Section 75(2), is especially problematic in the spatial planning arena. Successive policies have attempted to more precisely define and promote good relations, particularly ideas around sharing, integration and desegregation. The *Shared Future* policy, launched in 2005 by direct rule Ministers during the suspension of the Assembly, committed departments and public agencies to a more explicit integrationist agenda in education, housing and employment (OFMDFM 2005). The strategy also envisaged strengthening the good relations aspect of the Northern Ireland Act 1998 by extending the scope of EQIAs to cover Section 75(2) in order to create more comprehensive *Good Relations Impact Assessments*. In this respect, the strategy also considered legislating for an exemption from protections preventing sectarian discrimination to allow for 'mixed housing' schemes. The Campaign for the Administration of Justice (CAJ) argued at the time that given existing community differentials in housing, such a move would perpetuate inequalities and potentially undermine the good relations it was intended to support (CAJ 2013).

Ultimately, the proposal was not legislated for, overall progress was slow in developing firm *Shared Future* proposals and there was weak evidence that these translated into area-based projects in particular (Knox 2016). Subsequently, the restored Executive launched its own consultation framework for *Cohesion, Sharing and Integration (CSI)* which repeated the need to address interfacing, promote shared neighbourhoods and eliminate the waste caused by the duplication of segregated services (OFMDFM 2011). Again, Knox (2011) was critical of the financial allocation to support CSI, the lack of connection with mainstream policies and the weak evidence that the two main political parties were ever committed to delivering such change.

Political pressure, especially from the United Kingdom, Irish and United States Governments to stabilise political institutions; to agree to measures to deal with parades, flags and the past; and to genuinely promote an integrated society then prompted the Northern Ireland Executive to produce *Towards Building a United Community* (TBUC) (OFMDFM 2013). This, inter alia, promised action to remove all interfaces within ten years, increase shared social housing schemes as well as considerably expand integrated education at both primary and post-primary levels. It also reorganised and renamed the Equality Commission to the *Equality and Good Relations Commission* and granted the new body additional powers, including the implementation of an augmented 'good relations' impact assessment. CAJ (2013: i) again raised concerns that the 'good relations duty would be harnessed by the opponents of rights and equality to obstruct equality initiatives' and pointed out that the Northern Ireland Act 1998 deliberately subordinated the good relations duty to its equality counterpart. As noted previously, the legislation stated that the 'good relations duty' is to be discharged 'without prejudice' to the equality duty and that public authorities 'should have regard to the desirability of promoting good relations', rather than the stronger 'due regard to the need to promote equality of opportunity' in Section 75(1). The implications for spatial planning policy, especially in a highly segregated environment were also set out:

> This, it was hoped, would be sufficient to prevent equality initiatives, for example new build housing provision rightly likely to benefit one section of the community more than another due to greater objective need, being derailed on the grounds that they could lead to 'community tensions'. There was also the risk that special measures for minorities would be objected to by a more powerful political constituency and be hence deemed bad for 'good relations' and discontinued.
>
> *(CAJ 2013: ii)*

The Department for Communities and the Northern Ireland Housing Executive have created actual shared housing schemes funded by the TBUC initiative. The first wave of ten schemes is summarised in Table 15.1. These were selected based on a range of criteria that prioritised places with a degree of mixing in the housing market (that is, where the proportion of Catholics and Protestants was not higher than 70 percent of the local catchment). Participating housing associations received a 10 percent uplift in Housing Association Grant, which funded a *Good Relations Plan,* that, in turn, supported the community to live in a mixed area. As part of this, tenants signed a *Community Charter,* which made a commitment to living by the ideals of a shared community, including restricting the display of flags and emblems in the estate or on individual properties.

There have been criticisms that such schemes are located in comparatively safe areas where mixing was already likely and that they have tended to prioritise contact at the expense of a more explicit focus on poverty (McDowell et al. 2017; Murtagh et al. 2019). However, Stevenson and Sagherian-Dickey (2016) did show positive impacts on attitudes and, in particular, how mixed estates encouraged the assimilation of groups from previously single-identity neighbourhoods

TABLE 15.1 Shared housing developments supported by *Towards Building a United Community* initiative

Scheme location	*Number of houses*
Ballynafoy Close, Ravenhill Road, east Belfast	67
Global Crescent, Ravenhill Avenue, east Belfast	86
Burnvale Crescent, Burn Road, Cookstown, Co Tyrone	58
Manse Court in Crossgar, Co Down	12
Felden, Newtownabbey, greater north Belfast	97
Main Street, Dundrum, Co Down	17
Market Road, Ballymena, Co Antrim	36
Antrim Road, Ballynahinch, Co Down	24
Dromore Street, Banbridge, Co Down	32
Embankment, Ballynafeigh, south Belfast	55
Total	**484**

as well as migrant communities. The initiative undoubtedly produced important opportunities for learning and transferring practice across the built environment arena. However, as Murtagh and Ellis (2011) argued, there remains a skills deficit in the management of contested space and the application of equality principles beyond statutory obligations under Section 75 of the Northern Ireland Act 1998. We return to this specific matter later in the chapter.

As noted earlier, the Agreement and Northern Ireland Act 1998 initially identified compulsory procedural obligations on the equality duty that did not apply to its 'good relations' counterpart, including the scope of EQIAs. Here, CAJ argued that the EQIA concept draws heavily on established concepts in non-discrimination law, defining, for example, 'adverse impact' in a similar manner to concepts such as discriminatory detriment (whilst the term 'good relations' was not defined under any part of Section 75). Notwithstanding the legislative definition of the scope of EQIAs, the Equality Commission proposed in 2007 a specific shift in policy in recommending that the seven-stage EQIA process should now also be applied to good relations using the same methodology. For the CAJ, this again undermines the principles of equality and the responsibilities of public authorities:

> Such methodology was now being applied to an entirely different concept. The risk in applying the EQIA process to good relations is that simple negative perceptions, 'impacts' or 'tensions' which do not actually objectively reach the threshold of adverse impact, could in a lay sense be considered as such. Consequently it could then be read that the public authority is 'required' to take measures against such an 'adverse impact' on good relations grounds. The ECNI has confirmed it did not pilot the likely impact of this new approach before its implementation.
>
> *(CAJ 2013: iii)*

For CAJ and the Northern Ireland Human Rights Commission (NIHRC), Section 75(2) should be supported by a Bill of Rights that, inter alia, would place a qualified duty on public authorities to respect, on the basis of equality of treatment, the identity and ethos of both main communities, along with protections for the rights of other minorities (NIHRC 2009; CAJ 2020).

Planning and the equality challenge

In their review of the impact of the policy, CAJ (2013: 22) concluded the problem is that 'the term "good relations" is one not found in international human rights or equality instruments and hence, unlike key equality concepts, an international body of knowledge cannot be readily referred to for definition.' In other words, we are restricted in the application of such concepts primarily if they have not already been developed, tested and filtered through superior legal tests. This offers a dangerously restrictive understanding of peace-building and a narrow legal discourse within which to understand and encourage complex conflict transformation processes. McCrudden (2004: 21) summed it up well in arguing that 'community relations activity that is not based on a notion of tackling inequality is community relations built on sand. For those who argue thus, the idea that community relations is in some way in constant tension with equality is a dangerous notion'. However, the community relations sector has also failed to offer specific guidance, methods or examples of practice that could help planners make sense of Section 75(2). How and where to pursue a good relations outcome and when to respect choices about segregation have been lost in the high-level rhetoric. The focus on personal prejudice and interpersonal contact misses the point about the way in which space is socially produced through a range of political, sectarian and, increasingly, economic structures. As the data presented later show, planners have been left to sort out these complex and often conflicting objectives with little in the way of viable guidance or relevant support.

The first *Regional Development Strategy* (RDS) for Northern Ireland did attempt to satisfy both equality and good relations objectives in a Strategic Planning Guideline (SPG-SRC) 3 that aimed 'to foster development, which contributes to better community relations, recognises cultural diversity, and reduces socio-economic differentials within Northern Ireland.' It set out a dual approach to 'facilitate the development of integrated communities where people wish to live together; and promote respect, encouragement and celebration of different traditions and encourage communication and social intercourse in areas where communities are living apart' (DRD 2001: 34-35). The RDS was thin on detail or supportive guidance on how the central tension in these ambitions would be resolved, and O'Neill (2012) argued that professional planners lacked the skills or decision-making criteria to implement the strategy in practice. Indeed, her research showed that SPG-SRC 3 was rarely mentioned in Development Plans and even EQIAs, despite its policy equivalence with strategic objectives dealing with housing growth, retailing and employment. When the RDS was revised in 2012, a more general statement on *Strengthening Community Cohesion* did not identify religion specifically but, instead, offered a more expansive approach to 'encourage mixed housing development. Neighbourhoods with homes in a range of sizes and tenures will allow heterogeneous populations to live together. Diverse populations lead to more stable communities and can help reduce social isolation' (DRD 2012: 38).

The *Strategic Planning Policy Statement for Northern Ireland* (SPPS) (Department of the Environment 2015) also accepted a broad understanding of shared space and how planning might deliver more inclusive places. The list of examples included mixed use developments, shared neighbourhood centres and public spaces, including paths and cycleways. The emphasis on interface locations and the need to develop an integrated approach reflects a range of studies calling for a more direct engagement between the planning system and contested space (Mulholland et al. 2019). Here, the SPPS acknowledges that to remove interface barriers by 2023, planning initiatives 'should be considered as part of a more comprehensive regeneration of wider neighbourhood environments'; that community consultation is vital; and that 'LDPs (Local Development Plans) should take account of any "good relations" policies of the Council where relevant' (Department of the Environment 2015: Section 4.17).

The *Belfast Local Development Plan Draft Plan Strategy 2035* did subsequently set out a new policy (SP4) to 'support development that maximises opportunities to build strong cohesive communities that make a positive contribution to good relations' (BCC 2019: 37). In particular, it aims to deliver the commitments in the Council's Community Plan *The Belfast Agenda* (BCC 2017) to ensure 'a diverse city that is open, safe and welcoming to all' (BCC 2017: 37). The *Statement of Progress* on the Community Plan (BCC 2020) concedes that progress has been slow with sixty-three interfaces remaining across the city. The *Belfast Local Development Plan Draft Plan Strategy 2035* does not make specific proposals nor designate priority areas but sets out a methodology for developing shared cross-community initiatives informed by collaborative consultation in five stages:

1. The survey and analysis stage aims to develop a shared vision between neighbourhoods, informed by early community involvement;
2. Safeguarding sites prioritises particular locations accommodated by a programme of relationship building between communities;
3. Temporary interventions are encouraged, including meanwhile (or temporary) uses at interfaces (noted above in the SPPS), public art and landscaping;
4. Permanent interventions are aimed at the removal of 'peace infrastructure', creating new connections and building mixed neighbourhood facilities; and
5. Shared space is the ultimate outcome shaped by the creation of lifetime neighbourhoods that will address the spatial legacies of segregation and community division (Based on BCC 2019: 37).

The Planning Act (Northern Ireland) 2011 also requires a Local Development Plan Draft Plan Strategy be examined at Independent Examination (IE) to establish whether it is 'sound' in terms of content, conformity (with regional plans and policies) and the processes by which it is produced. *Development Plan Practice Note (DPPN) 6: Soundness* (DfI 2017) identifies twelve tests of soundness based on three categories – procedural, consistency and coherence – in order to examine how the Development Plan Document (DPD) has been produced, aligns with relevant central government and Northern Ireland plans and guidance and the coherence, consistency and effectiveness of the strategy itself. It acknowledged that the concept of testing soundness is based on established practice in England and Wales and, whilst environmental sustainability is evaluated, equality, in the unique context of Northern Ireland, is not.

Skills, knowledge and equality

O'Neill (2012) surveyed 116 stakeholders actively involved in EQIAs in the broad field of spatial planning covering land use planning, urban regeneration, housing and rural development. Most respondents felt that policymakers largely treat equality as a necessary but ultimately ineffective bureaucratic process. Three-quarters (75 percent) felt that it was treated in a *tick-box way* and only 7 percent felt strongly that it had transformed policy. Moreover, 26 percent thought that policymakers made an attempt to understand the diversity of consultees to EQIAs and 69 percent that process was emphasised at the expense of content. A professional planner responded:

> EQIAs have become just part of the list of things that need to be done; the problem is that they are also abused from outside, often for sectarian reasons. It's a way of making sure our side gets what it needs and wants without much reference to the policy aims. In this way, they have lost their meaning for some civil servants; they just make sure that they are not tripped up by it rather than seeing it as a way of improving policy.

A community representative also acknowledged that the problems were not just one for planners and the tactical use of EQIAs and equality, in general, did not help create the type of interdisciplinary and inter-organisational learning necessary for progressive implementation:

> I think that they (EQIAs) have to be more than a paper exercise. Government and voluntary organisations work in silos and only think of how they will implement their own programme; they are not thinking about cohesion and what is best for the community.

Moreover, time and experience have not necessarily improved practice or commitment to equality objectives. Only 29 per cent of the stakeholders surveyed thought that the quality of EQIAs had improved over time and only 22 per cent that consultation techniques had been strengthened, despite acknowledged problems with participatory practices in this part of planning policy (Neill and Ellis 2008). Less than one-in-ten (8 per cent) respondents felt that a policy or decision had been changed because of impact assessments and 21 per cent pointed to a lack of training as at least part of the explanation for the comparatively weak performance. Planners, in particular, return to the way in which EQIAs are treated as a procedural necessity in explaining why they make little substantive difference to the design and delivery of policies:

> It would seem that the EQIA is a statutory duty which is then tick boxed after the policy has been prepared rather than beginning with the EQIA. There does not seem to be anybody to ensure that this is carried out.

Similarly, a housing official acknowledged that the reformist potential of equality has been dented by the way in which it is valued within professional and civil servant cultures, coupled with increasing public apathy about its overall effects:

> They should be a cornerstone of any development but, sadly, this is not always the case. The process is very bureaucratic and the level of public response and interest seems very low.

O'Neill (2012) also looked at eight Area Plans and interviewed fourteen practitioners in planning, the equality and community relations sector and NGOs and community groups to evaluate the impact of Section 75 in plan preparation and implementation. The analysis had a particular focus on the effects of EQIAs in an attempt to understand the poor perception of its role in the survey data. The plans were at various stages of preparation and the analysis, which took place between 2008 and 2012, examined the Ards and Down Area Plan; Belfast

Metropolitan Area Plan (BMAP); Magherafelt Area Plan; Northern Area Plan; Banbridge, Newry and Mourne Area Plan; Armagh Area Plan; Antrim, Ballymena and Larne Area Plan; and West Tyrone Area Plan. The evaluation showed that:

- The Issues Papers and even the draft plans rarely mentioned equality issues, even where there were obvious challenges of ethno-religious segregation as was the case in the Belfast Metropolitan Area.
- The BMAP EQIA did evaluate land use zonings against the spatial distribution of the main equality categories and provided statistical significance tests to identify deviance, although it did not find any adverse or discriminatory effects. Few other EQIAs used the same rigour in applying quantitative or qualitative research, geocoded data or consultations to evaluate the differential impact of the main planning policies.
- The consultation approaches also varied considerably with the NGO *Community Places* using more creative participatory techniques to engage young people but most processes relied on written responses to the draft EQIA and accompanying Issues Paper. O'Neill concluded that it was hardly surprising that the responses often lacked analytical depth and, in some cases, NGOs working in Section 75 sectors, such as ageing, community relations, race and disability, made no response at all. For their part, the NGOs pointed out that they were overburdened by such consultation processes and that they often felt they had little impact on the planning system.

This also related to other aspects of planning policy and governance, including the Examination in Public (EiP) held in 2006 into revised housing growth forecasts, which were based on new demographic projections for Northern Ireland to 2025. O'Neill pointed out that of the fifty submissions to the draft policy paper on the new targets (for the then twenty-six district councils), eleven raised equality or community relations issues. Here, equality and, in particular, alleged discrimination against Catholics was used to justify the need for over-zoning of designated housing land across a number of local authority areas. Surprisingly little evidence was presented on how the projections were discriminatory, what the differential effects were and how rezoning could redress variations in high-level housing forecasts. For O'Neill, equality was reduced to a tactic simply to justify a clientelist outcome, often with little or no regard for the evidence of the impact on the Section 75 groups. Drawing on the same EiP case, McKay (2010) showed how planning professionals were selective about the use and interpretation of research, especially in such adversarial settings where significant financial advantage rests on gaining or losing planning permission. Inevitably, the abuse of knowledge has created what he called both a 'crisis of trust' and a 'crisis of truth' and questions 'the motives, tactics, and practices employed by participants in the plan-evaluation process' (McKay 2010: 131). The profession is therefore not blameless in explaining why equality has made such little progress in the planning system, regardless of the lack of clarity or the practicability of implementing Section 75 in practice.

Conclusion

The Northern Ireland Act 1998, signed in the optimism of the peace process, introduced a formidable range of legislative protection, support programmes and evaluation methods to embed equality and good relations in mainstream policy. Planning, at least initially, responded enthusiastically with a new Strategic Planning Guideline in the RDS that expressed the dual need to promote more cohesive and integrated places and, at the same time, protect the rights of people to live in segregated areas. In some senses, this is the crux of the problem with equality and how it has been translated into spatial planning and related policy areas in urban regeneration, rural development and housing. It is far from clear where and how policymakers make the choice about which path to follow and how the bundle of rights embodied in Section 75 can be simultaneously protected without having an adverse effect on other interests. Murtagh et al. (2021) showed how the demography of Belfast has changed significantly, especially in the last two decades as the number of older people living in central locations has declined to be replaced by younger, professional and higher spending groups, stimulated by, and stimulating, apartment developments rather than family homes or sheltered options. Older people have moved to the periphery, coastal locations and park homes with often poor conditions and support services. The city is now branded around youth lifestyles and hipster imagery, all accelerated by studentification linked to investments by the Ulster University in particular. The market, not equality and good relations, increasingly dictates the population mix of the city and what Section 75 groups get access to housing sites, elite investments and the infrastructure of the urban core. Religion is now just one of the set of equality categories that needs to be accommodated if planning genuinely aims to create inclusive, diverse and sustainable places.

A version of equality that simply enables the poorest to compete for scarce housing is hardly likely to create a more sustainable and fairer city or a progressive planning system that can play an important role in the long-term development of the region. Of course, housing should go to those most in need, but creating a demographic, social and religious mix cannot be realistically achieved without producing the quantity and quality of social housing in the right places to meet growing demand. It is simply impractical to expect integrated living in the social housing sector when urgent need among Catholics so significantly outstrips supply (ECNI 2017). The aim of achieving more cohesive and integrated communities is an empty signifier without ensuring that the investment in social housing can create fair access for a whole range of identities and social groups.

The most segregated communities are the most deprived, cut off from services and facilities and alienated from education and labour markets. Creating more integrated communities should be about planning more inclusive places, especially for the people who suffered most in the conflict and benefit least from the peace. But planners and related professions need better support to interpret the Act and how to make its provisions work in reality. They have been let down by a lack of guidance, workable models and skills support; and some have corrupted the principles of equality in the service of clientelism and profit. There is a need for a reenergised ethic for the profession and the need for responsibility in the use and application of knowledge. However, practitioners also need support from the profession, academia and, critically, the equality sector to ensure that the worthy ideals of good relations can make a difference to the most disadvantaged and deprived communities in Northern Ireland and to other places emerging from conflict.

Legislation

Northern Ireland Act 1998
Northern Ireland Constitution Act 1973
Planning Act (Northern Ireland) 2011

References

Agocs, C. and R. Osborne. 2009. 'Comparing Equity Policies in Canada and Northern Ireland: Policy Learning in Two Directions?' *Canadian Public Policy* 35(2): 237–62.

Aughey, A. 2005. *The Politics of Northern Ireland*. London: Routledge.

Belfast City Council (BCC). 2017. *The Belfast Agenda*. Belfast: BCC.

Belfast City Council (BCC). 2019. *Belfast Local Development Plan: Draft Plan Strategy 2035*. Belfast: BCC.

Belfast City Council (BCC). 2020. *The Belfast Agenda: Statement of Progress 2017–19*. Belfast: BCC. https://www.belfastcity.gov.uk/Community/Community-planning/Statement-of-Progress-2017-2019 (accessed September 2021).

Birrell, D. 2009. *The Impact of Devolution on Social Policy*. Bristol: Policy Press.

Central Secretariat. 1993. *Policy Appraisal and Fair Treatment, Circular 5/93*. Belfast: Northern Ireland Office.

Coakley, J. 2009. *Implementation Issues and the Pursuit of a Settlement in Northern Ireland, IBIS Discussion Papers: Breaking the Patterns of Conflict Series No. 6*. Belfast: Queen's University Belfast.

Coakley, J. and J. Todd. 2020. *Negotiating a Settlement in Northern Ireland, 1969–2019*. Oxford: Oxford University Press.

Committee on the Administration of Justice (CAJ). 2013. *Unequal Relations? Policy, The Section 75 Duties and Equality Commission Advice: Has 'Good Relations' Been Allowed to Undermine Equality?* Belfast: CAJ.

Committee on the Administration of Justice (CAJ). 2020. *Written Evidence to the NI Assembly Ad-Hoc Committee on the Bill of Rights from the Equality Coalition Co-Conveners*. Belfast: CAJ.

Department for Infrastructure (DfI). 2017. *Development Plan Practice Note (DPPN) 6: Soundness*. Belfast: DfI.

Department for Regional Development (DRD). 2001. *Regional Development Strategy for Northern Ireland 2025*. Belfast: DRD.

Department for Regional Development (DRD). 2012. *Regional Development Strategy 2035*. Belfast: DRD.

Department of the Environment. 2015. *The Strategic Planning Policy Statement for Northern Ireland*. Belfast: Department of the Environment.

Dickson, B. and R. Osborne. 2007. 'Equality and Human Rights since the Belfast Agreement', in P. Carmichael, C. Knox and R. Osborne (eds) *Devolution and Constitutional Change in Northern Ireland*. Manchester: Manchester University Press, pp. 125–37.

Donaghy, T. 2004. 'Mainstreaming: Northern Ireland's Participative Democratic Approach', *Policy and Politics* 32(1): 49–62.

Ellis, G. 2001. 'Social Exclusion, Equality and the Good Friday Peace Agreement: The Implications for Land Use Planning', *Policy and Politics* 29(4): 393–411.

Equality Commission Northern Ireland (ECNI). 2005. *Section 75 of the Northern Ireland Act 1998 Practical Guidance on Equality Impact Assessment*. Belfast: ECNI.

Equality Commission for Northern Ireland (ECNI). 2014. *Gaps in Equality Law between Great Britain and Northern Ireland*. Belfast: ECNI.

Equality Commission for Northern Ireland (ECNI). 2017. *Statement on Key Inequalities in Housing and Communities in Northern Ireland*. Belfast: ECNI.

Gale, R. and H. Thomas. 2018. 'Race at the Margins: A Critical Race Theory Perspective on Race Equality in UK Planning', *Environment and Planning C: Politics and Space* 36(3): 460–78.

Hughes, J. 2004. 'Attitudes towards Equality in Northern Ireland: Evidence of Progress?', in B. Osborne and I. Shuttleworth (eds) *Fair Employment in Northern Ireland: A Generation On*. Belfast: Blackstaff Press, pp. 166–83.

Hughes, J. and C. Donnelly. 2004. 'Attitudes to Community Relations in Northern Ireland: Signs of Optimism in the Post Cease-Fire Period?' *Terrorism and Political Violence* 16(3): 567–92.

Knox, C. 2011. 'Cohesion, Sharing and Integration in Northern Ireland', *Environment and Planning C* 29(3): 548–66.

Knox, C. 2016. 'Northern Ireland: Where is the Peace Dividend?', *Policy and Politics* 44(3): 485–503.

Knox, C. and P. Carmichael. 2005. 'Devolution - The Northern Ireland Way: An Exercise in 'Creative Ambiguity'', *Environment and Planning C: Government and Policy* 23(1): 63–83.

Knox, C. and P. Carmichael. 2010. 'Debate: Devolution in Northern Ireland', *Public Money and Management* 30(2): 82–83.

McCrudden, C. 1998. 'Mainstreaming Equality in the Governance of Northern Ireland', *Fordham International Law Journal* 22(4): 1696–775.

McCrudden, C. 2004. 'Mainstreaming Equality in Northern Ireland 1998–2004: A Review of Issues Concerning the Operation of the Equality Duty in Section 75 of the Northern Ireland Act 1998, Annex b, Vol.2', in E. McLaughlin and N. Faris (eds) *The Section 75 Equality Duty - An Operational Review*. London: Northern Ireland Office.

McCrudden, C., R. Ford and A. Heath. 2004. 'Legal Regulation of Affirmative Action in Northern Ireland: An Empirical Assessment', *Oxford Journal of Legal Studies* 24(3): 363–415.

McDowell, S., M. Braniff and J. Murphy. 2017. 'Zero-Sum Politics in Contested Spaces: The Unintended Consequences of Legislative Peacebuilding in Northern Ireland', *Political Geography* 61: 193–202.

McKay, S. 2010. 'In Pursuit of Parrhesia: An Investigation into Knowledge Selection in Inquisitorial Planning Practice', *Environment and Planning A* 42(1): 116–33.

McLaughlin, E. 2007. 'From Negative to Positive Equality Duties: The Development and Constitutionalisation of Equality Duties in the United Kingdom', *Social Policy and Society* 6(1) 111–21.

McLaughlin, E. and N. Faris. 2004. *The Section 75 Equality Duty - An Operational Review*. London: Northern Ireland Office.

Morrow, D. 2016. 'Acknowledging Religious and Cultural Diversity in an Antagonistic Society: The Challenge of Northern Ireland', in I. Honohan and N. Rougier (eds) *Tolerance and Diversity in Ireland, North and South*. Manchester: Manchester University Press, pp. 211–31.

Mulholland, C., K. Sterrett, M. Morrissey, C. Karelse and F. Gaffikin. 2019. *Making Space for Each Other: North Belfast*. Belfast: Queen's University Belfast.

Murtagh, B., C. Cleland, S. Ferguson, G. Ellis, R. Hunter, C.R. Rodriguez Añez, L.A. Becker, A.A.F. Hino and R.S. Reis. 2021. 'Age-Friendly Cities, Knowledge and Urban Restructuring', *International Planning Studies* 1–15. https://doi.org/10.1080/13563475.2021.1920374.

Murtagh, B. and G. Ellis. 2011. 'Skills, Conflict and Sustainable Communities', *Planning Theory and Research* 12(3): 1–22.

Murtagh, B., A. Grounds, P. Boland and L. Fox-Rogers. 2019. 'Social Economics, Poverty and Violence after Peace', *Third World Thematics: A TWQ Journal* 4(2–3): 220–38.

Murray, M., G.H. Fagan and P. McCusker. 2009. 'Measuring Horizontal Governance: A Review of Public Consultation by the Northern Ireland Government between 2000 and 2004', *Policy and Politics* 37(4): 553–71.

Neill, W. and G. Ellis. 2008. 'Spatial Planning in a Contested Territory: The Search for a Place Vision in Post-Troubles Northern Ireland', in C. Coulter and M. Murray (eds) *Northern Ireland after the Troubles*. Manchester: Manchester University Press, pp. 88–109.

Northern Ireland Human Rights Commission (NIHRC). 2009. *Summary Advice to the Secretary of State for Northern Ireland*. Belfast: NIHRC.

Office of the First and Deputy First Minister (OFMDFM). 2005. *A Shared Future: Policy and Strategic Framework for Good Relations in Northern Ireland*. Belfast: OFMDFM.

Office of the First and Deputy First Minister (OFMDFM). 2011. *Programme for Cohesion, Sharing and Integration*. Belfast: OFMDFM.

Office of the First and Deputy First Minister (OFMDFM). 2013. *Towards Building a United Community*. Belfast: OFMDFM.

O'Neill, D. 2012. *Collaborative Planning and Equality in Northern Ireland*. Unpublished PhD Thesis, Queen's University, Belfast.

Osborne, R. 2007. 'Evidence and Equality in Northern Ireland', *Evidence and Policy: A Journal of Research, Debate and Practice* 3(1): 79–97.

Rafferty, G. 2012. 'Embracing the Creation of Shared Space: Considering the Potential Intersection between Community Planning and Peace-Building', *Space and Polity* 16(2): 197–213.

Ruane, J. and J. Todd. 2012. 'Beyond Inequality: Assessing the Impact of Fair Employment, Affirmative Action and Equality Measures on Northern Ireland', in G. Brown, A. Langer and F. Stewart (eds) *Debating Affirmative Action*. Basingstoke: Palgrave Macmillan, pp. 182–208.

Standing Advisory Committee on Human Rights (SACHR). 1997. *Employment Equality: Building for the Future*. Belfast: SACHR.

Stevenson, C. and T. Sagherian-Dickey. 2016. 'Collectively Coping with Contact: The Role of Intragroup Support in Dealing with the Challenges of Intergroup Mixing in Residential Contexts', *British Journal of Social Psychology* 55(4): 681–99.

Tonge, J. and J. Evans. 2020. 'Northern Ireland: From the Centre to the Margins?' *Parliamentary Affairs* 73(1): 172–88.

16

PROPERTY LAW AND PLANNING

Thomas Murphy, John McCord and Peadar Davis

Introduction

The state plays a considerable role in the affairs of property owners in Northern Ireland, from the outworking of public policy under the auspices of the *Regional Development Strategy*, through to the operations of central government departments, local government and statutory undertakers, in delivering infrastructure and services. The state also plays a role in both regulating and facilitating the operations of the private sector where there is perceived to be a public interest. These interventions are achieved via a number of mechanisms. The planning system constitutes perhaps the most obvious and comprehensive of these mechanisms, by which the state can regulate what property owners can legitimately do with land. Planning decisions may affect property rights in three main respects. First, to give effect to planning policy or a particular planning decision, it may be necessary to acquire property by means of compulsory purchase. Second, by controlling the use to which land may be put, as well as the design and form of any development, planning decisions may directly restrict the freedom of property owners to exercise the normal rights of ownership. Third, by permitting development in the vicinity, a planning decision may have an adverse impact on surrounding land, a phenomenon known as planning blight. Additionally, it may also be argued that planning permission (once granted) is itself a property right.[1] In each of these cases, there will be two key matters that may result in contention. First, the decision in question (and subsequent actions, such as compulsory purchase of land) can be challenged on grounds of validity: in particular, was there a power to act in this manner in the case in question and was the correct procedure followed? Second, following on from the question of validity will be if, and on what basis, the affected property owner is entitled to be compensated, or to receive some other remedy, in respect of the interference with property rights?

In many respects, these are issues that arise in all jurisdictions and the manner in which they will be dealt with is often broadly similar. In particular, in the Northern Ireland context, there is much in the case law of England and Wales to draw upon. However, there are also striking differences, not only in matters of detail but also in some fundamental respects, for example in regard to the procedure for compulsory acquisition of land (Vesting Orders as opposed to

DOI: 10.4324/9781003218319-16

Compulsory Purchase Orders) and aspects of compensation. In addition, Northern Ireland has a particular history that has influenced the law relating to interference with individual property rights in the jurisdiction. That said, compulsory purchase compensation law is an enormously complex yet important area of law that is not well understood. In Northern Ireland, there is a paucity of authoritative reference material relating to this notoriously difficult area, apart from the various pieces of legislation, decisions from the Lands Tribunal and the systematic yet comprehensive operational guidance provided by Land & Property Services for affected parties. This chapter will, therefore, outline and discuss the main aspects of Vesting Orders and compensation in the Northern Ireland context.

Acquisition of land for planning purposes

In Northern Ireland an array of statutory provisions enables public authorities to acquire land for a variety of purposes, including for urban regeneration, redevelopment and infrastructure provision. This chapter adopts the expression 'acquiring authority' to describe the relevant entity utilising vesting powers. In the context of planning, Article 87(1) of the Planning (Northern Ireland) Order 1991 enables the acquiring authority to acquire most[2] land either compulsorily or by agreement where it is satisfied:

- that the land is required in connection with a development scheme; or
- that it is expedient in the public interest that the land should be held together with land so required; or
- that the land is required for development or redevelopment, or both, as a whole for the purpose of providing for the relocation of population or industry or the replacement of open space in the course of the redevelopment or improvement of another area as a whole; or
- that it is expedient to acquire the land for a purpose which it is necessary to achieve in the interests of the proper planning of an area in which the land is situated.[3]

Any public authority seeking to exercise a power to acquire land must comply with any specific procedural requirements related to the exercise of that particular power. In respect of Article 87, the acquiring authority is the relevant Department, which must initially consult with the district council within whose area the land is situated.[4] If the acquisition is to be 'otherwise than by agreement' it may proceed concurrently with proceedings for the adoption of the development scheme under Article 86, but any Vesting Order must not be made prior to the making of the order adopting the development scheme. Where acquisition is to be by compulsory means then the acquiring authority must also comply with the general procedural framework laid out in Schedule 6 of the Local Government Act (Northern Ireland) 1972 (as modified by Article 87(3) and Schedule 2 of the 1991 Order) for making a Vesting Order.[5] Specifically, paragraph 6(1) of Schedule 6 provides as follows:

> a Vesting Order shall operate, without further assurance, to vest in the [Department], as from the date on which the Vesting Order becomes operative (in this Schedule referred to as 'the date of vesting'), an estate in fee simple or such other estate (if any) in, to or over the land to which it relates as is therein specified, freed and discharged from all claims or estates whatsoever (except as is specified in the order).

Notably, this procedural framework differs somewhat from the Compulsory Purchase Order procedure that is operative in England and Wales and has been the focus of challenges to the exercise of compulsory acquisition powers in Northern Ireland. Indeed, a key difference between the Vesting Order and Compulsory Purchase Order approaches is detailed by Roots et al. (2018), who outline that the vesting procedure automatically vests title in the acquiring authority conferring an immediate right of entry against all those interested in the land, except certain minor tenants without any conveyance.[6] Consequently, as title is acquired without investigation, interests in land are transferred into rights to compensation. As the general vesting declaration procedure does not rely on participation by the landowner, formalities as to notification are prescribed and can be onerous.

Notices

The first requirement of Schedule 6 is that anyone potentially affected by a proposed Vesting Order receives notice that vesting is intended to take place. Such notice:

a) shall be published by the [Department] on at least two occasions in the locality in which the land is situated;
b) shall be served by the [Department] on every person appearing to the [Department] to have an estate in the land;
c) shall also be served by the [Department] on such government departments and public bodies as the [Department shall think fit].[7]

This notice requirement, in essence, marks the beginning of a period in which objections can be raised and, if any representations are received, and are not subsequently withdrawn; not concerned solely with the amount of compensation and (the acquiring authority is satisfied) not of a purely frivolous or vexatious nature, then a public local inquiry must be held.

Upon the expiry of one month from the last publication of the Notice of Intention to Vest and, having considered any relevant representations and (if applicable) any report of the person who held the public inquiry, the acquiring authority may:

(i) make a Vesting Order, which may contain any modifications of the [Department's] proposal that the [Department] thinks proper, vesting in the [Department] all or any part of the land for an estate in fee simple or for such other estate as may be specified in the Vesting Order; or
(ii) refuse to make the Order.

Where a Vesting Order is made, it must contain such provisions as the acquiring authority thinks necessary or expedient for carrying it into effect[8] and a further Notice must be published and served on every person who, having provided notice of an objection to the acquiring authority, appeared at the local inquiry in support of their objection.[9] The purpose is to provide anyone with an interest in affected land with a final opportunity to challenge the vesting before it takes effect. Any such challenge is strictly limited in terms of time, form and grounds.

Challenges: time and form

Issues of time and form are intertwined. According to Paragraph 5(1)(b) any person wishing to challenge the making of the Order 'may within one month from the publication of the

Notice of the making of the Vesting Order, make an application for the purpose to the High Court in accordance with rules of court'. This seems fairly clear cut, but a number of challenges to the Vesting Order made in respect of the lands acquired for the construction of the Victoria Square Shopping Complex in central Belfast demonstrate that this is not necessarily so. What these challenges demonstrate is that while the one-month period is rigid, there may, nevertheless be debate as to when it began and ended. The first question is when was the relevant Notice published? Again this seems straightforward. However, in *Re Scalene Investments Ltd*, the respondent Department placed Notices in three newspapers on 11 November 2003. Given that the applicant did not issue an originating notice of motion (in accordance with the rules of court) until 17 December 2003, the Department argued that the application should be struck out as being out of time. However, the Department had, in this case, undertaken a 'publication scheme' whereby the same Notice was again published in the same newspapers on 18 November 2003.

While the acquiring authority is required, under Paragraph 2, to publish Notice of the Intention to Make a Vesting Order on at least two occasions, Paragraph 5 does not make this necessary in respect of notice of the actual *making* of the Order and, furthermore, The Planning (Northern Ireland) Order 1991 amended the original wording of Paragraph 5 to remove any specific obligations in respect of the form and manner that notice must take.[10] Consequently, the form of the Notice is at the discretion of the acquiring authority, according to what it considers to be appropriate. That leaves it unclear what the minimum requirement in any given case would be but, in the *Scalene* case, the court considered that the initial publication in the three newspapers in question would have sufficed.

While there had thus been no obligation on the respondent to publish the Notice again on 18 November, the court nevertheless considered that 'publication' was not necessarily a single event and that in order to, for example, reach a sufficient audience using newspapers published on different days the publication in all of the newspapers would be, collectively, 'publication' for the purposes of Schedule 6. Accordingly, in this case, the respondent had clearly considered it appropriate to publish on two separate occasions in order to reach a sufficient audience and, consequently, in terms of determining the time limits for any challenge by the applicant, the later date – 18 November – was the date 'from' which the one-month period should be measured and thus the application was deemed to be within the statutory time limit[11].

Likewise, it seems clear that within one month, a challenge must have been brought according to the rules of court. A challenge under Schedule 6 is by way of *statutory* review and thus should be by way of originating notice of motion under Order 55 of the rules of court. However, in *Re Bowden* the applicant instead (within the statutory time limit) applied for leave to apply for judicial review under Order 53 and thus, in response to arguments that the applicant should be permitted to amend the application, the respondent argued that the statutory time limit had by this time (the date on which the court considered the matter) elapsed. Nevertheless, given that proceedings (albeit in the wrong form) had been issued within time, Weatherup J was prepared to exercise his discretion to treat the application as if it had been correctly formulated:

> In this case the proceedings were issued within time; they were also served within that time; the content of the application made under Ord 53 is the same content as that which would have applied had the application been properly formulated in the prescribed manner under Ord 55. Therefore, it seems to me there is no prejudice to the

> concept of legal certainty, or to good public administration, or any other prejudice to the proposed respondent or others between the present situation, and that which would have prevailed had the proceedings been in the proper form as an Ord 55 originating notice of motion. In those circumstances, I propose to exercise my discretion to treat the application as if it had been formulated by originating notice of motion under Ord 55. Accordingly, I will make an order that will require the applicant to amend the proceedings to comply with an Ord 55 application. If the application is converted into an originating notice of motion under Ord 55, leave is not required, and therefore when the matter is properly constituted, there will be no need for an application for leave.[12]

However, in the application of *JGS Services Ltd and Others*[13] the argument that there was any discretion to extend the actual time limit per se was rejected, as was the notion that the language of Paragraph 5(1)(b) (that a challenger '*may*' – as opposed to *must* within one month bring proceedings) in some way implies a right of challenge after a period of one month. Likewise, the argument that a one-month time limit was inherently disproportionate and thus contrary to Article 6 of the European Convention on Human Rights (ECHR) was rejected.

Grounds for challenge

Schedule 6, Paragraph 5 provides that a challenge may be made, and the High Court may suspend the operation of a Vesting Order[14] and/or quash the order[15], on the basis that the making of the order is outside the powers conferred by Article 87[16] and/or due to non-compliance with the procedures specified in Schedule 6. In respect of procedural non-compliance, a Vesting Order will be quashed only if it results in substantial prejudice to the interests of the applicant. In determining the matter the court is not concerned with the merits of the decision to make the Vesting Order[17], nevertheless, investigating whether there has been substantial prejudice (in the context of procedural non-compliance) will almost inevitably involve an examination of the background to the case and, in the context of any challenge, Article 1 of Protocol 1 (A1P1) of the ECHR applies. Determining whether A1P1 has been breached involves a balancing exercise. As Nicholson LJ noted in *Keys and another v Department for Regional Development*[18] any vesting engages A1P1 and 'It can only be justified if it is in the public interest and a fair balance has been struck between the demands of the general interest of the community and the need to protect individual rights'.[19] In that case[20], while there had been no improper exercise of statutory powers, there had been a clear failure to comply with the Schedule 6 procedural framework. Due to an administrative mistake, preliminary notice of the intention to acquire the relevant land compulsorily was not served, as required by Paragraph 2(b), on the applicants as owners of an estate in the land. Rather, the Department corresponded with the Londonderry and Lough Swilly Railway Company (a former owner), which, perhaps unsurprisingly, raised no objections. Nicholson LJ concluded, on the facts, that the Department should have ascertained that the applicants in fact owned an estate in the land and consequently there was a clear breach of Schedule 6. The question remained whether this breach was (in the context of A1P1) one that substantially prejudiced the interests of the applicants. Having noted the merits of the scheme and the relatively limited amount of the applicant's land that had been vested, Nicholson LJ essentially boiled it down to one question: 'A right to compensation is provided for but is that enough?'[21]. In this case, while there had been a procedural breach, which was consequently deemed to result in a beach of A1P1, compensation was considered to be sufficient and the Vesting Order was upheld. In essence, it was concluded that the Department had the requisite power to vest the land and:

> I cannot accept that any representations which they would have made following Notice of Intention to Vest would have had any effect on the outcome. I am satisfied that in the public interest the Skeoge Road Link would have been approved. If I quashed the Vesting Order so far as the Applicants are concerned and a Notice of Intention to Vest was served on them and they made representations … they would be unsuccessful and they would be in the same position as they are today. I would be doing them no favours. I would have needed evidence from a roads engineer or some such expert to satisfy me that their portion of the Skeoge Road Link was not needed or that some steps could be taken to prevent severance of one part of their land from the other.[22]

What was questionable in this case, however, was the level of compensation and the need to take into account the procedural irregularity that had occurred:

> the compensation must be fair. I have no evidence before me that it will not be fair. But it is important that compensation is assessed on the basis that the Keys family were denied the opportunity of negotiating with developers. If there is evidence that they would have obtained increased compensation, had they been served Notice of Intention to Vest, they must be compensated accordingly.[23]

Exclusion of other challenges and effect of vesting

Aside from the procedure discussed above, a Vesting Order or the making of that Order 'shall not be challenged in any legal proceedings whatsoever'[24] and the Order will become operative at the end of the period of one month from the date of publication of the Notice of the making of the Order.[25] The acquiring authority must either serve a copy of the Vesting Order or Notice[26] that the Order has become operative on every person appearing to it to have an estate in the land and notice must be served on such government departments and public bodies as the acquiring authority thinks fit. In respect of unregistered land, the fee simple estate (or such other estate as specified by the Order) becomes vested in the Department from the date on which the Vesting Order became operative (the date of vesting) and is freed and discharged from all claims and estates (except as provided by the Order).[27] In respect of registered land, the acquiring authority is required to lodge the Vesting Order with the Land Registry and, consequently, the date of vesting will be the date of registration (unless the date on which the Order would otherwise have become operative is later).[28]

Once the Vesting Order is operative, the interests of any former owners are converted to a right to be compensated. This is the end of the matter, almost, but perhaps not quite. While the Vesting Order, or indeed the making of the Order, cannot be challenged, the decision in *Re Bowden* indicates the possibility of a limited challenge to the actual taking of possession of the land in question in pursuance of a Vesting Order. In *Re Bowden* itself such a challenge was not necessary: the decision as to when the Department should actually take possession was made as part of the Vesting Order and was thus (to some extent) successfully challenged in the course of normal statutory review. However, if no decision with regard to the timetable for possession was taken in the context of the Vesting Order per se, then any dispute in this respect might be dealt with in one of two possible ways.

The first possible approach might be to conclude that the absence of such a timetable is simply part of the Vesting Order, which, if it causes anyone concern, should be challenged in the normal way (as discussed above). Any subsequent concerns in respect of the timetable, once presented, would thus be irrelevant. The fact that Weatherup J in *Re Bowden* considered the actual taking of possession as being part of the vesting and therefore challengeable under Paragraph 5 of Schedule 6 might be cited as justification for such an approach. The other possibility is that the court might entertain an application for judicial review. An applicant would, presumably, argue that this would be a challenge to the exercise of the acquiring authority's discretion under Paragraph 9 of Schedule 6, not a challenge to the Vesting Order (or the making thereof) per se and thus not contrary to Paragraph 5(1)(c) of Schedule 6 (excluding any other challenges). In *Re Bowden* Weatherup J specifically declined to comment on the correct approach in such a case.

As *Re Bowden* demonstrates, disputes surrounding the timetable for actual possession can raise significant issues. In essence, public sector occupiers had been given a longer period in which to remain in possession because the acquiring authority envisaged that they would all vacate by agreement whereas, conversely, it was anticipated that three months 'eviction time' would be needed to remove the non-public sector occupiers. In a number of applications considered by the court the businesses in question were to be extinguished and thus this was considered to be merely a matter of compensation. However, in respect of the *Bowden* application the business (Harrison's) was to relocate and the relocation to suitable alternative premises (a matter that could not be resolved purely by compensation) could not be reasonably completed in the relevant time. Having accepted that this was the case and, bearing in mind the more generous time period offered to the public sector occupiers, Weatherup J, while not quashing the Vesting Order, considered that it breached A1P1 ECHR and required assurance that Harrison's would be allowed to remain in possession for the same period of time as the public sector occupiers. The key factor in this case was that the continued presence of the public sector occupiers demonstrated that the site was not immediately required and, in the context of a private sector occupier who could not reasonably be expected to relocate premises in the time allocated, the same time frame had to be applied. Consequently, whilst the circumstances of the applicant are relevant in respect of having the matter examined, they will not be decisive in terms of determining the date for possession: in *Re Bowden* Harrisons reasonably required until November 2004 to relocate; possession was merely delayed to April 2004 as the conduct of the Department (as opposed to the needs of the applicant) demonstrated that it did not require possession until then.

Development and disposal

Having acquired the land, the acquiring authority may develop the land, erecting, constructing or carrying out any building or work, may enter into an agreement to develop the land and may repair, maintain and insure any buildings or works on the land and generally manage the land in a proper manner.[29] Alternatively, they may use the land for any other (non-planning related) purpose for which it has the power to acquire land or dispose of the land to such persons as:

> may appear to it to be expedient in order to secure the best use of that or other land and any buildings or works which have been, or are to be, erected, constructed or

carried out thereon, whether by itself or by any other person, or to secure the erection, construction or carrying out thereon of any buildings or works appearing to it to be needed for the proper planning of the area in which the land is situated.[30]

In disposing of the land, the acquiring authority must take into consideration ('as far as practical') the desire of any person who had been living, or carrying on business or other activities on the land to be accommodated thereon and to provide any such person, who is willing to comply with the requirements of the acquiring authority, with the opportunity to obtain suitable accommodation[31].

Reverse compulsory acquisition: Purchase Notices and Blight Notices

The operation of planning law can place restrictions on the possible use to which land is put and, consequently, may have a detrimental impact on the value of the land. In limited circumstances (considered below) there may be a right to compensation, but in two particular types of scenario the local authority/relevant body may be required to actually purchase the relevant estate in land. First, an adverse planning decision may mean that the land is incapable of reasonably beneficial use, for example, if a building has been accidentally destroyed and planning permission is refused for reconstruction: in this type of case a Purchase Notice may be served. Second, land may be blighted – effectively rendered unsaleable on the open market – if identified as being included in a plan or scheme that could potentially result in that land being compulsorily acquired at some point in the future. In this type of case a Blight Notice may be served.

Purchase Notices

Section 191 of The Planning Act (Northern Ireland) 2011[32] provides that the 'owner'[33] of land may, within six months or such longer period as the council may allow[34], for the following planning decisions (and if the owner claims that the relevant circumstances apply), serve a Purchase Notice in writing, requiring the local council to buy his/her interest in the land:

1. An application for planning permission is refused, or is granted subject to conditions, or existing planning permission is revoked or modified by the imposition of conditions and the owner claims that: the land has become incapable of reasonably beneficial use in its existing state; and, if conditions have been imposed, the land cannot be rendered capable of reasonably beneficial use through development in accordance with those conditions; and the land cannot be rendered capable of reasonably beneficial use through any other development in accordance with planning permission that has been granted[35];
2. An application for listed building consent is refused or is granted subject to conditions or existing listed building consent is revoked or is modified by the imposition of conditions and the owner claims that: the land[36] has become incapable of reasonably beneficial use in its existing state; and, if conditions have been imposed, the land cannot be rendered capable of reasonably beneficial use through works carried out in accordance with those conditions; and the land cannot be rendered capable of reasonably beneficial use by carrying out any other works in accordance with listed building consent that has been granted[37];

3. An order is made under section 73 requiring that a use of land should be discontinued, or should only be continued subject to conditions, or that any buildings or works should be altered or removed and the owner claims that: the land is incapable of reasonably beneficial use in its existing state; and the land cannot be rendered capable of reasonably beneficial use through any development in accordance with planning permission that has been granted.[38]

In situations 1 and 2, a Purchase Notice can be effective only if the land 'has become incapable of reasonably beneficial use in its existing state'. In respect of situation 3, the wording is slightly different: a Notice can be effective where the land 'is' thus incapable. The rationale for the difference, and any practical effect, is unclear: in *Purbeck District Council v Secretary of State for the Environment and Another*[39] while Woolf J declined to go so far as to conclude (in the context of corresponding English legislation[40]) that 'becomes' simply meant 'is' he nevertheless considered:

> All that is required … is that the situation that exists at the time of the Notice to purchase is considered. If the planning permission had been granted and not refused, the situation that then exists would not have arisen and therefore in that sense the land as it then exists can be said to have 'become incapable of reasonably beneficial use' when the application is made after planning permission has been refused. (I have only referred to the refusal of planning permission but the same principles apply where planning permission is granted subject to conditions[41]).
>
> When it is borne in mind that section 180(1)(a) refers to the existing state of the land, it becomes obvious that performing an exercise that requires one to investigate what was the state of the land at some stage in its history is inconsistent with the intent of the section[42].

Consequently, in all three situations the court will be concerned only with the present condition of the land and it is unnecessary to consider whether there has been any change in this respect, or indeed to consider whether the land was ever capable of reasonably beneficial use.[43] A possible exception, identified by Woolf J in *Purbeck*, is where the land has become incapable of reasonably beneficial use because of the unlawful actions of the owner, specifically the failure of the owner to comply with planning law. However, limits on the application of this principle were established by the Court of Appeal in the subsequent case of *Balco Transports Services Ltd v Secretary of State for the Environment (No 2)*.[44] Ruling that there should be no wholesale borrowing of the private law maxim that a person cannot take advantage from their own wrong, Glidewell LJ considered that the conduct of the owner would result in the conditions for a Purchase Notice not being fulfilled in the following limited circumstances:

1. The relevant land is in a state caused by development carried out without planning permission;
2. A valid Enforcement Notice in respect of that development has been/could be served;
3. Such an Enforcement Notice could require the owner/occupier to take steps to restore the land to its condition prior to the development taking place; and
4. In such a condition the land would be capable of reasonably beneficial use.[45]

Essentially, in these (narrow) circumstances an effective Purchase Notice cannot be served until the time limit for an Enforcement Notice has elapsed for the reason that the owner could be forced (up until that point) to return the land to a state capable of reasonably beneficial use.

In considering whether land has a reasonably beneficial use, section 191(5) specifically provides that 'no account shall be taken of any prospective use of that land which would involve the carrying out of new development as defined by section 43(1) of the Land Development Values (Compensation) Act (Northern Ireland) 1965 or of any works requiring listed building consent which might be executed to the building'. Similarly, in this context, the conditions referred to in sections 61[46], 62[47] and 94[48] are to be disregarded. Prospective development of a class mentioned in Schedule 1 of the 1965 Act can, however, be taken into consideration.[49]

In *Allen v Department of the Environment for Northern Ireland*,[50] planning permission in relation to part of a building was only granted subject to a condition. A Purchase Notice contended that the condition effectively prevented development, that it was difficult to attract tenants to the building in its current state (due to the likelihood that the building might be subject to compulsory acquisition in the near future) and thus that the building was incapable of reasonably beneficial use. Upholding the Purchase Notice, Lord Lowry LCJ stated that:

> The phrase 'reasonably beneficial use' is a straightforward expression made up of ordinary words. The use must be the owner's use, because … it is the owner who claims that the Department should purchase his estate in the land which, he claims 'has become incapable of reasonably beneficial use'. To say that 'reasonably beneficial' simply means something more than minimally beneficial attributes to that phrase a most unnatural meaning. It must, I consider, mean 'not only beneficial, in the sense that a benefit of some kind is to be derived, but reasonably beneficial, in the sense that a reasonable degree of benefit is to be derived'.[51]

In *R v Minister of Housing and Local Government, ex parte Chichester RDC*[52] it was considered that land is not rendered incapable of reasonably beneficial use simply because the land is of less value in its current state than if it were developed according to the owner's wishes: the refusal of permission to build houses on seventeen caravan sites did not render the land incapable of reasonably beneficial use. In *Brookdene Investments Ltd v Minister of Housing and Local Government*[53] it was established that a significant difference in value as a consequence of refusal of planning permission did not determine the matter, but was a relevant factor to be taken into consideration. In England this issue is covered by policy guidance originally contained in paragraphs 13 and 16 of Circular 13/83[54]:

> 13. In considering what capacity for use the land has, relevant factors are the physical state of the land, its size, shape and surroundings, and the general pattern of land-uses in the area; a use of relatively low value may be regarded as reasonably beneficial if such a use is common for similar land in the vicinity … Profit may be a useful comparison in certain circumstances, but the absence of profit (however calculated) is not necessarily material: the concept of reasonably beneficial use is not synonymous with profit;
> 16. The Secretary of State … may take into account, among other things, whether there is a reasonable prospect of the server's selling or letting the land for any

> purpose, were its availability to be made known locally. He would normally expect to see some evidence to show that the server has attempted to dispose of his interest in the land before he could be satisfied that the land had become incapable of reasonably beneficial use.

This guidance was considered without criticism by the Court of Appeal in *Colley v Secretary of State for the Environment and Canterbury City Council*[55] so, while it does not apply in Northern Ireland and there has been no equivalent Northern Ireland guidance issued on this matter, it may nevertheless be instructive. In *Allen* the decision of the Lands Tribunal (upheld by the Northern Ireland Court of Appeal) was based significantly on (loss of) rental income:

> In their existing state the premises were let at rents which were substantially lower than rents which were obtainable in other parts of Great Victoria Street unaffected by the scheme. Even taking into account general economic circumstances (which might depress rents generally to some extent) if rents had merely kept pace with the rise in the Retail Price Index between 1971 and 1981 the total income would have been more than double what is actually received. The threatened acquisition had a serious and adverse effect on the quality of the tenancies as well as their quantity. It was almost impossible to attract substantially long-term tenancies.[56]

A final issue raised in *Allen* was the fact that the planning condition in question related only to *part* of the premises, while the Purchase Notice related to the whole building. In *Wain v Secretary of State for the Environment*[57] it was held, in the context of the relevant English legislation, that a Purchase Notice should be served only in respect of the part of the land rendered incapable of reasonably beneficial use (not the whole land). In *Allen,* while Lord Lowry LCJ ultimately did not have to decide this matter, it was noted that the Northern Ireland legislation does not contain provisions equivalent to those on which the judgment in *Wain* was based[58].

Within two months of the date on which the Purchase Notice was served[59], the council must respond to the person by whom the Purchase Notice was served, by:

- Serving notice that the council is willing to comply; or
- Serving a counter-notice that the council is not, for stated reasons[60], willing to comply; or
- Serving a counter-notice, stating that planning permission or listed building consent (as relevant in the particular case) might reasonably be expected to be granted for any other development/works which, in the opinion of the council would, if carried out render the land capable of reasonably beneficial use.[61]

Within two months of the date of receipt of any counter-notice the recipient may refer the matter to the Lands Tribunal, which will consider the matters raised in the Purchase Notice and (only) the reasons specified in the counter-notice.[62] Where the council does not object, or subsequently withdraws an objection, or if an objection is not upheld by the Lands Tribunal or where the two-month period for response by the council has elapsed then, unless the Purchase Notice is withdrawn,[63] the claimant and the council are deemed to have entered into a contract for the sale of the claimant's estate in the land.[64] The council are liable to pay the amount which would have been paid had the estate been compulsorily acquired.[65]

Blight Notices

In certain situations, an owner-occupier may serve on the appropriate authority a Blight Notice requiring the authority to purchase their 'blighted' interest in property as defined in planning law. The main purpose of a Blight Notice is to compel authorities to purchase properties in advance of potential or intended compulsory acquisition to prevent economic hardship where the value of the interest may be significantly reduced by certain categories of planning or proposed development. The methods and procedures are governed by The Planning Blight (Compensation) (Northern Ireland) Order 1981[66], which provides for various circumstances of public development, including those where blight arises as a result of the inclusion of a property in a Vesting Order. The classes of blighted land to which the 1981 Order applies are specified in full in Article 3[67] and generally include land allocated for public authority functions in development plans, for housing redevelopment, the construction, improvement or alteration of roads or within compulsory acquisition. Only certain persons qualify to serve a Blight Notice, specifically those with an interest in the land the blight will effect:

- A resident owner-occupier of a private dwelling[68];
- An owner-occupier of an agricultural unit[69];
- An owner-occupier of any hereditament, including business owners, where:
  - the net annual value of the land does not exceed £19,685[70];
  - the capital value of the hereditament does not exceed £2,100,000[71];
- Certain mortgagees[72] and personal representatives.[73]

In this context, an owner-occupier is the owner of a freehold interest (whether legal or equitable) or legal tenancy with at least three years unexpired term who has been in occupation of the whole or substantial part of the hereditament or the six months preceding the service of a Blight Notice.[74] If the property is vacant when the Blight Notice is served, they must have been in occupation for six months ending not more than one year before the date of service of the Blight Notice[75] and, if the latter applies, the hereditament or dwelling must have been unoccupied since the owner-occupier vacated. Notably, where a property is let it cannot be regarded as occupied[76] and an investment property owner is not entitled to serve a Blight Notice.

A person who has an interest qualifying for protection under Article 4 and has land within the circumstances of planning blight detailed in Article 3 may be entitled to serve a Blight Notice on the body responsible requiring them to purchase the claimant's property interest at its untainted value.[77] Amongst other things, in order to succeed the claimant must, before serving the Blight Notice in the prescribed form[78], demonstrate that they have made reasonable endeavours to sell that interest[79] and that, in consequence of the blight, has been unable to sell that interest except at a price substantially lower than that for which it might reasonably have been expected in the absence of blight.[80] The 1981 Order does not specify what constitutes reasonable endeavours and it is a question of fact and degree in each case. In *Churches' Youth Welfare Council v Department of the Environment for Northern Ireland*[81], the Tribunal acknowledged that there may be circumstances where little evidence is required to demonstrate that reasonable endeavours to sell the relevant interest have been made, or that its value has been substantially reduced. Also in *Jeffers v Department of the Environment for Northern*

Ireland[82] the Tribunal stated that blight is a matter of degree and, in extreme cases, especially where the date for compulsorily vesting is close at hand, it may be obvious that the price will be gravely depreciated. In *Rhodia International Holdings Ltd v Huntsman International LLC*[83] the court instructively concluded that the qualification of reasonable endeavour only required the performing party to take one reasonable course, not all of those courses available. The facts of each case will therefore dictate what is considered reasonable endeavour to sell; however, it is not acceptable to make no attempts to sell on the basis that the planning blight would result in no purchase.[84]

It is not possible to serve a valid Blight Notice in respect of a part of a hereditament[85] and an objection must be raised on counter-notice by the appropriate authority.[86] Where the authority proposes to acquire compulsorily part only of the property and a Blight Notice is served in respect of the whole property, it may be difficult to compel them to take the whole.[87] Instead, the owners must show that the part cannot be acquired without causing material detriment to the house. Where the Blight Notice has been served and the acquiring authority either accepts it or fails to reject it within two months, the Notice automatically takes effect and the authority is deemed to have entered into a statutory contract to purchase the interest of the claimant on that date which becomes the valuation date.[88] In such circumstances, the principles of valuation and assessment of compensation are the same as those which apply had the authority compulsorily acquired the interest.[89] Any dispute as to the amount of compensation for the interest shall be determined by the Lands Tribunal.[90] However, compensation following acceptance of a Blight Notice does not extend to abortive marketing costs incurred prior to service of the Notice[91] or to the costs of preparing and serving the Blight Notice. In contrast, if the acquiring authority does not wish to purchase the property under the blight provisions, they may serve a counter-notice within the two-month period[92] objecting on one or more specified grounds[93] set out in Article 6(2) of the 1981 Order. They include the following:

- No part of the hereditament or agricultural unit is comprised in a relevant category of blighted land;[94];
- The appropriate authority does not propose to compulsorily acquire any of the land;[95]
- The acquiring authority only proposes to acquire part of the land referred to in the Blight Notice (although if there is material detriment to any part not acquired they can be made to acquire the whole of a person's property – see subsection 'Material detriment' below);[96]
- On the date of the Blight Notice, the claimant was not entitled to an interest in any part of the land;[97]
- The interest of the claimant does not qualify (which must be stated on the counter-notice);[98]
- That the person who made the Blight Notice has not made reasonable endeavours to sell (Article 5(1) (b)), or he could only sell at a value that was depreciated by the blight (Article 5(1) (c));[99]
- That in the case of land falling within Article 3(1) (b), (d), (e), (g), (h), (i), (j), or (l), the appropriate authority does not propose to acquire any of the land within the next 15 years (or such longer period as specified in the counter-notice).[100]

Article 6(3) further stipulates that the authority cannot rely on paragraph 2(g), that in appropriate cases it does not propose requiring the land within the next 15 years, if ground (b) is appropriate.[101] This provision is to deter authorities from making improper use of these objections.

Where a counter-notice has been served objecting to the Blight Notice on any of the above grounds, the owner-occupier may refer the matter to the Lands Tribunal within a period of two months and it will determine the matter.[102] On such a reference, and where the objection is not withdrawn, the Lands Tribunal shall consider the matters set out in the Blight Notice served by the claimant and *only* the grounds of objection specified in the counter-notice of the authority[103]. Unless it is shown to the Tribunal that the objection is not well founded, the Tribunal shall uphold the objection thus essentially placing the burden on the claimant[104].Notably, the position is different for an objection on the ground mentioned in Article 6(2) (b), (c) or (g) which will not be upheld unless it is shown to the satisfaction of the Tribunal that the objection by the authority is well founded.[105]

Where the Lands Tribunal does not uphold the objection, it will declare the Blight Notice to which the counter-notice relates as valid.[106] Alternatively, where the Lands Tribunal upholds an objection on the ground that the authority only proposes to acquire part of the land covered by the Blight Notice[107], it shall declare that the Notice is a valid Notice in relation to that part which the appropriate authority proposes to acquire compulsorily, but not in relation to any other part[108]. Moreover, the Tribunal shall not uphold an objection on the ground mentioned in Article 6(2) (c) if the part of the hereditament which the appropriate authority proposes to acquire is, or includes, part of a house, building or factory and the Tribunal is satisfied that the part cannot be acquired without causing material detriment to the house, building or factory.[109] In determining whether part of a house, building or factory can be taken without material detriment to the house, building or factory, the Lands Tribunal shall take into account not only the effect of the severance but also the use to be made of the part proposed to be acquired; and in a case where the part is proposed to be acquired for works or other purposes extending to other land, the effect of the whole of the works and the use to be made of the other land.[110]

A number of cases have considered aspects of the principles governing Blight Notices in relation to public works. In the matter of *Reilly & Reilly v Department for Regional Development*[111] the Department for Regional Development announced a proposed scheme for a high-speed dual carriageway, which included a proposed vesting of a portion of the claimant's front garden. Following subsequent unsuccessful attempts to sell the property, the claimants served a Blight Notice on the Department claiming they had made reasonable endeavours to sell their interest and in consequence of the proposed scheme they had been unable to do so. They further claimed that the part of the garden that the Department proposed to acquire could not be taken without material detriment to the house. The Department served a counter-notice objecting, on the grounds of Article 6(2) (c) and (f), to the Blight Notice. The claimant's referred the counter-notice objections of the authority to the Lands Tribunal for determination, and, prior to the hearing, the Department published a Notice of intention to make a Vesting Order. At hearing, the parties agreed the claimants had made a reasonable endeavour to sell their property at an appropriate marketing price pursuant to Article 5(1) (b). However, they disagreed as to whether the inability to sell was a result of the blight (Article 5(1) (c)). They also disagreed as to whether there would be material detriment to the remainder of the property, which was not to be vested. Essentially, the case pivoted on the question of whether the inability to sell was a result of the proposed scheme. Taking into account evidence of the difficult market conditions that prevailed at the time, the existing property footprint and attributes and comparators, the Tribunal was not persuaded, on balance, that there was sufficient evidence that the property would have been sold in this difficult market but for the proposed scheme. For this reason, and upholding the Department's objection to the Blight Notice to that part of land to be acquired, the Tribunal concluded that the claimants had not shown that, in consequence of the proposed scheme, they had been unable to sell their interest except at a price substantially lower than that for which it might reasonably have been expected to sell if there were no scheme. That said, the claimants did show that the part to be acquired could not be acquired without causing material detriment to the house – meaning the authority could be made to acquire the whole of the property (see 'Material detriment' below). This case demonstrates that, in some circumstances, the inability to sell will very much depend on the degree of material detriment, but that may not always be the case. Indeed, there may be other reasons why a property has not sold, and it is necessary to demonstrate that this is clearly related to the blight.

In the matter of *O'Neill & O'Neill v Department for Regional Development*[112], the claimants were the owners of a complex of buildings and land that fell within a proposed road dualling scheme and was scheduled for vesting. Following unsuccessful attempts to market the properties, the claimants served a Blight Notice[113] in which they claimed they had made reasonable endeavours to sell and in consequence of the proposed scheme they had been unable to do so. The Department served counter-notices objecting to the Blight Notices on the grounds of Articles 5(1) (b) and (c) and also on the ground that the interest of the claimants for one property was not that of an owner-occupier (Article 4), although this was withdrawn prior to the hearing. Following expert evidence pertaining to the nature, reasonableness and efficacy of the marketing approach and methods employed, with reference to market demand and enquiries in light of the blighted land, the Tribunal concluded that the properties were so blighted by the road scheme that no amount of marketing would have generated a sale and indeed that the evidence of no follow-up interest from buyers supported the view that the properties could not be sold at any price as a consequence of the effect of the blight. As such, the Tribunal was satisfied that reasonable endeavours had been made to sell the hereditaments subject to Blight Notices and these endeavours satisfied the conditions of Article 5(1) (b) and (c) of the order. The objections in the counter-notice were therefore not upheld and the Blight Notices declared valid. The case again emphasises that the Tribunal must be satisfied that reasonable endeavour to sell has been made and that the blight is the critical aspect of failure in that regard.

Compensation for interest in land

Both the Vesting Order and the Compulsory Purchase Order approaches share a common historical lineage found in disparate and varied private Acts from the late 18th and 19th centuries authorising the construction of railways, canals and harbours. Such private Acts ordinarily contained powers of compulsory acquisition and provided standard clauses and procedures for implementation and determining compensation, which were consolidated in the Land Clauses Consolidation Act 1845. Whilst this statute did not confer the power of compulsory acquisition, in the period post-1845, any Act authorising compulsory purchase was conventionally treated as incorporating this Act. In the early 20th century, the advent of the First World War and improvements in science and technology meant that the provision of public services for, inter alia, housing, transport schemes and public health assumed greater societal cognisance and importance. In anticipation of the postwar reconstruction period, the Scott Committee was established in 1918 to conduct an urgent review of the compensation laws in light of the need for public authorities to acquire large quantities of land. Unsurprisingly, the Committee noted the prioritisation and emphasis was on acquisition for wider public purposes, and it made a number of recommendations, which provided for less sympathetic treatment in the assessment of compensation. This was manifest in the six rules of compulsory purchase compensation provided in Section 2 of the Acquisition of Land Act 1919[114], as replaced, in England and Wales, by Section 5 of the Land Compensation Act 1961, which effectively was a consolidation statute. This has been subject to minor amendments over the years, most recently through the Planning and Compulsory Purchase Act 2004 and the Localism Act 2011. In England and Wales most of the rules governing procedure and compensation are found in the Compulsory Purchase Act 1965, supplemented by the Compulsory Purchase (Vesting Declarations) Act 1981, which enabled orders to be implemented by a vesting declaration.

In Northern Ireland, the rules of compulsory purchase compensation are found in the Land Compensation (Northern Ireland) Order 1982, which essentially consolidates the 1919 Act. However, compulsory purchase compensation law in Northern Ireland is spread over various other enactments, including the Lands Tribunal and Compensation Act (Northern Ireland) 1964, the Planning and Land Compensation Act (Northern Ireland) 1971, the Land Acquisition and Compensation (Northern Ireland) Order 1973 and the Land Acquisition and Compensation (Amendment) Act (Northern Ireland) 2016.

Clearly, the rights to compensation and assessment procedures are derived from an assortment of acts of Parliament, case law and established valuation practice. Understandably, the notorious complexity and conflicting application of compensatory rules means that the law can lack clarity and its application can lack obvious consistency. Indeed, the law has been described as 'a tortuous journey, through obscure statutes and apparently conflicting case law'[115] that is in dire need of reform or repeal. However, despite strong and repeated calls by senior judiciary[116] and the Law Commission[117] for much needed intelligibility and precision to compulsory purchase law, the appetite from Parliament to implement reform has been limited.

Compensation rules and heads of claim

In Northern Ireland, statutory provisions relating to compulsory purchase compensation are set out in Article 6 of the Land Compensation (Northern Ireland) Order 1982[118],The six core 'rules' for assessing compensation are as follows:

1. No allowance shall be made on account of the acquisition being compulsory;
2. The value of land shall, subject to rules 3 to 6, be taken to be the amount which the land if sold in the open market by a willing seller might be expected to realise;
3. The special suitability or adaptability of the land for any purpose shall not be taken into account if that purpose is a purpose to which it could be applied only in pursuance of statutory powers, or for which there is no market apart from the special needs of a particular purchaser or the requirements of any authority possessing compulsory acquisition powers;
4. Where the value of the land is increased by reason of the use of it or of any premises on it in a manner which could be restrained by any court, or is contrary to law, or is detrimental to the health of the occupants of the premises or to the public health, the amount of that increase shall not be taken into account;

5. Where land is, and but for the compulsory acquisition would continue to be, devoted to a purpose of such a nature that there is no general demand or market for land for that purpose, the compensation may, where reinstatement in some other place is *bona fide* intended, be assessed on the basis of the reasonable cost of equivalent reinstatement;
6. The provisions of Rule (2) shall not affect the assessment of compensation for disturbance or any other matter not directly based on the value of land.

Depending upon facts and peculiar circumstances of each case, compensation can be claimed under 'Heads of claim' categories, which include: (1) the value of the land taken; (2) severance and injurious affection; (3) disturbance; and (4) fees.

Compensation when land taken

Two principles underpin the value of the land taken: the principle of equivalence and the market value rule.

1. PRINCIPLE OF EQUIVALENCE

The underlying principle is that compensation following a compulsory acquisition of land is the principle of equivalence. In simple terms, this principle states that the effects of the Vesting Order on the value of a property should be ignored when assessing compensation.[119] This principle has been developed through case law spanning hundreds of years. Traditionally, the courts have determined compensation by ensuring that the property owners are in no better and no worse position financially than they would have been if the interest had not been compulsorily acquired[120]. In *Director of Buildings and Lands v Shun Fung Ironworks*[121], the Privy Council enunciated the principle, with Lord Nicholls opining:

> No allowance is to be made because the resumption or acquisition was compulsory; and the land is to be valued at a price it might be expected to realise if sold by a willing seller, not an unwilling seller. But subject to these qualifications, a claimant is entitled to be compensated fairly and fully for his loss. Conversely, and built into the concept of fair compensation, is the corollary that a claimant is not entitled to receive more than fair compensation: a person is entitled to compensation for losses fairly attributable to the taking of his land, but not to any greater amount.

Therefore, where land is compulsorily acquired, the court will try to ensure that owners are fairly and fully compensated for the loss arising. It is also important to highlight the fact that the basis of compensation is the value to the owner of the land being acquired; not the value to the public authority acquiring the land.[122]

2. MARKET VALUE

Generally, compensation payable in respect of land and property compulsorily acquired is valued on the basis of its open market value. Indeed, four of the six basic rules of valuation under the 1982 Order concern market value.

Rule 2, the market value rule, is perhaps the most commonly used principle. This envisages a hypothetical sale of the property in the open market by a willing seller[123] with no allowance for the fact that the property is being compulsorily acquired. An authoritative interpretation of the 'market value' principle is set out in *Vyricherla Narayana Gajapatiraju v Revenue Divisional Officer, Vizagapatam*[124], where Lord Romer opined:

> The compensation must be determined, therefore, by reference to the price which a willing vendor might reasonably expect to obtain from a willing purchaser. The disinclination of the vendor to part with his land and the urgent necessity of the purchaser to buy must alike be disregarded. Neither must be considered as acting under compulsion. This is implied in the common saying that the value of the land is not to be estimated at its value to the purchaser. But this does not mean that the fact that some particular purchaser might desire the land more than others is to be disregarded. The wish of a particular purchaser, though not his compulsion, may always be taken into consideration for what it is worth.[125]

In this context of compulsory acquisition, whilst the transaction is hypothetical, 'there is nothing hypothetical about the open market in which it is supposed to take place'[126] which should be kept as near as possible to reality.[127] In this statutory 'no scheme world', the vendor must determine the value of the land on the open market in conditions that would invite an offer, enhance value and obtain the best price reasonably obtainable, even though there will be no sale.[128] Moreover, the price will be assessed according to the prevailing market conditions at the time. This open market hypothesis should also be applied to the material condition and existence of the property, even where the vendor may have made improvements prior to marketing the property[129]. Indeed, the condition of the property on the valuation date is the relevant condition. Furthermore, the hypothetical vendor and purchaser should reflect the reality of reasonable people engaged in a transaction for the property.

Consequently, it is necessary to value the land on the basis of its open market value; without any increase or decrease attributable to the scheme of development connected to the Vesting Order. This was set out in the Court of Appeal case of *Pointe Gourde Quarrying and Transport Company Limited v Sub-Intendent of Crown Lands*[130], commonly referred thereafter as the Pointe Gourde principle, as codified in Rule (3) of Article 6. Applying the Pointe Gourde rule to Rule 2, the 'market value rule', means that the landowner would not be entitled to enhanced value attributable solely to the particular use proposed to be made of the land under a scheme for which compulsory acquisition of the land is an integral part. This element of value is not part of market value because the owner could not realise it in the open market. However, in applying this rule, Lord Nicholls[131] offered the following useful insights:

- It should not be applied too severely but in a manner that achieves a fair and reasonable result;
- A result is not fair and reasonable where it requires a valuation exercise that is unreal or virtually impossible;
- A valuation result should be viewed with caution when it will lead to a gross disparity between the amount of compensation payable and the market values of comparable adjoining properties that are not being acquired.

Market value does not, however, mean simply the existing use value. Whilst it is assumed that a willing seller would only be willing to sell at the best price which they could reasonably achieve in the open market, a willing seller would expect to receive recognition of any additional value due to the development potential of the land, which, in practice, will not be negligible simply because there is only one buyer. Where market value reflects development value this may be taken into account.[132] The lack of planning permission will not matter where the open market value reflects any 'hope value'. This can be evidenced by reference to, inter alia, land sales comparators and the planning pipeline which would support the likelihood of permissions being granted[133].

With regards to determining any possible development value, a claimant can rely to an extent on any existing planning permission or any permission that is generally permitted or in accordance with the zoning and other policy in a development plan or, indeed, the use intended by the scheme, save for uses requiring statutory powers to undertake. To provide clarity in this regard, the affected owner can obtain a certificate of acceptable development that would list acceptable changes of use of the land. It is important to note that the availability of planning permission does not, in itself, imply development value, which would be required to be demonstrated to exist in a 'no scheme' scenario.

The market value basis for the assessment of compensation can include any additional ransom or marriage value elements resulting from the combination of the affected land with other parcels of land or due to the potential to merge interests in the land, should such additional value be able to be proven to exist in a 'no scheme world'. This is necessary to properly reflect the market value of the affected land.

It should be noted that Rule 3 in the 1919 Act (and the equivalent Northern Ireland clauses) eliminates value attributable to special suitability of land for a purpose for which there is no market other than the special needs of a *particular* (indicating singular) purchaser. This 'particular purchaser' part of Rule 3 may act to exclude a marriage or ransom value from the assessment of compensation. This part of the rule was repealed in Great Britain by Section 70 of the Planning and Compensation Act 1991. It is interesting to note that this has *not* been repealed in Northern Ireland. This albeit rare and perhaps hard to prove aspect nevertheless has the potential to produce considerable disparity between otherwise similar cases in the different jurisdictions and is an aspect that review and reform activity would do well to consider.

In keeping with the 'market value' basis of compensation, unusual properties, such as schools, churches or public conveniences, for which there is no effective market, can be valued on the basis of equivalent reinstatement of a similar building or modern alternative should the current use be continued in the future. This mirrors the 'contractors test' valuation approach, which would be adopted for such properties for other purposes.

Mortgages and negative equity

It is common for dwellings to be subject to a mortgage and both mortgagees and mortgagors have an interest in land for which there is an entitlement to compensation.[134] However, when the outstanding loan on the mortgage exceeds the value of the property, there is no compensation to the mortgagor in respect of the property. Indeed, the acquiring authority will pay the value of the property to the lender. The borrower will remain liable to the lender for the amount of any outstanding sum plus interest due from the operative date of vesting until the mortgage is redeemed and this amount is not compensatable. However, the borrower may still have an entitlement to other heads of compensation such as a home loss payment. There is also an entitlement to compensation if losses occur as a result of having to take out a new mortgage (see 'Disturbance', considered below).

Following the advent of the global financial crisis and subsequent property market crash in 2010, circumstances where the outstanding loan on the mortgage exceeded the value of the property ('negative equity' as a result of falling property value) became a more prominent matter of consideration in compulsory acquisition. In such a situation, the value of the property being acquired is agreed between the acquiring authority, the lender and the borrower.[135] A number of recent decisions have considered the issue of mortgage arrears and payments. In *O'Neill v Northern Ireland Housing Executive*[136] the Lands Tribunal considered the issue of 'negative equity' where the mortgage debt exceeded the market value of the lands and the claimants sought to recover some or all of that negative equity as compensation pursuant to the Land Compensation (Northern Ireland) Order 1982 and Local Government Act (Northern Ireland) 1972. Primarily, the Lands Tribunal was asked to consider a number of preliminary questions relating to:

- Whether the Northern Ireland Housing Executive, as a promoter pursuant to the Land Clauses Consolidation Act 1845, was bound to discharge all or any of the mortgaged debts secured on each property where the assessed compensation falls below the current redemption figure?
- Whether the determination of compensation payable upon the vesting of mortgaged lands is the value of the lands settled by agreement of the mortgagee, the party entitled to the equity of redemption and the promoters of the undertaking and in default of such agreement that amount may be determined by the Lands Tribunal? and
- Whether the determination of compensation payable upon the vesting of mortgaged lands is the sum of money required to redeem the mortgage or mortgages upon the lands assuming such sum is greater than that as agreed or determined by question (2) above?

The Tribunal noted that under Schedule 6 of the 1972 Act, the mortgagor and mortgagee have no residual property rights and that they must claim from the compensation fund and, in regard to any rights against each other, the terms of their mortgage agreement. In rejecting the claimant's argument, the Tribunal held that where the assessed compensation falls below the current redemption figure, the Northern Ireland Housing Executive was not bound to discharge all of the mortgaged debts secured on each property. Pertinently, the Tribunal found that the amount of compensation payable upon the vesting of mortgaged lands is not measured by the sum of money required to redeem the mortgage or mortgages upon the lands. It further held that the compensation payable upon the vesting of mortgaged lands is not limited to the value of the lands, as provided under Article 6(1)(2) of the 1982 Order, meaning it can include compensation for disturbance or any other matter not directly based on the value of land (Article 6(1)(6)) and severance and injurious affection (Article 8).

More recently, in *Kerr v Northern Ireland Housing Executive*[137] the Lands Tribunal provided further clarification with regard to level of compensation payable when a property is vested for regeneration and the owner-occupier is in negative equity. In this case, the applicant purchased a property which was subject to a Vesting Order. In line with compensation policy, the applicant was offered £91,000 in compensation for the open market value. In addition to this the applicant was also entitled to a home loss payment of 10 percent of the market value (£9,000) and a disturbance payment of £800. The total claim for compensation amounted to around £100,900, resulting in a shortfall of around £45,000 for which the applicant remained liable to pay under the conditions of his mortgage. In considering the matter, the Tribunal also considered two issues: (1) whether the applicant's loss and liability to his mortgage lender could be recovered as compensation under 'disturbance'; and (2) the implications of Article 1 of the First Protocol of the European Convention on Human Rights. The Attorney General for Northern Ireland argued that, if not compensated, the severe financial hardship that the applicant would suffer would result in an excessive burden due to the operation of the Vesting Order. This, it was contended, violated A1P1 unless the authority demonstrated that the compensation offered was *proportionate to the aim pursued in the public interest*. However, the Tribunal ruled that negative equity is a situation 'directly based' on the value of land and that, in the instant case, the loss resulted from the collapse of the property market which had occurred prior to, and independent of, the Vesting Order. Furthermore, the Tribunal also held that the statutory framework could not provide for every anomaly and hardship but provided a natural balance between public and private interests including urban regeneration and housing development schemes and the individual requirements of those adversely affected. In arriving at this decision, the Tribunal also considered that the vast majority of properties affected by the Vesting Order were likely to be in 'positive equity' and would benefit from the vesting. Accordingly, it was held that the applicant remained liable for the mortgage shortfall.

In the conjoined cases of *Northern Ireland Housing Executive v Magee and another*[138] and *Northern Ireland Housing Executive v Rodgers and another*[139] the Lands Tribunal also considered and provided further clarification as to the statutory authority of the Tribunal to award the compensation monies to a respondent as mortgagee where the existence of charges exceeded the market value of the properties. Whilst this appears a straightforward issue, the Tribunal had to consider the correct amounts to be awarded as there was no agreement as to the amount of compensation payable and whether the mortgagee was entitled to receipt of the compensation monies directly. In determining the matter, the Tribunal[140] noted that the existence of charges exceeding the market value of the properties engaged Section 110 of the Land Clauses Consolidation Act 1845, which provides:

> Sum to be paid when mortgage exceeds the value of the lands
>
> 110. If any such mortgaged lands shall be of less value than the principal, interest, and costs secured thereon, the value of such lands, or the compensation to be made by the promoters of the undertaking in respect thereof, shall be settled by agreement between the mortgagee of such lands and the party entitled to the equity of redemption thereof on the one part, and the promoters of the undertaking on the other part; and if the parties aforesaid fail to agree respecting the amount of such value or compensation, the same shall be determined as in other cases of disputed compensation; and the amount of such value or compensation, being so agreed upon or determined, shall be paid by the promoters of the undertaking to the mortgagee, in satisfaction of his mortgage debt, so far as the same will extend; and upon payment or tender thereof the mortgagee shall convey or release all his

> interest in such mortgaged lands to the promoters of the undertaking, or as they shall direct.

Also, Paragraph 17 of Schedule 6 of Schedule 6 of the Local Government Act (Northern Ireland) 1972 where relevant provides:

> 17(1) Where the amount of compensation has been determined but for some reason it is not possible for the council to obtain a good discharge therefore, -
>
> b) if the total amount of the compensation exceeds £1,000, the amount payable by the council shall be paid, applied and dealt with in accordance with the provisions of the Lands Clauses Consolidation Act 1845 [1845 c.18] with respect to the purchase money or compensation coming to parties having limited interests, or prevented from treating, or not making title, and those provisions shall have effect accordingly.

Section 111 of the Land Clauses Consolidation Act 1845 also dictates how the amount payable by the applicant should be paid, applied and dealt with in accordance with the Act.[141] Considering this, the Tribunal held that Section 110 of the Land Clauses Consolidation Act 1845 confers on the Tribunal the statutory authority to award the compensation monies to a mortgagee.

Compensation for partial vesting of a holding

The situation is complicated somewhat should the Vesting Order be for part only of a land holding for a scheme. In this case, in addition to compensation for the piece of land taken, there is also the possibility of additional compensation based on any detrimental effect the loss of land has had on the remainder of the claimant's affected land, described as compensation for severance or injurious affection to retained land.

Article 8(1) of the 1982 Order provides for this compensation for severance or injurious affection:

> In assessing compensation to be paid … regard shall be had not only to the value of the land acquired but also to the damage, if any, sustained or which may be sustained by that person by reason of the severing of the land from other lands of that person held with that land, or otherwise injuriously affecting such other lands by the exercise of powers conferred on the acquiring authority by any transferred provision.

Severance relates to the loss of land which gave value to the remainder[142], such as farmland to a farm, with injurious affection relating to the general adverse effect of the scheme on the value of land retained[143], such as road noise or other pertinent loss of amenity or utility, which unlike the case where no land is taken can relate to both the carrying out of public works and their subsequent operational use. The established test for compensation is found in *Cowper Essex v Action Local Board*[144], where the Law Lords determined that a claim could be made for severance or injurious affection, where, notwithstanding the fact that the parts of land are not adjoining, the claimant has an interest in separate parts of land, the parts of land are in sufficient proximity to each other where possession and control of each confers an enhanced aggregated value. The Privy Council also held in *Holditch v Canadian Northern Ontario Railway*[145] that, where compulsorily acquired, the lands injuriously affected by the severance must be related to the remaining lands to the extent that the owner of the land is prejudiced in his disposition or use by virtue of the severance. The tenor of case law also suggests that common occupation of the land acquired and retained does not constrain compensation for the effects of severance and injurious affection.[146] This includes situations where the land acquired is held under different interests[147] or not held by the same interest or title. Therefore, where the proximity relationship is established and the owner with a legal and/or equitable interest in the lands can demonstrate there is enhanced value, there is an entitlement to compensation.

Determining the value of the acquired land and the effect of severance and injurious affection should be performed on a consistent basis where the claimant receives no compensation for injurious affection above or below his loss[148]. From a practical perspective the approach to assessing the actual loss involves a calculation of the total value of the holding before and after the effect of the scheme, with compensation forming the difference between the two figures. This 'before and after' methodology is also adopted to ascertain whether there is any 'net gain' to an affected party's gross contiguous holdings, to determine if there is, in fact, any net gain due to the scheme. This may occur when a scheme is having a positive effect on values and the dispossessed owner stands to gain from this in value uplift in other retained land. Should this be the case, then any compensation due for any aspect of compensation can be reduced by the amount of the value uplift or 'betterment' received, although there is authority to suggest that this should only be deducted from compensation where land has been acquired under statutory provision[149]. Of course, there is a natural limit to this logic, with an affected party not expected to make any net payment, regardless of the size of the net benefit enjoyed. Therefore, while this method has been subject to some criticism,[150] it is generally considered to be a convenient method of valuation and has advantages in cases of severance as it may reflect the true loss to a claimant[151].

Material detriment

There will be instances where the vesting proposal of an acquiring authority will sever units of occupation and ownership boundaries. Clearly, this can be problematic for affected owners as property amenities and characteristics are inherently related to value. Therefore, the partial/full loss of a front or back garden or taking of only part of the land can be inconvenient. In a situation where the authority has acquired only part of the land, the owner has two main options: (1) sell only that land the authority requires and claim compensation for the value of the land lost and reduced value of the land retained; (2) try to compel the acquiring authority to acquire the whole interest in the land. This latter course of action relates to a claim for 'material detriment' and the owner must show that the part cannot be acquired without causing material detriment to the house[152] in that the remaining part is 'less useful or less valuable in some significant degree'.[153] Such a claim must be made within six weeks of the operative date of the Vesting Order and, in the case of a dispute, the issue will be determined by reference to the Lands Tribunal. In determining whether there is material detriment, the Lands Tribunal shall take into account not only the effect of the severance but also the use to be made of the part proposed to be acquired and, in a case where the part is proposed to be acquired for works or other purposes extending to other land, the effect of the whole of the works and the use to be made of the other land.[154]

The tests for determining 'material detriment' have been the subject of some debate in the past and include considering whether the part could be taken without seriously affecting the amenity or convenience of the remainder or whether the extent of the effect on the value of the remainder would be material or serious.[155] This was also considered in the matter of *Reilly v Department for Regional Development*[156], where the claimants served a Blight Notice on the Department claiming, in addition to reasonable endeavours to sell, that the part of the garden that the Department proposed to acquire could not be taken without material detriment to the house. In this context, the Department served a counter-notice objecting on the grounds of Article 6(2) (c) of the 1973 Order. The claimants referred the counter-notice objections of the authority to the Lands Tribunal for determination. In the instant case, whilst it was clear that there were likely to be both advantages and disadvantages flowing from the scheme as it affected this property and the net effect on the value of many aspects was debatable, the Tribunal asserted that it must take into account the effect of the whole scheme. The Tribunal accepted that the effect of this scheme on this property would be exceptional, the part required for the scheme could not be taken without seriously affecting the amenity or convenience of the remainder and the effect on value of the remainder probably would be material or serious. For these reasons, the Tribunal concluded that there would be material detriment to the remainder of the property which is not to be vested. Importantly, consistent with the language of the legislation, the Tribunal emphasised that the preferred approach in consideration of material detriment is to assess the nature and impact of the scheme as a whole, as opposed to an assessment of the effect on the value of discrete aspects.

Compensation for decisions

Until relatively recently, compensation could potentially be claimed where planning permission was refused or was granted subject to conditions or, in the case of a listed building, where consent to alter the building was refused or granted subject to conditions. These rights were considered necessary in early planning legislation in order to encourage engagement with the planning system and also, arguably, reflected a traditional view that planning law was an imposition upon property rights, the operation of which necessitated the payment of compensation. In Northern Ireland a right to compensation in such cases was removed by the Planning (Compensation, etc.) Act (Northern Ireland) 2001.[157] This change is reflective of the approach of the European Court of Human Rights (ECtHR) to planning decisions, which, unless they result in a 'deprivation' of possessions (as opposed to mere 'control'), has been to generally assume that no compensation should be due. In *Baner v Sweden* it was stated that '[e]veryone must be prepared to accept a certain interference in the public interest without compensation'.[158] Moreover, in the specific context of planning, the ECtHR in *Buckley v United Kingdom* stated that in view of the 'multitude of local factors … inherent in the choice and implementation of planning policies … the margin of appreciation accorded to national authorities in the planning context is necessarily wide'.[159]

However, as noted above, the ECtHR in *Pine Valley v Ireland*[160] considered that planning permission (once granted) is itself a property right that may be protected under A1P1 against any act of deprivation by the state (such as revocation or modification). This is reflected by the (potential) provision of compensation in a number of situations under Part 6 of the Planning Act (Northern Ireland) 2011.

1 *Planning permission is revoked or modified*

Where a council or the acquiring authority revokes or modifies any planning permission under, respectively, section 68 or 72 of the Planning Act (Northern Ireland) 2011, section 179 provides that compensation may be payable under section 26 of the Land Development Values (Compensation) Act (Northern Ireland) 1965 to any person 'interested in the land' to which the permission related who: (a) has incurred expenditure in carrying out work which is rendered abortive by the revocation or modification; or (b) has otherwise sustained loss or damage which is directly attributable to the revocation or modification.[161]

Compensation will not be payable in respect of any expenditure, loss or damage incurred prior to the grant of permission (which has been revoked or modified) or arising out of anything done or omitted to be done before that grant of permission.[162] In addition, where it appears to the council that planning permission could have been granted for development other than that which gave rise to the claim then, for the purposes of assessing compensation, it may be assumed that such permission would be granted (either unconditionally or conditionally).[163]

Specific provisions apply to any claim in respect of buildings, plant or machinery in respect of the revocation or modification of permission relating to the winning or working of minerals.[164]

2 *Listed building consent is revoked or modified*

Where listed building consent is revoked or modified under, respectively, section 98 or 101 of the Planning Act (Northern Ireland) 2011, section 181 provides that compensation may be payable on the same basis and in respect of the same factors as specified above in respect of the modification or revocation of planning permission.[165]

3 *An order is made under section 73, 75 or 112*

Where an order is made under section 73 or 75 requiring a use of land to be discontinued (or imposing conditions upon continuance) or requiring any building or works on land to be altered or removed, or where an order is made under section 112(3) revoking or modifying a hazardous substances consent, then the council will be liable to pay compensation for any damage suffered as result of any consequent depreciation of the value of any interest in the land, or as a result of any person being disturbed in their enjoyment of the land.[166]The council will also be liable to pay compensation in respect of any expenses reasonably incurred in carrying out any works in compliance with such an order.[167] Any claim for compensation must be made in writing to the council within six months from the date of the order, or within such extended period as the council may allow.[168] No compensation will be payable where an estate in the land is acquired by the council following the service of a Purchase Notice,[169] and any compensation payable will be reduced by the value of any timber, apparatus or other minerals removed for the purposes of complying with the order.[170]

4 *Operation of a Tree Preservation Order*

Under section 183 a Tree Preservation Order may provide for the payment of compensation if consent is refused to fell, lop or top a tree that is the subject of the order, or if permission is granted subject to conditions that result in consequent loss or damage. Any assessment of compensation will take into consideration the compensation (if any) paid to any person other than the claimant in respect of the same tree and any injurious effect on the land of the claimant that would result from the felling of the tree.[171]

5 *Hazardous substances consent is revoked or modified under section 116*

Where land is subject to a hazardous substances consent and there is a change in the person in control of *part* of the land and a council revokes or modifies the consent upon an application for its continuation under section 116(2) then compensation is payable to the person previously in control of the whole land in respect of any loss or damage sustained by that person that is directly attributable to the revocation/modification.

6 *Loss incurred due to a Stop Notice or a Temporary Stop Notice*

A council[172] will be liable to pay compensation to a person who, at the time when a Stop Notice was served, had an estate in or occupied the land in question in any of the following circumstances:

- the Enforcement Notice to which the Stop Notice relates is quashed on any ground other than on the grounds mentioned in section 143(3)(a) (planning permission granted for those items contained in the Stop Notice on appeal);
- the Enforcement Notice is varied, otherwise than on the grounds specified in section 143(a), so that the activity prohibited by the Stop Notice ceases to be relevant;
- the Enforcement Notice is withdrawn for reasons other than the grant of planning permission for the development to which the Stop Notice relates; or
- the Stop Notice is withdrawn.[173]

Similarly, compensation will be payable to such a person if a Temporary Stop Notice is issued and any of the following circumstances apply:

- the activity specified in the Notice is authorised either by a planning permission or development order;[174]
- a certificate of lawfulness of existing use or development is issued in respect of the activity under section 169 or granted by virtue of an appeal against a decision not to issue a certificate under section 173;[175] or
- a council withdraws the Temporary Stop Notice.[176]

Whether the Stop Notice is temporary or otherwise, such a person is entitled to be compensated for any loss or damage directly attributable to the prohibition contained in the Notice and this will include any loss due to a breach of contract caused by the taking of action necessary to comply with the prohibition.[177] Compensation is not payable in respect of the prohibition of an activity that would, in any case, have constituted, or contributed to, a breach of planning control, nor in respect of any loss or damage that could have been avoided by the claimant providing information (or otherwise cooperating) as required under section 133 or 240.[178] Any claim must be made in writing to the council within six months from the date of the decision in respect of which the claim is made, or within such extended period as the council may allow (Article 30 of the Planning (General Development Procedure) Order (Northern Ireland) 2015).

7 *Loss or damage caused by service of a Building Preservation Notice*

Where a Building Preservation Notice ceases to have effect without the building being included on the list of buildings of special architectural or historic interest compiled by the Department under section 80, then compensation is payable to any person who, at the time the Notice was served, had an estate in the building.[179] Compensation shall be in respect of loss or damage directly attributable to the effect of the Notice[180] and includes any loss due to a consequent breach of contract caused by discontinuing or countermanding any works to the building.[181]

Compensation when no land is taken

A right to compensation may also arise in limited circumstances when no land is taken but when statutory powers are exercised. In general, compensation can be claimed for depreciation in the value of land caused by the use of public works or the injurious effects caused by public works, including compensation for the adverse effects of development. It is generally acknowledged that compensation under these circumstances is inferior to that available should even a single square foot of land be vested, presenting the possibility of perceived inequity and/or unfairness in the treatment of affected parties who may be deemed to be very similarly situated in relation to public works. The compensation available where no land is taken is tightly controlled and limited to losses relating to the use of public works and losses relating to the carrying out of public works. Each of these will be examined in turn.

Reduction in value caused by the use of public works

Part II of the 1973 Order[182] provides a right to certain owners to claim compensation where land is depreciated in value by the use of certain public works,[183] although, there are a number of restrictions on the measure of compensation. Indeed, owners are usually better compensated, where they can show land is taken through a claim for severance or injurious affection. In this event, compensation for depreciation by use of public works is not payable where such has been paid under injurious affection in connection with the acquisition of the land.[184]

A person may claim compensation for the depreciation in value of an interest in land due only to seven specific physical factors caused by the use of the public works.[185] Article 4(2) details that the seven specified factors are: (1) noise, (2) vibration, (3) smell, (4) fumes, (5) smoke, (6) artificial lighting and (7) the discharge on to the land in respect of which the claim is made of any solid or liquid substance. It is sufficient to show depreciation in value due to the physical factors. However, any depreciation in value that is attributable to reasons other than these seven specific factors is not compensatable.[186]

The noise, vibration or other associated physical factors occasioned by aircraft arriving at or departing from an aerodrome are treated as caused by the use of the aerodrome whether or not the aircraft is within the boundaries of the aerodrome. However, in all other circumstances, the source of the physical factors must be situated on or in the public works.[187] Indeed, this latter point is a debilitating restriction on the right to claim and the important distinction between this and a claim for the execution of works is that it is the use of the works that must cause the depreciation. For example, if a motorway is constructed in close proximity to a dwelling, any claim must relate to the depreciation in value caused by the physical factors associated with the traffic using the road, not to the physical existence of the highway. Similarly, where a claim arises from noise due to alterations to a public road, the physical factors must emanate from that altered stretch. However, where increased traffic using an altered stretch of road is the consequence of new works elsewhere, depreciation due to that increased traffic causing more noise on the altered stretch of road is compensatable.[188] It must be noted, however, that an intensification of use occasioning depreciation in a non-altered stretch of road is not compensatable, again raising the spectre of unfairness and/or inequity in similarly affected properties. Notably, pursuant to Article 4(7), no compensation is payable in respect of physical factors caused by accidents involving vehicles on a road or accidents involving aircraft.

To be able to submit a claim a person must own a qualifying interest in the dwelling or land before the relevant date.[189] Where the public works in question is a highway, no compensation will be paid prior to the date on which it was first open to public traffic.[190] With regard to any other public works, it is payable only on the date on which they were first used after completion.[191] A qualifying interest is a freehold interest or a tenancy in a dwelling house or flat with at least three years unexpired at the date of notice of claim.[192] The claimant must be in residential occupation of the property at the date of notice of the claim,[193] entitling both landlord and tenant, if otherwise qualified, to claim.[194] Where the qualifying interest is in land other than a dwelling, the claimant must have an owner's interest and be an owner-occupier at the same time. An owner-occupier in relation to land in a hereditament, means a person who occupies, in right of an owner's interest, the whole or a substantial part of the land and; in relation to and in an agricultural unit, it means a person who occupies the whole of that unit and is entitled while so occupying it to an owner's interest in the whole or in any part of that land.[195]

A claim is made by serving on the responsible authority a Notice which contains the following particulars[196]:

- the land in respect of which the claim is made;
- the claimant's interest and the date on which, and the manner in which, it was acquired;
- the claimant's occupation of the land (in cases where occupation is a qualification for a claim);
- any other interests in the land so far as known to the claimant;
- the public works to which the claim relates;
- the amount of compensation claimed;
- details of other any land contiguous or adjacent land owned by the claimant.

No claim shall be made otherwise than in the claim period, that is to say, the period of two years beginning on the expiration of twelve months from the relevant date. This provision has been amended in England and Wales as it may be slightly unjust where owners have suffered depreciation on account of the physical factors but are unaware of the right to compensation. Where a claimant has entered into a contract during the twelve months preceding the first claim day to dispose of the interest or to grant a tenancy (where land not a dwelling), he may still submit a claim. However, the claim should be submitted during that twelve-month period and before he disposes of the interest or grants a tenancy, otherwise he will lose rights to compensation.[197] This can potentially lead to unjust situations where a new purchaser has acquired the property at a depreciated value and had no incentive to pursue a claim with the original owner, who, selling at the depreciated price, has no entitlement to pursue a claim.

Compensation is based upon prices current on the first claim day, which is twelve months after use of the public works first commenced.[198] In assessing depreciation due to the physical factors caused by the use of any public works, account shall be taken of the use of those works as it exists on the first day of the claim period and of any intensification that may then be reasonably expected of the use of those works in the state in which they are on that date.[199] Moreover, it is the depreciation of the value of the interest by the physical factors that is assessed; proximity to public works, unattractiveness or effect on a view may indeed cause depreciation but this is not compensatable. The benefit of any soundproofing work, or work to mitigate the effects of a road or other public works will be taken into account in assessing the extent of the depreciation.[200]

The compensation payable on a claim will also be reduced, inter alia, where other land has benefited, attributable to the existence of or the use or prospective use of the public to works to which the claim relates. The reduction will be equal to an increase in the value of the claimant's interest in the land in respect of which the claim is made and any interest in other land contiguous or adjacent to the land to which the claimant was entitled in the same capacity on the relevant date.[201]

Compensation for the adverse effects of development

In addition to decreasing the value of land, the nature of certain public development works can cause great inconvenience and discomfort to people living nearby; again, the obvious example is the development of a new dual carriageway or motorway or an aerodrome. The detrimental character of such works may depreciate the value of affected land meaning that compensation for injurious affection is payable. However, such compensation is often limited to land compulsorily acquired and those with a proprietary interest in the land and will not eradicate the basis of the inconvenience. Part III of the 1973 Order, therefore, confers a range of discretionary powers to mitigate the injurious or adverse effect that the construction, alteration, existence or use of any public works may have. The main purpose is to enable an authority to assuage the effects of the works through purchasing affected land and, where necessary, soundproofing and screening work on such land. The following powers are noted:

1. Acquisition of land in connection with public works.

   Under Article 27 of the 1973 Order, a responsible authority may acquire land by agreement for the purpose of mitigating the adverse effect of public works.[202] Such land can be either: (1) land the enjoyment of which is seriously affected by the carrying out of works by the authority for the construction or alteration of any public works; or (2) land the enjoyment of which is seriously affected by the use of any public works, if the interest of the vendor is one which falls within the Planning Blight (Compensation) (Northern Ireland) Order 1981.[203] The compulsory power of acquisition in 1 is not exercisable unless the acquisition is begun before that date; the power in 2 must be exercised before the expiration of one year after the date.[204]

2. Execution of works in connection with public works.

   Article 28 of the 1973 Order empowers the responsible authority to carry out on land, acquired under Article 27 or owned by them, works for mitigating the adverse effect which the construction, alteration, existence or use of any public works has or will have on the surroundings of the works.[205] The works of mitigation include landscaping activities such as the planting of trees, shrubs or plants of any other description and the laying out of any area as grassland.[206] Indeed, the authority may develop or redevelop the land for the purpose of improving the surroundings of public works in any manner which they think desirable by reason of the construction, alteration, existence or use of the works;[207] and can also dispose of any land acquired by them under Article 27.[208]

3. Power to pay expenses of persons moving temporarily during construction works.

Article 29 of the 1973 Order provides that if the enjoyment of a dwelling adjacent to a site where any public works construction or alteration (other than a road) are being carried out is affected to the extent that continued occupation is not reasonably practicable,[209] the acquiring authorities may pay reasonable expenses incurred by the occupier in providing alternative rental accommodation for the whole or any part of the period.[210] The appropriate authority must agree on the need to seek alternative accommodation and the amount entitled is the amount by which the claimant's expenses in the temporary accommodation exceed what their expenses would have been had they continued to occupy the original property. Moreover, the agreement of the authority must be received before any expense is incurred.[211] Nonetheless, the operation of this provision is clearly narrow and restricted by a number of conditions: (1) it relates only to the responsible authority; (2) there is no statutory duty to discharge expenses; (3) only adjacent properties qualify; and (4) agreement is required prior to expenses being claimed.

4. Compensation for the loss of some right in land due to the execution of public works.

   Compensation is payable in some (notoriously limited) circumstances due to the execution of public works interfering with a right in land, such as a right of way, without actually interfering with the land itself. The compensation allows for recompense where an activity would have been actionable in law, but for the statutory authorisation. Proving such a case is notoriously difficult, requiring to satisfy the so called 'McCarthy Rules': that it relates to a physical interference with a right in land, that such action would be actionable in law save for the statutory power and that it relates strictly to the *execution* rather than the *use* of the works. Compensation exists only for losses to the value of land and not to any other losses, such as business revenue. It is commonly agreed that the existence of such specific circumstances are difficult to prove.

Additional compensation payments

Traditionally, the Land Acquisition and Compensation (Northern Ireland) Order 1973, as amended by the Home Loss Payments (Northern Ireland) Order 1992,[212] makes provisions for benefit of persons displaced from land. Under these articles of the 1973 Order and, subject to qualifying conditions and a minima and maxima range, there is provision for entitlement to home loss payments, farm loss payments, disturbance payments, and payments for certain housing executive tenants and for rehousing. More recently, the Land Acquisition and Compensation (Amendment) Act (Northern Ireland) 2016 ('the 2016 Act') was introduced to provide for addition payments for loss following the compulsory acquisition of land and, therefore, provide parity for land and property owners in Northern Ireland at the enhanced compensation amounts available in England and Wales. In addition, the provisions of the 2016 Act adumbrate supplementary payments for loss payments[213] and the owner will generally be paid reasonable expenses including his legal costs and other professional fees incurred in preparing a claim for compensation or moving property or purchasing a new property.

1. Basic Loss Payment

   Section 1 of the 2016 Act inserts a mechanism for basic loss payment.[214] This clause provides for the payment of basic loss to a person who has a qualifying interest in the land[215] and whose land has been acquired compulsorily[216] to the extent that he is not entitled to a home loss payment in respect of any part of the interest.[217] The enhanced payment is limited to whichever is the lower of either 7.5 percent of the value of his interest[218] or £75,000.[219] Also, where a home loss payment has been made, this amount must be deducted from the basic loss entitlement, to avoid double payment.[220] Where the compensation is being paid on the basis of equivalent reinstatement (Rule 5 of Article 6(1) Land Compensation (Northern Ireland) Order 1982) the value of the interest is nil and a payment is not due. Any dispute as to the amount of a payment under this provision must be determined by the Lands Tribunal.[221] A basic loss payment is also payable where an interest in land is acquired in consequence of the service of a Purchase Notice or Blight Notice.[222]

2. Occupier's loss payment for agricultural or other land

   Section 2 of the 2016 Act introduced provision for occupier's loss payment for agricultural and other land.[223] Article 33B provides for occupiers loss payment for agricultural land and applies to a person if he has a qualifying interest in the land,[224] the land is agricultural land,[225] the interest has been acquired compulsorily[226] and he has occupied the land for the period specified.[227] A person to whom this article applies is entitled to a payment of 2.5 percent of the value of his interest,[228] the land amount[229] and the building amount,[230] subject to a maximum payment of £25,000.[231] Article 33C relates to other land. This applies to a person if he has a qualifying interest in the land,[232] the land is not agricultural land,[233] the interest has been acquired compulsorily[234] and he has occupied the land for the period specified.[235] A person to whom this article applies is entitled to a payment of 2.5 percent of the value of his interest,[236] the land amount[237] and the building amount,[238] subject to a maximum payment of £25,000.[239]

3. Repeal of the farm loss payments.

   Section 6 of the 2016 Act repealed provisions for farm loss payment.[240] For both payment types, one aspect of the payment that remains open to interpretation is whether the percentage of the value of interest payable for the acquisition, up to the maximum value, applies to the value of the land or to the total compensation that comprises the value of land and any disturbance compensation. In *Hughes v Doncaster Metropolitan Borough Council*[241] the Law Lords appear to indicate that compensation for the acquisition of an interest in land includes all elements of the claim and, therefore, one amount, even though it consists of both the value of the land and disturbance. In contrast, it could be argued that the term 'value of his interest' relates to value in the context of 'market value' and is determined pursuant to rule 2 of Article 6(1) Land Compensation (Northern Ireland) Order 1982.

Conclusion

This chapter has provided a broad overview of the application of the core procedural and compensatory principles of compulsory acquisition in Northern Ireland. Compulsory acquisition and compensation law is a complex, but increasingly specialist and relevant area of law. Much of the legal theory, case law authority and vagaries of day-to-day practice of compensation are broadly similar to the situation in Great Britain. However, there are significant differences due to a different legal structure and a different development track over recent decades. This has created a situation whereby affected parties in Northern Ireland may be viewed as being materially disadvantaged in comparison to other UK residents in terms of compensation in some key areas. Whilst this has been acknowledged, and although it is to be welcomed that some of these issues have recently been the subject of reform, there still remain key differences in legal form, terminology, process and potential outcomes that may constitute the basis of future reform efforts to bring much needed clarity and consistency. Indeed, such differences need to be borne in mind by politicians, public officials and practitioners when considering the implications of both private and public investment and development activity that may require compulsory land assembly to achieve its aims.

Notes

1 *Pine Valley Developments Ltd and others v Ireland* App 12742/87. Series A No 222 (1992) 14 EHRR 319. By contrast, a driving license is not a possession: App 9177/80, *X v Federal Republic of Germany*, Decision of 6 October 1981, (1982) 26 DR 255.

2 An exception is provided by Article (1A): *The Department shall not acquire any estate in Crown land unless—(a) it is an estate which is for the time being held otherwise than by or on behalf of the Crown; and (b) the appropriate authority consents to the acquisition.*

3 Additionally, by virtue of Article 87(7) the Department may appropriate land originally acquired under other powers and/or for other purposes for these (planning) purposes and, conversely, by virtue of Article 89, may appropriate land acquired for these (planning) purposes 'for any purpose for which it is or may be authorised in any capacity to acquire land under any transferred provision'.

4 Article 87(5).

5 Schedule 6, in essence, provides a common procedural framework for compulsory purchase in Northern Ireland. It is specifically incorporated into (and specifically amended by) a range of legislation including, in the present context, the Planning (Northern Ireland) Order 1991.

6 Guy Roots et al., *The Law of Compulsory Purchase* (3 ed., Bloomsbury Professional 2018).

7 Paragraph 2 (as amended).

8 1972 Act, sch. 6, para. 4; 1991 Order, sch. 2, para. 6.

9 Paragraph 5(1)(a). Such notice must state that the Vesting Order has been made and name a place where a copy of the Vesting Order and of any map or plan referred to in it may be seen at all reasonable hours.

10 Schedule 2, para 7.

11 Incidentally, the respondent Department had written to the applicant (and relevant others) indicating that the Vesting Order had been made on 7 November 2003, that the Order would be first published on 11 November, would be published for a second time on 18 November and would become operative on 18 December. While the applicant had cited this as evidence of the respondent's belief that the second publication was the relevant date from which the time limit began, the courts dismissed this argument, noting that this was a question of law and not one to be determined by the respondent.

12 At 22.

13 [2004] NIQB 32.

14 Either generally or insofar as it affects any property of the applicant, until final determination of the proceedings.

15 Either generally or insofar as it affects any property of the applicant.

16 And similar powers.

17 *Re Bowden*, para 59, per Weatherup J.

18 [2006] NIQB 89.

19 Para 18.

20 [2006] NIQB 89.

21 Para 18.

22 Para 23.

23 Para 24.

24 1972 Act, sch. 6, para. 5(1)(b).

25 1972 Act, sch. 6, para. 5(1)(c).
26 Naming a place where a copy of the order and any relevant map or plan can be inspected.
27 1972 Act, sch. 6, para. 6(1).
28 1972 Act, sch. 6, para. 6(3).
29 Article 91.
30 Article 90(1).
31 Article 90(2).
32 Sections 191–197 reenact Articles 94–99 of the Planning (Northern Ireland) Order 1991. Where the land in question is Crown land then a Purchase Notice can only be served as provided in section 192.
33 Section 250 provides that: '"*owner*" in relation to land, means a person, other than a mortgagee not in possession, who, whether in his or her own right or as trustee for any other person, is entitled to receive the rack rent of the land, or where the land is not let at a rack rent, would be so entitled if it were so let'. For a discussion, see J.A. Dowling, *Planning Law in Northern Ireland* (Gill & MacMillan 1995, 257–58).
34 Section 191(4) and Article 30 of the Planning (General Development Procedure) Order (Northern Ireland) 2015. An exception is where a Repairs Notice has been served under section 202(4). This situation is governed by section 191(7).
35 Section 191(1).
36 Section 191(8) provides that 'the land' in this context means 'the building in respect of which listed building consent has been refused, or granted subject to conditions, or revoked or modified by the imposition of conditions, and in respect of which its owner serves a Notice under subsection (2), together with any land comprising the building, or contiguous or adjacent to it, and owned with it, being land as to which the owner claims that its use is substantially inseparable from that of the building and that it ought to be treated, together with the building, as a single holding'.
37 Section 191(2).
38 Section 191(3).
39 (1983) 46 P. & C.R. 1.
40 Section 180 of the Town and Country Planning Act 1971.
41 (1983) 46 P. & C.R. 1 at 6.
42 (1983) 46 P. & C.R. 1 at 6.
43 In *Purbeck* the fact that relevant land had always been marshland was irrelevant.
44 [1986] JPL 123.
45 [1986] JPL 123 at 127.
46 Relating to duration of planning permission.
47 Relating to duration of outline planning permission.
48 Relating to the duration of listed building consent.
49 Development that is not 'new'. This reflects the historical origin of modern Purchase Notice concept: in the aftermath of Second World War the owners of damaged buildings obviously could not make beneficial use of them in their bombed-out state but, equally, due to development plans for the area, might well be refused permission to rebuild.
50 [1985] NI 195.
51 Ibid.
52 [1960] 1 WLR 587.
53 (1969) 21 P & CR 545.
54 Now contained in Ministry of Housing, Communities and Local Government, *Guidance on Compulsory Purchase Process and the Crichel Down Rules* (2019), pages 131-132.
55 (1999) 77 P. & C.R. 190 at 197.
56 Ibid., 193.
57 (1982) 44 P & CR 289.
58 Section 183(3) of the Town and Country Planning Act 1971.
59 Section 193(1). Section 193(2) provides that in cases where an appeal under section, 58, 96, 143 or 159 is pending on the date on which the purchase notice is served, or is made within two months of that date, then the council must respond within two months of the date on which the appeal is disposed of.

60 Even where the council accepts that the land is incapable of reasonably beneficial use, section 194 allows the council to serve a counter-notice where land has a restricted use by virtue of being part of a larger area in respect of which a previous planning permission exists (and has not been revoked) and which contemplated that the land should remain undeveloped, preserved or laid out as amenity land to the remainder.
61 Section 193(1).
62 Section 195. Reasons not specified in the counter-notice will not be considered: *Allen v Department of the Environment for Northern Ireland* [1985] NI 195; *Essex CC v Essex Incorporated Congregational Church Union* [1963] 1 All ER 326.
63 Section 196(4). The person who has served the Purchase Notice may withdraw it at any point up until the amount to be paid has been agreed with the council or determined by the Lands Tribunal or at any time before the end of six weeks from the date of that agreement or determination. The effect of such a withdrawal is that any contract is deemed not to have been made. For a discussion of withdrawal in the context of the relevant English legislation, see *White v Herefordshire Council* [2007] EWCA Civ 1204 in which the Court of Appeal had to consider the effect of the council being served, somewhat bizarrely, with a series of four Purchase Notices by the respondent.
64 Section 196(1). Section 196(5) provides that the date of completion 'shall, unless the owner of the estate and the council otherwise agree, be 3 months from the date on which they agree upon the amount to be paid for that estate or, in default of agreement on that amount, from the date on which the Lands Tribunal determines that amount'. On this date the council is liable to pay full compensation and, until this is paid, is liable to pay interest as governed by section 196(6).
65 Section 196(2). Section 197 provides special provisions where, by virtue of section 26 of the Land Development Values (Compensation) Act (Northern Ireland) 1965, compensation is payable in respect of expenditure incurred in carrying out any work on land.
66 As amended: The Planning (Amendment) (Northern Ireland) Order 1982.
67 See Article 3(1)(a-l) for the detailed list of blighted land.
68 Article 4(1)(a).
69 Article 4(1)(b)(i).
70 Article 4(1) (b)(ii)(aa).
71 Article 4(1) (b)(ii)(bb).
72 See Article 12 which provides that a mortgagee has the right to serve a Blight Notice where he is entitled to exercise his powers to sell an interest in a hereditament, can give immediate vacant possession of the land and has been unable to sell the interest concerned except at a price substantially lower than he might reasonably be expected to achieve had the land not been within the circumstances of blight.
73 See Article 14, which states that a personal representative of a deceased owner-occupier has a similar right to serve a Blight Notice and must demonstrate that he has been unable to sell except at a substantially lower price than he might reasonably be expected to obtain.
74 Article 4(4)(a); Article 4(5).
75 Article 4(4)(b).
76 For example, see: *Aardvark SRE Ltd v Sedgefield BC* [2008] (Unreported 28/2/2008-BNO/205/2006).
77 Article 5.
78 A Blight Notice must be served in writing and state the interest in land and statutory ground pursuant to the Planning Blight Regulations (Northern Ireland) 1989.
79 Article 5(1)(b).
80 Article 5(1)(c).
81 (R/9/1977).
82 (R/7/1977).
83 (2007) EWHC 292.
84 See: *Perkins v West Wiltshire District Council* [1976] 1 EGLR 182.
85 Article 5(2).
86 See: *Binns v Secretary of State for Transport* [1986] 2 EGLR 207.
87 See: *Lake v Cheshire County Council* [1976] 32 P&CR 143, LT.
88 Article 8(1).
89 Article 8(2).

90 Article 8(3). Note that under Article 8(3A), a claimant may withdraw a Blight Notice at any time before the amount to be paid for the interest of the claimant has been agreed with the appropriate authority or determined by the Lands Tribunal or at any time before the end of six weeks beginning with the date on which that amount is so agreed or determined; and where a Blight Notice is withdrawn by virtue of this paragraph any contract deemed to have been made in consequence thereof shall be deemed not to have been made. A Blight Notice can also be withdrawn where a claimant sells the property: For intuitive authority see, *Carrel v London Underground* [1995]; *Bennet v Wakefield Metropolitan District* [1997] 37 RVR 32.
91 See: *Budgen v Secretary of State for Wales* [1985] 2 EGLR 203.
92 Article 6(1).
93 Article 6(4).
94 Article 6(2)(a).
95 Article 6(2)(b).
96 Article 6(2)(c).
97 Article 6(2)(d).
98 Article 6(2)(e).
99 Article 6(2)(f).
100 Article 6(2)(g).
101 Article 6(3) provides that an objection may not be made under paragraph (2) (g) if it could be made under paragraph (2)(b).
102 Article 7(1).
103 Article 7(2). See *Allen v Department of the Environment for Northern Ireland* [1985] NI 195; *Essex CC v Essex Incorporated Congregational Church Union* [1963] 1 All ER 326.
104 Ibid. The onus is on the claimant (Article 7(2)] to demonstrate that he has made reasonable endeavours to sell and that, in consequence of the proposed scheme, they have been unable to sell their interest except for a substantially lower price than what may be expected if there were no scheme.
105 Article 7(3).
106 Article 7(5).
107 Article 6 (2) (c) where the appropriate authority proposes to acquire compulsorily part of the hereditament, or (in the case of an agricultural unit) a part of the affected area specified in the counter-notice, but (unless compelled to do so by virtue of this order) does not propose to acquire compulsorily any other part of that hereditament or area.
108 Article 7(6).
109 Article 7(4).
110 Article 7(7).
111 R/10/2011.
112 R/21/2011; R/22/2011; R/23/2011.
113 In respect of four qualifying properties.
114 For discussion of the fundamental effects of the changes of the 1919 Act, see *Horn v Sunderland Corporation* [1941] 2 KB 26.
115 *Waters and others v Welsh Development* [2004] 1WLR 1304, per Carnwath LJ.
116 See for example: *Spirerose Limited v Transport for London* [2009] 1 WLR 1797.
117 The Law Commission, *Towards A Compulsory Purchase Code: (1) Compensation A Consultative Report* (Consultation Paper No 165 London: TSO, 2002).
118 Article 6(1) – Rules (1)–(6).
119 Rule (1) No allowance shall be made on account of the acquisition being compulsory.
120 See: *Horn v Sunderland Corporation* [1941] 2 KB 26, where Scott LJ stated 'The principle of equivalence … is at the root of statutory compensation, which lays it down that the owner shall be paid neither less nor more than his loss'.
121 [1995] 2 AC 111.
122 *Stebbing v Metropolitan Board of Works* [1870] LR 6 QB 37.
123 In this context a willing seller does not mean a person selling without reserve for any price he can obtain: See *Inland Revenue Commissioners v Clay & Buchanan* [1914] 3 KB 466. The rule does not mention the requirement of a willing buyer. This is because Rule 2 assumes a sale, which means a willing buyer is implied.
124 (the Indian case) [1939] AC 302.
125 Ibid.

126 *Lady Fox's Executors v Commissioners for Inland Revenue* [1994] 2 EGLR 185, per Hoffman LJ.
127 See *Trochette Property Co. v Greater London Council* [1974] 28 P&CR 408.
128 See *Inland Revenue Commissioners v Clay & Buchanan* [1914] 3KB 466.
129 See *Duke of Buccleuch v Inland Revenue* [1967] 1 AC 506.
130 [1947] AC 565.
131 See *Waters v Welsh Development Agency* [2004] 1WLR 1304.
132 *R v Brown* [1867] LR2 QB 630.
133 *Corrin v Northampton Borough Council* [1980] 1 EGLR 148.
134 Article 13 1973 Order.
135 See: Sections 110-111 of the Land Clauses Consolidation Act 1845.
136 (2009) R/49/2009.
137 (2013) R/37/2010.
138 R/16/2014.
139 R/17/2014.
140 Paragraph 11 of Schedule 6 of the Local Government Act (Northern Ireland) 1972 gives the Lands Tribunal the statutory authority to determine a dispute in relation to compensation arising from the operation of a Vesting Order.
141 Section 111 'Deposit of such sum when refused on tender' provides:

'If upon such payment or tender as aforesaid being made any such mortgagee fail so to convey his interest in such mortgage, or to adduce a good title thereto to the satisfaction of the promoters of the undertaking, it shall be lawful for them to pay into the Supreme Court the amount of such value or compensation and the making of payment to the mortgagee or into the Supreme Court shall be accepted by the mortgagee in satisfaction of his mortgage debt, so far as the same will extend, and shall be a full discharge of such mortgaged lands from all money due thereon; and it shall be lawful for the promoters of the undertaking, if they think fit, to execute a deed poll, duly stamped, in the manner herein-before provided in the case of the purchase of lands by them; and thereupon such lands, as to all such estate and interest as were then vested in the mortgagee, or any person in trust for him, shall become absolutely vested in the promoters of the undertaking, and they shall be entitled to immediate possession thereof in case such mortgagee were himself entitled to such possession; nevertheless all rights and remedies possessed by the mortgagee against the mortgagor, by virtue of any bond or covenant or other obligation, other than the right to such lands, shall remain in force in respect of so much of the mortgage debt as shall not have been satisfied by payment to the mortgagee or into the Supreme Court'.

142 *Holt v Gas, Light and Coke Company* [1872] 8 LR 728.
143 *Duke of Buccleuch v Metropolitan Board of Works* [1872] LR 5 HL 418.
144 [1889] HL, All ER Rep 901.
145 [1916] 1 AC 536.
146 *Holt v Gas, Light and Coke Company* [1872] 8 LR 728.
147 *Oppenheimer v Minister of Transport* [1942] KB 242.
148 *Horn v Sunderland Corporation* [1941] 2 KB 26; *BP Petroleum Developments Ltd v Ryder* [1987] 2 EGLR 233.
149 *Hughes v Doncaster Metropolitan Borough Council* [1991] 1 All ER 295.
150 *Abbey v Homesteads Group Ltd v Secretary of State for Transport* [1982] where support for the view that compensation should be assessed as a separate exercise. See also: *South Eastern Railway v London County Council* [1915]; *Hoveringham Gravels Ltd v Chiltern District Council* [1977] 2 EGLR 17.
151 *Hughes v Doncaster Metropolitan Borough Council* [1991] 1 All ER 295.
152 Article 54 Land Acquisition and Compensation (Northern Ireland) Order 1973.
153 *Ravenseft Properties Ltd v Hillingdon London Borough Council* [1986] EGD 315.
154 Article 54.
155 *Ravenseft Properties Ltd v Hillingdon London Borough Council* [1986] EGD 315.
156 R/10/2011 – See Blight Notice above for fuller discussion of the case. See also *Smith v Kent County Council* [1995] 2 EGLR 196.
157 Sections 1–3, revoking relevant provisions of the Land Development Values (Compensation) Act (Northern Ireland) 1965 and the Planning (Northern Ireland) Order 1972.
158 (1989) 60 DR 128.

159 (1997) 23 EHRR 101 at [75].
160 App 12742/87. Series A No 222 (1992) 14 EHRR 319.
161 Section 26(1). Section 26(2) provides that 'any expenditure incurred in the preparation of plans for the purposes of any work, or upon other similar matters preparatory thereto, shall be taken to be included in the expenditure incurred in carrying out that work'.
162 Section 26(3).
163 Section 188.
164 Section 180 of the Planning Act (Northern Ireland) 2011.
165 There are no special provisions relating to the winning or working of minerals in this context.
166 Section 182(3).
167 Section 182(3).
168 Section 182(6).
169 Where the order in question is made under section 73 or 75.
170 Section 182(4). Under section 182(7) any dispute about compensation shall be determined by the Lands Tribunal.
171 Under Section 183(3) any dispute about compensation shall be determined by the Lands Tribunal unless otherwise provided for by the Tree Preservation Order.
172 Section 185(7). Where the Notice was served by the Department under section 151 claims are made to the council deemed to have served it under that section.
173 Section 185(2).
174 Section 187(1)(a). Section 187(2) provides that this does not apply if the planning permission is granted on/after the date on which a copy of the Notice is first displayed as specified in section 135(5).
175 Section 187(1)(b).
176 Section 187(1)(c). Section 187(3) provides that this does not apply if the Notice is withdrawn following the grant of planning permission.
177 Section 185(1) and (4) (Stop Notices) and section 187(4) and (6) (Temporary Stop Notices).
178 Section 185(5) (Stop Notices) and section 187(7) (Temporary Stop Notices). Under section 185(6) and section 187(8) any dispute about compensation shall be determined by the Lands Tribunal.
179 Section 186.
180 Section 186(2).
181 Section 186(3).
182 Articles 4–21.
183 Article 4(3) provides that public works are any road; any aerodrome; and any works or land (not being a road or aerodrome) provided or used in the exercise of statutory powers.
184 See *Bannocks v Secretary of State for Transport* [1995] 2 EGLR 157.
185 Article 4(1).
186 See for example: *Blower v Suffolk County Council* [1994] 2 EGLR 204 where glare from streetlights was a physical factor. An example of a factor that is not compensatable is the loss of a view.
187 Article 4(5).
188 *Williamson v Cumbria County Council* [1994] 2 EGLR 206.
189 Article 5(1).
190 Article 4(9)(a).
191 Article 4(9)(b).
192 Article 5(4). If a tenant has less than three years unexpired, and therefore ineligible to claim, he may be regarded as having an owners interest where, before the relevant date, the tenant had served notice on his landlord to have the freehold or an extended lease under the Leasehold (Enlargement and Extension) Act (Northern Ireland) 1971.
193 Article 5(2).
194 See *Allen v Department of Transport* [1994]; *Nesbitt v National Assembly for Wales* [2003] RVR 302.
195 Article 5(5).
196 Article 6(1). A claim must be particularised: See *Methodist Church Purposes Trustees v North Tyneside Metropolitan Borough Council* (1979) 38 P & CR 665.
197 Article 6(3). See *Donaldson v Hereford and Worcester County Council* [1997] RVR 242.
198 Article 7(1). In *Davis v Mid-Glamorgan County Council* [1979] 2 EGLR 158 the altered first use of the airport occurred in three phases and the Tribunal decided that the claim could be made once all phases were complete as it was a single scheme for improvement. However, in *Price v Caerphilly*

County Borough Council [2004] 1 EGLR 157 this scheme concept was not applied to a road that had been opened in stages as the relevant highway was the section or phase of a road, not the whole road.

199 Article 7(2). However, there may be justification for assuming increased traffic flows where a new section or link is completed in the future. See *Dhenin v Department of Transport* [1990] 60 P&CR 349.

200 Article 7(3).

201 Article 9(1).

202 203 N.B. 'public works' does not include a road or any works forming part of the works of a statutory undertaker as defined in Article 2(2) of the Planning (Northern Ireland) Order 1972, as amended.

203 Taking references to the date of service of a Blight Notice under Article 5 of that order as references to the date on which the purchase agreement is made.

204 Article 27(3). The acquisition of any land shall be treated as 'begun' when the agreement for its acquisition is made: Article 27(4).

205 Article 28(1)(a)–(b).

206 Article 28(2).

207 Article 28(3)(a).

208 Article 28(3)(b).

209 Article 29(1).

210 Article 29(2).

211 Article 29(3).

212 See Articles 30–44 of the Home Loss Payments (Northern Ireland) Order 1992.

213 See: Section 3 of the Land Acquisition and Compensation (Amendment) Act (Northern Ireland) 2016.

214 Inserting Article 33A into Article 33 of the Land Acquisition and Compensation (Northern Ireland) Order 1973. This provision is analogous to Section 106 of the 2004 Act providing for basic loss payment.

215 Article 33A(1)(a) See Article 33A(4) which provides that 'An interest in land is a qualifying interest if it is a freehold interest (legal or equitable) or an interest under a tenancy and (in either case) it subsists for a period of not less than one year ending with whichever is the earlier of – (a) the date on which the Vesting Order is operative; (b) the date on which compensation is agreed between the person and the acquiring authority'.

216 Article 33A(1)(b).

217 Article 33A(1)(c).

218 Article 33A(2)(a).

219 Article 33A(2)(b).

220 Article 33A(6).

221 Section 3 of the Land Acquisition and Compensation (Amendment) Act (Northern Ireland) 2016, inserting Article 33G(1).

222 Section 3 of the Land Acquisition and Compensation (Amendment) Act (Northern Ireland) 2016, inserting Article 33H(2), where the interest is in consequence of the service of a Purchase Notice pursuant to Section 191 of the Planning Act (Northern Ireland) 2011 (right to require purchase of certain interests) or a Blight Notice under Article 5 of the Planning Blight (Compensation) (Northern Ireland) Order 1981 (purchase of blighted land).

223 Inserting Articles 33B and 33C into Article 33 of the Land Acquisition and Compensation (Northern Ireland) Order 1973. This provision is similar to Section 107 of the 2004 Act which inserted a new Section 33B providing for occupier's loss payment in respect of agricultural land and a new Section 33C into the 1973 Act providing for occupier's loss payment in respect of other land.

224 Article 33B(1)(a). 'Qualifying interest' is the same for the purposes of Article 33A.

225 Article 33B(1)(b).

226 Article 33B(1)(c).

227 Article 33B(1)(d). The period is specified in Article 33A(4).

228 Article 33B(2)(a).

229 Article 33B(2)(b).

230 Article 33B(2)(c).

231 Article 33B(3).
232 Article 33C(1)(a). 'Qualifying interest' is the same for the purposes of Article 33A.
233 Article 33C(1)(b).
234 Article 33C(1)(c).
235 Article 33C(1)(d). The period is specified in Article 33A(4).
236 Article 33C(2)(a).
237 Article 33C(2)(b).
238 Article 33C(2)(c).
239 Article 33C(3).
240 Repealing Article 2(2); Articles 34–36 of the Land Acquisition and Compensation (Northern Ireland) Order 1973.
241 [1991] 1 AC 382.

17

LIQUOR LICENSING LAW AND PLANNING

Michael Murray

Introduction

Licensing authorities across the United Kingdom exercise a large number of regulatory functions covering a wide range of activities that may include the sale of alcohol, entertainment, gambling, amusement arcades, taxi services, adult shops, street trading and houses in multiple occupation. This oversight is implemented for the wellbeing of the general public, say as consumers and residents, and seeks to prevent nuisance, crime and disorder. From a planning perspective, due consideration is given to the protection of environmental amenity. In all these matters local government plays a key role, but within Northern Ireland the courts exercise distinct decision making in regard to the opening of new premises for the sale of liquor and for bookmaking. Judicial involvement removes this aspect of land use regulation away from the moral scruples of elected representatives and their constituents and, when licence applications are objected to by those with *locus standi*, the arrangements provide for a rigorous adversarial examination of merits.

Northern Ireland has its own legislative provisions relating to the sale of liquor for consumption on or off the premises that involve both planning and licensing statutes. Kerr J *In the Matter of an Application by Philip Russell Limited for Provisional Grant of a Licence* (1993) has cogently set out that context: 'In this field of law judicial interpretation and application of the various statutes and associated regulations have played an important part in developing a comprehensive and cohesive code which governs the supply of intoxicating liquor to the public at large'. This chapter is primarily concerned with these matters and, by drawing on a considerable corpus of case law, gives particular attention to the involvement of the planner in assisting the courts in regard to the legal tests of vicinity and inadequacy. Put very simply, the relevant planning practice-related hurdle to be overcome by an applicant for a new liquor licence is evidential proof that the number of licences in the vicinity of the application premises is inadequate. To provide context, the narrative in this chapter considers, first, the changing character of supply and demand within the wider setting of evolving public policy. This is followed, second, by an examination of the interaction between planning and the liquor licensing statute in regard to the opening of new public houses and off-licences; particular

DOI: 10.4324/9781003218319-17

attention is given to the need for planning permission. Third, the chapter addresses the role of the planner as an expert witness within the court arenas and identifies significant principles derived from case law that are shaping the spatial distribution of facilities for the sale of alcohol within this jurisdiction. Litigation is central to the formation of this case law and, in the final section, the scope for objecting to licence applications is addressed.

The changing character of supply and demand

While the history of liquor licensing in Ireland can be traced as far back as the 17th century with the introduction of the Alehouses Act of 1635, significant attempts to control the problems of the increasing number of liquor outlets gained momentum during the 19th century. As recorded by McBrien (1997: 6) the Licensing (Ireland) Act 1833 provided the foundation of Irish licensing law and included a provision that new licences could be refused on the ground of the number of previously licensed premises in the neighbourhood. The outworking of that legislation needs to be set against the spilling over of the influence of the social reformers in Britain regarding health and sanitation during these years of city improvement (see Cherry 1988), along with the more particular activities of an emergent temperance movement in Ireland at that time (see Malcolm 1982). Restricting the grant of new licences was carried forward in the subsequent Licensing (Ireland) Act 1902, though discretion was given to licensing authorities to approve new licences where urban growth could be demonstrated and the location of the premises directly related to that population increase. Following partition, the Northern Ireland Parliament reenacted the provisions of the 1902 Act in the form of the Intoxicating Liquor Act (Northern Ireland) 1923 with some modifications (McBrien 1997). The mixed trading carried out by the then spirit-grocer was ended and these businesses were given the option of selling either alcohol or groceries but not both. Moreover, in an attempt to reduce the number of outlets, two existing licences had to be surrendered for each new one. The population growth criterion was retained as evidential proof of need and interestingly, as will be noted later in this chapter, it continues to inform the contemporary granting of new licences albeit the surrender requirement has changed to one subsisting licence.

From the early 1970s the presence of the off-licence shop on high streets gained momentum in line with amended licence surrender requirements for new outlets and changing consumer behaviour. The quest for regulatory review continued with the publication of the Blackburn Report on intoxicating liquor in Northern Ireland in 1979. It proposed four objectives impacting supply and demand as the basis for future legislation:

- to control the level and pattern of alcohol consumption so as to protect social welfare and the health of the community;
- to protect social amenity and the physical environment;
- to encourage a higher standard of licensed facilities; and
- to be fair to all sections of the community and the licensed trade.

The Licensing (Northern Ireland) Order 1987 was subsequently enacted on the back of that report with measures that were, in the main, operational in nature, for example, revised trading hours, guest accommodation in pubs and licences for places of public entertainment. The greater part of the parliamentary debate on the draft statute revolved around the deeply contested issue of the Sunday opening of public houses, related problems of enforcement tied to the availability of meals and an increasing number of clubs over which government had expressed a wish for tighter control. In successfully requesting the House of Commons to approve the Order, the then Parliamentary Under-Secretary of State for Northern Ireland Richard Needham commented:

> In 1985 the number of public houses in Northern Ireland stood at 1,832 and that is 39 less than the number of public houses in 1978. The number of off-licences has increased from 200 to 250; the number of hotels has gone down from 177 to 165. I hope that the House will be pleased to learn that the number of restaurants has risen by 82 to 208. Registered clubs, with which we are concerned, have risen from 483 to 625.
>
> *(Licensing (Northern Ireland) Hansard, 14 July 1987)*

The dark backdrop of 'the Troubles' in Northern Ireland has resonance with this analysis, albeit set within the context of a society painfully transitioning towards new norms of social behaviour and an acceptance of greater pluralism. Changing attitudes to alcohol consumption and expectations around availability have been part of that transformation and paved the way for another major review of licensing arrangements over the period from 1992. This led to the enactment of the Licensing (Northern Ireland) Order 1996. It extended the categories of licensed premises to include guest houses, conference centres and university campuses, increased permitted trading hours and allowed qualified access by children in adult company. However, for purposes of this chapter, the significant amendment related to the introduction of mixed trading that has allowed for the sale of alcohol for consumption off the premises within food stores. This is in line with the expanding provision of outlets by national multiples and symbol group franchises, which seek to offer an alternative to the stand-alone off-licence shop and greater customer convenience. Again, during the debate in Parliament on the draft Order, concern was expressed that this deregulation measure could herald a return to the spirit-grocer 'on every street corner' (Licensing Law (Northern Ireland) Hansard, 7 November 1996). A robust response by Parliamentary Under-Secretary of State Malcolm Moss rejected that assertion:

> I wish to make it clear that that cannot happen, for two reasons. First, if a licence were to be obtained from the Court, need must be justified. In other words, there should be inadequate provision in a locality before a licence is granted. Secondly, and most important ... we have not removed the requirement of surrender. Accordingly, there is a finite number of licences to allow alcohol to be sold in Northern Ireland. That number cannot increase. Corner shops will be able to obtain licences to sell alcohol only if other licences elsewhere are surrendered. A licence is trading at about £40,000, and is used by many off licences as security at the bank for loans. That being so, I do not believe that what we propose will lead to many shops taking up licences to sell alcohol.
>
> *(Licensing Law (Northern Ireland) Hansard, 7 November 1996)*

The licence surrender principle is a facet of Northern Ireland liquor licensing law as distinct from Great Britain where this does not apply; emphasis there is placed on the evidence-led criterion of overprovision. This requirement has drawn judicial comment, when, in *Re application for the provisional grant of In/Off Licences; Strike Four (Belfast) Limited v Lofty Inns Limited and Another; McCann v Lofty Inns Limited and Others* (1998), Nicholson LJ opined:

> Under the legislation which governed the issue of licences before 1971 the licence or licences which had to be surrendered were required to be in the same geographical area as the licence which was sought. This requirement had significance. Since 1971 the licence surrendered may come from any part of Northern Ireland. An artificial trade has grown up in the sale of such licences. Legal points have been taken as to what is a 'subsisting licence'. This is attractive to lawyers and to opponents of applicants. But I fail to see any purpose in requiring the surrender of such a licence and I regret that the 1996 Order did not make some amendment of the licensing law, so that the surrender of such a licence played no effective part in the control of licensing.

In January 1995 the Minister with responsibility for the sector had advanced the proposition of dropping the surrender requirement in Northern Ireland, and notwithstanding the consequent deregulatory impact had this happened, it was argued: 'about £100 million was invested in licences and that banks were greatly concerned at the prospect of so much being wiped off the balance sheets of publicans and off-licence owners, to devastating commercial effect' (Licensing Law (Northern Ireland) Hansard, 7 November 1996). However, this was not the end of the matter. A further review of liquor licensing law in Northern Ireland commenced in 2004 to take account of progress towards peace in a society moving out of conflict, tourism demands and the growth of a new night-time economy in cities and towns (Department for Social Development 2005). On 20 July 2006 Minister of State at the Northern Ireland Office David Hanson announced in a Parliamentary written statement that the surrender principle would be abolished, subject to the findings of a business impact assessment and that, furthermore, in regard to liquor licensing governance, decision making for granting new licenses would be transferred from the courts to local government in line with the agenda of public administration reform. The matter of the surrender licence was subsequently debated by the Transitional Northern Ireland Assembly on 29 January 2007 and a motion was passed, calling upon the Minister with responsibility for the Department for Social Development to ensure that no action would be taken with regard to furthering the proposed abolition of the 'surrender principle' for liquor licences, as proposed by the Northern Ireland Liquor Licensing review; the Transitional Assembly further called for this issue to be dealt with by a restored Northern Ireland Assembly. The proposals under direct rule were not taken forward as responsibility for liquor licensing law passed to the Northern Ireland Executive following the restoration of devolved government on 8 May 2007. Then in what seems to have drawn a line under abolishing the surrender principle, Northern Ireland Executive Minister for Social Development Margaret Ritchie announced to the Assembly on 17 November 2008 that no action would be taken at the present time on this matter given the absence of a robust evidence base (Northern Ireland Assembly Research and Library Services 2009).

However, in contrast, the transfer of responsibility for new liquor licences to local government was strongly canvassed by this Minister. Anecdotally, there would seem to have been little enthusiasm at that time among commercial interests or at the local authority level for that to happen due to perceived costs of operation, legal challenges, personal ethics and an overloaded council reform programme (Institute of Licensing 2014). Most significantly, in a 2012 public consultation on 'Proposed changes to the law regulating the sale and supply of alcohol in Northern Ireland' the focus was not on reshaping liquor licensing governance, but rather on potentially modifying customer purchasing arrangements in off-licence facilities and public houses (see Department for Social Development 2013). This 2012 consultation did lead into a legislation agenda that reached Committee Stage but which was then abandoned when the Northern Ireland Assembly was dissolved in 2017. The review process was subsequently revisited in 2019 with a further public consultation that, in turn, led to the passage of the Licensing and Registration of Clubs (Amendment) Act (Northern Ireland) 2021. Essentially this new legislation seeks to modernise operating practices within the sector while having regard to societal protection obligations. But significantly it provides for new licensing arrangements to facilitate sales by local alcohol producers. It also promises an independent review of the licensing system to include the surrender principle discussed above.

It is against this evolving social, commercial and policy / legal context that consideration can now be given to changes in the profile of licensed premises in Northern Ireland. Table 17.1 summarises complete time series data available for the period 31 December 1997 to 2020 following the enactment of the Licensing (Northern Ireland) Order 1996. The data demonstrate a marked decline of some 25 percent in the number of public houses across Northern Ireland from 1,632 establishments in 1997 to 1,224 establishments by the end of 2020. Over the same period, off-licences have increased by almost 84 percent, albeit off a much smaller base, from 335 premises in 1997 to 615 premises by the end of 2020. The operation of the surrender principle is clearly in evidence in providing for the expansion of off-licence facilities. When these data are disaggregated by Petty Sessions Districts, there is evidence that both urban and rural areas have been shedding public house provision (Table 17.2). For example, the combined Petty Sessions Districts of East Tyrone, Fermanagh, Omagh and Strabane in the rural west of the region experienced a reduction of eighty-one public houses between 1997 and 2020 and an addition of fifty-one new off-licences over the same period, suggesting not only an 'export' of licences to other parts of Northern Ireland but also a shift in purchasing behaviour. Geographical factors are playing out in all rural areas of Northern Ireland related

TABLE 17.1 Liquor licences in force in Northern Ireland, 1997 and 2020

Category	*1997*	*2020*
Public houses	1,632	1,224
Off-licences	335	615
Hotels	152	159
Guest houses	14	30
Restaurants	345	614
Conference centres	1	32
Higher education institutions	0	21
Places of public entertainment	11	21
Refreshment rooms	13	7
Seamen's canteens	2	1
Non-seagoing vessels	2	4
Indoor arena	0	1
Outdoor stadia	0	2

Source: Department for Social Development, data received 23 June 2015; Department for Communities, data received 9 July 2021. (It is possible for the courts to renew a licence up to twelve months after it expires and annual figures may incorporate adjustments due to successful late applications.)

TABLE 17.2 Public houses and off-licences by petty sessions districts, 1997 and 2020

Petty session district	*1997 Public houses*	*2020 Public houses*	*1997 Off-licences*	*2020 Off-licences*
Antrim	41	29	9	18
Ards	77	59	30	46
Armagh	78	51	7	20
Ballymena	61	43	7	22
Banbridge	54	33	4	10
Belfast	225	168	110	132
Craigavon	49	36	11	33
Down	102	76	6	23
East Tyrone	131	99	13	29
Fermanagh	91	77	19	33
Larne	26	15	4	9
Limavady	44	32	7	15
Lisburn	58	39	21	34
Londonderry	100	88	23	42
Magherafelt	62	52	4	16
Newry and Mourne	129	93	19	37
North Antrim	121	94	17	41
North Down	33	25	15	25
Omagh	80	63	4	20
Strabane	70	52	5	10
TOTAL	1,632	1,224	335	615

Source: Department for Social Development, data received 23 June 2015; Department for Communities, data received 28 July 2021. (It is possible for the courts to renew a licence up to twelve months after it expires and annual figures may incorporate adjustments due to successful late applications.)

to a dispersed settlement pattern, a plethora of villages and small towns and customer accessibility linked to the patronage of public houses. The conversion of former high street pubs to shops or residential development, and the presence of derelict licensed premises in the countryside, are most emblematic of changed times (Photograph 17.1). A similar trend has been observed across rural areas within the United Kingdom as a whole and is but part of the wider loss of services in rural communities that are facing challenges related to economic viability, urban competition, quality of provision, changing patterns of personal consumption and variations in digital connectivity (see, for example, Pratten and Lovatt 2002; Wilson 2019). And within Northern Ireland the capital value of a liquor licence, as discussed above, in the face of reduced profitability or proprietor retirement adds considerably to the dynamic of change in both rural and urban areas.

Planning and the liquor licensing statute in Northern Ireland

The premises for which liquor licences may be granted are set out in the Licensing (Northern Ireland) Order 1996. Public houses are denoted as Article 5(1)(a) premises 'in which the business carried on under the licence is the business of selling intoxicating liquor by retail for consumption either in or off the premises'. Off-licences are referred to as Article 5(1)(b) premises 'in which the business carried on under the licence is the business of selling intoxicating

PHOTOGRAPH 17.1 Former pub and residence converted to shop and storage, Ballynahinch. *Source*: Author 2017.

liquor by retail for consumption off the premises'. Article 7 deals with the grant of licences and the following paragraphs are key to planning practice:

> 7. – (1) An application for the grant of a licence shall be made to a county court…
> – (4) A court shall refuse an application for the grant of a licence unless it is satisfied—
>
> (e) where the premises are of a kind mentioned in Article 5(1)(a) or (b)—
>
> (i) subject to paragraph (6), that the number of licensed premises of the kind specified in the application which are in the vicinity of the premises is, and having regard to any licences provisionally granted under Article 9 or any sites approved under Article 10 will be, inadequate; and…
>
> (f) either—
>
> (i) that there is in force planning permission to use the premises as premises of the kind specified in the application for the period during which the licence would be in force; or
>
> (ii) that the premises may be used as such premises for that period without such permission.
>
> *(The Licensing (Northern Ireland) Order 1996)*

Accordingly, formal planning permission may be an essential proof. For new build premises and materially significant physical alterations, obtaining planning consent needs to be demonstrated to the court. Stand-alone off-licences are regarded as Class A1 Shops (for the retail sale of goods) under the Planning (Use Classes) Order (Northern Ireland) 2015 and do not attract scrutiny regarding that particular use at development management stage; indeed, if the former established use for a proposed off-licence has been a shop, then the provisions of Article 7(4)(f)(ii) are relevant within the court proceedings. One consequence of this arrangement is that local objectors may become aware and will be able to pursue an objection only when an application to the court for an off-licence (either new build or change of use) has been advertised in a local paper that circulates within the area. In contrast, change of use developments related to public houses are sui generis within the Planning (Use Classes) Order (Northern Ireland) 2015 (for the sale of food or drink for consumption on the premises) and require the determination of planning applications against the provisions of the development plan and other material considerations (see, for example, *Appeal to the Commission by J D Wetherspoon PLC in respect of the non-determination of application for full planning permission for change of use from retail unit to a J D Wetherspoon traditional ale, wine and food bar at 23–24 The Diamond, Londonderry*, Planning Appeals Commission Reference: 1999/A061). Issues of noise, disturbance, loss of residential amenity and traffic safety are important and Development Control Advice Note 7 (Department of the Environment for Northern Ireland 1983) expands on these material considerations in regard to both new build and change of use public house development:

> The effects of noise and disturbance and traffic safety are the main planning considerations in relation to public houses. There may be a number of suitable locations provided that problems of noise can be minimised and other amenity and traffic considerations are met. For example, a suitable location for a public house may be on the edge of a residential estate, in an area of mixed land uses or within a neighbourhood shopping centre. A public house may also be suitable in a town centre area or, indeed, in a rural area, although regarding the latter location, the proposal would, of course, be subject to the Department's policy for the control of development in rural areas. However, a public house will not generally be acceptable within a wholly residential area. This is because of loss of amenity, including the problems of noise and disturbance which may arise, and also the possible traffic hazards associated with heavy lorries delivering supplies, and customers arriving and leaving by car in the latter part of the evening.
>
> *(paragraph 9)*

Under the new planning governance arrangements in Northern Ireland, where a council adopts its Plan Strategy, this existing guidance contained within DCAN 7 - Public Houses will cease to have effect in the district of that council and shall not be material from that date, whether a planning application has been received before or after that date. In short, however, prior to a court hearing in regard to an application for a new liquor licence, it is necessary that matters that rest within the remit of development management are addressed and that appropriate certificates of planning consent, if required, are available to be tendered as statutory proofs.

For completeness, it is worth noting that Article 6(1) of the licensing statute prohibits the granting of a liquor licence for premises situated in a service area within the meaning of the Roads (Northern Ireland) Order 1993 and for premises in which the principal business carried on is the business of a garage retailing petrol or derv, or the sale or maintenance of motor vehicles, or premises which form part of such premises. Important clarification on this latter prohibition is provided in *Re an application by Glenshane Tourist Services Limited for the grant of a licence* (1999) when Nicholson LJ in granting a restaurant licence, considered matters relating to main and ancillary uses:

> I do not consider that the restaurant comprises premises in which the business of a garage is carried on ... I consider that it is probably premises which form part of a garage business premises because the shop is essential to the running of the garage. But in my opinion the principal businesses carried on are those of a restaurant and a supermarket and the garage business is ancillary to the other businesses, albeit an important ancillary business ... (P)remises are to be refused a licence if they are ancillary to or where the principal business is a petrol filling station.

The planner as expert witness

Advising on planning applications and giving evidence in planning appeals comprise an important part of the workload of planning consultants operating within the licensing arena. But planners have also become central to the determination of liquor licence applications before the courts (on initial application before the County Court and on appeal before the High Court), where their role is that of expert witness. The key matters that planners seek to address in their evidence relate to vicinity and inadequacy as discussed below, but it is valuable to note that the standing of the professional planner in addressing these matters has been the subject of judicial comment:

> In my opinion the test of adequacy is an objective test, between the character of the area as a whole and having regard to the number of persons in that area, the number of existing licensed premises and indeed, in some instances, transient persons who may be entitled to consideration and the adequacy of the facilities being considered. It is in a sense the sort of test a planner would adopt in determining the potential requirements of an area. Examples can be seen in New Town and redevelopment areas where provision is made for the siting of certain types of premises having regard to the area and the proposed development.
>
> (McGonigal J in *Magill and Another v Bell and Others (1972)*)

Case complexity, however, may well require a multidisciplinary approach and thus it is not unusual, though not always necessary, for additional technical evidence to be presented to the court that encompasses traffic engineering and market surveys. In regard to the latter, applications for liquor licences in Northern Ireland have resorted, on occasion, to the use of shopper and local resident questionnaire surveys to help provide evidential proof on the matter of inadequacy. Judicial comment has also provided helpful insight on the added value or otherwise of that empirical work as, for example, in the case *Lidl (Northern Ireland) GmbH and Winemark the Wine Merchants Ltd* (2008), where Morgan J stated:

> Survey evidence in this type of case can be helpful in establishing patterns of behaviour and assisting in understanding the reasons for those ... As well as helping in relation to patterns of behaviour, surveys may also be of assistance in demonstrating that there is an unusual factor at play which it is necessary for the court to take into account. Where, however, the matter at issue concerns the way in which ordinary people are likely to make judgments about how to proceed with their everyday activities it seems to me unlikely that a survey will have much to contribute.

But looked at in the round, the planner's first task as expert witness is to analyse and set the scene, in line with the sentiments of McGonigal J above, and any additional evidence, as deemed appropriate, will seek to complement that perspective. Additionally, it has become an element of good practice in contested applications for the planning evidence of applicants and objectors to be shared in advance of the hearing and for a note on points of agreement and disagreement to be prepared for the court. Keegan J *On Appeal from the Decision of the County Court for the Division of Craigavon in the Matter of an Application for the Provisional Grant of an Off licence Pursuant to Article 5(1)(B) of the Licensing (Northern Ireland) Order 1996: Winemark (The Wine Merchants Limited) v Hagan (James Aidan), Hagan (Mary Bridget) and The Tunnel Bar Limited* (2016 NIQB 90) has commented on the value this arrangement:

> I also asked the parties at the outset whether or not they would consider sharing the expert evidence prior to hearing. I am grateful to both senior counsel for the practical approach they took in this matter. Counsel agreed to share expert reports and an expert meeting took place prior to the hearing. This approach accords with the sentiments expressed by Gillen J in the matter of an application by Sainsbury's [2012] NIQB 45 and it led to a more efficient use of court time. The expert meeting took place on 20 April 2016 and an agreed note of the meeting sets out the following under a number of headings: The Proposed Site … The Vicinity … The Demand … The Existing Facilities … This agreed minute was a useful template upon which the evidence progressed.

The matter of vicinity

The liquor licensing legislation circumscribes the granting of new public houses and off-licences by the geography of 'vicinity', and within contested cases it can often be the case that different vicinities are contended for by the parties. The statute does not define the meaning of vicinity and thus its interpretation has been left to the courts to deal with on a fact-sensitive, case-by-case basis. In bookmaking licence applications, the comparable geographical descriptor is 'locality' and again the onus has fallen on judicial opinion to give guidance. McGonigal J in *Magill and Another v Bell and Others* (1972) and later in *McFerran v Casey and Others* (unreported 1975) has held, for example, that vicinity is a smaller area than locality. McDermott LJ in *Re Samuel A McClean* (unreported 1994) has added to this by stating: 'In my opinion vicinity is not to be confused with locality which defines an area but means a physical proximity best indicated by the sense of neighbourhood'. These more fine-grained geographies of belonging are central to liquor licensing with the meaning and extent of neighbourhood examined closely by Carswell LJ in *Re Boots the Chemists Application* (1994):

> In my opinion the concept of a neighbourhood has the same limitation as that of a vicinity—its etymological parent—of which McGonigal J said in *Magill v Bell* (1972), 164 that it is limited to premises in the neighbourhood in the sense in which one speaks of being a neighbour of another. In *Cormican*'s case (1975, unreported) Judge JP Higgins QC expressed the principles as follows:
>
>> 'I think it is impossible to lay down any general rule as to the extent of the area indicated by the word "vicinity". In country districts people are said to be neighbours, that is to live in the same neighbourhood, who live many miles apart. The same cannot be said of dwellers in a city, where a single square may constitute a

> neighbourhood. Physical features may determine the boundary or boundaries of a neighbourhood as, for example, a river, a railway or a range of hills. In an urban area lacking such physical features the layout of the streets and the nature, character and use of the buildings need to be looked at, as well as the size and distribution of the population, whether residing or working in the area.'

I would perhaps qualify that in the way in which I expressed it in *Donnelly v Regency Hotel Ltd* (1985):

> 'I think that it is of importance to look both at the physical features of an area and any natural boundaries, and also at the established dwelling patterns and any geographical allegiances of those who live, work, or shop there. A vicinity accordingly seems to me to connote more than the area plotted on a map; its determination has to take into account the habits and movements of people in the area, and the directions in which those habits take them in the course of their daily lives.'

In my view two important conclusions should be drawn: (1) There is no sensible distinction between a vicinity and a neighbourhood. The authorities in the field of licensing law in which the proper extent of a vicinity are discussed are therefore relevant in the present context. (2) A neighbourhood is an area defined by physical and social factors, and its extent remains the same whatever the context in which one comes to consider it.

Gillen J has usefully drawn together relevant criteria that have helped shape the concept of vicinity and which have stood the test of legal authority over time. *In the matter of an application by Sainsbury's Supermarket LTD and Winemark the Winemerchants Limited, Wine Inns Limited, Phillip Russell Limited and Toby Inns Limited* (2012) he has stated:

> Vicinity is not limited to premises immediately surrounding the proposed premises. Whilst it is impossible to lay down any general rule as to the extent of the area indicated by the word vicinity, it is limited to the premises in the neighbourhood in the sense in which one speaks of being a neighbour of another. This is not to be confused with locality. It means a physical proximity best indicated by a sense of neighbourhood. People are within a neighbourhood if they or the neighbourhood are near enough to be affected in some way if the application is granted. In arriving at a fact sensitive conclusion in any particular case, it is necessary for the court to consider the following affective tributaries that flow into any definition of vicinity:
>
> - The physical features of an area e.g. a river, railway, range of hills, and layout of the streets, nature, character and use of buildings;
> - Any natural boundaries;
> - The size and distribution of the population;
> - Established dwelling patterns;
> - Geographical allegiances of those who live work or shop there;
> - The habits and movements of people in the area;
> - The direction in which these habits take them in the course of their daily lives;
> - One test is to stand at the applicant's premises and look out to determine the neighbourhood;

- The functional relationship and mixes of uses and whether these are consistent across the vicinity.

> In the context of this case I make it clear that whilst usually the concept of neighbourhood embraces residential estates, there is no reason why it inevitably must do so. I see no logical reason why a shopping centre cannot constitute a neighbourhood in the same way that that concept is recognised in pharmaceutical decisions…Equally, I see no reason why in appropriate circumstances vicinity cannot be defined by virtue of a road system network taking into account ease of movement and travel patterns. In short, the first step is logically to determine what licensed premises are in the vicinity of the proposed premises. One might approach that in either of two ways, by defining with more or less precision the area which one determines to be the vicinity or by looking at each of the other licensed premises near to the proposed premises and deciding if each counts as being in the vicinity of the proposed premises. I agree with the approach of Carswell J in *Donnelly's* case where he was inclined to the latter approach as containing the necessary flexibility and elasticity of the concept of vicinity.

The delineation of a vicinity in regard to an application for a liquor licence fits well with the townscape analysis training and expertise of planners. The seminal book by Kevin Lynch (1960) *The Image of the City* endures as a text in planning schools that facilitates this examination of urban morphology through a set of criteria comprising: (1) *paths*, the streets, pavements and other channels in which people travel; (2) *edges*, perceived boundaries such as walls, buildings and shorelines; (3) *districts*, relatively large sections of the city distinguished by some identity or character; (4) *nodes*, focal points, intersections or loci; and (5) *landmarks*, readily identifiable objects that serve as external reference points. In more prosaic terms, what the planner is seeking to contribute to the court by way of evidence is an appreciation of how local people understand and define their area that draws on these physical analysis factors. Judicial comment, however, has sought to temper overly prescriptive definitions of vicinity. In *F.A. Wellworth & Co Ltd v Phillip Russell Ltd and Others* (1997) Girvan J stated:

> There is an ever present danger that parties will seek to define vicinity by a rigid delineation of its boundaries and lose sight of the reality that in a large conurbation like Belfast a neighbourhood or vicinity can rarely be rigidly defined. An area peripheral to the heart of the centre of the vicinity may fall within more than one neighbourhood. The present case exemplifies the approach of experts seeking to define a vicinity by rigid lines on a map.

The principle advanced here is that part of any vicinity may well overlap with adjacent areas. A further observation by Nicholson LJ in *re Application for the provisional grant of In/Off Licences; Strike Four (Belfast) Limited v Lofty Inns Limited and Another; McCann v Lofty Inns Limited and Others* (1998) is that vicinities may not be rigidly fixed over time and that while a 'consistency of approach is helpful … times and circumstances change and what was valid in 1993 may not be valid in 1998'. As previously indicated, the identification of the vicinity and the relevant licensed premises within it (if any) constitutes the initial matter of concern to the court. The follow-on test comprises the extent to which the existing facilities within the vicinity of the application site are inadequate.

The matter of inadequacy

The concept of inadequacy has generated considerable explication arising from litigation in the County Court, High Court and Court of Appeal. The first question here may well be 'inadequate for whom?' and, in that regard, there is useful guidance provided by McDermott J in *Belfast Co-Operative Society v Tohill* (1975):

(1) I am concerned with the requirements of the public at large – not with one particular section of it. In granting this licence the learned County Court Judge in large measure based his decision on the fact that such a licence would be appreciated by and a benefit to the members of the Society. I am sure this is so but with all due respect I believe he erred in law in emphasising the requirements and benefits of the Society members as distinct from the public at large.
(2) In referring to the requirements of the public I have regard to the reasonable requirements of the public. In evidence it was said, and I accept, that at busy times, for instance, coming up to Christmas, a customer would have to wait to be served in either of the existing off licence premises: I am sure this problem is not peculiar to those premises and such evidence would not in my view render an adequate number of premises inadequate.
(3) The concept of the reasonable requirements of the public is not met simply by the existence of a number of off licence premises in a particular area. For the number of premises to be adequate the public must have reasonable opportunities of selection and purchase at competitive prices. For instance a chain of several off licence premises under one ownership with no competition as to price or selection might well not be an adequate number. It would be idle to conjecture as to all the matters which might have to be considered under the heading of reasonable requirements but in addition to considerations of choice and price, service and situation could be relevant matters.

In terms of numbers, while public houses with off-sales facilities would be taken into account in an off-licence application within the same vicinity, it would not seem appropriate that the converse should apply given the very different manner of consumption as between public houses and off-licences. This matter picks up on the phrase in the statute 'the number of licensed premises of the kind specified in the application' and was considered in *Hynes v McAlinden and Others* (1974) where the key issue revolved around whether or not a court could have regard to public houses with off-sales facilities in an off-licence application situation. In this instance the Court of Appeal held by a majority that public houses with off-sales facilities could indeed be taken into account. Lowry LCJ ruled:

> the question still has to be answered to what extent the number of on-licence premises in the vicinity is relevant to the determination of the question whether there is an adequate number of off-licence premises, since both kinds of premises provide facilities for the sale of intoxicating liquor for consumption off the premises…
>
> One is forced to recognise the logic of saying that, if there are many on-licences in the vicinity, there must be less demand for off-licences. Another point of view is that a court must take off-licenses by themselves and must ask whether the number of such licenses in the vicinity is adequate, having regard to the demand or probable demand

> for off-licences as such, and not merely for premises with facilities for buying and taking away intoxicating liquor. Indirectly, one may say, the number of on-licences will affect the situation, since their presence in large numbers will reduce the trade of the off-licences and weaken the case for increasing the number of such licences. But it can be argued that it is at the off-licence premises, as affected by circumstances, that the court must look directly.
>
> I test the matter by considering it from the applicant's point of view. It would be difficult, in my opinion, to prevent him from giving evidence, as part of his case, of the scarcity of any kind of facilities for the sale of liquor to be consumed off the premises and, conversely, I think an objector must also be allowed to point to the large number of on-licences with such facilities, and therefore consider that the court is not precluded in such a case from taking into account the number of on-licences in the vicinity.
>
> One thing, however, ought to be made clear for the benefit of courts hearing applications for off-licences: they are quite distinct in character and the Legislature so treats them. The mere fact that there are many on-licence premises with facilities similar to those provided by off-licence premises is not a complete answer to an application for an off-licence. The public is entitled to an adequate supply of off-licence premises and not only to facilities similar to those which they provide.

Of course, 'taking into account' is a different proposition from 'giving determining weight', as commented by Curran, LJ in the same ruling:

> The Act requires the court to decide the question of inadequacy, but it gives no directions as to how that question is to be determined. I see no reason why objectors should be precluded from giving evidence as to the number of on and off-licensed premises in the vicinity. It would be for the court to decide what weight, if any, should be given to such evidences in the circumstances of the particular case.

The key requirement for planning practice that flows from this Court of Appeal judgment is that, within the vicinity of an application for an off-licence, all public houses with their off-sales facilities and off-licences need to be mapped in order to inform the court.

Additionally, it has to be borne in mind that the absence of demand within a vicinity may cast doubt on the inadequacy of licensed premises even if no licensed premises exist. His Honour Judge Smyth in *Moyle Supermarkets v Wineflair (Belfast) Ltd, CoOp Group (CWS) Ltd and Gerard A Connolly trading as Anzac Bar* (2008) stated:

> The number of premises in the vicinity has to be shown, on balance to be inadequate. There are no premises inside the vicinity but I do not regard that as determinative. If there is no, or very little, demand then it would be difficult to say that the number of premises was inadequate and a finding on vicinity does not dispose of the case.

The general principles around inadequacy that can be applied to facts on the ground have again been usefully synthesised by Gillen J in *Phillip Russell Limited and D W Retail* (2013):

> The restrictive effect of the concept of inadequacy is a key component in licensing legislation. The legal curb is thus the obligation placed on the applicant to prove

> inadequacy. Demand can be generated from both inside and outside the vicinity. I can take into account the demand not only from persons residing in the vicinity but also from persons who work there, who resort to the vicinity for the purpose of recreation or shopping etc. The statutory test is inadequacy of numbers, not suitability or convenience. The fact that the public would find the presence of an off licence in a particular location convenient is not proof of such inadequacy. That is not to say that convenience conceptually cannot be a facet of adequacy. Thus in the context of accessibility it is relevant. If other premises in the vicinity are difficult to reach because of distance or the topography, car parking around them is poor, the character of the area where they are found is not conducive to all members of the public, or they are well off the beaten track etc. a court can take these into account. In particular, in the context of this case, distance between off licences is fact sensitive. There is no maximum or minimum set distance which will determine whether or not there is an inadequate supply in the vicinity. The nature of the accessibility together with the general topography and character of the area will all influence the impact that distance between premises will have on any case. In short, convenience in terms of accessibility and location are proper considerations and may have a particular bearing where members of the public are attracted into a particular location from outside the vicinity. Similarly, issues of appearance, range of goods, pricing, competition, service and choice are all proper components of the judicial probe of the concept of inadequacy in each case. The need for a proper competitive supply and range, choice and service does not, of course, mean that every village must have more than one off licence. To so conclude would drive a coach and horses through the concept of inadequacy. As in this case, each application must be judged on its merits. The advantages of integration in a mixed trading shop cannot by itself be a factor pointing towards the grant of an application for an integrated off-sales facility. A significant level of trade and large numbers of customers at particular premises do not necessarily lead to a conclusion that inadequacy of off licence provision has been proven. The court must go back to the legislation and recognise that the test is inadequacy in the vicinity. Demand for an off licence cannot be created simply by illustrating that large numbers of people come to a particular store. If this was to be the case, virtually every large filling station or supermarket with evidence of increasing sales could successfully apply for an off licence.

Within this guidance set out by Gillen J there are matters that clearly go well beyond the planning brief as they cross over into commercial/operational considerations relating to the nature of the business proposed. From a land use planning perspective, two key components in a proof of evidence comprise the multifaceted dimensions of demand and accessibility. An assessment of demand usually has regard to the overall number and change over time of the adult population residing within the vicinity. The analysis may additionally make reference to the primary catchment area of the vicinity to take account of those drawn into the vicinity, say for purposes of shopping, social activities or for employment. A land use analysis can throw light on these activities. For licensed facilities proposed on the edge of an urban area it would be logical to also include the adult population within the countryside adjacent to an application site as this may well act as a potential component of demand. Additionally, passing traffic may be an element of demand and thus examination of traffic volume and direction of movement can be relevant. The second and related matter is accessibility and in that regard ease of walking, car ownership, road infrastructure and level of service, car parking and public transport may be relevant to the determination of a liquor licence application. Inevitably this brings into play the issue of convenience and, while this is discussed in *Phillip Russell Limited and D W Retail* (2013), the comments of Morgan J in *Lidl (Northern Ireland) GmbH and Winemark the Wine Merchants Ltd* (2008) give added direction:

> I conclude, therefore, that convenience in terms of accessibility and location are entirely proper considerations in the assessment of inadequacy and that these may have a particular bearing where members of the public are attracted into a particular location from outside the vicinity.

Within the legislation, there are, finally, a number of provisions that give exception to the burden of proving inadequacy. Under conditions set out in Schedule 2 of the Licensing (Northern Ireland) Order 1996 a subsisting licence that is proposed to be surrendered can effectively be 'relocated' to premises in the vicinity of the surrender licence premises. The relevant circumstances where inadequacy need not be addressed include the destruction of the surrender licence premises, the termination of a tenancy or compulsory acquisition.

Licence application objections

The considerable number of authorities generated over time on matters of vicinity and inadequacy has depended in the main on commercial litigants coming forward to object to liquor licence applications. Paragraph 4 of Schedule 1 of the Licensing (Northern Ireland) Order 1996 identifies those who may appear at the hearing of the application and object to the grant of a licence as comprising the sub-divisional commander of the police sub-division in which the premises are situated, the relevant local authority or 'or any person owning, or residing or carrying on business in, premises in the vicinity of the premises for which the licence is sought'. As noted by Kerr J *In the Matter of an Application by Philip Russell Limited for Provisional Grant of a Licence* (1993):

> The provision of the right to local residents to object, is an unequivocal but unsurprising acknowledgement of the contribution which inhabitants of the area in the vicinity of the proposed licensed premises can make to the debate as to the propriety of their being permitted in the intended location.

The grounds of objection set out in Article (4)(a) to (e)(i) also include mention of procedural compliance, the fitness of the applicant to hold a licence and the suitability of the premises. Evidence from an architect is usually called to deal with the latter point if it arises in a case and when the matter relates to structure and design issues. Nonetheless, there is judicial opinion that the effect on the area of locating a licensed premises in a particular place is also a legitimate aspect of the inquiry as to the suitability of the premises (see *In the Matter of an Application by Philip Russell Limited for Provisional Grant of a Licence*, 1993). Accordingly, community-based objections by residents living within the vicinity of the licence application premises are permitted to be heard by the court in line with the provisions above and can, on occasion, be persuasive in an application being refused. In *Wine*

Inns Ltd v Haggerty and others (1988), Nicholson J, having found no need for such premises, ruled as follows:

> I would have exercised my discretion to refuse to grant a licence on the following grounds: 1. A substantial majority of those residing in the vicinity do not want an off-licence premises on reasonable grounds. They are the public whose needs are to be provided for. 2. I do not consider that a Court ought to impose on the public what they do not want, provided that their grounds are reasonable.

Moral objections may of course come before the court and, in that regard, useful guidance has been provided by His Honour Judge Curran in *Application for Provisional Grant of Licence (Off-Licence): Applicants – John James Robinson, Harriett Margaret Ann Robinson and Andrew Robinson* (2003):

> I have to exercise the Court's discretion in this matter, but it must be exercised within the confines of previous authority and on a legal basis. My sympathy for the proper concerns of the residents is so strong that I have to guard against making a decision founded, not upon law, but upon that sympathy. I have weighed up the various criteria binding not just this Court but all Courts and while I would sooner have seen an off-licence further away from the Church and Church Hall round the corner near the main (area) of business premises (it is only 45 metres from the boundary of the Church grounds), proximity to church premises while relevant is not an overriding factor. I cannot conclude that the justified concerns of the Church Vestry and the residents outweigh the substantial indeed formidable case made on behalf of the applicants.

Conclusion

The liquor licensing system in Northern Ireland is purposefully designed as restrictive and places an obligation on an applicant for a new public house licence or an off-licence to prove inadequacy within the vicinity of the application site. This chapter places planning practice at the heart of liquor licensing in Northern Ireland in regard to assisting the courts on these matters of vicinity and inadequacy. Again a point made in other chapters in this book rings true here in that case law has provided helpful interpretations of the statute content when set against the skeletal nature of the primary legislation. The courts have given direction on guiding principles over a long period of time that have carefully facilitated implementation of evolving policy preferences and legislative adjustments, and which, in turn, seek to accommodate changing public expectations and demands. No doubt there will be ongoing legislative revisions in liquor licensing in Northern Ireland, but the evidence presented in this chapter unequivocally conveys the caution, fairness and rigour with which justice is dispensed in this sphere of public policy.

Legislation

Alehouses Act 1635
Intoxicating Liquor Act (Northern Ireland), 1923
Licensing Act (Northern Ireland) 1971

Licensing and Registration of Clubs (Amendment) Act (Northern Ireland) 2021
Licensing (Ireland) Act, 1833
Licensing (Ireland) Act 1902
The Licensing (Northern Ireland) Order 1987
The Licensing (Northern Ireland) Order 1996
The Planning (Use Classes) Order (Northern Ireland) 2015
The Roads (Northern Ireland) Order 1993

Cases cited

Appeal to the Commission by J D Wetherspoon PLC in respect of the non-determination of application for full planning permission for change of use from retail unit to a J D Wetherspoon traditional ale, wine and food bar at 23–24 The Diamond, Londonderry (Planning Appeals Commission Reference: 1999/A061)

Application for provisional Grant of Licence (Off-Licence): Applicants – John James Robinson, Harriett Margaret Ann Robinson and Andrew Robinson (2003 County Court decision)

Belfast Co-Operative Society v Tohill (1975 2 NIJB)

Donnelly v Regency Hotel Ltd (1985 NI 144)

F.A. Wellworth & Co Ltd v Phillip Russell Ltd and Others (1997 NI 172)

Hynes v McAlinden and Others (1974 NI 166)

In the matter of an application by Philip Russell Limited for Provisional Grant of a Licence (1993 KERF1159)

In the matter of an application by Sainsbury's Supermarket LTD and Winemark the Winemerchants Limited, Wine Inns Limited, Phillip Russell Limited and Toby Inns Limited (2012 NIQB 45)

Lidl (Northern Ireland) GmbH and Winemark the Wine Merchants Ltd (2008 NIQB 146).

Magill and Another v Bell and Others (1972 NI 159)

McFerran v Casey and Others (1975 NIQB unreported judgment)

Moyle Supermarkets v Wineflair (Belfast) Ltd, CoOp Group (CWS) Ltd and Gerard A Connolly trading as Anzac Bar (2008 County Court decision)

On Appeal from the Decision of the County Court for the Division of Craigavon in the Matter of an Application for the Provisional Grant of an Off licence Pursuant to Article 5(1)(B) of the Licensing (Northern Ireland) Order 1996: Winemark (The Wine Merchants Limited) v Hagan (James Aidan), Hagan (Mary Bridget) and The Tunnel Bar Limited (2016 NIQB 90)

Phillip Russell Limited and D W Retail (2013 NIQB 56)

Re an application by Glenshane Tourist Services Limited for the grant of a licence (1999 NIQB unreported judgment)

Re application for the provisional grant of In/Off Licences; Strike Four (Belfast) Limited v Lofty Inns Limited and Another; McCann v Lofty Inns Limited and Others (1998 NIQB unreported judgment)

Re Boots the Chemists Application (1994 NIJB 11)

Re Samuel A McClean, (1994 NIQB unreported judgment)

Wine Inns Ltd v Haggerty and others (1988 4 NIJB 1)

References

Cherry, G.E. 1988. *Cities and Plans: The Shaping of Urban Britain in the Nineteenth and Twentieth Centuries.* London: Edward Arnold.

Department for Social Development. 2005. *Liquor Licensing – The Way Forward: Government Proposals to Reform Liquor Licensing Law in Northern Ireland – Consultation Document*. Belfast: Department for Social Development.

Department for Social Development. 2013. *Department for Social Development (DSD) Consultation on Proposed Changes to the Law Regulating the Sale and Supply of Alcohol in Northern Ireland – Outcome Report*. Belfast: Department for Social Development.

Department of the Environment for Northern Ireland. 1983. *DECAN 7: Public Houses*. Belfast: Department of the Environment for Northern Ireland.

Institute of Licensing. 2014. *Entertainments and Liquor Licensing*. Presentation at Antrim Borough Council, 5 June.

Lynch, K. 1960. *The Image of the City*. Cambridge, MA: MIT Press.

Malcolm, E. 1982. 'The Catholic Church and the Irish Temperance Movement, 1838–1901', *Irish Historical Studies* 23(89): 1–16.

McBrien, E.J.D. 1997. *Liquor Licensing Laws of Northern Ireland*. Dublin: Gill and Macmillan Ltd.

Northern Ireland Assembly Research and Library Services. 2009. *Liquor Licensing Reform in Northern Ireland*. Research Paper 61/09. Belfast: Northern Ireland Assembly.

Pratten, J. and C. Lovatt. 2002. 'Can the Rural Pub Survive? A Challenge for Management or a Lost Cause', *Management Research News* 25(1): 60–72.

Wilson, B. 2019. *State of Rural Services 2018*. Craven Arms: Rural England CIC.

18

PLANNING AND PROFESSIONAL PRACTICE

Gary McGhee, William Orbinson QC, Stephen McKay and Michael Murray

Introduction

As previous chapters of this book have demonstrated, town and country planning requires a multidisciplinary perspective that most commonly engages with the legal profession. The contributions of the solicitor and barrister are key to successful case management and presentation and thus these matters merit consideration in a book devoted to planning law and practice in Northern Ireland. In the sections that follow, Gary McGhee gives particular attention to the changing workload of the planning and environmental solicitor. William Orbinson QC addresses the specialist skillset that a planning barrister can bring to the development arena. The planning profession operates within a framework of ethical standards and it is appropriate that their contribution to the integrity of planning practice should also be explored. Accordingly, Stephen McKay and Michael Murray explicate principles for planning practitioners derived from the Royal Town Planning Institute's (RTPI) Code of Conduct and conclude by setting out the roles and responsibilities of the planner as expert witness.

The role of the planning and environmental solicitor *(Gary McGhee)*

The role of the planning and environmental solicitor has changed beyond all recognition over the course of the last two decades. In the early 1990s, the number of Northern Ireland law firms with significant experience in planning or environmental law could be counted on two or three fingers.

The mid-1990s witnessed the arrival of the supermarket giants, Tesco and J. Sainsbury, to Northern Ireland and their ambitious plans to obtain permission for stores across the country triggered a series of lengthy planning inquiries. These inquiries provided many of Northern Ireland's leading planning solicitors and barristers with their first significant experience of planning law as local supermarkets retained them to slow down and resist the arrival of the English supermarket giants. However, while the supermarket inquiries gave local planning lawyers a first opportunity in this area of the law, the challenges facing local solicitors aspiring to forge a career out of planning and environmental law were considerable.

DOI: 10.4324/9781003218319-18

With lengthy planning inquiries into food stores being commonplace in Great Britain, major companies such as Tesco and J. Sainsbury were, even in the mid-1990s, well aware of the different roles played by solicitors and barristers in the planning process and the added value that each brings to a major development proposal. The same could not be said about large sections of the development community in Northern Ireland at that time. Whilst many local developers readily accepted and understood the need to retain a barrister to act as an advocate at a planning appeal or inquiry, the benefits of instructing a solicitor specialising in these areas enjoyed limited understanding locally. As such, the first major obstacle facing local planning and environmental solicitors was a near-constant requirement to explain their role in the planning process and how that role would work in concert with those of a planning consultant and barrister. The position in Great Britain at this time could not have been more different. In that marketplace, any debate as to the added value and role of planning and environmental solicitors had long been settled with solicitors specialising in these areas of the law present in all of the major law firms. In particular, City of London firms had already established large teams of planning and environmental lawyers to cater to the needs of their clients – needs that were, and still are, replicated by clients in Northern Ireland. Accordingly, the local market had to be educated as to the role of the planning and environmental solicitor.

The second major obstacle facing aspiring local planning and environmental solicitors was, and remains, the ability of planning consultants and architects to access and instruct planning barristers directly and without recourse to a solicitor. This entitlement – known as 'direct access' – distinguishes a solicitor advising on planning or environmental matters from most other solicitors in a law firm and fundamentally alters the relationship between the solicitor and the barrister. With direct access, the referral of work is a two-way street with both lawyers being in a position to refer work to the other. Having received instructions directly from a planning consultant or architect, it is the barrister who may feel compelled to advise a client that a solicitor specialising in planning or environmental law should be retained (the planning consultant or architect having already chosen not to do so). This situation is in stark contrast to the vast majority of areas of law that do not have direct access and in which the referral of work is almost exclusively one direction – from solicitor to barrister – with the latter rarely in a position to reciprocate. Planning and environmental matters that elevate to a level involving court proceedings necessitate the involvement of a solicitor, but, these aside, direct access effectively compels solicitors specialising in these areas to bring added expertise, coordination, professionalism and knowledge to a planning case if counsel, consultants and clients are not to progress without them.

The final obstacle confronting the pioneering local planning and environmental solicitors was the decision by the Planning Appeals Commission to abolish the entitlement of a planning applicant, when exercising his/her right of appeal to the Planning Appeals Commission, to choose for the appeal to be conducted formally. Formal hearings provided the parties with an opportunity to cross-examine witnesses and were popular with appellants who frequently instructed both counsel and solicitors to appear on their behalf at the appeals and to prepare their cases, respectively. Prior to the removal of this entitlement to a formal hearing, local planning solicitors could expect to be instructed in relation to multiple appeals per month. However, once the appeals process moved to informal procedures, lawyer participation in the process declined significantly, with planning and environmental solicitors being hit hardest of all. The Planning Appeals Commission's ability to interrogate the merits and demerits of certain types of development proposal has, in the opinion of this contributor, sadly declined as a result of the decision to informalise the appeals process. The last five years have not witnessed a single formal hearing before the Planning Appeals Commission. While it is acknowledged that there is an onus on the appellant to request and justify the need for a formal hearing, can it credibly be argued that no cases have come before the Planning Appeals Commission over the last five years which were not appropriate for a formal hearing? The balance has, arguably, shifted too far in favour of informal hearings and the time is ripe for the Planning Appeals Commission to consider the reinstatement of the right of an appellant to obtain a formal hearing.

The aforementioned obstacles have, to a large extent, framed the role of solicitors specialising in planning and environmental law in Northern Ireland today. The local market is now educated as to the role of such lawyers, who, in 2021, number around a dozen. Legally complex enforcement appeals aside, the legacy of the demise of formal hearings is that appearances at the Planning Appeals Commission by solicitors are increasingly confined to appeals pertaining to larger proposals and public inquiries. Judicial reviews (untouched by direct access) remain an important component of a planning and environmental solicitor's practice augmented with developing areas of environmental law, such as permitting, waste law, contaminated land law and environmental due diligence. Every high-profile planning or environmental judicial review in Northern Ireland seems to trigger complaints from some politicians about the quantum of such challenges, but, in reality, there is no evidence that, proportionately, more challenges are instigated here than in Great Britain. In any event, the complaints of some local politicians sit uncomfortably with the unending drive at an international level by both the United Nations and the European Union to provide individual citizens with increased access to environmental justice. International treaties and directives have brought down the cost of planning and environmental judicial reviews considerably with the availability of protective costs orders for an increasing number of such challenges, doing much to ensure that the courts will continue to be kept busy with these types of cases in the coming years. The transfer of most planning powers to local councils has also resulted in a growing area of work for local planning and environmental solicitors as a number of the local councils have opted to retain private sector planning and environmental solicitors to provide legal advice that had previously been provided to the Department of the Environment by the Departmental Solicitor's Office. A further outworking of the transfer of planning powers has been the increased use of planning agreements by local councils as a means of securing planning gain and increased revenue. This has provided additional work for local planning solicitors given the contractual nature of such documents.

So, to conclude, with the passage of over twenty-five years since the first planning inquiries into 'supermarket wars' in Northern Ireland, the role of the planning and environmental solicitor locally is now well established, and the local market largely now understands the added value that such lawyers can bring to a project. This is evidenced by the growth in the number of local planning and environmental lawyers and the fact that almost all of the leading law firms in Belfast now consider such solicitors to be an essential part of the services that they offer to their clients. The obstacles of direct access and the demise of formal appeal hearings, however, remain and have compelled practitioners to reposition themselves in the marketplace. This repositioning, allied with the opportunities for growth, that continue to be provided by international law, in particular, ensure that the role of the planning and environmental solicitor is assured for years to come.

The role of the planning barrister *(William Orbinson QC)*

'Northern Ireland's first planning Silk?', said an amused criminal practitioner in 2006 at a seminar for aspiring QC's. '*Why not?*', I said. Six months on, when I took Silk and he didn't, he graciously wrote to me with apologies and congratulations. But it struck me then that being a pure planning barrister – in the sense of having an exclusive specialism, at least – was well off the beaten track, and for the local Bar a bit of an oddity. Fifteen years on from taking Silk and three decades after my first big planning case against formidable Queen's Counsel opposition what strikes me now is that writing this chapter contribution proved a bit of a challenge. What, exactly, do I do for a living as a specialist planning barrister, and how, exactly, is it different from being a barrister practising in other areas of law? How to explain that distinction? Am I still a bit odd?

Well, forget much of what you know about barristers. The role of today's planning barrister is not, primarily, the classic advocate's role. It is not, primarily, about the cut and thrust of cross-examination because encountering cross-examination in today's planning world is like a celebrated naturalist stumbling upon a rare and endangered species. So, if today's planning barrister is not Rumpole in muddy wellies, what is he, what is she?

As the foundation for everything else, the planning barrister's role is to ensure compliance with the increasingly complex regimen of legal regulation surrounding planning practice, or, failing that, to at least identify the risks of non-compliance and recommend ways to reduce and mitigate those risks. Difficult in a European Union context, that is apt to become more difficult as the United Kingdom teases out how its planning and environmental laws will relate to European norms. Particularly in the enforcement field, but not exclusively, the planning barrister's role is to identify possible legal arguments to ground the client's case. Done right, that requires lateral thinking, creativity, and a preparedness to 'think outside the box', to ask where the client needs to reach, and how legal principle steers that journey, or be developed to smooth the way.

If anything is the key role of the planning barrister beyond navigating pure legalities, it is policy interpretation and application. That is because what policy means is a matter of law for ultimate determination by the courts, not a matter of discretion or of what might feel right. So, the foundational task for the planning barrister, in close collaboration with the planning consultant, is to divine the proper meaning of policy and then to shape the client's case around that meaning, to deliver a policy-compliant result. Legally driven policy interpretation and application require distinct skills, posing a series of questions for close consideration. Is the policy itself lawful, both substantively and in terms of process? What does the adopted development plan have to say about the proposal, and why? Given the primacy of the adopted development plan, what relative weight should be given to it and to other material considerations, including other policies? What relative weight should be given to conflicting regional policies, and why? Is the nature, purpose and content of policy properly understood? Has the policy been read as a whole, in context, and as part of the overall policy document? Has it been read commonsensically, and by reference to the objectives it serves, and the objectives the policy document containing it serves? Is it appreciated that supposed requirements in policy are not mandatory requirements, to be construed with the strictness applied to legislation, so that there is no obligation to adhere mechanistically to each point but rather freedom to make a departure where that is considered justified? What material considerations might reasonably be said to outweigh non-compliance with policy? Is it unreasonable in the

Wednesbury sense to reach a determination on the basis of the policy, or to interpret the facts to sustain a particular application of the policy? Is the weight proposed to be given to the policy unreasonable in the *Wednesbury* sense? And so on.

Particularly with major and regionally significant applications but also in every case, a central function of the planning barrister is to maintain from the outset a broad, strategic focus while ensuring that the detail is right, so that the team works together towards a clear objective, achieving what the client wants to achieve. Winning even big battles is no use if the war is lost. And no planning barristers worth their salt will allow a client blindly to launch into, or stick with, a case that will likely fail or be a pyrrhic victory. Rather, frank advice on the merits, and demerits, must be given, from the get-go.

Particularly with major and regionally significant applications and in every case, there is a danger in having a hive of experts buzzing around doing their own thing, not moving in the same direction or, worse, still moving in contradictory directions. So, the planning barrister needs to give direction from the outset on what needs to be done, by whom and to what end, so that the entire team moves as one to a common goal. And that is a dynamic process as the case develops, not a 'starter for ten', with no ongoing coordination and direction.

Increasingly, the planning barrister acting for the planning applicant is engaged from the earliest stages in moulding the form of the proposal around legal principle, policy, guidance and supporting evidence, and shielding it from legal challenge. That links to the vital role that the planning barrister plays in ensuring that assessments prepared for purposes of Environmental Impact Assessment and Habitats Regulations are legally compliant and robust. As time has gone on, that has become a bigger and bigger part of my practice, and usually an onerous part, given the sheer volume of material involved (often, highly technical) and the need to scrutinise it all closely to detect, and mend, vulnerabilities. I have lost years at my Mac, grinding through seemingly endless draft chapters, but it needs to be done, and done with dedication, with precision.

With the return of everyday planning to district councils after 2015, the importance of making strong representations for or against an application came to the fore, and with that an increasing tendency for planning barristers to be instructed to influence the content of those representations to ensure robustness and persuasiveness, in accessible and concise language: to scratch where it itches. That extends to appearing before Planning Committees: an art in itself. Selling, or damning, a proposal succinctly and in a way that appeals to hard-pressed Members, often fatigued by a slew of planning applications before your slot, is in many ways harder than persuading a High Court judge. Enforced brevity breeds extreme focus, an acute need to press the right buttons precisely before the lights go out. A week may be a long time in politics, but you can see your life flash by in the three or five minutes you have to persuade a Planning Committee.

And then there are appeals, hearings, inquiries and the examination in public. I lump them together because essentially the planning barrister's role in all of them is the same, especially in a more or less post-cross-examination age. Rigorous identification of the issues and the arguments, rigorous construction of the arguments and the evidence package, rigorous formulation of rebuttal and counter-rebuttal points. All of it in advance, none of it on the hoof. All of it shaped by a keen understanding grounded in long experience of how the Planning Appeals Commission approaches its task, as reflected in my *Planning Appeals Principles: The Decision-Making Approach of the Planning Appeals Commission*, fifth edition, 2020 Planning Online. And, critically, all of it geared towards delivering the strategic goal for the client.

Last but not least, a major component in my work is judicial review. I remember London's most famous planning Silk saying when he led me in promoting the Victoria Square development twenty years ago that finding yourself in the JR court was a sign of something having gone wrong with your proposal. That may have been true then, but it is not true now. As I found when writing my *Planning and Judicial Review* 2009 SLS and *Planning and Judicial Review Case Summaries 1976 – 2019* Planning Online, planning judicial review has gone from a trickle in the 1970s, 1980s and 1990s to a constant flow from around 2005 onwards, in part due to Aarhus costs protection, with challenges varying in merit from hopeless to compelling. Because of that, the planning barrister promoting a proposal has to be scrupulous in detecting, and closing down as far as possible, potential grounds of challenge and, when seeking to challenge an approval, has to have *Line of Duty* levels of sleuthing skill to spot points of weakness, things that do not add up. Either way, above all, the planning barrister must not come up 'the Lagan in a bubble'.

And that is true of every facet of the planning barrister's role.

Ethics and the planning profession *(Stephen McKay and Michael Murray)*

The purpose of the planning system is to regulate the development and use of land in the public interest thereby ensuring that all development is carried out in a way that would not cause demonstrable harm to interests of acknowledged importance. It is, therefore, important to distinguish those matters that planning can influence from those which are outside its control. In this context, planning determines what kind of development is appropriate, how much is desirable, where it should best be located and what it should look like. The corollary of this is that a huge responsibility is placed on the shoulders of those involved in the process to act with honesty and integrity. A key question is, therefore, what is the public interest? The public interest in the planning arena is explained by Campbell and Marshall (2002) as being manifested in development management and land use planning and, in this context, Howe (1994) perceives its consideration as an ethical prerequisite of professional planning practitioners. In the case of both democratically elected representatives and professional planners, in both the public and the private sectors, there is a powerful resonance of ethical consequentialism underpinned by utilitarian principles whereby the actions of each should be in pursuit of maximising positive consequences when pitted against negative, optimising impacts upon human welfare and producing the maximum net benefit for the majority in the long term (Taylor 2009).

For professional planning practitioners, the ethical benchmark is delineated by the Code of Conduct of the RTPI, which aims to 'advance the science and art of town planning for the benefit of the public' (RTPI 2016: 1). Sheridan (2010) explains how expertise and knowledge are 'makers of RTPI membership' (p. 24), but planning professionals are constantly challenged by ethical issues extending beyond the remit of expertise and knowledge. In this context, the seminal work of Marcuse (1976) scrutinises the ethical implications for planners at a more substantive level and poses challenges to practitioners that might help them to better understand what is demanded by the profession. It is, for example, incumbent upon them to have a fundamental understanding of public sector decision-making processes and with this comes a responsibility to provide advice with honesty, integrity and, of course, in the public interest, no matter what the ramifications of doing so for their personal financial gain or career development. While exemplary behaviour such as this might reasonably be considered as a manifestation of fearless speech (Foucault 1983), it is demanded in all walks of the profession from private sector practitioner to local government planner and right to the highest level of government adviser. While allegiance is imperative, with planners unreservedly respecting client confidentiality in delivering their duties with due diligence (RTPI 2016), advice must be of the highest integrity albeit that it may not be what the client wants to hear. Mizzoni (2010) indicates that while this relationship is of paramount importance when the employer is legitimised through the democratic process, it should not diminish in any way the requirement upon the professional to dissent when it is evident that inappropriate interpretation of evidence that may be injurious to the public interest has occurred. The code of conduct for the planning profession is, therefore, driven by a series of core principles, which include competence; honesty and integrity; independent professional judgement; due care and diligence; equality and respect; and professional behaviour.

Whilst the ethical principles for planning practitioners are clearly established in the Code of Conduct (2016) for the RTPI, a parallel set of planning principles has been developed for councillors to reflect their own set of responsibilities to the public interest (Department of the Environment 2015). Part 9 of the Northern Ireland Code of Conduct for Councillors (the Code), which specifically applies to planning matters, came into force on 1 April 2015. In this context, the Code demands that councillors must provide a declaration of interests whereby they must declare direct or indirect pecuniary interest and withdraw from any planning proceedings and must not act as an agent pursuing planning matters on their own council. Acting in the public interest means that they must have regard for planning matters and not individual constituents or specific interests. This means that they must act fairly and openly, determining applications on their own merits and reporting approaches by lobbyists to the relevant planning officer. It is strictly forbidden to attempt to pressurise or influence planning officers to provide a particular recommendation on a planning application contrary to their professional view and similarly coerce them to either take or refrain from taking enforcement action against unauthorised development. The same principles underpin the development plan-making process. Councillors must not attempt to influence planning officers to include policies and proposals, such as the zoning of land in the local development plan, that would be to their own advantage or the advantage of any individual or group or make public statements about pending policies and proposals in the draft plan, prior to these being placed in the public domain. It is imperative that they must not take payment or reward of any kind for planning favours on their council. In terms of decision making, the Code demands that councillors keep an open mind and be willing to listen to all arguments, representations and evidence presented at the planning committee meeting and must not decide how they are going to vote before the planning committee meeting has taken place.

For those bound by the codes of conduct, sanctions can be taken. In the case of Chartered Town Planners (members of the RTPI) procedures are in place for the conduct of disciplinary proceedings in respect of any complaint made against a member of any action contrary or prejudicial to the aims, objects and interests of the Chartered Institute or of conduct unbecoming to a member. The RTPI has powers in place that conform to the principles of natural justice, enabling it to warn, reprimand, suspend or expel members as it sees fit. In the case of councillors, punitive measures are also in place whereby they can be censured or even suspended or partially suspended from being a councillor for a period of up to one year. In extreme circumstances, there is provision to disqualify a person from being or becoming a councillor for a period not exceeding five years. Perhaps, almost perversely, for those who are neither chartered planners nor democratically elected councillors there are no sanctions and, even if a chartered planner is expelled from the RTPI, there is nothing to stop him/her from continuing to work in planning practice.

The planner as expert witness *(Stephen McKay and Michael Murray)*

The arenas within which a planner may be involved in submitting expert evidence range from court proceedings to planning appeals and public inquiries and latterly may include appearances before local authority planning committees by way of predetermination hearings. The duties and responsibilities of expert witnesses more generally in civil cases have been set out by the courts with established practice commonly regarded as being derived from the 'Ikarian Reefer' case (*National Justice Compania Naviera S.A. v Prudential Assurance Co Ltd.* 1993) in which Cresswell J advised:

> 1. Expert evidence presented to the Court should be and should be seen to be the independent product of the expert uninfluenced as to form or content by the exigencies of litigation … 2. An expert witness should provide independent assistance to the Court by way of objective unbiased opinion in relation to matters within his expertise … An expert witness in the High Court should never assume the role of advocate. 3. An expert witness should state the facts or assumptions on which his opinion is based. He should not omit to consider material facts which detract from his concluded opinion … 4. An expert witness should make it clear when a particular question or issue falls outside his expertise. 5. If an expert's opinion is not properly researched because he considers that insufficient data is available then this must be stated with an indication that the opinion is no more than a provisional one … 6. If after exchange of reports, an expert witness changes his view on a material matter … such change of view should be communicated … to the other side without delay and when appropriate to the Court … 7. Where expert evidence refers to photographs, plans, calculations … survey reports or other similar documents these must be provided to the opposite party at the same time as the exchange of reports,

These considerations considerably amplify the RTPI (2016) Code of Conduct and are in line with the RTPI (2018) Practice Advice on the role of planners more generally as expert witnesses when preparing written representations, participating in a discussion at hearings or giving evidence at an inquiry. They have also informed Practice Direction preparation issued by the High Court of Justice in Northern Ireland. Under the terms of Practice Direction No 01 of 2015, incorporated as Appendix D into Practice Direction No 01 of 2019, experts are required to give a declaration that encompasses the following considerations:

> (1) I understand that my primary duty in furnishing written reports and giving evidence is to assist the court and that this takes priority over any duties which I may owe to the party or parties by whom I have been engaged or by whom I have been paid or am liable to be paid. I confirm that I have complied and will continue to comply with this duty; (2) I have endeavoured in my reports and in my opinions to be accurate and to have covered all relevant issues concerning the matters stated, which I have been asked

> to address, and the opinions expressed represent my true and complete professional opinion; (3) I have endeavoured to include in my report those matters of which I have knowledge and of which I have been made aware which might adversely affect the validity of my opinion; (4) I have indicated the sources of all information that I have used; (5) I have where possible formed an independent view on matters suggested to me by others including my instructing lawyers and their client; where I have relied upon information from others, including my instructing lawyers and their client, I have so disclosed in my report. (6) I will notify those instructing me immediately and confirm in writing if, for any reason, my existing report or opinion requires any correction or qualification; (7) I understand that: *(a)* My report, subject to any corrections before swearing as to its correctness, will form the evidence which I will give under oath or affirmation; *(b)* I may be cross-examined on my report by a cross-examiner assisted by an expert; *(c)* I am likely to be the subject of public adverse criticism by the judge if the court concludes that I have not taken reasonable care in trying to meet the standard set out above. (8) I confirm that I have not entered into any arrangement whereby the amount or payment of my fees, charges or expenses is in any way dependent upon the outcome of this case.

Judicial comment has drawn attention to the limits of expert witness opinion and *In the Matter of an Application by Mooreland and Owenvarragh Residents' Association for Judicial Review* (2022 NIQB 40) Humphreys J states in a postscript to this judgment that allegations that a planning permission is unlawful is "an issue which is demonstrably for the court to determine, not a planning consultant". Planning education places great weight on familiarising students with the RTPI Code of Conduct, but the curriculum would certainly be further enriched by introducing the obligations placed on professional practice by the courts.

Conclusion

As this and previous chapters have demonstrated, the planning practitioner will invariably be operating within a multidisciplinary setting that requires an engagement with members of the legal profession. Solicitors and barristers have a vital role to play in effective case management and presentation. It is within this context that planners are often called to give evidence and, as demonstrated above, there is a serious onus to act in accordance with principles of ethical behaviour that are formalised through a code of professional conduct and which carry over to the spirit and purpose of rules governing expert witnesses. Planners also require a toolkit of methodology and a capacity to articulate complex ideas about space and place with a soundness of opinion. In representing the public interest that lies at the heart of the planning system, the planning profession must continue to embrace this trinity of appropriate method, soundness of opinion and ethical practice that are the hallmarks of its distinctive and valuable contribution to society.

Cases cited

In the Matter of an Application by Mooreland and Owenvarragh Residents' Association for Judicial Review (2022 NIQB 40)

National Justice Compania Naviera S.A. v Prudential Assurance Co Ltd. 1993. Lloyd's Law Reports 2: 68.

References

Campbell, H. and R. Marshall. 2002. 'Utilitarianism's Bad Breath? A Re-evaluation of the Public Interest Justification for Planning', *Planning Theory* 1(2): 165–89.

Department of the Environment. 2015. *The Northern Ireland Code of Conduct for Councillors*. Belfast: Department of the Environment.

Foucault, M. 1983. 'Discourse and Truth: The Problematization of Parrhesia', *Six Lectures Given by Michel Foucault at the University of California at Berkeley, Oct–Nov. 1983*. Berkley, CA: California University of California Press.

High Court of Justice in Northern Ireland. 2019. *Practice Direction No 01 of 2019*. Belfast: The High Court of Justice in Northern Ireland.

Howe, E. 1994. *Acting on Ethics in City Planning*. The State University of New Jersey. New Brunswick, NJ: CUPR Rutgers.

Marcuse, P. 1976. 'Professional Ethics and Beyond: Values in Planning', *Journal of the American Planning Association* 42(3): 264–74.

Mizzoni, J. 2010. *Ethics: The Basics*. Oxford: Wiley Blackwell.

RTPI (Royal Town Planning Institute). 2016. *Code of Professional Conduct*. London: RTPI.

RTPI (Royal Town Planning Institute). 2018. *Planners as Expert Witnesses: RTPI Practice Advice*. London: RTPI.

Sheridan, C. 2010. 'Expertise and Knowledge are Makers of RTPI Membership', *Planning* 26 (March): 1861.

Taylor, N. 2009. 'Consequentialism and the Ethics of Planning Research', in Lo Piccolo, F. and H. Thomas (eds) *Ethics and Planning Research*. Farnham: Ashgate, pp. 13–28.

APPENDIX

Northern Ireland Judicial Decisions

Page references in this book are noted in italics.

2022 In the matter of an application by Alpha Resource Management Limited for leave to apply for judicial review and in the matter of a decision of Belfast City Council (NICA 27) Keegan LCJ and Treacy LJ

2022 In the matter of an application by Mooreland and Owenvarragh Residents' Association for judicial review (NIQB 40) Humphreys J...*24, 424*

2022 In the matter of an application by Noeleen McAleenon for judicial review (NIQB 39) Humphreys J

2022 In the matter of an application by Gordon Duff (Re Glasdrumman Road, Ballynahinch) for judicial review and in the matter of a decision of Newry, Mourne and Down District Council (NIQB 37) Scoffield J

2022 In the matter of an application by ABO Wind (NI) Limited for judicial review and in the matter of a decision of the Planning Appeals Commission (NIQB 3) Scoffield J...*294*

2021 In the matter of an application by Alpha Resource Management Limited for leave to apply for judicial review and in the matter of a decision of Belfast City Council (NIQB 122) Scoffield J

2021 In the matter of an application by Martin Tracey for leave to apply for judicial review (NIQB 104) Humphreys J

2021 In the matter of an application by ABO Wind NI Limited and Energia Renewables Company 1 Limited for judicial review and in the matter of the Chief Planner's Update 7 (NIQB 96) Humphreys J...*213*

2021 In the matter of an application by Hartlands (NI) limited for judicial review and in the matter of a decision of Derry City and Strabane District Council (NIQB 94) Scoffield J

2021 In the matter of an application by Safe Electricity A&T Limited and Patrick Woods for judicial review and in the matter of decisions of the Minister for the Department for Infrastructure and the Executive Office (NIQB 93) Scoffield J...*23, 292*

2021 In the matter of an application by Fernmount Trading (NI) Limited and Sharp (NI) Limited for leave to apply for judicial review and in the matter of a decision by Belfast City Council (NIQB 89) Scoffield J

2021 In the matter of an application by Rural Integrity (Lisburn 03) Limited for leave to apply for judicial review and in the matter of a decision of Lisburn and Castlereagh City Council (NIQB 32) Scoffield J

2020 In the matter of an application by Patrick Lynch for judicial review (NICA 32) Horner J...*209*

2020 In the matter of an application by Rural Integrity (Lisburn 01) Limited for judicial review (NICA 12) Morgan LCJ delivering the judgement of the Court...*204*

2020 In the matter of Belfast City Council and Madame Zhang Meifang and The Consul General of the People's Republic of China, Belfast (2020 NICh 12) McBride J...*163*

2020 In the matter of an application by William Donnelly for leave to apply for judicial review (NIQB 35) Humphreys J...*49, 148*

2020 In the matter of 32 applications by Rural Integrity (Lisburn 01) Limited and related Limited Companies for leave to apply for judicial review (NIQB 25) McCloskey J...*204*

2020 In the matter of an application by Coulters Hill Residents Ltd for leave to apply or judicial review v Planning Appeals Commission (NIQB 1) McCloskey J...*190, 203*

2019 In the matter of an application by James Hugh Allister and Robert Edwin Agnew for judicial review v Causeway Coast and Glens Borough Council (NIQB79) McCloskey J...*48, 60, 62*

2019 In the matter of an application by James Allister and Robert Agnew for judicial review (NIQB 66) McCloskey J

2019 In the matter of an application by Rural Integrity (Lisburn 01) Limited for judicial review v Planning Appeals Commission - protected costs and security for costs (NIQB 40) McCloskey J...*204*

2019 In the matter of an application by Stuart Knox for judicial review v Causeway Coast and Glens Borough Council (NIQB 34) McCloskey J...*66*

2019 In the matter of an application by Greencastle Rouskey Gortin Concerned Community Limited for judicial review v Department for Infrastructure (NIQB 24) McCloskey J...*39*

2019 In the matter of an application by Greencastle Rouskey Gortin Concerned Community Limited for judicial review v Department for Infrastructure - Judgement Summary (NIQB 12) McCloskey J...*39*

2018 In the matter of an application by James Allister and Robert Agnew for judicial review (NIQB 103) McCloskey J

2018 In the matter of an application by Dean Blackwood, representing River Faughan Anglers Limited for judicial review v Derry City and Strabane District Council (NIQB 87) McCloskey J...*191*

2018 In the matter of an application by Andrea Kells, Chief Commissioner of the Planning Appeals Commission, for leave to apply for judicial review (NIQB 84) McCloskey J

2018 In the matter of an application by Colum Sands for judicial review v Newry, Mourne and Down District Council (NIQB 80) McCloskey J...*66, 275*

2018 In the matter of an application by Dean Blackwood, representing River Faughan Anglers Limited for leave to apply for judicial review v Derry City and Strabane District Council (NIQB 73) McCloskey J

2018 In the matter of an application by Christine Alexander for judicial review v Causeway Coast and Glens Borough Council (NIQB 55) McCloskey J...*48*

2018 In the matter of an application by Elizabeth Conlon for judicial review v Belfast City Council (NIQB 49) McCloskey J...*62, 180*

2018 In the matter of an application by Colin Buick as Chairperson of NOARC21 for judicial review. In the matter of a decision by the Department for Infrastructure dated 13 September 2017 (NICA 26) Morgan LCJ, Stephens LJ and Treacy LJ...*22*

2018 In the matter of an application by Colin Buick for judicial review and in the matter of a decision of the Department for Infrastructure made on 13 September 2017 under Section 26 of the Planning Act (Northern Ireland) 2011 (NIQB 43) Keegan J...*21*

2018 In the matter of an application by Oasis Retail Services Limited for leave to apply for judicial review and in the matter of decisions made by Belfast City Council (Licensing Committee) on 19 March 2014 granting an amusement permit to Belfast Leisure Company Ltd at 24-28 Bradbury Place, Belfast, 6 October 2014 granting an amusement permit to Hazeldene Enterprises Ltd for premises at 25-41 Botanic Avenue, Belfast (NICA 32) Deeny LJ delivering the judgement of the Court

2018 In the matter of an application by Rural Integrity for judicial review (McC 10820) McCloskey J

2018 In the matter of an application by Paul McNamara for leave to apply for judicial review v Lisburn and Castlereagh City Council (NIQB 22) McCloskey J...*199*

2018 In the matter of an application by Belfast City Council for judicial review v the Planning Appeals Commission (NIQB 17) McCloskey J...*40, 45, 61, 62, 67, 70*

2018 In the matter of an application by RES UK and Ireland Limited for judicial review and in the matter of decisions of the Planning Appeals Commission dated 26 June 2017 (References 2015/A012, 2015/A0168 and 2015/A 0169) (NIQB 16) Keegan J ...*283*

2018 In the matter of an application by Stewart Mills, as representative of Ringhaddy Area Residents, for leave to apply for judicial review v Newry, Mourne and Down District Council (NIQB 14) McCloskey J

2017 In the matter of an application by Rural Integrity (Lisburn 01) Limited for leave to apply for judicial review and in the matter of an application by Dr Theresa Donaldson for judicial review and in the matter of a decision by Lisburn and Castlereagh City Council to issue planning permission reference S/2014/0908F (NIQB 133) Keegan J

2017 In the matter of an application by Dermott Nesbitt for leave to apply for judicial review v Newry, Mourne and Down District Council (Choice Housing Ireland Limited - interested party) (NIQB 116) McCloskey J

2017 In the matter of an application by Frank McGurk for judicial review and in the matter of a decision of the Department of the Environment for Northern Ireland made on 23rd April 2014 (NIQB 108) McBride J

2017 In the matter of an application by William Donnelly for judicial review and in the matter of a decision by the Department of the Environment for Northern Ireland made on 27 July 2015 (NIQB 84) McBride J...*49, 277, 279*

2017 In the matter of an application by Mr Richard Irwin for judicial review and in the matter of the decisions of Mid and East Antrim Borough council on prospecting for hydrocarbons at land at Woodburn Forest, near Carrickfergus, County Antrim (NIQB 75) Colton J

2017 In the matter of an application by Ards and North Down Borough Council for judicial review and in the matter of an application by Gordon Duff for judicial review and in the matter of a decision by the Planning Appeals Commission dated 27 July 2016 (NIQB 73) Maguire J

2017 In the matter of an application by Christine Gibson for leave to apply for judicial review and in the matter of a decision of Newry, Mourne and Down District Council made on 29 June 2015 and 25 February 2016 and in the matter of a decision of the Department of Agriculture, Environment and Rural Affairs (Marine and Fisheries Division) made on 3 June 2016 (Respondents); Frazer Ferries Limited (Notice Party) (NICA 77) Gillen LJ delivering the judgement of the Court...*195, 199, 278*

2017 An application by Friends of the Earth Limited for judicial review (NICA 41) Weatherup LJ delivering the judgement of the Court...*160*

2017 In the matter of an application by Mr Chris Murphy to apply for judicial review in the matter of a decision by the Department of Infrastructure on 17 August 2016 to proceed with the Castledawson to Toome Dualling Scheme (NIQB 35) Keegan J...*198, 278*

2017 In the matter of an application by the Minister of Enterprise, Trade and Investment for judicial review and in the matter of a decision by the Minister for the environment dated 3 September 2014 between: Belfast City Council Appellant / Notice Party and the Minister of Enterprise, Trade and Investment Respondent / Applicant (NICA 28) Weir LJ delivering the judgement of the Court...*21*

2017 In the matter of an application by Patrick Heffron for judicial review and in the matter of a decision by the Department of the Environment for Northern Ireland (Planning Service) on 23 March 2015 to grant planning permission (NIQB 25) Keegan J

2017 In the matter of an application by Oasis Retail Services Limited for leave to apply for judicial review and in the matter of decisions made by Belfast City Council (Licensing Committee) on 19 March 2014 granting an amusement permit to Belfast Leisure Company Ltd at 24-28 Bradbury Place, Belfast, 6 October 2014 granting an amusement permit to Hazeldene Enterprises Ltd for premises at 25-41 Botanic Avenue, Belfast (NIQB 13) Maguire J

2016 In the matter of an application by Friends of the Earth Limited for judicial review and in the matter of a decision of the Department of the Environment not to issue a Stop Notice to mandate the cessation of sand and gravel extraction at Lough Neagh pending the outcome of an enforcement appeal (NIQB 91) Maguire J

2016 On Appeal from the Decision of the County Court for the Division of Craigavon in the matter of an application for the Provisional Grant of an Off licence Pursuant to Article 5(1)(B) of the Licensing (Northern Ireland) Order 1996: Winemark (The Wine Merchants Limited) v Hagan (James Aidan), Hagan (Mary Bridget) and The Tunnel Bar Limited (NIQB 90) Keegan J...*406*

2016 Gibson's (Christine) application (NIQB 82) Colton J ...*195*

2016 The Minister of Enterprise, Trade and Investment's application (NIQB 26) Treacy J ...*20*

2016 O'Hare's application (NIQB 20) Maguire J...*50*
2015 Newry Chamber of Commerce and Trade's application (NIQB 65) Treacy J...*49, 189*
2015 Roberts' (Gregory) application (NIQB 20) O'Hara J...*207*
2014 Mooreland and Owenvarragh Residents' Associations' application (NIQB 130) Horner J...*189, 213*
2014 Doyle's (Ellen) application (NIQB 82) Treacy J...*197*
2014 River Faughan Anglers Limited's application (NIQB 34) Treacy J...*207*
2014 McLean Bookmakers and North West Bookmakers applications (NIQB 32) Treacy J
2014 Ulster Architectural Heritage Society's application (NIQB 21) Treacy J...*53, 92*
2014 Department of the Environment's application (NIQB 4) Treacy J...*50, 197*
2014 Lamont's (David John Stewart and Elaine) Application (NIQB 3) Treacy J...*53*
2013 Sandale Developments Limited's application (NICA 58) Morgan LCJ, Girvan LJ and Coghlin LJ
2013 The Planning Service of Northern Ireland v Young & Anor (NICA 29) Morgan LCJ, Girvan LJ and Sir John Sheil
2013 Crusaders Football Club's application (NIQB 68) Treacy J...*192*
2013 National Trust's application (NIQB 60) Weatherup J...*197, 209, 276*
2013 Phillip Russell Limited and D W Retail (NIQB 56) Gillen J...*410, 412*
2013 In the matter of an application by the Northern Ireland Environment Agency for judicial review and in the matter of a decision of the Planning Appeals Commission made on 7 June 2012 (NIQB 43) Treacy J
2013 Sandale Developments application (NIQB 38) Treacy J
2013 Alternative A5 Alliance's application for judicial review (NIQB 30) Stephens J ...*197, 206, 214*
2013 In the matter of an application by Brian Quinn and Michael Quinn (NIQB 24) Treacy J...*197, 200, 207*
2012 Musgrave Retail Partners (NI) Ltd's application (Leave stage) (NIQB 109) Maguire J ...*193*
2012 Corbo Properties' application (NIQB 107) Horner J...*192, 202*
2012 In the matter of an application by Sainsbury's Supermarket LTD and Winemark the Winemerchants Limited, Wine Inns Limited, Phillip Russell Limited and Toby Inns Limited (NIQB 45) Gillen J...*407*
2012 Young's (William) application (Leave stage) (NIQB 15) Treacy J...*191*
2011 Judgement of the Court (Fourth Chamber) of 20 October 2011 (reference for a preliminary ruling from the Court of Appeal in Northern Ireland (United Kingdom)) – Department of the Environment for Northern Ireland v Seaport (NI) Ltd, Magherafelt District Council, F P McCann (Developments) Ltd, Younger Homes Ltd, Heron Brothers Ltd, G Small Contracts, Creagh Concrete Products Ltd (Case C-474/10)...*127, 233*
2011 Tesco Stores Ltd v Department of the Environment Planning Service (NICA 23) Morgan LCJ, Higgins LJ and Coghlin LJ...*212*
2011 In the matter of an application by Central Craigavon Limited for judicial review (NICA 17) Higgins LJ, Girvan LJ and Coghlin LJ
2011 Warner Chilcott UK Ltd and Caridian BCT NI Ltd's application (NIQB 137) Treacy J...*46*
2011 LS Banbridge Phase 2 Ltd's application (NIQB 106) Treacy J

2011 Taggart's (Michael) application (NIQB 82) Treacy J
2011 Belfast International Airport Ltd's application (Leave stage) and Belfast City Airport Watch Ltd's application (Leave stage) (NIQB 34) McCloskey J...*214*
2010 Thompson's (Ciara) application (NIQB 120) Treacy J...*204*
2010 In the matter of an application by House of Fraser Ltd & Ors for leave to apply for judicial review (NIQB 105) Girvan LJ...*215*
2010 An application for judicial review by Central Craigavon Limited (NIQB 102) Treacy J...*211*
2010 Central Craigavon Limited's application (NIQB 73) Morgan LCJ...*197*
2010 Blackburn's (Louisa) application (Leave Stage) (NIQB 71) Treacy J...*193*
2010 Sandale Developments Limited's application (NIQB 43) Weatherup J
2010 In the matter of an application by Ciara Patricia Thompson for judicial review and in the matter of a decision of the Department of the Environment for Northern Ireland (Planning Service) (NIQB 38) Morgan LCJ...*203*
2010 Tesco Stores Limited's application (NIQB 15) Weatherup J...*59, 65, 193*
2010 The Planning Service and William Young, Roberta Young (NIMag 5) White DJ (MC)
2009 Boswell's application (NIQB 95) Weatherup J...*201*
2009 Anderson Haulage Limited's application (NIQB 90) Weatherup J
2009 Thompson's application (NIQB 79) Weatherup J
2009 Heron Property Ltd's application for judicial review (NIQB 75) Morgan LCJ
2009 Woods' application (Peter, Patrick, Mary and Peter Gary) (NIQB 34) Weatherup J
2009 Vico Kent's application (Leave hearing) (NIQB 16) Morgan J
2009 Ace Bates Skip Ltd and Samuel's application (NIQB 13) Weatherup J
2008 Lidl (Northern Ireland) GmbH and Winemark the Wine Merchants Ltd (NIQB 146) Morgan J...*405, 412*
2008 Re Central Craigavon Limited (NIQB 46) Gillen J...*43*
2008 In the matter of an application by the applicant Sheridan Millennium Limited for judicial review and in the matter of a decision by the Department for Social Development and in the matter of a decision by Laganside Corporation (NIQB 8) Gillen J
2008 Moyle Supermarkets v Wineflair (Belfast) Ltd, CoOp Group (CWS) Ltd and Gerard A Connolly trading as Anzac Bar (County Court decision) His Honour Judge Smyth...*410*
2007 In the matter of an application by William Young for judicial review (NICA 32) Kerr LCJ, Campbell LJ and Higgins LJ...*197, 214*
2007 In the matter of an application by HM, a Minor, by PM, her Father and Next Friend for judicial review (NICA 2) Kerr LCJ, Nicholson LJ and Campbell LJ...*209*
2007 In the matter of an application by John Hill for leave to apply for judicial review (NICA 1) Kerr LCJ, Campbell LJ and Higgins LJ...*190*
2007 (1) Seaport Investments Ltd and (2) Magherafelt District Council & Ors applications for judicial review (remedies) (NIQB 103) Weatherup J...*233*
2007 An application by John Hill for judicial review (NIQB 97) Weatherup J
2007 In the matter of an application by Sheridan Millennium Limited for judicial review and in the matter of a decision by the Department for Social Development and in the matter of a decision by Laganside Corporation (NIQB 92) Gillen J...*200*
2007 In the matter of an application by Kinnegar Residents' Action Group, Park Road and District Residents' Association, Old Stranmillis Residents' Association, Belfast

Holyland Regeneration Association and Cultra Residents' Association for judicial review (NIQB 90) Girvan LJ...*199*

2007 (1) An application by Seaport Investments Limited (2) An application by Magherafelt District Council, F P McCann (Developments) Limited, Younger Homes Limited, Herron Bros Limited, G Small Contracts and Creagh Concrete Products Limited (NIQB 62) Weatherup J...*126, 233*

2007 In the matter of an application by Omagh District Council for judicial review (NIQB 61) Gillen J...*197, 202*

2007 Re Young's application for judicial review (NIQB 29) Weatherup J

2007 In the matter of an application by Sheridan Millennium Limited for leave to apply for judicial review and in the matter of the decision by the Department for Social Development and in the matter of the decision by Laganside Corporation (NIQB 27) Gillen J...*189*

2007 In the matter of an application by James Stewart Moore for judicial review and in the matter of a planning application to the Planning Service (Downpatrick) by the Northern Ireland Housing Executive (Newtownards) (NIQB 23) Gillen J...*197*

2007 In the matter of an application by Francis McHenry for leave to apply for judicial review in the matter of a decision of the Planning Appeals Commission dated 22 November 2006 (NIQB 22) Gillen J...*193*

2006 Between James Keys and Bernadette McGinley and Department of Regional Development (NIQB 89) Nicholson LJ...*362*

2006 In the matter of an application by Liam Ward for leave to apply for a judicial review (NIQB 67) Girvan J

2006 Application by Friends of the Earth for judicial review (NIQB 48) Weatherup J ...*197, 201*

2006 In the matter of an application for judicial review by: (1) Bow Street Mall Limited (2) Lisburn Chamber of Commerce (3) Multi-development UK Limited (4) Central Craigavon Limited (5) Belfast City Council (6) Belfast Chamber of Trade and Commerce and in the matter of a planning decision made by the Department of the Environment (NIQB 28) Girvan J...*53, 207, 212*

2006 In the matter of applications by: (1) AM Developments UK Limited (2) Belfast City Council and Belfast Chamber of Trade and Commerce (3) Bow Street Mall Limited and Lisburn Chamber of Commerce (4) Central Craigavon Limited for judicial review and in the matter of a decision of the Rt Hon Lord Rooker, Minister with responsibility for the Department of the Environment (NIQB 9) Girvan J...*200*

2005 In the matter of an application by Misbehavin Limited for judicial (NICA 35) Kerr LCJ, Sheil LJ and Hart J

2005 Windsor Securities Ltd re an application for judicial review (NICA 34) Kerr LCJ, Campbell LJ and Sheil LJ

2005 In the matter of an application by FP McCann Developments Limited and Maghera Presbyterian Church for judicial review (NIQB 78) Weatherup, J...*57*

2005 In the matter of an application by the Logan Rodgers Partnership for judicial review and in the matter of a decision by the Planning Appeals Commission dated 24 March 2004 (NIQB 4) Girvan J...*56*

2004 In the matter of an application by HM, a Minor, by PM, her Father and Next Friend for judicial review (NIQB 85) Weatherup J

2004 In the matter of an application by Misbehavin' Limited for judicial review; in the matter of an application by Ian Brown for judicial review (NIQB 61) Weatherup J

2004 Re Department of the Environment for Northern Ireland's application for judicial review (NIQB 51) Weatherup J

2004 Re Moore's application for judicial review (NIQB 49) Weatherup J...*214*

2004 In the matter of an application by Windsor Securities Ltd for leave to apply for judicial review and in the matter of a decision of the Planning Appeals Commission dated 17 December 2003 (NIQB 43) Girvan J

2004 In the matter of an application by Bowden & Oths (practicing as Harrisons), Beattie, Prabera & Sabherwal. In the matter of an application by JGS Services Ltd, Simon NC Jones and Nigel TJ Brown. In the matter of an application by Scalene Investments Limited (NIQB 32) Weatherup J...*361, 362*

2004 In the matter of an application by Ronald Foster for judicial review (NIQB 1) Kerr J

2004 57 Developments and the Department of the Environment for Northern Ireland (NICh 3) Weatherup, J...*202*

2003 In the matter of an application by SOS (NI) Limited for leave to apply for judicial review (NICA 15) Carswell LCJ, Campbell LJ and Coghlin J

2003 In the matter of an application by 'D' for judicial review (NICA 14) Carswell LCJ, Nicholson LJ and McCollum LJ...*196*

2003 In the matter of an application by Barry Gilligan and others for judicial review (NICA 10) Carswell LCJ, Nicholson and McCollum LJJ...*90*

2003 In the matter of an application by James Stewart for judicial review (NICA 4) Carswell LCJ, Campbell LJ and Kerr J...*69*

2003 Re Rowsome's application for judicial review (NIQB 61) Weatherup J

2003 In the matter of an application by Patrick Joseph Green for judicial review (NIQB 54) Kerr J

2003 Application for provisional Grant of Licence (Off-Licence): Applicants – John James Robinson, Harriett Margaret Ann Robinson and Andrew Robinson (County Court decision) His Honour Judge Curran...*413*

2002 In the matter of an application by UK Waste Management Limited for judicial review (NICA 8) Carswell LCJ

2002 Killultah Estates Ltd (NIQB 62) Girvan J

2002 In the matter of an application by Barry Gilligan, Valerie Allen, Anthony Cox, Maeve Fee, Henry Glover and Malone Park Residents' Association for judicial review and in the matter by Planning Service of the Department of the Environment for Northern Ireland dated 16 January 2002 (NIQB 53) Coghlin J...*89*

2002 In the matter of an application by D for judicial review and in the matter of decisions of the Chief Constable of the Royal Ulster Constabulary and the Department for Regional Development (NIQB 51) Coghlin J

2002 In the matter of an application by James Stewart for judicial review (NIQB 38) Gillen J...*69*

2001 In the matter of an application by Belfast Chamber of Trade and Commerce, Belfast City Council and North Down Borough Council for judicial review (NICA 6) Carswell LCJ

2001 BP Amoco PLC and John Kelly Ltd and Glenshane Tourist Services Limited (NICA 3) Carswell LCJ...*78*

2000 Re Lisburn Development Consortium's application for judicial review (Queens Bench Division – Crown Side) Kerr J...*55*

2000 In the matter of an application by Belfast Chamber of Trade and Commerce, Belfast City Council and North Down Borough Council for judicial review and in the matter of a decision by the Minister for the Environment for Northern Ireland to approve planning permission sought under planning application Z95/1088 ('D5') and any subsequent grant and planning permission on foot of the Minister's press announcement (Queen's Bench Division – Crown Side) Coghlin J

1999 In the matter of an application by UK Waste Management Limited for judicial review (CARF2798) Carswell LCJ...*60, 201*

1999 In the matter of an application by Ava Leisure Limited for judicial review (NICA 4) Carswell LCJ

1999 In re UK Waste Management Limited (Application for judicial review) (Queen's Bench Division – Crown Side) Coghlin J

1999 Re Aquis Estates Ltd's and others' application for judicial review (Queen's Bench Division – Crown Side) Kerr J

1999 In re Nicholson and another (Application for judicial review) (Queen's Bench Division – Crown Side) Coghlin J

1999 Re an application by Glenshane Tourist Services Limited for the grant of a licence (NIQB unreported judgement) Nicholson LJ...*404*

1998 In re Interface Europe Limited and Another (Queen's Bench Division – Crown Side) Coghlin J

1998 In re Ava Leisure Limited (Application for judicial review) (Queen's Bench Division – Crown Side) Kerr J

1998 Re application for the provisional grant of In/Off Licences; Strike Four (Belfast) Limited v Lofty Inns Limited and Another; McCann v Lofty Inns Limited and Others (NIQB unreported judgement) Nicholson LJ...*399, 408*

1997 In the matter of an application by Queen's University of Belfast for judicial review (Queen's Bench Division – Crown Side) Kerr J...*197*

1997 In the matter of an application by Callan for judicial review (Queen's Bench Division – Crown Side) Coghlin J...*42*

1997 Re an application by Bray Investments Limited for judicial review (Queen's Bench Division – Crown Side) Kerr J

1997 F.A. Wellworth & Co Ltd v Phillip Russell Ltd and Others (NI 172) Girvan J...*408*

1996 Re FA Wellworth & Co Ltd's application and related application (J Sainsbury plc, notice party) (Court of Appeal) Hutton LCJ, Nicholson LJ and Kerr J

1996 In the matter of an application by Wellworth & Co Limited and by Boots the Chemist Limited for judicial review (Queen's Bench Division – Crown Side) Carswell LJ

1995 In the matter of an application by Meekatharra (Northern Ireland) Limited for judicial review (Queen's Bench Division – Crown Side) Kerr J...*47*

1995 R v Secretary of State ex parte Blair (Queen's Bench Division – Crown Side) Kerr J ...*59*

1995 R v Planning Appeals Commission ex parte Bailie (Queen's Bench Division – Crown Side) Kerr J...*35*

1994 Re Samuel A McClean (NIQB unreported judgement) McDermott LJ...*406*

1994 Re Boots the Chemists Application (NIJB 11) Carswell LJ...*406*

1993 In the matter of an application by Philip Russell Limited for Provisional Grant of a Licence (KERF1159) Kerr J...*397, 412*

1992 Re an application by the Very Reverend McCann for judicial review (Queen's Bench Division – Crown Side) Carswell J

1992 Application by McCann for judicial review (Queen's Bench Division – Crown Side) Carswell J

1992 Re Thompson's application (Queen's Bench Division – Crown Side) Carswell J

1992 R v Secretary of State ex parte Genova Ltd and another (Queen's Bench Division – Crown Side) Carswell J

1991 Re an application by Bangor Flagship Developments Ltd (Queen's Bench Division – Crown Side) Carswell J...*59*

1990 R v Secretary of State ex parte McGuigan (Queen's Bench Division – Crown Side) Campbell J

1990 R v Secretary of State ex parte Ulster Estates Limited (Queen's Bench Division – Crown Side) Carswell J...*59*

1988 In re Magill's application (Queen's Bench Division) Carswell J

1988 Wine Inns Ltd v Haggerty and others (4 NIJB 1) Nicholson J...*412*

1986 In re North Down Borough Council's application (Queen's Bench Division) Carswell J...*201*

1985 R v Secretary of State for Northern Ireland ex parte Blair (Queen's Bench Division – Crown Side) Lowry LCJ...*189*

1985 Allen v Department of the Environment for Northern Ireland (NI 195) Lowry LCJ...*367*

1985 Donnelly v Regency Hotel Ltd (NI 144) Carswell LJ...*407*

1984 R v Secretary of State for Northern Ireland ex parte Johnston (Queen's Bench Division – Crown Side) Kelly LJ...*201*

1984 R v Secretary of State for Northern Ireland ex parte Kane and others (Queen's Bench Division – Crown Side) MacDermott J

1982 R (Thallon) v Department of the Environment (Queen's Bench Division) Hutton J

1976 Morelli v Department of the Environment (N.I. 159) Murray J...*45*

1975 McFerran v Casey and Others (NIQB unreported judgment) McGonigal J...*406*

1975 Belfast Co-Operative Society v Tohill (2 NIJB) McDermott J...*409*

1974 Hynes v McAlinden and Others (NI 166) Lowry LCJ...*409*

1972 Magill and Another v Bell and Others (NI 159) McGonigal J...*405, 406*

1959 Belfast Corporation v O.D. Cars Ltd (1960 1 All ER 65)...*3*

INDEX

For Product Safety Concerns and Information please contact our EU representative GPSR@taylorandfrancis.com
Taylor & Francis Verlag GmbH, Kaufingerstraße 24, 80331 München, Germany

www.ingramcontent.com/pod-product-compliance
Lightning Source LLC
LaVergne TN
LVHW080803011026
849160LV00015B/1519
* 9 7 8 1 0 3 2 1 1 0 7 3 8 *